Routing Algorithms in Networks-on-Chip

Maurizio Palesi • Masoud Daneshtalab
Editors

Routing Algorithms in Networks-on-Chip

Springer

Editors
Maurizio Palesi
Facoltà di Ingegneria
Kore University of Enna
Cittadella Universitaria
Enna, Italy

Masoud Daneshtalab
Department of IT
University of Turku
Turku, Finland

ISBN 978-1-4614-8273-4 ISBN 978-1-4614-8274-1 (eBook)
DOI 10.1007/978-1-4614-8274-1
Springer New York Heidelberg Dordrecht London

Library of Congress Control Number: 2013950182

Printed on acid-free paper

Springer is part of Springer Science+Business Media (www.springer.com)

Foreword

In the year 2000 when the idea of Networks on Chip (NoC) was proposed, many people looked at it sarcastically as "bizarre," "too complicated," and "unacceptably inefficient." In a few years they were proved wrong. But even the greatest optimists did not predict that this new paradigm, for designing inter-core connections for multi-core systems using packet-switched communication, will get unanimous endorsement from academicians as well as semiconductor industries in just a matter of 10 years. But it has happened with the announcement of NoC-based 48-core chip by Intel. Now annually there are many international workshops and special track sessions held in important international conferences specifically dealing with issues related to NoC architecture. Within the NoC architecture area, new ideas in routing algorithm design continue to dominate the research publications. During the earlier years, routing algorithm proposals attempted communication performance improvement by maximizing routing adaptivity (while avoiding any possibility of deadlock) and by reducing congestion in general or application-specific contexts. Recently, researchers have expanded the scope of routing algorithm design by including fault tolerance and lowering power consumption as added objectives along with high performance. Motivated by advances in new technologies, proposals of routing algorithms for 3D architectures and mixed electro-optical or pure optical NoCs have also started appearing in literature.

As is obvious from research publications in various conferences and workshops, NoC is becoming an important topic of research and postgraduate teaching in universities all over the globe. Routing algorithm design is a challenging topic for researchers since it provides the possibility of graph theoretic analysis of any proposed new solution. Thanks to the availability of free NoC simulators, this area provides the possibility of concrete and speedy experimental evaluation of new ideas in NoC routing. Availability of ASIC design tools and FPGA prototyping tools also allows evaluation of cost and power consumption implications of the new ideas. Research in the area of routing algorithms is still flourishing and by no means has reached the saturation point. This book, entitled *Routing Algorithms in Networks-on-Chip*, is a collection of papers describing representative solutions to important aspects and issues related to routing algorithms. This collection does not claim to include best solutions for any aspect of routing algorithms, nor does it claim a complete coverage of topics related to this important area of NoC architecture. But the book does provide a good source of reference to postgraduate students and researchers getting started in this exciting area.

For many years, both Maurizio Palesi and Masoud Daneshtalab have been very active in research related to various aspects of NoC architecture design in general, and design of routing algorithms in particular. They have made significant and distinctive contributions in the area of routing algorithms. Their contributions, through the organization of NoC-related workshops and special sessions in international conferences, as well as through special issues for various international journals, are well known and highly appreciated by the NoC community. The contacts and knowledge gained by them through these experiences have placed them in a unique position to put together this excellent collection of papers in a book form.

The book is organized in six logical parts such that each part contains papers related to a common theme. For example, Part I contains papers proposing ideas to improve routing performance in NoC platforms. Similarly, Part II collects ideas related to multicast routing in NoC platforms. There are parts, each with multiple chapters, dealing with fault tolerance routing in NoC, power/energy-aware routing, and routing for 3D and optical NoC. The single chapter in the last part describes an industrial case study of routing algorithm in a tera-scale architecture. This organization makes the book directly useable as a reference or as a textbook in a special topic graduate course.

I recommend this book to all those who are new to the area of NoC architecture and NoC routing and want to understand the basic concepts and learn about important research issues and problems in this area. The book will also be useful as a reference source to established research groups as well as industry involved in NoC research. I feel this book will make an important contribution in promoting education and research in NoC architecture.

Jönköping University,
Jönköping, Sweden

Shashi Kumar

Preface

Modern Systems-on-Chip (SoCs) today contain hundreds of Intellectual Properties (IPs)/cores, including programmable processors, coprocessors, accelerators, application-specific IPs, peripherals, memories, reconfigurable logic, and even analog blocks. We have now entered the so-called many-core era. The International Technology Roadmap for Semiconductors foresees that the number of Processing Elements (PEs) that will be integrated into an SoC will be in the order of thousands by 2020. As the number of communicating elements increases, there is a need for an efficient, scalable, and reliable communication infrastructure. As technology geometries shrink to the deep submicron regime, however, the communication delay and power consumption of global interconnections become the major bottleneck. The Network-on-Chip (NoC) design paradigm, based on a modular packet-switched mechanism, can address many of the on-chip communication issues, such as performance limitations of long interconnects and integration of a large number of PEs on a chip.

The overall performance of a network depends on several network properties such as topology, routing algorithm, flow control, and switching technique. This book is focused on routing algorithms. The routing algorithm has a strong impact on several nonfunctional requirements of an NoC-based system. Performance, reliability, energy consumption, power dissipation, thermal aspects, and fault tolerance represent just a short list of the major common metrics affected by the routing algorithm.

The scientific literature related to NoC architectures and design methodologies is mostly dominated by works that address issues concerning the routing algorithms. Unfortunately, although the topic is so important and so widely discussed in the NoC community, there is a lack of structured resources (i.e., books, articles, etc.) aimed at organizing the great deal of literature and information on routing algorithms used in the NoC-based systems. The goal of this book is to provide a unified platform for students (master's, Ph.D.), researchers (from both academy and industry), and practitioners for building the basis of routing algorithms for NoC-based systems as well as providing in-depth discussions on advanced solutions applied (and to be applied) in the current and the next generation of NoC-based many-core SoCs. After a brief introduction on the basic concepts of on-chip networks, routing algorithms for NoC architectures will be presented and discussed at different abstraction levels – starting from the algorithmic level to their actual circuital implementation. The impact on current and future key design objectives, namely, power dissipation, energy consumption, thermal aspects, reliability, and performance, will be analyzed and discussed.

This book is organized based on the key problems which affect the current and (may also affect) the next generation of many-core SoCs by putting emphasis on the role played by the routing algorithm. The book is organized into six parts, each of them aimed at presenting a selection of routing algorithms specifically designed for addressing the key issues which characterize the many-core era. The six parts will cover the following macro-topics: Performance Improvement, Multicast Communication, Fault Tolerance and Reliability, Power/Energy and Thermal Issues, Emerging Technologies, and Industrial Case Study.

The book can be used as a reference for university (post)graduate courses and by Ph.D. students, as well as for advanced courses on SoCs. The audience of the book is not limited to various NoC-related research groups; it also might be attractive for faculty and students from other fields when high-performance computing, supercomputers, and many-core systems become the key elements. In fact, routing algorithms play an important role for these elements. As there is much literature related to this subject, we attempt to collect and adjust the state-of-the-art research results, which have been recently published, into the chapters.

Chapter 1 provides the basic concepts of on-chip networks, including network topologies, switching techniques, and routing algorithms. Such topics represent the conceptual bases exploited by the strategies, the mechanisms, and the methodologies discussed in the rest of the book.

The first part of the book consists of a set of chapters related to performance improvement. Chapter 2 presents a system-level framework for designing minimal deterministic routing algorithms for NoCs that are customized for a set of applications. Chapter 3 studies deadlock detection and recovery strategy in NoCs as opposed to deadlock avoidance. Chapter 4 discusses the abacus turn model which allows to keep the network deadlock-free by dynamically applying forbidden turns. Chapter 5 investigates routing algorithms based on learning approaches for balancing the traffic over the network.

The second part of the book focuses on multicast communication. Chapter 6 presents a new method and concept for implementing efficient and deadlock-free tree-based multicast routing algorithms. Chapter 7 addresses how to implement unicast and multicast routing efficiently in 2D and 3D mesh networks.

Fault tolerance and reliability are discussed in the third part of the book. Chapter 8 presents a fault-tolerant routing algorithm that keeps the negative effect of faulty components on the NoC power and performance as low as possible. Chapter 9 provides extensive knowledge on how to develop a fault-tolerant routing algorithm based on the characteristics of a system.

The fourth part of the book addresses power, energy, and thermal issues. Chapter 10 presents the use of bufferless deflection routing for removing input buffers which are responsible for a significant fraction of the total power budget. Chapter 11 proposes a routing algorithm to reduce the hotspot temperature for application-specific NoCs.

Emerging technologies are explored in the fifth part of the book. Chapter 12 introduces some design concepts of traffic- and thermal-aware routing algorithms in 3D NoC architectures, which aim at minimizing the performance impact caused by the run-time thermal managements. A new architecture for nanophotonic NoCs which consists of optical data and control planes is proposed in Chap. 13.

Finally, in the last part of the book (Chap. 14), an industrial case study illustrates a comprehensive approach in architecting (and micro-architecting) a scalable and flexible on-die interconnect and associated routing algorithms that can be applied to a wide range of applications in an industry setting.

Enna, Italy — Maurizio Palesi
Turku, Finland — Masoud Daneshtalab

Acknowledgments

"Life is like riding a bicycle. To keep your balance you must keep moving."
"Imagination is more important than knowledge. Knowledge is limited. Imagination encircles the world."
– Albert Einstein

A collective work of this nature is not possible without a considerable amount of mutual support. For that, we would like to express our gratitude to all the contributing authors for their hard work and cooperation. We would also like to thank the Springer staff involved in the publication of this book.

Maurizio Palesi would like to pay special thanks to his wife, Cristina Galvagno, for constant support and encouragement and for taking care of Alberto (their son) especially as the deadlines approached.

Masoud Daneshtalab would like to express his deepest gratitude to his wonderful wife, Masoumeh Ebrahimi, for her excellent guidance and patience. Without her love, encouragement, and patience, he would not be able to finish this work. She is always the first one he would go to whenever he needs support, and the first one to share his happiness on his success. Finally, Masoud would like to thank his mother, Mahnaz, for her constant love, support, and prayers.

"Let yourself be drawn by the stronger pull of that which you truly love"
"Two there are who are never satisfied – the lover of the world and the lover of knowledge"
– Rumi

Contents

Chapter 1
Basic Concepts on On-Chip Networks

Masoud Danashtalab and Maurizio Palesi

Abstract As the number of cores integrated into a System-on-Chip increases, the role played by the communication system becomes more and more important. The Network-on-Chip design paradigm is today recognised as the most viable communication infrastructure to deal with the scalability issues which characterise the ultra-deep sub-micron silicon era. In this chapter, some of the most important concepts in the context of on-chip networks will be reviewed. Basic concepts including, network topologies, switching techniques, and routing algorithms will be recalled. Such topics represent the conceptual bases exploited by the strategies, the mechanisms, and the methodologies discussed in the subsequent chapters.

1.1 Introduction

Programmable chips, like microprocessors, were originally developed with a single computational core. Such chips could execute program instructions in sequence and with a speed proportional to the clock frequency. Thus, increasing the clock frequency was the most common way to improve the overall performance of a chip. Unfortunately, increasing the clock frequency of a chip has several drawbacks. One of the most important concern is related to the increase in power dissipation which in turn determines the operational temperature of the chip. High clock frequency results in high temperature that has to be dissipated in a small surface of a few square centimeters. Temperature also impacts the system reliability as every 10 °C increase in operating temperature roughly doubles failure rate of the components.

M. Danashtalab (✉)
University of Turku, Turku, Finland
e-mail: masdan@utu.fi

M. Palesi
Kore University, Enna, Italy
e-mail: maurizio.palesi@unikore.it

M. Palesi and M. Daneshtalab (eds.), *Routing Algorithms in Networks-on-Chip*,
DOI 10.1007/978-1-4614-8274-1_1,

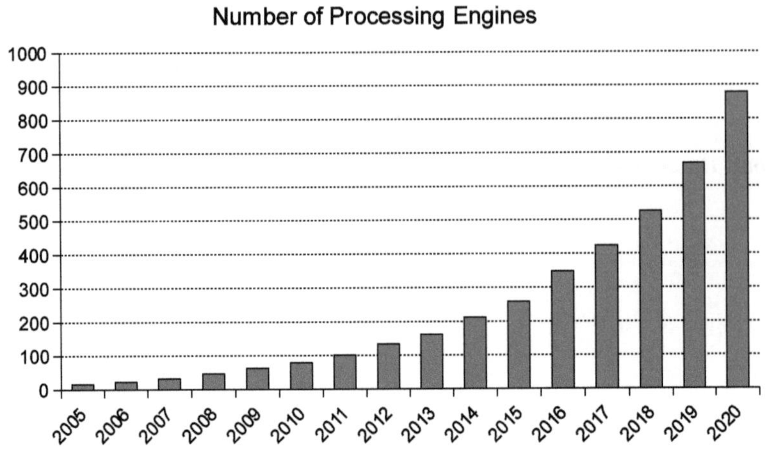

Fig. 1.1 Expected number of processing elements into a SoC [20]

The increasing demand in computational power of current and next generation applications cannot be "simply" provided by increasing the clock frequency. To overcome the limitation in increasing the clock frequency, modern chips are designed with multiple parallel computational cores able to work in parallel but with a reduced clock frequency. Thus, although instructions are executed slower than in a single core microprocessor running at higher clock frequency, they can be executed in parallel.

As the number of cores integrated into a single silicon chip increases (Fig. 1.1), the on-chip communication related issues are becoming more and more relevant than the computational related issues. That is, we can have all the computational horsepower we need (simply by increasing the number of computational cores) but, at the same time, there is a need for an efficient communication infrastructure for orchestrating the interaction between them.

The Network-on-Chip (NoC) design paradigm is currently considered as the most viable solution to deal with communication issues which affect the next generation of many-core System-on-Chip (SoC). A NoC can be imagined as a computer network in which computers are replaced by cores into the chip. In which information travels through the micronetwork in the form of packets traversing on-chip links and routed by on-chip routers.

NoCs are emerged as a highly scalable, reliable, and modular interconnect fabric for MPSoCs [2,8,21,22]. However, the network fabric takes up a substantial portion of system power budget [26]. As power is one of the most important issues in billion-transistor chips, in addition to network delay and area, the interconnect power consumption should be taken into consideration. Therefore, NoCs should be accommodated into the limited silicon area using efficient topology, routing algorithm, and router implementation.

In this chapter, basic concepts of NoCs including network topologies, switching techniques, flow control mechanisms, virtual channels, output selection, routing algorithms, and a general NoC architecture are presented.

1.2 Network Topology

The network topology is the study of the arrangement and connectivity of the routers. In other words, it defines the various channels and the connection pattern that are available for the data transfer across the network. Performance, cost, and scalability are the important factors in the selection of the appropriate topology. Shared-Bus, Crossbar, Butterfly Fat-Tree, Ring, Torus, and 2D-Mesh are the most popular topologies for on-chip interconnects which have been commercially used [14, 21].

Direct networks have at least one processing element (PE) attached to each router of the network so that routers may regularly spread between PEs. This helps to simplify the physical implementation. The shared-bus, ring, and 2D mesh/torus topologies (Fig. 1.2) are examples of direct networks, and provide tremendous improvement in performance, but at a cost of hardware overhead, typically increasing as the square of the number of PEs. On the other hand, indirect

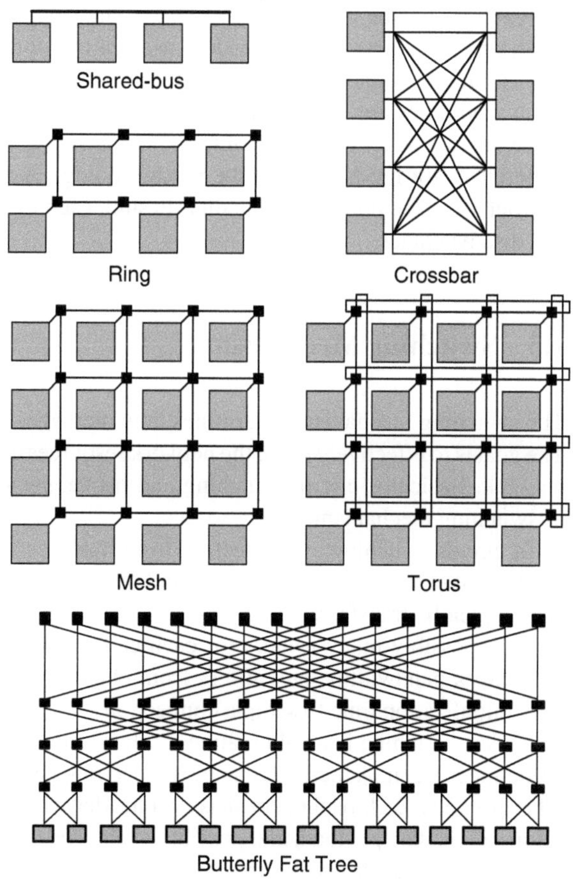

Fig. 1.2 Network topologies of Shared-bus, Ring, Crossbar, Mesh, Torus, and Butterfly

networks have a subset of routers which are not connected to any PE. All tree-based topologies where PEs are connected only to the leaf routers (e.g., the butterfly topology) as well as crossbar switch (Fig. 1.2) are indirect networks.

The shared-bus topology is the simplest using a common shared link common to all PEs where they compete for exclusive access to the bus. For communication intensive applications, it is necessary to overcome the bandwidth limitations of the shared-bus topology and move to scalable networks. However, this topology scales very poorly as the number of PEs increases. A small modification to the shared-bus topology to allow more concurrent transactions is to create the ring topology where every PE has exactly two neighbors. In this topology, messages hop along intermediate PEs until they arrive at the final destination. This causes the ring to saturate at a low injection rate for most traffic patterns. The crossbar topology is a fully connected one which allows every PE to directly communicate with any other PE.

The fat-tree topologies suffer from the fact that the number of routers exceeds the number of PEs, when the amount of PEs increases. This incurs an important network overhead. For the on-chip interconnects the network, overhead is more critical than for the off-chip networks and the design scalability is more essential. Mesh and torus networks are widely used in multiprocessor architectures because of their simple connection and easy routing provided by adjacency. Both torus and mesh topologies are fully scalable. Although torus provides a better performance, the regularity, better utilization of links, and lower network overhead are some of the preferences for mesh. That is, the mesh topology is more economic scheme since the routers on the borders are smaller. In sum, each topology has its own advantages and disadvantages.

1.3 Switching Mechanism

The switching mechanism determines how messages traverse a route in a network. The goal is to effectively share the network resources among messages traversing the network. Basically, circuit switching and packet switching form the two extremes of switching mechanisms.

In circuit switching, a connection from a source to a destination is established prior to the transmission of data and exclusively reserved until the message is completely transferred (e.g., as in telephone networks that set up a circuit through possibly many routers for each call). This mechanism has low delay and guarantees bandwidths, but suffers from channel utilization, low throughput, and long initialization time to setup a connection.

Packet switching is an alternative mechanism where data is not transmitted on a predefined circuit. A message can be divided into packets which share channels with other packets. Each packet consists of a header which contains routing and control information, data payload, and possibly a tail. The data payload follows the channel reserved by a header while the tail releases the channel reservation. Packets are individually and independently routed through the network, and at the destination the packets are assembled into the original message. If a message is divided into several packets, the order of packets at arrival PE must be the same as departure. Therefore, in-order delivery is an essential part that should be supported by on-chip networks. The packet switching mechanism improves channel utilization and network throughput.

In the packet switching domain, buffered flow control defines the mechanism that deals with the allocation of channels and buffers for the packets traversing between source and destination. The flow control mechanism is necessary when two or more packets compete to use the same channel, at the same time. Commonly three different buffered flow control strategies are used: store-and-forward, virtual cut through, and wormhole. When these mechanisms are implemented in on-chip networks, they have different performance metrics along with different requirements on hardware resources.

1.3.1 Store-and-Forward

Store-and-forward is the simplest flow control mechanism [14]. In this approach, each router along the path stores the entire packet in the buffer and then, the packet is forwarded to a selected neighboring router if the chosen neighboring router has enough empty buffering space available to hold the whole packet. This mechanism requires a large amount of buffering space (at least the size of the largest packet) in each router of the network, which can increase the cost dramatically. On top of that, network latency increases significantly because a packet cannot be forwarded to the next router until the whole packet is received and stored in the current router. Consequently, the store-and-forward approach is impractical in large-scale NoCs.

1.3.2 Virtual Cut-Through

The virtual cut-through mechanism [14] was proposed to address the large network latency problem in the store-and-forward strategy by reducing the packet delays at each routing stage. In this approach, one packet can be forwarded to the next stage before it is entirety received by the current router which reduces the store-and-forward delays. However, when the next stage router is not available, similar to the store-and-forward mechanism, the virtual cut-through approach also requires a large buffering space at each router to store the whole packet.

1.3.3 Wormhole

In this mechanism, a packet is divided into smaller segments called FLITs (FLow control digIT) [27]. Then, the flits are routed through the network one after another in a pipelined fashion. The first flit in a packet (header) reserves the channel of each router, the body (payload) flits will then follow the reserved channel, and the tail flit will later release the channel reservation. The wormhole mechanism does not require the complete packet to be stored in the router while waiting for the header flit to route to the next stages. One packet may occupy several intermediate routers at the same time. That is, the wormhole approach is similar to the virtual cut-through, but here the channel and buffer allocation is done on a flit-basis rather than packet-basis. Accordingly, the wormhole approach requires much less buffer space, thus, enabling small, compact and fast router designs. Because of these advantages, the wormhole mechanism is an ideal flow control candidate for on-chip networks.

1.4 Virtual Channels

There is a possibility of blocking in the wormhole network when a packet reserves a channel along a path which is prevented to be used by other packets. The use of Virtual Channels (VCs) overcomes the problem of blockages in the wormhole network via allowing blocked packets to be passed by other packets. This is accomplished by assigning several VCs, each with a separate flit queue, to each physical channel. For each VC, when the header flit arrives, a buffer will be assigned to the incoming packet, and is reserved until the trailer flit is transmitted. If a packet holding a VC gets blocked, other packets from other VCs can still traverse the physical channel.

As depicted in Fig. 1.3, at an input port the incoming flits are stored in distinct channel buffers which are multiplexed together again onto the output ports. If one of the channels is blocked, the other channels can access the outputs. VCs were introduced to solve the deadlock avoidance problem and to improve network latency

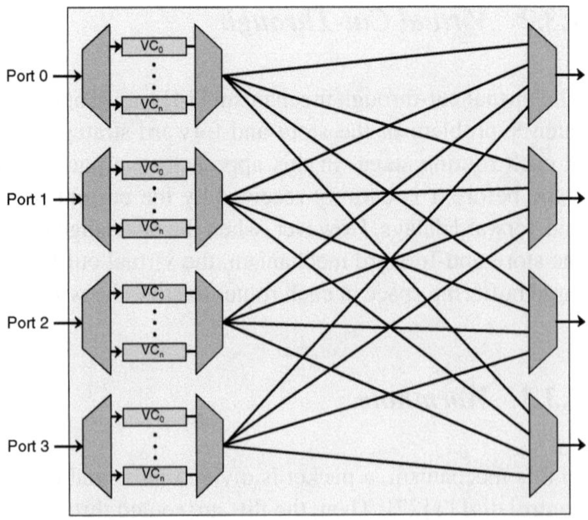

Fig. 1.3 A typical router using VCs

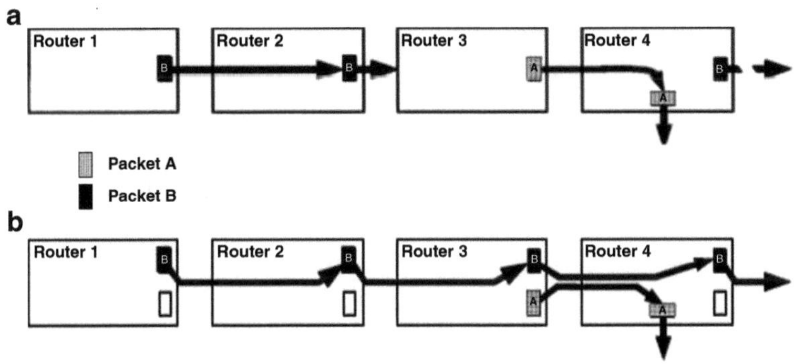

Fig. 1.4 Using VC for avoiding deadlock

and throughput. Figure 1.4a shows how a packet *A* blocked between routers 3 and 4 which also blocks the packet *B* when the network is not equipped with VCs. As illustrated in Fig. 1.4b, using VCs allows dual utilization of the physical channel between routers 3 and 4 where the packet *B* can pass the router 3. However, although employing VCs improves the performance and reduces Head-of-Line blocking (HoL) efforts in the network, it increases design complexity of the link controller and flow control mechanisms.

1.5 Network Dimension

The NoC design is commonly discussed in the form of two-dimensional (2D) and three-dimensional (3D) architectures. As shown in Fig. 1.5a, in 2D NoCs design, all switches are laid down in a single layer and connected to each other via intra-layer connections. In 3D NoCs (Fig. 1.5b), layers are stacked on top of each other via inter-layer connections instead of being spread across a 2D plane [4, 17, 31]. Each layer can use different technology, topology, clock frequency, etc. In recent years, through-silicon-via (TSV) has attracted a lot of attention to be employed for the inter-layer connections (vertical channels). TSVs enable faster and more power efficient inter-layer communication across multiple stacked layers. Figure 1.5 illustrates a 2D and 3D network with almost similar numbers of cores.

1.6 Output Scheduling

When multiple packets request for an output port, the need of an output scheduling algorithm that determines the priority order of candidate packets to advance emerges. In fact, the scheduler gives a priority order to each packet, and then the

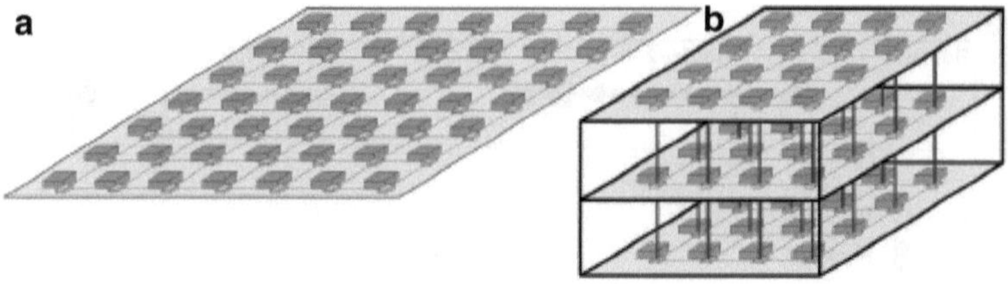

Fig. 1.5 (**a**) 2D NoC (**b**) 3D NoC

output ports select the highest-ordered packets to be forwarded. There is a variety of solutions with different implementation complexity and different performance characteristics e.g. round-robin, first-come first-served. The starvation prevention is the main concern that must be considered in the scheduler. Starvation is the situation in which low priority packets cannot get access to output ports as high priority packets always win the arbitration. The arbitration policy can be based on the round-robin or the packet priority scheme. Round-robin provides a fair scheduling by serving packets in a circular manner, so that it is starvation-free. However, packet prioritization might be needed in the network for the quality of service or it can be used for increasing the overall performance. On the other hand, prioritization schemes may lead to starvation in the network. A proper resource assignment can avoid starvation while packets have different priorities.

1.7 Routing Algorithm

Routing is the process that is used to forward the packets along appropriate directions in the network between a source and a destination.

In general, a routing algorithm can be seen as the cascade of two main blocks which implement the *routing function* [5, 6, 18, 32] and the *selection function* [1, 19, 29, 34] as shown in Fig. 1.6. First, the routing function computes the set of admissible output channels towards which the packet can be forwarded to reach the destination. Then, the selection function is used to select one output channel from the set of admissible output channels returned by the routing function. In a router implementing a deterministic routing algorithm (Sect. 1.7.2), the selection block is not present since the routing function returns only a single output port (see Fig. 1.6a). In a router implementing an oblivious routing algorithm (Sect. 1.7.2) the selection block takes its decision based solely on the information provided by the header flit (see Fig. 1.6b). Finally, network status information (e.g., link utilization and buffer occupation) are exploited by the selection function of a router implementing an adaptive routing algorithm (Sect. 1.7.2) (see Fig. 1.6c).

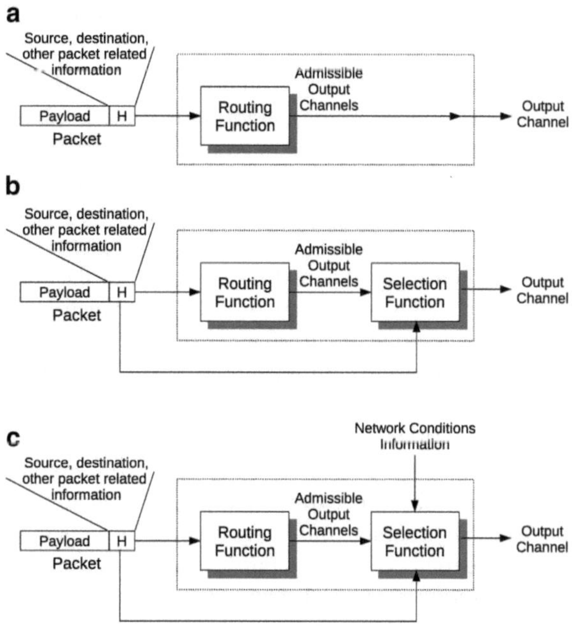

Fig. 1.6 Routing function and selection fucntion

Routing algorithms not only affect the transmission time but also can impact the power consumption and congestion conditions in the network.

1.7.1 Source Versus Distributed Routing

Routing can be utilized either at the source router or with a distributed manner by routers along the path. In the source routing scheme the entire route of a packet is decided by the source router stacking the exact router-to-router itinerary of a packet in the header. As the packet traverses in the network this information is used by each router on the path to navigate the packet towards the destination. This scheme is a simple solution for on-chip networks while the problem of the routing information overhead is the drawback of this scheme, i.e. for a network with a diameter of k, each packet requires at most k routing information stacked on the header of the packet. Accordingly, if the network grows, the header overhead becomes significant which is impractical for on-chip networks. In contrast, in the distributed routing approach the routing decision is taken by the individual routers depending on different parameters while the header of a packet has to include only the destination address. Each intermediate router examines the destination address (sometimes source address is also needed) and decides along which channel to forward the packet. However, the router complexity of the latter scheme is higher than the former scheme.

1.7.2 Deterministic Versus Adaptive Routing

Distributed routing scheme can be classified as *deterministic*, *oblivious*, and *adaptive*. Deterministic routing algorithms route packets in a fixed path between the source and destination routers. Implementations of deterministic routing algorithms are simple but they are not able to balance the load across the links in non-uniform or bursty traffic [3, 9]. Oblivious and adaptive routing algorithms are proposed to address these limitations. In oblivious routing, multiple paths from the source to the destination are possible and the selection of the routing path does not depend on the current status of the network. By better distributing load across links, adaptive algorithms improve the network performance and also provide tolerance if link or router failure occurs. In adaptive routing algorithms, the path of a packet from the source to the destination is determined by network conditions. An adaptive routing algorithm decreases the probability of passing a packet from a congested or malfunction link. While deterministic routing algorithms are the best choices for uniform or regular traffic patterns, the adaptive schemes are preferable in the presence of irregular traffic (non-uniform or bursty traffic) or in networks with unreliable routers and links. Furthermore, since packets may arrive at the destination from different paths and with different latencies, an adaptive routing could not guarantee the order of packets. To achieve in-order delivery property, a reordering module is required. These requirements increase both design complexity, and likely communication latency.

1.7.3 Minimal Versus Non-minimal Routing

Adaptive routing algorithms can be either minimal or non-minimal [15]. Minimal routing algorithms allow only the shortest paths to be chosen, while non-minimal routing algorithms also allow longer paths. Besides, a minimal fully adaptive routing algorithm can route packets along any shortest path adaptively; and a minimal partially adaptive routing algorithm cannot route packets along every shortest path.

1.7.4 Unicast and Multicast Routing Protocols

The communication in on-chip networks can be either unicast (one-to-one) or multicast/broadcast (one-to-many) [10, 16, 24]. In the unicast communication, a packet (message) is sent from a source router to a single destination router, while in the multicast communication a packet is transmitted from a source router to an arbitrary set of destination routers. Thus, the former is a special case of the latter.

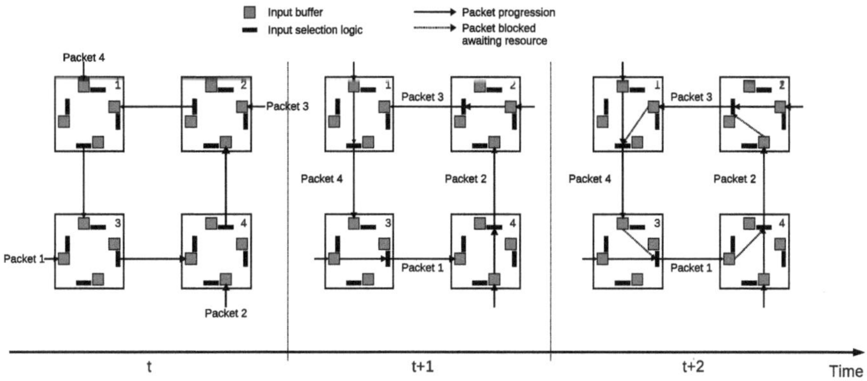

Fig. 1.7 An example of deadlock

The multicast communication is frequently employed in many applications of MPSoC such as replication [25], barrier synchronization [33], cache coherency in distributed shared-memory architectures [23], and clock synchronization [11]. Although the multicast communication can be implemented by multiple unicast communications, it produces a significant amount of unnecessary traffic increasing the latency and congestion in the network [24].

1.7.5 Deadlock and Livelock

Figure 1.7 shows an example of deadlock situation [28] occurring when, at the same time t, four packets Packet 1, Packet 2, Packet 3, and Packet 4 present at the west port of router 3, south port of router 4, east port of router 2, and north port of router 1 respectively. Destinations of the packets are two hops counterclockwise: Packet 1 is destined to node 2, Packet 2 to node 1, Packet 3 to node 3, and Packet 4 to node 4. Assuming a routing function which returns the output port to reach the destination in a minimal hop count favouring counterclockwise direction, at time $t+1$ the input logic of router 3 selects the west input creating a shortcut between the west input port and the east output port. Based on wormhole rules, such a shortcut is maintained until all the flits of Packet 1 have traversed router 3. At the same time, a shortcut from south input port of router 4 with its north output port is created. Similarly, a shortcut between east input port to west output port, and from north input port to south output port are created at time $t+1$ in router 2 and router 1 respectively. At time $t+2$, the first flit of Packet 1 is stored into the west input buffer of router 4. The routing function determines that the flit has to be forwarded to the north output port, but this port is already assigned for forwarding the flits of Packet 2. Thus, the first flit of Packet 1 is blocked in the west input buffer of router 4. Similarly, the first flit of Packet 2 is blocked in the south input port of router 2, the first flit of Packet 3 is blocked in the east input port of router 1, and the first flit of Packet 4 is blocked in the north input port of router 3. Assuming 1 flit input buffer size, the flits of the four packets cannot advance and, consequently, there is a deadlock.

A common way to verify the deadlock freedom property of a routing function is by means of Duato's theorem [12] which is an extension of Dally and Seitz theorem [7] for adaptive routing functions. It is based on the analysis of the channel dependency graph (*CDG*) associated to the routing function and the network topology. Specifically, the following definitions are needed to introduce the theorem.

Definition 1. A *Topology Graph* $TG = G(P,L)$ is a directed graph where each vertex p_i represents a node of the network and each directed arc $l_{ij} = (p_i, p_j)$ represents a physical unidirectional link connecting node p_i to node p_j.

Let $L_{in}(p)$ and $L_{out}(p)$ be the set of input links and output links for node p respectively. Mathematically:

$$L_{in}(p) = \{l|\ l \in L \wedge dst(l) = p\},$$
$$L_{out}(p) = \{l|\ l \in L \wedge src(l) = p\}.$$

where $src(l)$ and $dst(l)$ indicate the source and the destination network node of the link l.

Definition 2. A *Routing Function* for a node $p \in P$, is a function $R(p) : L_{in}(p) \times P \rightarrow \wp(L_{out}(p))$. $R(p)(l,q)$ gives the set of output links of node p that can be used to send a message received from the input link l and whose destination is $q \in P$.

The $\wp$ indicates a power set. We indicate with R the set of all routing functions: $R = \{R(p) :\ p \in P\}$.

Definition 3. Given a topology graph $TG(P,L)$ and a routing function R, there is a *direct dependency* from $l_i \in L$ to $l_j \in L$ if l_j can be used immediately after l_i by messages destined to some node $p \in P$.

Definition 4. A *Channel Dependency Graph* $CDG(L,D)$ for a topology graph TG, and a routing function R, is a directed graph. The vertices of *CDG* are the links of TG. The arcs of *CDG* are the pair of links (l_i, l_j) such that there is a direct dependency from l_i to l_j.

Based on the above definitions, the following theorem gives a sufficient condition for deadlock freedom.

Theorem 1 (Duato's Theorem [12]). *A routing function R for a topology graph TG is deadlock-free if there are no cycles in its channel dependency graph CDG.*

Livelock is a condition where a packet keeps circulating within the network without ever reaching its destination. It is the result of using a non-minimal adaptive routing algorithm. A livelock free routing algorithm has to guarantee forward progress of each packet, where after each hop the packet is in one step closer to its destination.

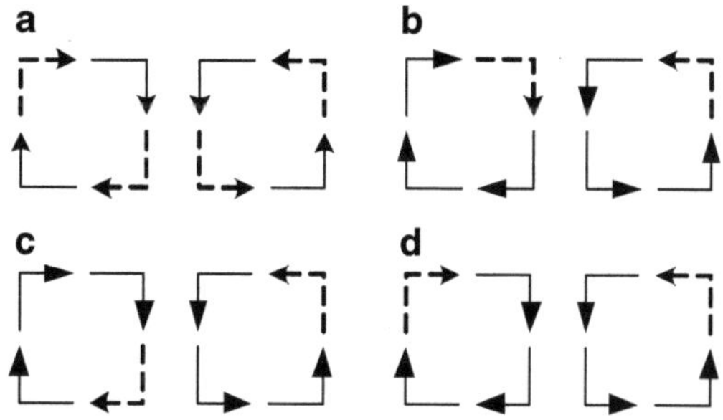

Fig. 1.8 All possible turns in (**a**) XY routing (**b**) Negative-First (**c**) West-First (**d**) North-Last (The *solid lines* indicate the allowable turns and the *dash lines* indicate the unallowable turns)

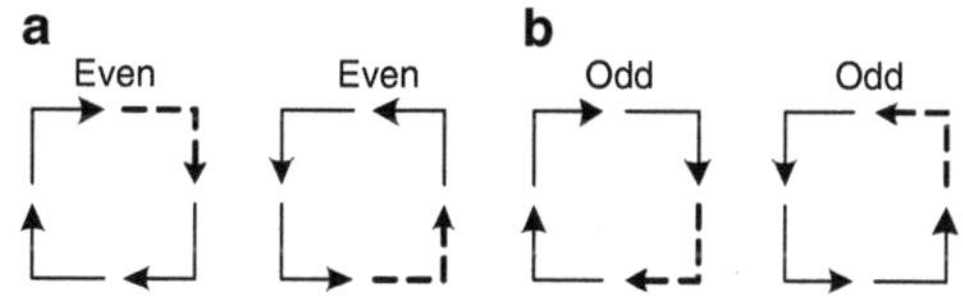

Fig. 1.9 The Odd-Even turn model rules: (**a**) prohibited turns in even columns (**b**) prohibited turns in odd columns

1.7.6 Turn Model Routing

Turn Model routing scheme based on wormhole switching mechanism provides deadlock and livelock freedom in the two-dimensional mesh topology [6, 18]. This model is also chosen as a representative of minimal and partial adaptive routing. In the turn model, deadlock can be avoided by prohibiting just enough turns to break all the cycles. Four well-known turn models are XY, Negative-First (NF), West-First (WF) and North-Last (NL) as shown in Fig. 1.8. Although the XY routing algorithm prohibits four turns to avoid deadlock, the other models avoid only two turns out of eight turns.

The Odd-Even model is one of the most popular partial adaptive wormhole routing algorithms in 2D mesh on-chip interconnection network [6] without virtual channels. Unlike the turn model which prohibits certain turns in all locations of the network, in the Odd-Even model some turns are restricted only in even columns and some other turns are prohibited in odd columns. Therefore, the degree of adaptiveness provided by this model is higher than the other turn models. Odd-Even rules can be described by the following rules:

Rule 1: East turns cannot be taken in even columns (Fig. 1.9a).
Rule 2: North turns cannot be taken in odd columns (Fig. 1.9b).

1.7.7 Application Specific Routing Algorithms

In the embedded systems scenario, unlike general networks, the computational as well as communication requirements of the applications can be very well characterized. Specifically, it is known which pairs of cores in the NoC system

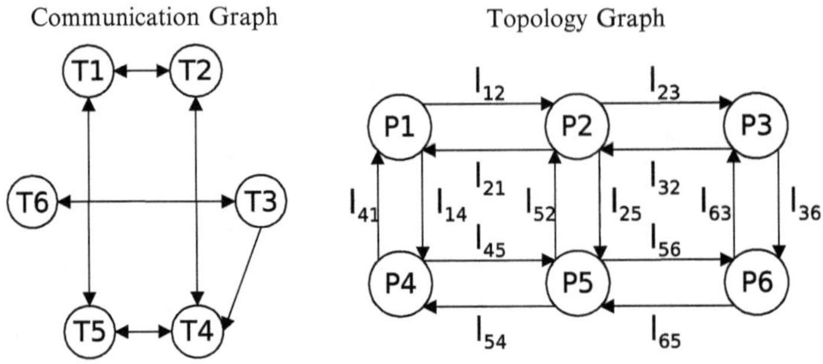

Fig. 1.10 Communication graph and topology graph

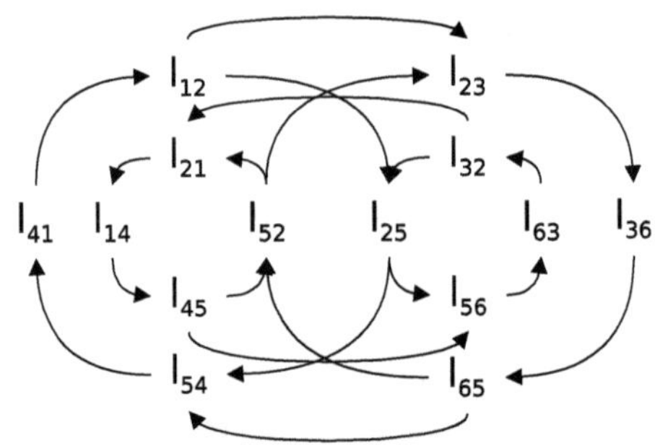

Fig. 1.11 Channel dependency graph for the network topology in Fig. 1.10, given a minimal fully adaptive routing algorithm

communicate and which do not. By off-line profiling and analysis, one can also estimate some quantitative information like communication bandwidth requirements between communicating pairs. After the applications have been mapped and scheduled on the NoC platform, information about communications which are never concurrent is also available. The APplication Specific Routing Algorithms (APSRA) methodology [30] allows a specific application to generate high efficient routing algorithms tailored. The basic idea behind APSRA is computing the channel dependency graph (*CDG*) [13] by considering just the direct channel dependencies generated by the communicating cores. Such *CDG*, called application specific *CDG* (*ASCDG*), will contain less cycles than the *CDG*. For this reason, there will be less cycles to be removed (i.e., less prohibited turns) with a consequent reduction of the impact on the adaptivity of the routing function.

Let us consider the communication graph and the topology graph depicted in Fig. 1.10. For the sake of simplicity, let us consider that the task T_i is mapped on node P_i, $i = 1,2,\ldots,6$.

The *CDG* [13] for a minimal fully adaptive routing algorithm is shown in Fig. 1.11. Since it contains several cycles, Duato's theorem [13] cannot assure deadlock freedom for minimal fully adaptive routing for this topology. To make the routing deadlock free, it is necessary to break all cycles of the *CDG*. Breaking a cycle by means of a dependency removal results in restricting the routing function and consequently loss of adaptivity. As the cycles to be removed are many, the adaptivity of the resulting deadlock free routing algorithm will be strongly reduced.

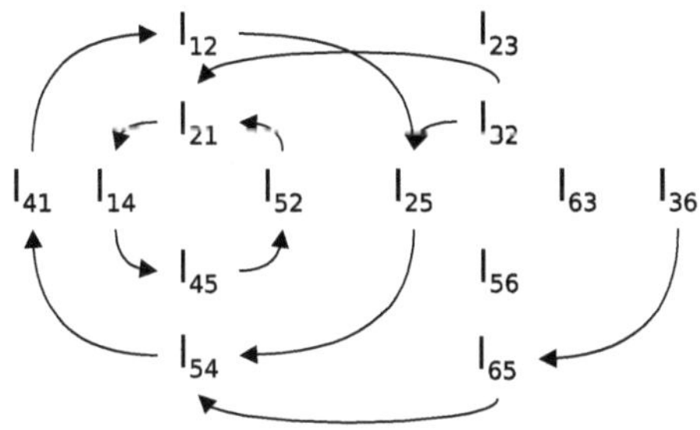

Fig. 1.12 Application specific channel dependency graph for the network topology and the communication graph in Fig. 1.11, given a minimal fully adaptive routing algorithm

However, if communication information is considered, several cycles of the *CDG* can be safely removed without any impact on the adaptivity of the routing algorithm. For example, study the dependency $l_{1,2} \to l_{2,3}$. Such a dependency is present in the *CDG* but it is not present in the *ASCDG*. In fact, channels $l_{1,2}$ and $l_{2,3}$ can be used in sequence only for the communications $T_1 \to T_3$, $T_1 \to T_6$, and $T_4 \to T_3$ which are not present in the communication graph. If we analyze the rest of the dependencies taking into consideration in the communication graph, we find that several dependencies can be safely removed without the need of restricting the routing function. The resulting *ASCDG* is shown in Fig. 1.12. It has been obtained from the *CDG* by removing all the dependencies which cannot be activated by the communication graph.

Although the number of cycles is reduced to two for the *ASCDG*, we still have a possibility of deadlock. To handle this, we can simply break the cycles as follows. The application specific channel dependency $l_{4,1} \to l_{1,2}$ is due to the communication $T4 \to T2$. Such communication can be realized by both paths $P4 \to P5 \to P2$ and $P4 \to P1 \to P2$. If the routing function is restricted in such a way that the latter path is prohibited, the application specific channel dependency $l_{4,1} \to l_{1,2}$ does not exist any longer. In a similar way it is possible to break the second cycle, for instance, by removing the dependency $l_{1,4} \to l_{4,5}$ due to communication $T1 \to T5$.

1.8 Performance Metrics

Many metrics have been used for estimating, evaluating and comparing the performance of NoCs. These metrics include different versions of latency (e.g., spread, minimum, maximum, average and expected), various versions of throughput, jitter in latency, jitter in throughput etc. The most common metrics used are the *average delay* and *average throughput*. The average delay, is the mean of the average communication delay over all the communications. The average communication delay is the average delay experimented by the packets of the communication to reach their destinations. The packet delay is the time elapsed from the instant in which the header of the packet is injected into the network, to the instant in which the tail of the packet (i.e., the tail flit) reaches the destination. The average throughput is the mean of the throughput over all the destination nodes. That is, the average number of packets received by the destination in the time unit.

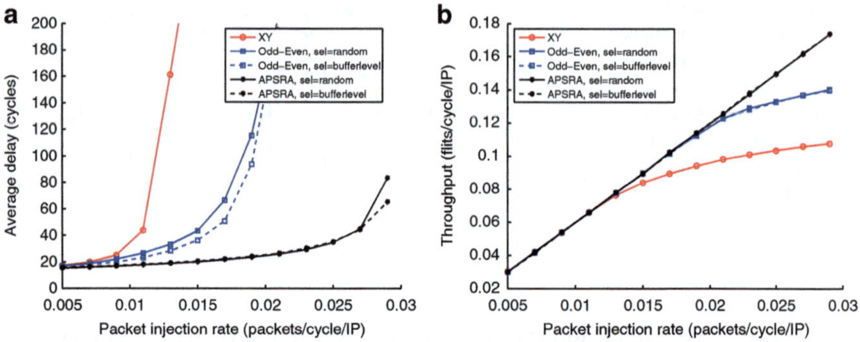

Fig. 1.13 Delay variation (**a**) and throughput variation (**b**) under *Transpose 2* traffic

Usually, the average delay and the average throughput are reported by means of a diagram for different packet injection rates. An example of such a diagrams is shown in Fig. 1.13. The figure shows the delay variation and throughput variation under *Transpose 2* [9] traffic scenarios for different routing functions (XY, Odd-Even [6], and APSRA [30]) and different selection policies (random, buffer-level [30]).

1.9 Summary

In this chapter, several important concepts in the domain of NoC design were presented. We have discussed various topologies for direct and indirect networks. Different switching, flow control mechanisms along with using virtual channels, routing schemes, output selection technique, and a general network-on-chip architecture were also described. These concepts presented here are further mentioned in various places in the rest of this book.

References

1. G. Ascia, V. Catania, M. Palesi, D. Patti, Implementation and analysis of a new selection strategy for adaptive routing in networks-on-chip. IEEE Trans. Comput. **57**(6), 809–820 (2008)
2. L. Benini, G.D. Micheli, Networks on chips: a new SoC paradigm. IEEE Comput. **35**(1), 70–78 (2002)
3. D.P. Bertsekas, R.G. Gallager, *Data Networks* (Prentice Hall, Englewood Cliffs, 1992)
4. C.-H. Chao, K.-Y. Jheng, H.-Y. Wang, J.-C. Wu, A.-Y. Wu, Traffic- and thermal-aware run-time thermal management scheme for 3D NoC systems, in *ACM/IEEE International Symposium on Networks-on-Chip*, Grenoble, 2010, pp. 223–230
5. A.A. Chien, J.H. Kim, Planar-adaptive routing: low-cost adaptive networks for multiprocessors. J. ACM **42**(1), 91–123 (1995)

6. G.-M. Chiu, The odd-even turn model for adaptive routing. IEEE Trans. Parallel Distrib. Syst. **11**(7), 729–738 (2000)
7. W.J. Dally, C. Seitz, Deadlock-free message routing in multiprocessor interconnection networks. IEEE Trans. Comput. **C**(36), 547–553 (1987)
8. W.J. Dally, B. Towles, Route packets, not wires: on-chip interconnection networks, in *ACM/IEEE Design Automation Conference*, Las Vegas, 2001, pp. 684–689
9. W.J. Dally, B. Towles, *Principles and Practices of Interconnection Networks* (Morgan Kaufmann, San Francisco, 2004)
10. M. Daneshtalab, M. Ebrahimi, T.C. Xu, P. Liljeberg, H. Tenhunen, A generic adaptive path-based routing method for MPSoCs. Elsevier J. Syst. Archit. **57**(1), 109–120 (2011)
11. M.M. de Azevedo, D. Blough, Fault-tolerant clock synchronization of large multicomputers via multistep interactive convergence, in *International Conference on Distributed Computing Systems*, Hong Kong, 1996, pp. 249–257
12. J. Duato, A new theory of deadlock-free adaptive routing in wormhole networks. IEEE Trans. Parallel Distrib. Syst. **4**(12), 1320–1331 (1993)
13. J. Duato, A necessary and sufficient condition for deadlock-free routing in wormhole networks. IEEE Trans. Parallel Distrib. Syst. **6**(10), 1055–1067 (1995)
14. J. Duato, S. Yalamanchili, L. Ni, *Interconnection Networks: An Engineering Approach* (Morgan Kaufmann, San Francisco, 2002)
15. M. Ebrahimi, M. Daneshtalab, F. Fahimeh, P. Liljeberg, J. Plosila, M. Palesi, H. Tenhunen, HARAQ: congestion-aware learning model for highly adaptive routing algorithm in on-chip networks, in *ACM/IEEE International Symposium on Networks-on-Chip*, Copenhagen, May 2012, pp. 19–26
16. M. Ebrahimi, M. Daneshtalab, P. Liljeberg, J. Plosila, J. Flich, H. Tenhunen, Path-based partitioning methods for 3d networks-on-chip with minimal adaptive routing. IEEE Trans. Comput. **99**, pp. 1, doi: 10.1109/TC.2012.255
17. M. Ebrahimi, M. Daneshtalab, P. Liljeberg, J. Plosila, H. Tenhunen, Cluster-based topologies for 3D networks-on-chip using advanced inter-layer bus architecture. Elsevier J. Comput. Syst. Sci. **79**(4), 475–491 (2013)
18. C.J. Glass, L.M. Ni, The turn model for adaptive routing. J. Assoc. Comput. Mach. **41**(5), 874–902 (1994)
19. J. Hu, R. Marculescu, DyAD – smart routing for networks-on-chip, in *ACM/IEEE Design Automation Conference*, San Diego, 7–11 June 2004, pp. 260–263
20. ITRS 2011 edition, *International Technology Roadmap for Semiconductors* (2011). http://www.itrs.net/
21. A. Jantsch, H. Tenhunen (eds.), *Networks on Chip*, chapter 1 (Kluwer Academic, Boston, 2003)
22. S. Kumar, A. Jantsch, J.-P. Soininen, M. Forsell, M. Millberg, J. Oberg, K. Tiensyrja, A. Hemani, A network on chip architecture and design methodology, in *IEEE Computer Society Annual Symposium on VLSI*, Pittsburg, p. 117, 2002
23. K. Li, R. Schaefer, A hypercube shared virtual memory, in *International Conference on Parallel Processing*, University Park, 1989, pp. 125–132
24. X. Lin, L.M. Ni, Multicast communication in multicomputer networks. IEEE Trans. Parallel Distrib. Syst. **4**, 1105–1117 (1993)
25. P.K. McKinley, H. Xu, E.T. Kalns, L.M. Ni, CompaSS: efficient communication services for scalable architectures, in *International Conference on Supercomputing*, Washington, D.C., 1992, pp. 478–487
26. G.D. Micheli, L. Benini, Powering networks on chips: energy-efficient and reliable interconnect design for SoCs, in *International IEEE Symposium on Systems Synthesis*, Montréal, 2001, pp. 33–38
27. P. Mohapatra, Wormhole routing techniques for directly connected multicomputer systems. ACM Comput. Surv. **30**(8), 374–410 (1998)
28. L.M. Ni, P.K. McKinley, A survey of wormhole routing techniques in direct networks. IEEE Comput. **26**, 62–76 (1993)

29. E. Nilsson, M. Millberg, J. Oberg, A. Jantsch, Load distribution with the proximity congestion awareness in a network on chip, in *Design, Automation and Test in Europe*, Washington, D.C., 2003, pp. 1126–1127
30. M. Palesi, R. Holsmark, S. Kumar, V. Catania, Application specific routing algorithms for networks on chip. IEEE Trans. Parallel Distrib. Syst. **20**(3), 316–330 (2009)
31. D. Park, S. Eachempati, R. Das, A. Mishra, Y. Xie, N. Vijaykrishnan, C.R. Das, MIRA: a multi-layered on-chip interconnect router architecture, in *International Symposium on Computer Architecture*, Beijing, 2008, pp. 251–261
32. J. Upadhyay, V. Varavithya, P. Mohapatra, A traffic-balanced adaptive wormhole routing scheme for two-dimensional meshes. IEEE Trans. Comput. **46**(2), 190–197 (1997)
33. H. Xu, P.K. McKinley, E.T. Kalns, L.M. Ni, Efficient implementation of barrier synchronization in wormhole-routed hypercube multicomputers. J. Parallel Distrib. Comput. **16**, 172–184 (1992)
34. T.T. Ye, L. Benini, G.D. Micheli, Packetization and routing analysis of on-chip multiprocessor networks. J. Syst. Archit. **50**(2–3), 81–104 (2004)

Part I
Performance Improvement

Chapter 2
A Heuristic Framework for Designing and Exploring Deterministic Routing Algorithm for NoCs

Abbas Eslami Kiasari, Axel Jantsch, and Zhonghai Lu

Abstract In this chapter, we present a system-level framework for designing minimal deterministic routing algorithms for Networks-on-Chip (NoCs) that are customized for a set of applications. To this end, we first formulate an optimization problem of minimizing average packet latency in the network and then use the simulated annealing heuristic to solve this problem. To estimate the average packet latency we use a queueing-based analytical model which can capture the burstiness of the traffic. The proposed framework does not require virtual channels to guarantee deadlock freedom since routes are extracted from an acyclic channel dependency graph. Experiments with both synthetic and realistic workloads show the effectiveness of the approach. Results show that maximum sustainable throughput of the network is improved for different applications and architectures.

2.1 Introduction

Thanks to high performance and low power budget of ASICs (application specific integrated circuits), they have been common components in the design of embedded systems-on-chip. Advances of semiconductor technology facilitate the integration of reconfigurable logic with ASIC modules in embedded systems-on-chip. Reconfigurable architectures are used as new alternatives for implementing a wide range of computationally intensive applications, such as DSP, multimedia and computer vision applications [1]. In the beginning of the current millennium, network-on-chip (NoC) emerged as a standard solution in the on-chip architectures [10, 11].

A.E. Kiasari (✉) • A. Jantsch • Z. Lu
KTH Royal Institute of Technology, Stockholm, Sweden
e-mail: kiasari@kth.se

M. Palesi and M. Daneshtalab (eds.), *Routing Algorithms in Networks-on-Chip*,
DOI 10.1007/978-1-4614-8274-1_2, © Springer Science+Business Media New York 2014

In network-based systems, the performance of the communication infrastructure is critical, as it can represent the overall system performance bottleneck. The performance of networks depends heavily on the routing algorithm effectiveness, since it impacts all network metrics such as latency, throughput, and power dissipation.

Routing algorithms are generally categorized into deterministic and adaptive. A deterministic routing algorithm is oblivious of the dynamic network conditions and always provides the same path between a given source and destination pair. In contrast, in adaptive routing algorithms, besides source and destination addresses, network traffic variation plays an important role for selecting channels to forward packets. However, adaptive routing may cause packets to arrive out-of-order since they may be routed along different paths. The re-order buffers needed at the destination for ordering the packets impose large area and power on system [18]. Deterministic routers not only are more compact and faster than adaptive routers [5], but also guarantee in-order packet delivery. Therefore, it is not surprising that designers would like to use deterministic routing algorithms in the NoCs which suffer from limited silicon resources. However, in deterministic routing a packet cannot use alternative paths to avoid congested channels along its route; this leads to degraded performance of the communication architecture at high levels of network throughput.

A well-designed routing algorithm utilizes the network resources uniformly as much as possible and avoids the congested channels, even in the presence of non-uniform traffic patterns, which are usual in the embedded systems. In this chapter, we propose a system-level Latency-Aware Routing (LAR) framework for designing minimal deterministic routing algorithms for network-based platforms. Especially, LAR is appropriate for reconfigurable embedded systems-on-chip which host several applications with high computational requirements and static workloads. To the best of our knowledge, the proposed framework is the first one to deal with traffic burstiness. Before the execution of a new application, the routing tables are configured with pre-computed routes, as well as other components in the system. After selecting the route and adding it to the packet, no further time is needed on routing at the intermediate nodes along the path. Due to advantages of table-based routing, it is one of the most widely used routing methods for implementing deterministic routing algorithm, e.g., IBM SP1 and SP2 [5].

LAR uses a recently proposed analytical model in [14] to calculate the average packet latency in the network. The results obtained from simulation experiments confirm that the proposed routing framework can find efficient routes for various networks and workloads.

The rest of the chapter is organized as follows. We start by reviewing previous studies in Sect. 2.2. The proposed heuristic framework is proposed in Sect. 2.3. Experimental results in Sect. 2.4 show that our proposed approach can improve the system performance. Finally, concluding remarks are given in Sect. 2.5.

2.2 Related Work

Turn model for designing partially adaptive routing algorithms for mesh and hypercube networks was proposed in [9]. Prohibiting minimum number of turns breaks all of the cycles and produces a deadlock-free routing algorithm. Turn model was used to develop the Odd-Even adaptive routing algorithm for meshes [4]. This model restricts the locations where some turns can be taken so that deadlock is avoided. In comparison with turn model, the degree of routing adaptivity provided by the Odd-Even routing is more even for different source-destination pairs.

DyAD routing scheme, which combines deterministic and adaptive routing, is proposed in [12] for NoCs, where the router works in deterministic mode when the network is not congested, and switches to adaptive mode when the network becomes congested. In [23] the authors extend routers of a network to measure their load and to send appropriate load information to their direct neighbors. The load information is used to decide in which direction a packet should be routed to avoid hot-spots. Recently, the authors in [19] present APSRA, a methodology to develop adaptive routing algorithms for NoCs that are specialized for an application or a set of concurrent applications. APSRA exploits the application-specific information regarding pairs of cores that communicate and other pairs that never communicate in the NoC platform to maximize communication adaptivity and performance.

Since all of these approaches are based on adaptive routing, they suffer from out-of-order packet delivery. Our proposed routing framework overcomes this problem while it minimizes the average packet latency across the network.

An application-aware oblivious routing is proposed in [14] that statically determines deadlock-free routes. The authors presented a mixed integer-linear programming approach and a heuristic approach for producing routes that minimize maximum channel load. However, in case of realistic workload, they did not study the effect of task mapping on their approach. Also, we have addressed the congestion-aware routing problem in [15]. With the analysis technique, we first estimated the congestion level in the network, and then embedded this analysis technique into the loop of optimizing routing paths so as to find deterministic routing paths for all traffic flows while minimizing the congestion level in the network. Since this framework cannot capture the traffic burstiness, in this work we utilize an analytical model [14] to deal with traffic burstiness.

2.3 LAR Framework

The LAR framework consists of five steps as its flowchart is shown in Fig. 2.1. At first, we represent the architecture and application using *topology graph* (TG) and *communication graph* (CG), respectively. Then we construct the *channel dependency graph* (CDG) based on TG and CG. In the third step, an acyclic CDG is extracted by deleting some edges from CDG to guarantee the deadlock freedom.

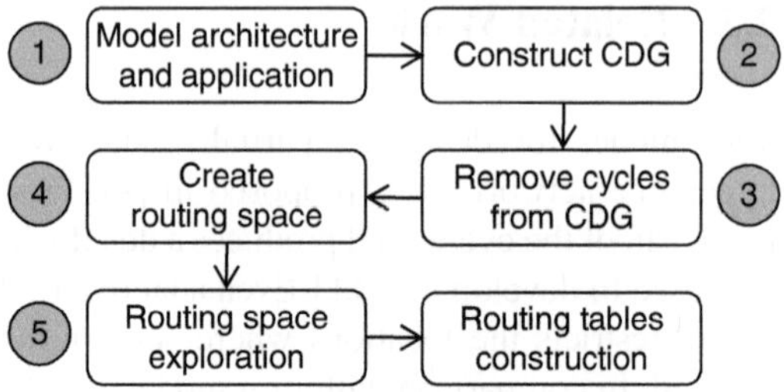

Fig. 2.1 The flowchart of LAR framework

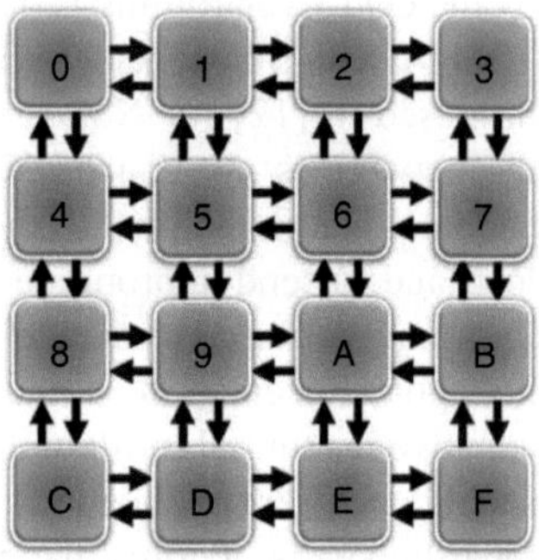

Fig. 2.2 TG of a 4 × 4 mesh network

After that, we find all possible shortest paths for each flow to create the routing space. Finally, we formulate an optimization problem over the routing space and solve it. In the following subsections, each step is described in detail.

2.3.1 Model Architecture and Application

In order to characterize network performance, a network model is essential. As shown in Fig. 2.2, a directed graph, which is called *Topology Graph* (*TG*), can represent the topology of an NoC architecture. Vertices and edges of TG show nodes and links of the NoC, respectively. Every node in TG contains a core and a wormhole-switched router. Such cores are local computing or storage regions, which may contain a processor, a DSP core, a configurable hardware, a high-bandwidth I/O controller, or a memory block. Each core is equipped with a Resource Network Interface (RNI). The RNI translates data between cores and routers by packing/unpacking data packets and also manages the packet injection process. Packets are injected into the network on injection channel and leave the network from ejection channel. We consider the general reference architecture for routers [7], where a routing logic determines the output channel over which the packet advances. Routing is only performed with the head flit of a packet. After routing phase, a crossbar switch handles the connections among input and output channel.

An application can be modeled by a graph called *Communication Graph* (CG). CG is a directed graph, where each vertex represents one selected task, and each

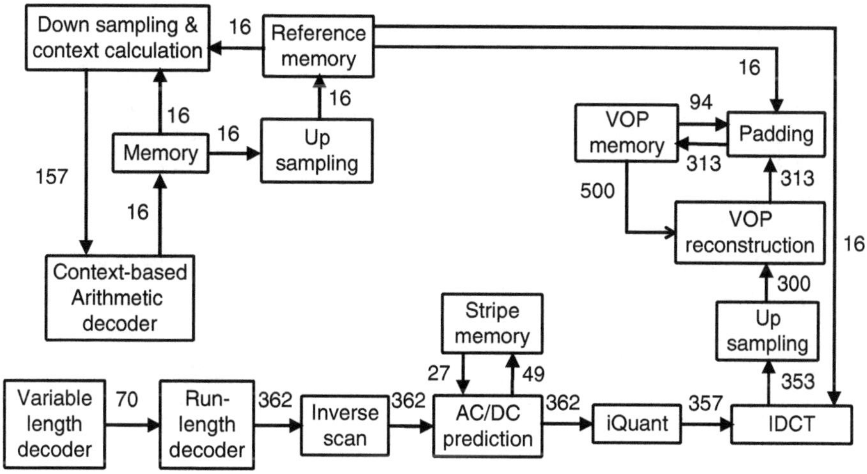

Fig. 2.3 CG of a video object plane decoder (VOPD) application [24]

directed arc represents the communication volume from source task to destination task. As an example, the CG of a video object plane decoder (VOPD) is shown in Fig. 2.3 [24]. Each node in the CG corresponds to a task and the numbers near the edges represent the bandwidth (in MBytes/s) of the data transfer, for a 30 frames/s MPEG-4 movie with $1{,}920 \times 1{,}088$ resolution [24].

2.3.2 *Construct Channel Dependency Graph*

Dally and Seitz simplified designing deadlock-free routing algorithms with a proof that an acyclic channel dependency graph (CDG) guarantees deadlock freedom [6]. Each vertex of the CDG is a channel in TG. For instance, vertex 01 in Fig. 2.4 corresponds to the channel from node 0 to node 1 in Fig. 2.2. There is a directed edge from one vertex in CDG to another if a packet is permitted to use the second channel in TG immediately after the first one. To find the edges of a CDG, we use the *Dijkstra's algorithm* to find all shortest paths between source and destination of any flows in corresponding TG. CDG of a 4×4 mesh network (Fig. 2.2) under minimal fully adaptive routing is shown in Fig. 2.4a, when any two nodes have the need to communicate such as in the uniform traffic pattern.

2.3.3 *Remove Cycles from CDG*

Traditional routing algorithms, such as *dimension-order routing* (DOR) and turn model, extract an acyclic CDG by systematically removing some edges from the CDG regardless of the traffic pattern. This may result in poor performance of

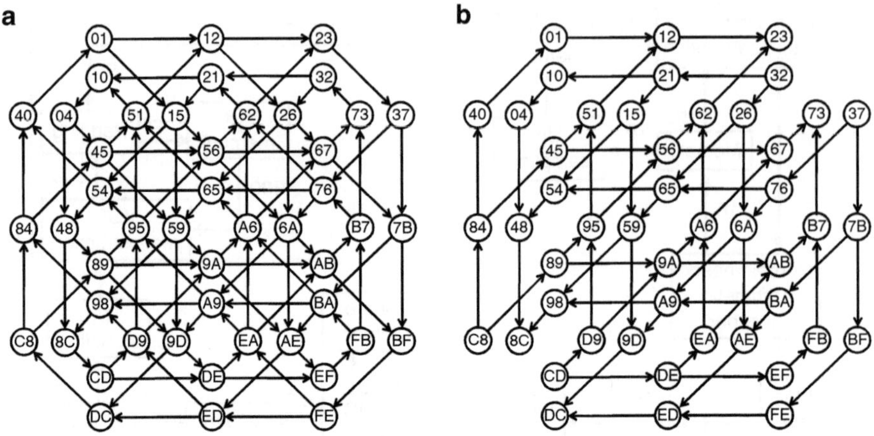

Fig. 2.4 The CDG of 4 × 4 mesh network for minimal fully adaptive routing under (**a**) uniform and (**b**) transpose traffic patterns

Table 2.1 Number of cycles in CDG of mesh networks

TG	Number of cycles in corresponding CDG
Mesh (2 × 2)	2
Mesh (2 × 3)	8
Mesh (3 × 3)	292
Mesh (3 × 4)	14,232
Mesh (4 × 4)	6,982,870
Mesh (4 × 5)	3,656,892,444

routing algorithm due to prohibition of unnecessary turns. For instance, as shown in Fig. 2.4b, there is no cycle in CDG of 4 × 4 mesh network under transpose traffic pattern, which the node in row i and column j sends packets to the node in row j and column i. However, traditional routing algorithms conservatively remove some edges from the CDG.

We modify the *depth-first-search* (*dfs*) algorithm to find cycles in a given CDG. Since we want to remove minimum number of edges, we delete an edge from the CDG which is shared among more cycles. Note that, this edge is removed if the reachability of all flows is guaranteed. For example, in a CDG of 4 × 4 mesh network, shown in Fig. 2.4a, there are 6,982,870 cycles and the edge from vertex 40 to vertex 01 is shared among 5,041,173 cycles. Thus by removing this edge from the CDG, the number of cycles is considerably reduced to 1,941,697. These steps are repeated again while there is no cycle in the CDG. Table 2.1 shows the numbers of cycles found by LAR in the CDG of different mesh networks. As it can be vividly seen, number of cycles is exponentially grown with the size of the TG and it takes a long time to find all cycles in the CDG. Hence, we find cycles in the CDG till certain number of cycles, and then remove an edge from the CDG which is shared among more cycles.

2.3.4 Create Routing Space (RS)

In this step, we apply Dijkstra's algorithm to the acyclic CDG to find all shortest paths between source and destination of flows in corresponding TG and create a set of f flows $RS = \{F_1, F_2, \ldots, F_f\}$ where f is the number of all flows in the system. $F_i = (\lambda_i, C_{A_i}, n_i, P_i)$, where λ_i is the average packet generation rate and C_{A_i} is the coefficient of variation (CV) of packet interarrival time for flow i. We remind that the relationship between CV of random variable X and its moments is represented by $C_X^2 = \overline{x^2}/\bar{x}^2 - 1$. In [14], we show that CV of a random variable reflects the burstiness intensity very well. n_i is the number of available shortest paths for flow i and P_i is itself a set and includes all n_i routes for flow i.

Usually more than one shortest path is available between two nodes ($n_i > 1$) in the routing space RS, so it is reasonable to choose a path such that the average packet latency is minimized. In the next subsection, we formulate an optimization problem over RS to find a suitable route for each flow and then use the simulated annealing heuristic to solve this problem.

2.3.5 Routing Space Exploration

2.3.5.1 Define Optimization Problem

In this subsection, we define an optimization problem to explore the routing space of RS. It is essential to define *decision variables* and *objective functions* in formulating an optimization problem. Our goal is to select a path for flow i ($1 \leq i \leq f$) among n_i available paths to minimize the average packet latency. Therefore, we define $X = \{x_1, x_2, \ldots, x_f\}$ as decision variables in the space of RS where x_i refers to a path number for flow i ($1 \leq x_i \leq n_i$) and the average packet latency as objective function.

The use of simulation experiments makes the task of searching for efficient designs computationally intensive and does not scale well with the size of networks since the search space of such a problem increases dramatically with the system size. Therefore, it is simply impossible to use the simulation in optimization loops. In the following subsection, we use an efficient analytical model to find nearly optimal solutions in reasonable time.

2.3.5.2 Analytical Latency Model

If the performance of a routing algorithm is measured in terms of average packet latency, then maximizing the performance means, in fact, minimizing the end-to-end packet latency. In this section, we briefly review a recently proposed analytical performance model which estimates the average packet latency in on-chip networks [14].

In a wormhole switched network, the end-to-end delay of a packet consists of two parts: the latency of the head flit and the latency of the body flits which follow the header flit in a pipelined fashion. The average latency of the head flit can be computed as the sum of delays at each hop, clearly, the link delays the head flit experienced and the residence times of the head flit in each of the routers along the path. Therefore, generally the only unknown parameter for computing the average packet latency is the mean waiting time for a packet from input channel i to output channel j in router N ($W_{i\to j}^{N}$). Using a G/G/1 priority queueing model, we estimated this value by [14]

$$W_{i\to j}^{N} = \begin{cases} \dfrac{\rho_j^N\left(C_A^2+C_{S_j^N}^2\right)}{2\left(\mu_j^N-\lambda_{1\to j}^N\right)}, & i=1, \\[2ex] \dfrac{\lambda_j^N\left(C_A^2+C_{S_j^N}^2\right)}{2\left(\mu_j^N-\sum_{k=1}^{i-1}\lambda_{k\to j}^N\right)^2}, & 2\le i\le p \end{cases} \tag{2.1}$$

where the variables are listed in Table 2.2 along with other parameters used in this chapter. Therefore, to compute the $W_{i\to j}^{N}$ we have to calculate the arrival rate from IC_i^N to OC_j^N ($\lambda_{i\to j}^N$), and also first and second moments of the service time of OC_j^N $\left(\bar{S}_j^N, \overline{\left(s_j^N\right)^2}\right)$. In the following two subsections, packet arrival rate and channel service time are computed.

Assuming the network is not overloaded, the arrival rate from IC_i^N to OC_j^N can be calculated using the following general equation

$$\lambda_{i\to j}^N = \sum\nolimits_S \sum\nolimits_D \lambda^S \times P^{S\to D} \times R\left(S\to D, IC_i^N \to OC_j^N\right) \tag{2.2}$$

In Eq. 2.2, the routing function $R(S\to D, IC_i^N \to OC_j^N)$ equals 1 if a packet from IP^S to IP^D passes from IC_i^N to OC_j^N; it equals 0 otherwise. Note that we assume a deterministic routing algorithm, thus the function of $R(S\to D, IC_i^N \to OC_j^N)$ can be predetermined, regardless of topology and routing algorithm. After that, the average packet rate to OC_j^N can be easily determined as

$$\lambda_j^N = \sum\nolimits_i \lambda_{i\to j}^N \tag{2.3}$$

After estimating the packet arrival rates, now we focus on the estimation of the moments of channel service times. At first, we assign a positive integer index to each output channel. Let D_j^N be the set of all possible destinations for a packet which passes through OC_j^N. The index of OC_j^N is equal to the maximum of distances among N and each M where $M \in D_j^N$. Obviously, the index of a channel is between 1

Table 2.2 Parameter notation

t_r	Time spent for packet routing decision (*cycles*)
t_s	Time spent for switching (*cycles*)
t_w	Time spent for transmitting a flit between two adjacent routers (*cycles*)
m	Average size of packets (*flits*)
L_b	The latency of body flits
$L^{S\to D}$	Average packet latency from IP^S to IP^D (*cycles*)
L	Average packet latency in the network (*cycles*)
IP^N	The IP core located at address N
R^N	The router located at address N
IC_i^N	The *i*th input channel in router R^N
OC_j^N	The *j*th output channel in router R^N
IB_i^N	Capacity of the buffer in IC_i^N (*flits*)
OB_j^N	Capacity of the buffer in OC_j^N (*flits*)
$P^{S\to D}$	Probability of a packet is generated in IP^S and is delivered to $IP^D \left(\sum_S\sum_D P^{S\to D}=1\right)$
λ^N	Average packet injection rate of IP^N (*packets/cycle*)
$\lambda_{i\to j}^N$	Average packet rate from IC_i^N to OC_j^N (*packets/cycle*)
λ_j^N	Average packet rate to OC_j^N (*packets/cycle*) $\left(\lambda_j^N=\sum_i\lambda_{i\to j}^N\right)$
$P_{i\to j}^N$	Probability of a packet entered form IC_i^N to be exited from OC_j^N
μ_j^N	Average service rate of the OC_j^N (*packets/cycle*)
$C_{S_j^N}$	Coefficient of variation (CV) for service time of the OC_j^N
C_A	CV for interarrival time of packets
ρ_j^N	The fraction of time that the OC_j^N is occupied
$W_{i\to j}^N$	Average waiting time for a packet from IC_i^N to OC_j^N (*cycles*)

and diameter of the network. In addition, the index of all ejection channels is supposed to be 0. After that, all output channels are divided into some groups based on their index numbers, so that group k contains all channels with index k.

Determination of the channel service time moments starts at group 0 (ejection channels) and works in ascending order of group numbers. Therefore, the waiting time from lower numbered groups can then be thought of as adding to the service time of packets on higher numbered groups. In other words, to determine the waiting time of channels in group k, we have to calculate the waiting time of all channels in group $k-1$. This approach is independent of the network topology and works for all kinds of deterministic routing algorithm, whether minimal or non-minimal.

In the ejection channel of R^N, the head flit and body flits are accepted in t_s+t_w and L_b cycles, respectively. Therefore, we can write $\bar{S}_1^N=t_s+t_w+L_b$ and since the standard deviation of packet size is known, we can easily compute $C_{S_1^N}$. Now, by using Eq. 2.1, the waiting time of input channels for ejection channel, $W_{i\to j}^N$, can be determined for all nodes in the network, where $2\le i\le p$.

Although the moments of service time can be computed simply for all ejection channels, service time moments of the other output channels cannot be computed in a direct manner by a general formula, and we have to use a more complicated approach. Now, we can estimate the first moment or average service time of ${OC_i}^M$ as

$$\bar{S}_i^M = \sum_{k=1}^{q} P_{j\to k}^N \left(t_s + t_w + t_r + W_{j\to k}^N + \bar{S}_k^N - \left(IB_j^N + OB_k^N\right) \times max\left(t_s, t_w\right)\right) \quad (2.4)$$

$$\overline{\left(s_i^M\right)^2} = \sum_{k=1}^{q} P_{j\to k}^N \left(t_s + t_w + t_r + W_{j\to k}^N + \bar{S}_k^N - \left(IB_j^N + OB_k^N\right) \times max\left(t_s, t_w\right)\right)^2 \quad (2.5)$$

where $P_{i\to j}^N$ is the probability of a packet entered form ${IC_j}^N$ to be exited from ${OC_k}^N$ and equals

$$P_{j\to k}^N = \lambda_{j\to k}^N / \lambda_i^M \quad (2.6)$$

Here, we should remind that to calculate $\bar{S}_i^M$ and $\overline{\left(s_i^M\right)^2}$ all values of $\bar{S}_k^N$ $(1 \le k \le q)$ must be computed before. Finally, the CV of channel service time for ${OC_i}^M$ can be given by

$$C_{S_i^M}^2 = \overline{\left(s_i^M\right)^2} / \left(\bar{S}_i^M\right)^2 - 1 \quad (2.7)$$

Now, we are able to compute the average waiting time of all output channels using Eq. 2.1. After computing $W_{i\to j}^N$ for all nodes and channels, the average packet latency between any two nodes in the network, $L^{S\to D}$, can be calculated. The average packet latency is the weighted mean of these latencies.

$$L = \sum_S \sum_D P^{S\to D} \times L^{S\to D} \quad (2.8)$$

where $P^{S\to D}$ is the probability of a packet is generated in IP^S and is delivered to IP^D. LAR framework uses the simulated annealing heuristic to minimize the average packet latency L as described briefly in the next subsection.

2.3.5.3 Simulated Annealing

Simulated Annealing is a stochastic computational method for solving the global optimization problem in a large search space. It is often used when the search space is discrete. For instance, simulated annealing has been applied to some computer-aided design (CAD) problems such as module placement [21] and packet routing [15]. For such problems, simulated annealing may be more efficient than exhaustive enumeration. While simulated annealing is unlikely to find the optimum solution, it can often find an acceptably good solution in a fixed amount of time. Simulated annealing was independently proposed as an optimization technique in 1983 [17] and 1985 [3]. This technique stems from thermal annealing in metallurgy which aims to increase the size of crystals and reduce their defects by heating a material and then slowly lowering the temperature to give atoms the time to attain the lowest energy state.

To simulate the physical annealing process, the simulated annealing algorithm starts with an initial solution and then at each iteration, a trial solution is randomly generated. The algorithm accepts the trial solution if it lowers the objective function (better solution), but also, with a certain probability, a trial solution that raises the objective function (worse solution). Usually the Metropolis algorithm [2] is used as the acceptance criterion in which worse solution are allowed using the criterion that

$$e^{-\Delta E/T} > R(0,1), \tag{2.9}$$

where ΔE is the difference of objective function with current and trial solutions (negative for a better solution; positive for a worse solution), T is a synthetic temperature, and $R(0,1)$ is a random number in the interval $[0,1]$. Typically this step is repeated until the system reaches a state that is good enough for the application, or until a given computation budget has been exhausted. By accepting worse solutions, the algorithm avoids being stuck at a local minimum in early iterations and is able to explore globally for better solutions. Detailed information about simulated annealing approach can be found in [17].

As mentioned in Sect. 2.3.5.1, objective function is the average packet latency and decision variables are represented by the routing set $X = \{x_1, x_2, \ldots, x_f\}$ where x_i is the path number for flow i $(1 \le x_i \le n_i)$. Let $X = \{x_1, x_2, \ldots, x_r, \ldots, x_f\}$ be the initial routing set. To choose a trial routing set $X_{new} = \{x_1, x_2, \ldots, x_r{}^{new}, \ldots, x_f\}$, we generate a random number r where $1 \le r \le f$ to choose a flow, and then generate another random number $x_r{}^{new}$ where $1 \le x_r{}^{new} \le n_r$ and $x_r{}^{new} \ne x_r$ to choose another path for flow r. Using analytical model describe in Sect. 2.3.5.2, we estimate the average packet latency for current and trial routing set.

2.4 Experimental Results

To evaluate the capability of the proposed framework, we developed a discrete-event simulator that mimics the behavior of routing algorithm in the networks at the flit level. Due to the popularity of the mesh network in NoC domain, our analysis focuses on this topology but LAR framework can be equally applied for other topologies without any change. Through all the experimental results, DOR routing is considered as the initial solution for the simulated annealing algorithm. We compare the performance of LAR with DOR which becomes XY routing algorithm in 2D mesh networks.

To achieve a high accuracy in the simulation results, we use the batch means method [20] for simulation output analysis. There are ten batches and each batch includes 1,000 up to 1,000,000 packets depending on the workload type, packet injection rate, and network size. Statistics gathering was inhibited for the first batch to avoid distortions due to the startup transient. The standard deviation of latency measurements is less than 1.8% of the mean value. As a result, the confidence level and confidence interval of simulation results are 0.99 and 0.02, respectively.

For the sake of comprehensive study, numerous validation experiments have been performed for several combinations of workload types and network size. In what follows, the capability of LAR will be assessed for both synthetic and realistic traffic patterns. Since their applications differ starkly in purpose, these classes of NoC have substantially different traffic patterns.

2.4.1 Synthetic Traffic

Synthetic traffic patterns used in this research include *uniform, transpose, shuffle, bit-complement,* and *bit-reversal* [5]. After developing models describing spatial traffic distributions, we should use an appropriate model to model the temporal traffic distribution. In the case of synthetic traffics, we use the Poisson process for modeling the temporal variation of traffic. It means that the time between two successive packet generations in a core is distributed exponentially. The Poisson model widely used in many performance analysis studies, and there are a large number of papers in many application domains that are based on this stochastic assumption.

The average packet latencies in the 4×4 and 8×8 mesh networks are plotted against offered load in the network in Figs. 2.5 and 2.6, respectively. We observe that under uniform and bit-complement traffic patterns LAR converges to DOR, because in such traffic patterns the average packet latency is minimum for DOR. It means that the simulated annealing algorithm is not able to find better routes and the final solution is the same as initial solution. This result is consistent with other results reported in [4, 9, 12, 19]. The main reason is that the DOR distributes packets evenly in the long term [9]. Previous works, Odd-Even [4], turn model [9], DyAD [12], and APSRA [19] indicate that in the case of uniform traffic, their proposed approaches underperform DOR. However, as can be seen in Figs. 2.5a and 2.6a, our proposed framework has the same performance as DOR for different traffic loads.

Figure 2.5b, c compare the latency of DOR and LAR in 4×4 mesh network under transpose and bit-reversal workloads, respectively. It can be vividly seen that LAR considerably outperforms DOR. Also, in the case of 8×8 mesh network, LAR has better performance than DOR as shown in Fig. 2.6b, c.

Figures 2.5d and 2.6d reveal that under shuffle traffic pattern LAR slightly outperforms DOR. Table 2.3 shows the maximum sustainable throughput of the network for each workload and for each routing algorithm in 4×4 and 8×8 mesh networks. It also shows the percentage improvement of LAR over DOR and reveals that on average LAR outperforms DOR. The maximum load that the network is capable of handling using LAR is improved by up to 205%.

Also, the performance of LAR framework is compared against DyAD routing scheme [12] which combines deterministic and adaptive routing algorithms. We

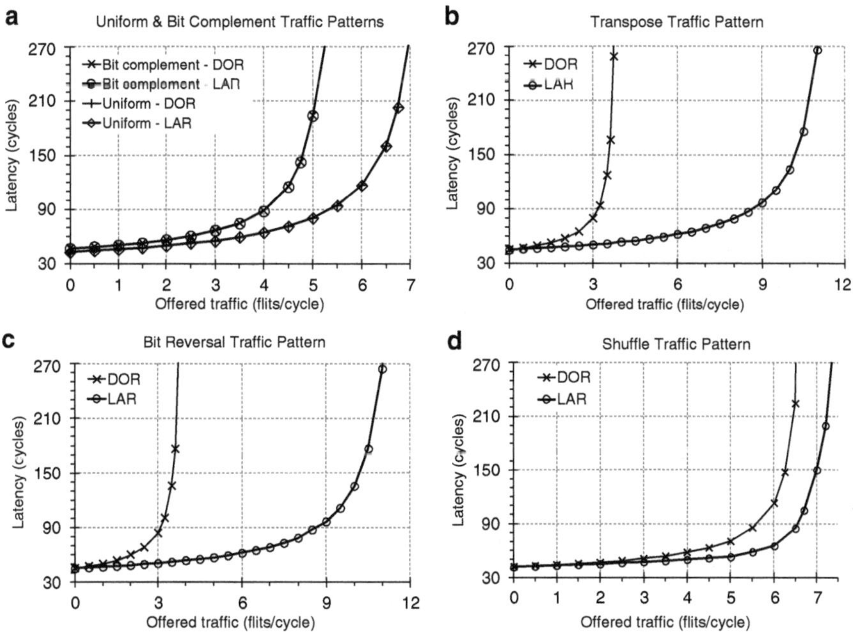

Fig. 2.5 Average packet latency under (**a**) uniform and bit-complement, (**b**) transpose, (**c**) bit-reversal, and (**d**) shuffle traffic patterns in 4×4 mesh network

Table 2.3 Improvement in maximum sustainable throughput of LAR as compared to DOR for different synthetic workloads

	4×4 mesh network			8×8 mesh network		
Workload	DOR	LAR	Impr.	DOR	LAR	Impr.
Uniform	7.4	7.4	0	15.9	15.9	0
Transpose	3.8	11.6	205%	7.7	10.5	36%
Bit-comp.	5.6	5.6	0	8.8	8.8	0
Bit-rev.	3.8	11.6	205%	7.6	9.2	21%
Shuffle	6.6	7.4	12%	12.2	13.4	10%

simulate the uniform and transpose workloads on the similar architecture (6×6 mesh network) and compare their improvement over DOR. Table 2.4 shows the percentage improvement of DyAD and LAR over DOR. In case of uniform workload, DyAD underperforms DOR while LAR has the same performance as DOR. In case of transpose traffic pattern, DyAD and LAR give about 62% and 60% improvement over DOR, respectively. This means that our deterministic routing policy can compete with adaptive routing policies (DyAD switches to adaptive mode under high traffic load) and meanwhile guarantees in-order packet delivery.

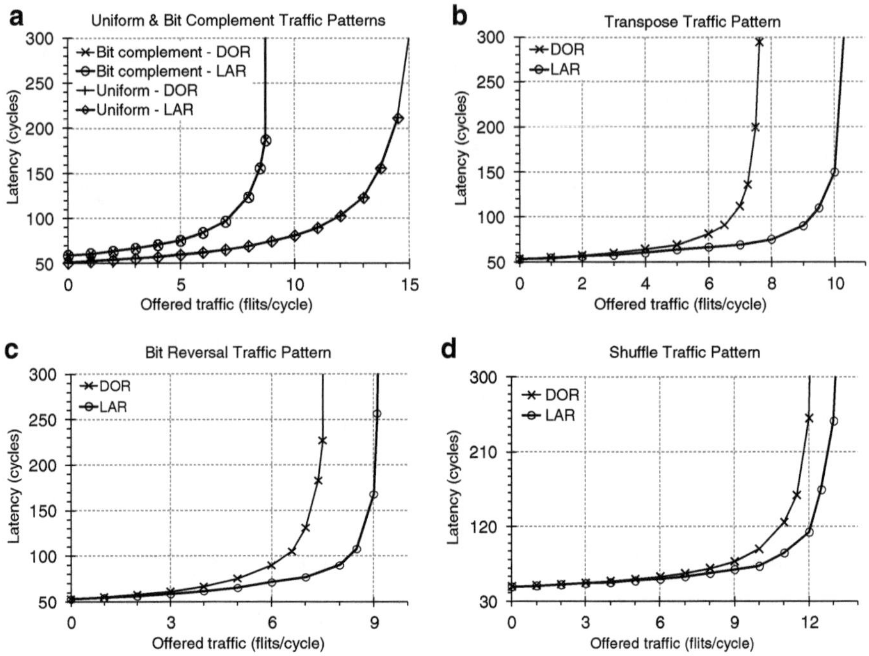

Fig. 2.6 Average packet latency under (**a**) uniform and bit-complement, (**b**) transpose, (**c**) bit-reversal, and (**d**) shuffle traffic patterns in 8×8 mesh network

Table 2.4 Improvement in maximum sustainable throughput of DyAD and LAR over DOR

	Improvement over DOR	
Workload	DyAD	LAR
Uniform	−21%	0
Transpose	62%	60%

2.4.2 *Realistic Traffic*

In case of realistic traffic, we consider two virtual channels for links to show the consistency of proposed framework with multiple virtual channel routing. As realistic communication scenarios, we consider a generic multimedia system (MMS) and the video object plane decoder (VOPD) application. MMS includes an H.263 video encoder, an H.263 video decoder, an mp3 audio encoder, and an mp3 audio decoder [13]. The communication volume requirements of this application are summarized in Table 2.5. VOPD is an application used for MPEG-4 video decoding and its communication graph is shown in Fig. 2.3. Several studies reported the existence of bursty packet injection in the on-chip interconnection networks for multimedia traffic [22, 25].

Table 2.5 MMS application traffic requirement [13]

src	dst	vol. (bytes)	src	dst	vol. (bytes)
ASIC1	ASIC2	25	DSP2	DSP1	20,363
ASIC1	DSP8	25	DSP3	ASIC4	38,016
ASIC2	ASIC3	764	DSP3	DSP6	7,061
ASIC2	MEM2	640	DSP3	DSP5	7,061
ASIC2	ASIC1	80	DSP4	DSP1	3,672
ASIC3	DSP8	641	DSP4	CPU	197
ASIC3	DSP4	144	DSP5	DSP6	26,924
ASIC4	DSP1	33,848	DSP6	ASIC2	28,248
ASIC4	CPU	197	DSP7	MEM2	7,065
CPU	MEM1	38,016	DSP8	DSP7	28,265
CPU	MEM3	38,016	DSP8	ASIC1	80
CPU	ASIC3	38,016	MEM1	ASIC4	116,873
DSP1	DSP2	33,848	MEM1	CPU	75,205
DSP1	CPU	20,363	MEM2	ASIC3	7,705
DSP2	ASIC2	33,848	MEM3	CPU	75,584

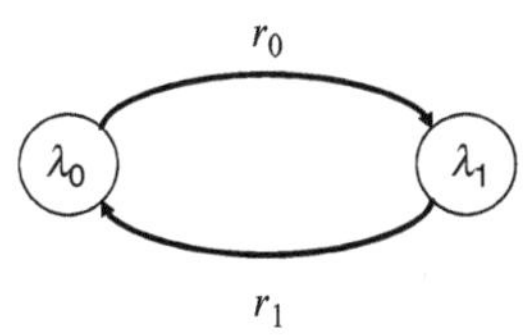

Fig. 2.7 Two-state MMPP model

Poisson process is not the appropriate model in case of bursty traffic; consequently, we used Markov-modulated Poisson process (MMPP) model as stochastic traffic generators to model the bursty nature of the application traffic [5, 8]. MMPP has been widely employed to model the traffic burstiness in the temporal domain [8]. Figure 2.7 shows a two-state MMPP in which the arrival traffic follows a Poisson process with rate λ_0 and λ_1. The transition rate from state 0 to 1 is r_0, while the rate from state 1 to state 0 is r_1.

Since in such systems, there are various types of cores with different bandwidth requirements, placement of tasks on a chip has strong effect on the system performance. To find a suitable mapping of these applications, we formulate another optimization problem to prune the large design space in a short time and then again use the simulated annealing heuristic to find a suitable mapping vector. Initially, we map task i to node i and then try to minimize the average packet latency through the simulated annealing approach. Figure 2.8a shows that in the case of MMS application and DOR, for the initial mapping M1, average packet latency equals 87 and after a certain number of tries, the mapping vector converges to the mapping M4 with average packet latency = 41. Furthermore, average packet latency values for mappings M2 and M3, which are two local minimum points in simulated annealing process, are shown in the figure.

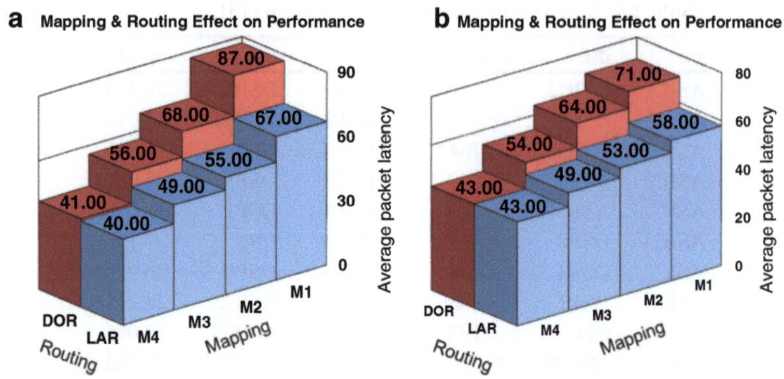

Fig. 2.8 The effect of mapping and routing on the performance of (**a**) MMS application and (**b**) VOPD application

After the mapping phase, we apply the LAR framework to these four mapping vectors. Figure 2.8a reveals that in case of mapping M1, LAR can significantly reduce the average packet latency from 87 to 67. However, for more efficient mapping vectors (M2, M3, and M4), we achieve less improvement. Specially, in the case of best mapping (M4), average packet latency is reduced insignificantly from 41 to 40. It is reasonable that DOR is latency-aware for the best mapping, because during the mapping problem solving process, we fix the routing policy to DOR and strive to minimize average packet latency for this routing policy. Likewise, as shown in Fig. 2.8b, for the VOPD application, the analysis result is the same as MMS application.

Figure 2.8 reveals that in case of application-specific traffic patterns, the improvement in the performance of the routing schemes highly depends on how the application tasks are mapped to the topology. This fact was not considered in the related works such as [16]. Nowadays, in embedded systems-on-chip there are several different types of cores including DSPs, embedded DRAMs, ASICs, and generic processors which their places are fixed on the chip. On the other hand, such a system hosts several applications with completely different workload. Furthermore, modern embedded devices allow users to install applications at run-time, so a complete analysis of such systems is not feasible during design phase. As a result, it is not feasible to map all applications such that the load is balanced for all of them with specific routing algorithm and we should balance the load in routing phase.

In this section we used the LAR framework to find low latency routes in the mesh network. Due to simplicity, regularity, and low cost merits of 2D mesh topology, it is the most popular one in the field of NoC. However, for large and 3D NoCs, which will be popular in the future, the communication in mesh architecture takes a long time. In the next subsection we use LAR to find deadlock-free paths in an arbitrary topology.

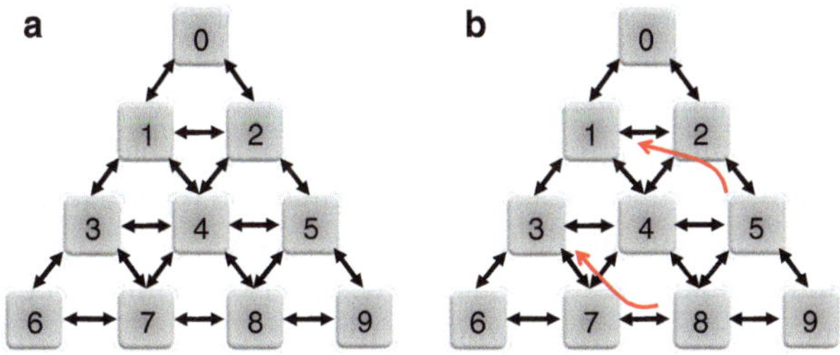

Fig. 2.9 (**a**) A custom topology and (**b**) prohibited turns

Table 2.6 Routing table for node 0 of topology in Fig. 2.8a

dst.	route	dst.	Route
0	No packet	5	SE, SE, EJ
1	SW, EJ	6	SW, SW, SW, EJ
2	SE, EJ	7	SE, SW, SW, EJ
3	SW, SW, EJ	8	SW, SE, SE, EJ
4	SW, SE, EJ	9	SE, SE, SE, EJ

2.4.3 Find Routes in an Arbitrary Topology

To show the capability of LAR framework to find deadlock-free routes in an arbitrary topology, we consider the topology shown in Fig. 2.9a. LAR reports that under uniform traffic pattern there are two cycles in the corresponding CDG and by prohibiting turns 52–21 and 87–73 (shown in Fig. 2.9b) the deadlock-freedom is guaranteed.

Table 2.6 shows the routing table for node 0 of the topology in Fig. 2.9a. Each route in the table specifies a path from node 0 to a given destination as channels name. SE, SW, and EJ specify South East, South West, and ejection channels, respectively. To route a packet, the routing table is indexed by destination address to look up the pre-computed route by LAR. This route is then added to the packet. Since there are seven channels in this network (E, S, NE, NW, SE, SW, and EJ), they can be encoded as 3-bit binary numbers. Also, there are techniques to reduce the size of routing tables [5, 19].

2.5 Conclusion

On-chip packet routing is extremely crucial because it heavily affects performance and power. This calls for a great need of routing optimization. However, due to the diverse connectivity enabled by a network and the interferences in sharing network buffers and links, determining good routing paths, which are minimal and deadlock free for traffic flows, is nontrivial. In this chapter, we have addressed the latency-aware routing problem. Using an analytical model, we first estimate the average packet latency in the network, and then embed this analysis technique into the loop of optimizing routing paths so as to quickly find deterministic routing paths for all traffic flows while minimizing the latency.

The proposed framework is appropriate for reconfigurable embedded systems-on-chip which run several applications with regular and repetitive computations on large set of data, e.g., multimedia and computer vision applications. LAR can not only design minimal and deterministic routing, but also can implement non-minimal routing without virtual channels in arbitrary topology.

References

1. K. Bondalapati, V.K. Prasanna, Reconfigurable computing systems. Proc. IEEE **90**(7), 1201–1217 (2002)
2. O. Catoni, Metropolis, simulated annealing, and iterated energy transformation algorithms, theory and experiments. J. Complex. **12**(4), 595–623 (1996)
3. V. Cerny, Thermodynamical approach to the traveling salesman problem: An efficient simulation algorithm. J. Opt. Theory. Appl. **45**, 41–45 (1985)
4. G.-M. Chiu, The odd-even turn model for adaptive routing. IEEE Trans. Parall. Distr. Syst. **11**(7), 729–738 (2000)
5. W.J. Dally, B. Towles, *Principles and practices of interconnection networks*, 1st edn. (Morgan Kaufmann, San Francisco, 2004)
6. W.J. Dally, C.L. Seitz, Deadlock-free message routing in multiprocessor interconnection networks. IEEE Trans. Comput. **36**(5), 547–553 (1987)
7. J. Duato, C. Yalamanchili, L. Ni, *Interconnection Networks: An Engineering Approach* (IEEE Computer Society Press, Los Alamitos, Morgan Kaufmann, San Francisco, 2003)
8. W. Fischer, K. Meier-Hellstern, The Markov-Modulated Poisson Process (MMPP) cookbook. Perform. Evaluat. **18**(2), 149–171 (1993)
9. C.J. Glass, L.M. Ni, The turn model for adaptive routing. J. Assoc. Comput. Mach. **41**(5), 874–902 (1994)
10. P. Guerrier, A. Greiner, A generic architecture for on-chip packet-switched interconnections, in *Proceedings of the Design, Automation, and Test in Europe* (Paris, France, 2000), pp. 250–256
11. A. Hemani et al., Network on a chip: An architecture for billion transistor era, in *Proceedings of the IEEE NorChip* (Turku, Finland, 2000), pp. 166–173
12. J. Hu, R. Marculescu, DyAD – smart routing for networks-on-chip, in *Proceedings of the Design Automation Conference* (San Diego, 2004), pp. 260–263
13. J. Hu, R. Marculescu, Energy- and performance-aware mapping for regular NoC architectures. IEEE Trans. Comp. Aid. Des. Integr. Circuit. Syst. **24**(4), 551–562 (2005)
14. A.E. Kiasari, Z. Lu, A. Jantsch, An analytical latency model for networks-on-chip. IEEE Trans. VLSI Syst. **21**(1), 113–123 (2013)
15. A.E. Kiasari, A. Jantsch, Z. Lu, A framework for designing congestion-aware deterministic routing, in *Proceedings of the 3rd International Workshop on Network-on-Chip Architectures (NoCArc), Held in conjunction with the 43rd Annual IEEE/ACM International Symposium on Microarchitecture (MICRO-43)* (Atlanta, 2010), pp. 45–50
16. M.A. Kinsy et al., Application-aware deadlock-free oblivious routing, in *Proceedings of the ISCA* (Austin, 2009), pp. 208–219
17. S. Kirkpatrick, C.D. Gelatt, M.P. Vecchi, Optimization by simulated annealing. Science **220**(4598), 671–680 (1983)
18. S. Murali et al., Analysis of error recovery schemes for networks on chips. IEEE Des. Test Comp. **22**(5), 434–442 (2005)

19. M. Palesi et al., Application specific routing algorithms for networks on chip. IEEE Trans. Parall. Distr. Syst. **20**(3), 316–330 (2009)
20. K. Pawlikowski, Steady-state simulation of queueing processes: A survey of problems and solutions. ACM Comput. Surv. **22**(2), 123–170 (1990)
21. C. Sechen, A. Sangiovanni-Vincentelli, The TimberWolf placement and routing package. J. Sol. State Circuit. **SC-20**, 510–522 (1985)
22. V. Soteriou, H. Wang, L.-S. Peh, A statistical traffic model for on-chip interconnection networks, in *Proceedings of the MASCOTS* (Monterey, 2006), pp. 104–116
23. W. Trumler et al., Self-optimized routing in a network-on-a-chip. IFIP World Comp. Cong. **268**, 199–212 (2008)
24. E.B. van der Tol, E.G. Jaspers, Mapping of MPEG-4 decoding on a flexible architecture platform. SPIE **4674**, 1–13 (2002)
25. G. Varatkar, R. Marculescu, Traffic analysis for on-chip networks design of multimedia applications, in *Proceedings of the Design Automation Conference* (New Orleans, 2002), pp. 795–800

Chapter 3
Run-Time Deadlock Detection

Ra'ed Al-Dujaily, Terrence Mak, Fei Xia, Alex Yakovlev, and Maurizio Palesi

Abstract Relentless technology downscaling provides a promising prospect to realize heterogeneous system-on-chip (SoC) and homogeneous chip-multiprocessor (CMP) based on the networks-on-chip (NoCs) paradigm with augmented scalability, modularity and performance. In many cases in such systems, scheduling and managing communication resources are the major design and implementation challenges instead of the computing resources. Moreover, adaptive packets routing in such systems are susceptible to routing deadlock, which could lead to performance degradation or system failure. Past research efforts were mainly focused on complex design-time approaches to avoid deadlock or simple heuristic run-time approaches to detect and recover from deadlock. The later, however, consider only local or partial information about the network which may produce substantial false detections, especially with the network close to saturation where blocked packets could be fagged as deadlocked.

This chapter studies deadlock detection and recovery strategy in NoCs as opposed to deadlock avoidance. It presents a deadlock detection method that utilizes run-time transitive-closure (TC) computation to discover the existence of deadlock-equivalence sets, which imply loops of requests in NoCs. This detection scheme guarantees the discovery of all true-deadlocks without false alarms in contrast with

R. Al-Dujaily (✉)
University of Southampton, Southampton SO17 1BJ, UK
e-mail: r.al-dujaily@ecs.soton.ac.uk

T. Mak
The Chinese University of Hong Kong, Ho Sin-Hang Engineering Building, Shatin, Hong Kong
e-mail: stmak@cse.cuhk.edu.hk

F. Xia • A. Yakovlev
Newcastle University, Newcastle Upon Tyne, NE1 7RU, UK
e-mail: fei.xia@newcastle.ac.uk; alex.yakovlev@newcastle.ac.uk

M. Palesi
Kore University, Enna, Italy
e-mail: maurizio.palesi@unikore.it

M. Palesi and M. Daneshtalab (eds.), *Routing Algorithms in Networks-on-Chip*,
DOI 10.1007/978-1-4614-8274-1_3,

state-of-the-art approximation and heuristic methods. A distributed TC-network architecture, which couples with the network-on-chip (NoC) infrastructure, is also presented to realize the detection mechanism efficiently. Detailed hardware realization architectures and schematics are also discussed. The captured results based on a cycle-accurate simulator demonstrate the effectiveness of the proposed method. It drastically outperforms timing-based deadlock detection mechanisms by eliminating false detections and, thus, reducing energy wastage in retransmission for various traffic scenarios including real-world application.

3.1 Introduction

Networks-on-Chip (NoCs) emerged as an on-chip communication architectural paradigm that applies networking theory and methods to very large scale integration (VLSI) systems incorporated on a single chip [7, 13]. Such an architecture consists of a network constructed from multiple point-to-point data links (channels) interconnected by routers. The routers are connected to a set of distributed intellectual properties (IP) cores, where each IP core can be an implementation of processor cores, memory modules, digital signal processing (DSP) blocks, etc. Communication among these distributed IP usually uses a packet-switching method where messages are divided into suitably-sized blocks, called packets and flits. The packets can be relayed from any of the source-destination communication pairs, over several data links, by making routing decisions at routers. NoCs bring a remarkable improvement in performance, flexibility, scalability and power efficiency over conventional bus interconnections. However, in NoCs communication deadlocks may appear at the routers network and cause an impasse in the communications, hence leading to performance degradation or system failure.

A deadlock is a situation in NoCs wherein two or more packets (agents) are waiting for one another to release channels (resources) and are unable to make progress. The network is deadlocked if a chain of waiting packets forms a dependency cycle. As a simple deadlock example, consider the dependency cycle of four packets shown in Fig. 3.1. Each of the four packets in the figure holds some channels, but it cannot proceed further until it acquires another channel currently held by one of the other of the packets. However, no packet can release the channel needed by another until it acquires its own requested channel. The packets are deadlocked and will remain in this state, halting all the occupied channels, until some intervention brings them back to life by breaking the cycle. Deadlocks can paralyze network communications by halting the resources occupied by the deadlocked packets which, in turn, could increase the likelihood of other packets blocking [33]. Hence the removal of deadlocks is essential. There are two strategies to deal with deadlocks: deadlock avoidance [14] and deadlock recovery [15].

In deadlock avoidance, resources are granted to packets in a way that the overall network is deadlock free. This can be based on a turn model which prohibits the routing algorithm from making certain turns in the network [9, 10] or based on

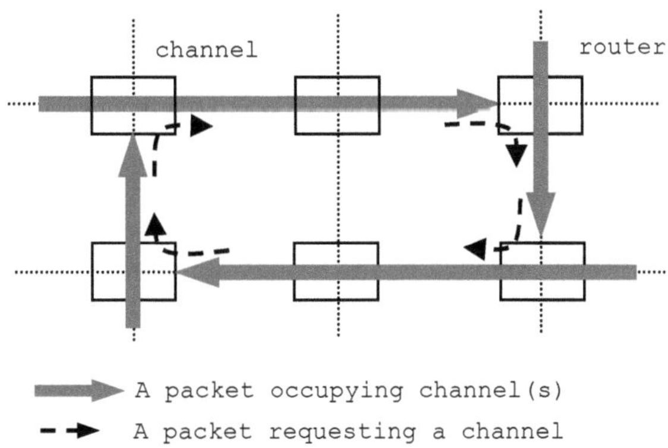

Fig. 3.1 An example of simple routing deadlock cycle, packets in the network are waiting for one another and forming a deadlock

the strict ordering of virtual channels (VCs) [15]. In general, avoidance techniques require restricted routing functions or additional resources, e.g., VCs. Due to its simplicity the turn model technique is popular in NoCs, even though it limits the routing alternatives and diminishes fault tolerance capabilities [21]. Moreover, it is not applicable to arbitrary network topologies.

Alternatively, deadlock recovery implies that channels are granted to packets without any routing restrictions, potentially outperforming deadlock avoidance [3, 25]. Deadlocks may occur and efficient detection and recovery mechanisms are required to intervene. However, detecting deadlock in a network is challenging because of the distributed nature of deadlocks. Heuristic approaches, such as timeout mechanisms, are often employed to monitor the activities at each channel for deadlock speculations. These techniques may produce a substantial number of false detections, especially with the network close to saturation where blocked packets could be flagged as deadlock. Several techniques have been proposed for reducing the number of false detections in general computer networks [22,24,30], nonetheless they are all based on the timeout idea and finding the best threshold values for different network settings is not an easy matter.

Unlike general computer networks, where internode information can only be exchanged through packets, on-chip networks can take advantage of additional dedicated wires to transmit control data between routers. This chapter exploits this NoCs-specific capability and proposes a new deadlock detection method which guarantees true deadlock detection. A run-time transitive-closure (TC) computation scheme is employed to discover the existence of deadlock-equivalence sets which imply loops of requests. Also, the proposed detection scheme can be realized using a distributed architecture, which closely couples with the NoC infrastructure, to speed up the necessary computation and to avoid introducing traffic overhead in the communication network. Initial results of this study and a sketch of the proposed architecture were presented in [1]. In [2], a complete theoretical framework of the TC computational approach is presented. Hardware architecture for the framework realization is detailed and experimental results on real-life applications are included.

3.2 Related Work

The study of deadlock recovery in the context of NoCs is rare and, as a consequence, the following information has been mainly acquired from the field of general computer networks. In [27, 33], the authors conclude that deadlocks can be highly improbable in interconnection networks when sufficient routing freedom is provided and fully exploited by the routing algorithm. Thus it is not favorable to limit the adaptivity of the routing algorithm to avoid a rare event, e.g., using the turn model [9, 10], nor to complicate the routers' design by devoting VCs specifically to prevent deadlocks [14, 15]. Since then, deadlock recovery has gained recognition for its potential for outperforming deadlock avoidance provided efficient detection and recovery mechanisms exist [22, 24, 25, 30]. Deadlock *detection* and *recovery* are the two important stages of any deadlock recovery scheme [15].

In the detection stage, the network must discover at run-time any deadlock dependency cycle. However, detecting deadlocks at run-time is challenging because of their highly distributed characteristics. Thus deadlock detection is usually implemented in a distributed way using a timeout mechanism [3,4,18,21,22,24,25,30]. In its simple form, a packet occupying a channel is suggested to be in a deadlock if the channel has been inactive for a given threshold time value [3] and thus the recovery stage is started. The minimum hardware components for each physical channel in the router to implement such a scheme are: a counter, comparator, latch and a register to store the threshold value, in case a programmable threshold is required. Accurate detections of deadlock in this mechanism are very sensitive to network load and message length transmitted over network channels. A long threshold time leads to more accurate deadlock detections, but takes a longer time to discover deadlock and thus increases packets blocking and dramatically degrades network performance [25]. Conversely, a short threshold value detects deadlock faster, but with a higher probability of false detections (false positives) and thus could saturate deadlock recovery channels, in the case of a dedicated central channel used for recovery in each router [3], which again can reduce the network performance [25].

Several techniques have been proposed for interconnection networks to reduce the number of false positive deadlocks. In [24] a packet is suggested as being deadlocked if all requested channels by a blocked packet are inactive for a given timeout. To further reduce the number of false deadlock detections, the mechanism presented in [25] is intended to identify only one packet in a sequence of blocked packets as being deadlocked. The work of [25] is less vulnerable to false detections at the expense of extra hardware (two comparators, two latches and two threshold values). In [30] the author proposed a technique that employs special control packets (called probes) to cross along inactive channels for more accurate detection. At large, all these works are based on the timeout mechanism and pose a difficulty in tuning the threshold value, i.e., to select a unique threshold value for different traffic and network loads. Therefore, a method which accurately detects deadlock without false alarms is required. A method which quickly detects deadlock independent of the network's traffic and load is desirable.

In the recovery stage, there are two kinds of deadlock recovery schemes: *regressive* and *progressive*. A regressive recovery is based on the abort-and-retry mechanism [18] which eliminates the suspected packets from the network. While a progressive recovery resolves deadlocked configuration without removing suspected packets from the network. For example, the DISHA progressive recovery scheme utilizes additional hardware in each router (central buffer) to bypass the suspected packets to their destination sequentially [3] or concurrently [4]. The bandwidth of these central channels is a fraction of the original network bandwidth. Therefore if the detection technique is detecting a lot of false deadlocks this will saturate the recovery bandwidth and as a result will degrade the network's performance [25].

This chapter presents a new deadlock detection method which guarantees true deadlock detection for NoCs. The results of this study, based on a cycle-accurate simulator, demonstrate the effectiveness of the method. It drastically outperforms timing-based deadlock detection mechanisms by eliminating false detections in various traffic scenarios. In this chapter the emphasis is not on comparing with deadlock avoidance techniques. Hence, a simple abort-and-retry approach to recover from detected deadlock is employed. Thus the primarily target is to analyze the deadlock detection rate and trueness of each studied detection techniques.

3.3 Methodology for Deadlock Detection

The following sections present the proposed deadlock-equivalence set principle and the general methodology for the proposed deadlock detection method.

3.3.1 Background and Assumptions

This work assumes fully adaptive routing algorithm with minimal paths. 'Minimal' in this context means that the routing algorithm always chooses the shortest path between sender and receiver. A wormhole flow control technique [12] is employed, a method which has been widely used in NoCs [29]. The packets in wormhole flow control are divided into smaller flits. A header flit is routed first and the rest will follow it in a pipeline manner and this will allow a packet to occupy several channels simultaneously. Thus, wormhole reduces the number of buffers required in each router which is a desirable feature for on-chip networks. However, wormhole makes networks more prone to blocking and deadlock [26, 33].

In this chapter, NoCs are studied without using any deadlock avoidance techniques, but adopting deadlock recovery with an accurate deadlock detection method (proposed in the next section). In line with existing work on deadlock detection/recovery [3, 22, 24, 25, 30], this work assumes that a channel buffer cannot contain flits belonging to different packets. Moreover, a packet arriving at its destination is eventually consumed. In other words, the network is deadlock free at the protocol interaction between the NoC and IP cores. It is important to notice that this work focuses on studding deadlocks that caused by packets at the network routing level. Extending this work to detect deadlocks that may arise by message dependencies which caused by protocol interaction between the IP cores through their network interfaces (NIs) and the network (routers and channels) is one of the future work. Moreover, the used terminology for an *agent* that own and request *resources* (channels) in this work is a *flit*. However, it is equally true that an agent is a message or a packet.

3.3.2 Equivalence Set Criterion for Deadlock

This section introduces the proposed method for deadlock detection in NoCs. It first introduces some important definitions and then defines a deadlock-equivalence set (DES) criterion for detecting loops of packet requests. It uses the TC computation to determine whether there is a set of channels in the NoCs forming a DES. The following definitions lead to that of a deadlock in NoCs:

Definition 1. A SoC and/or CMP consists of communication infrastructure called a NoC and computation/storage cores called IPs.

Definition 2. A NoC $\mathscr{N}(\mathscr{V},\mathscr{C})$ is a strongly connected directed graph, where $\mathscr{V}$ is a set of elements called Vertices that represent a set of router nodes and $\mathscr{C} = \mathscr{V} x \mathscr{V}$ is a set of ordered pairs, $(c_i, c_j) \neq (c_j, c_i)$, called edges that represent a set of Channels that connect routers. A single channel is only allowed to connect a given pair of routers in one direction.

Definition 3. An $\mathscr{I}$ is a set of processing/storage elements represent the IP integrated on-chip. Each IP has one injection channel and one delivery channel directly connected to a router $(v \in \mathscr{V})$.

Definition 4. The routing function R in $\mathscr{N}$ is a function that return the output channel, $c_{out} \in \mathscr{C}$, for each current node, $v_c \in \mathscr{V}$, and destination node, $v_d \in \mathscr{V}$, so that $c_{out} \neq c_{in}$. In other words, deflection routing is not allowed and no channel has the same network node as both its source and destination.

The information routed and propagated in the NoC are packets/flits. These represent the agents that own and/or request the network resources (channels). The resource ownership and request in a NoC at any particular time can be expressed as a channel wait-for graph (CWG) [14].

Definition 5. A CWG is a directed graph $\mathscr{G} = (\mathscr{C}, \mathscr{E})$ where $\mathscr{C}$ is a set of vertices in the network $\mathscr{N}$ and represents the set of channels. $\mathscr{E}$ is a set of ordered pairs called edges which represents the set of channel occupation and requisition status. At any particular state of the NoC, there exists an edge $(u, v) \in \mathscr{E}$ either if (1) there is a head flit in channel u requesting channel v, or if (2) there is a flit in channel u and another flit in channel v and both of them belong to the same packet. Case (1) refers to the request status and is drawn as dashed arcs in the CWG, while case (2) refers to the ownership status and is drawn as solid arcs.

For instance, case (1) can be seen in Fig. 3.1, where a head flit occupying ch_1 and requesting ch_2. Also case (2) is shown in the figure where a data flit occupying ch_2 and its head flit in ch_3. A matrix representation of the CWG can be extracted based on the Adjacency Boolean matrix.

Definition 6. The Adjacency Boolean matrix A of a directed graph $\mathscr{G} = (\mathscr{C}, \mathscr{E})$ is an $n \times n$ matrix, where n is the cardinality of $\mathscr{C}$ and $a_{i,j} \in A, \forall_{i,j} \in n$. A is constructed as follows: (1) $a_{ij} = 1(True)$ iff $(i, j) \in \mathscr{E}$ (edge exists between vertex i and vertex j), (2) $a_{ij} = 0(False)$ otherwise (including when $i = j$, see Definition 4).

Given a directed graph $\mathscr{G}$, it is possible to answer reachability questions using the concept of transitive-closure. For example, can one get from node (vertex) u to node v in one or more edges (hops)?

Definition 7. The Transitive Closure (TC) of $\mathscr{G}$ is a derived graph, $\mathscr{G}^+ = (\mathscr{V}, \mathscr{E}^+)$, which contains an edge (u, v) if, and only if, there is a path from u to v in one or more hops. It means that if $\mathscr{G}$ contains the edges (u, w) and (w, v), then v can be reached from u (transitivity property).

The derived graph $\mathscr{G}^+$, the transitive-closure of $\mathscr{G}$, is the result of adding to $\mathscr{G}$ only the edges that cause $\mathscr{G}$ to satisfy the transitivity property and no other edges (i.e., not adding edges that do not represent paths in the original graph). Thus, the TC of a directed graph is obtained by adding the fewest possible edges to the graph such that it is transitive. It can be computed from A using Floyd-Warshall algorithm [11] (see Algorithm 1, *lines 12–18*) and will be denoted as T.

At this point, it is necessary to introduce the deadlock-equivalence set criterion for detecting a loop of packet requests.

Definition 8. Given a set of channels $\mathscr{C} = \{c_1, c_2, \ldots, c_n\}$ and $n = \mid \mathscr{C} \mid$, from the network $\mathscr{N}$ with n channels and consider a subset $\mathscr{M} = \{c_1, c_2, \ldots, c_m\}$ of channels for some $m \leq n$. $\mathscr{M}$ is a DES iff in all its channels there are flits waiting for one another in a cyclic manner to progress to their respective destinations, i.e., c_1 occupied by a flit and it requests c_2, c_2 occupied by a flit and it requests $c_3, \ldots, c_{m-1}$ occupied by a flit and it requests c_m and c_m occupied by a flit and it requests c_1.

In NoCs characteristics, the CWG, at a particular time, has for any node at most one outgoing arc. Hence, a node can appear in a DES only once and cannot appear in multiple DES's at the same time. Thus, members of a set of simultaneous DES's, $\mathscr{S} = \{S_i\}$, are pair wise disjoint; that is, S_i, $S_j \in S$ and $i \neq j$ implying $S_i \cap S_j = \phi$. The TC computation can be used to determine whether there is a set of channels in the network forming a DES. To demonstrate the idea so far, consider the following example:

Example 1. Given a channel i and a channel j, where $i, j \in \mathscr{C}$. Assume there is a flit occupies each of these channels. Also, each flit requests the other channel,

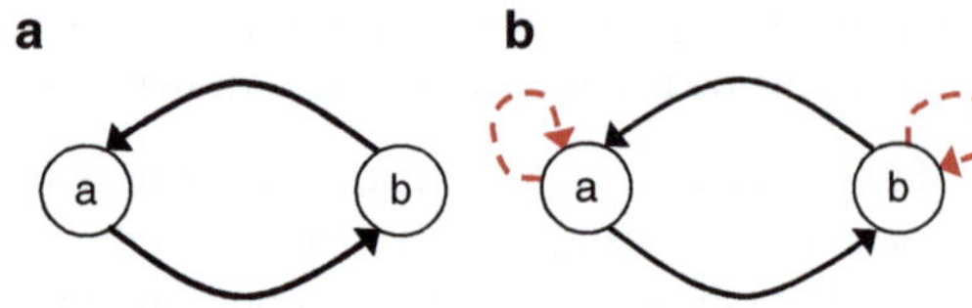

Fig. 3.2 Graphical representation of Example 1. (**a**) Original graph. (**b**) TC graph

i.e., the flit occupies channel i requesting channel j and the one occupies channel j requesting channel i. Then the corresponding adjacency Boolean matrix is:

$$A = \begin{bmatrix} 0 & 1 \\ 1 & 0 \end{bmatrix}.$$

The transitive closure of A is:

$$T = \begin{bmatrix} 1 & 1 \\ 1 & 1 \end{bmatrix},$$

i.e., $T_{1,2} = T_{2,1} = 1$ and $T_{1,1} = T_{2,2} = 1$. The last two imply a loop of channel requests as channel i requesting itself and channel j requesting itself. These will be shown as self-reflexive paths in the TC graph (see Fig. 3.2).

This can be extended to a subset $\mathscr{M} = \{c_1, c_2, \ldots, c_m\}$ of channels for some $m \leq n$, such that all pairs of elements in $\mathscr{M}$ meet the self-requesting condition:

$$DES(i,j) = \begin{cases} 1 & \text{if } T_{i,j} = T_{j,i} = T_{i,i} = T_{j,j} = 1, \\ 0 & \text{otherwise.} \end{cases} \quad \forall i,j \in \mathscr{M} \tag{3.1}$$

To recover from a deadlock situation, the dependency cycle formed by the DES should be resolved. A deadlock can be resolved if one or more of the packets forming the deadlock is removed from the network [18, 22, 25]. The next section presents the DES computational complexity for NoCs.

3.3.3 Equivalence Set Computational Complexity

As shown in the previous section (more details can be found in [2]), the deadlock-equivalence set (DES) provides a simple criterion for deadlock detection. The technique is to generate the CWG from the network and then derive the TC of the CWG and identify the channels that satisfy the criterion. Figure 3.3 illustrates this idea. Given a network at any particular state (Fig. 3.3a) the CWG can readily be drawn (Fig. 3.3b). The derived TC graph (Fig. 3.3c) clearly shows four vertices (channels) with self-reflexive paths and all pairs of these satisfy the condition of the DES (Eq. 3.1).

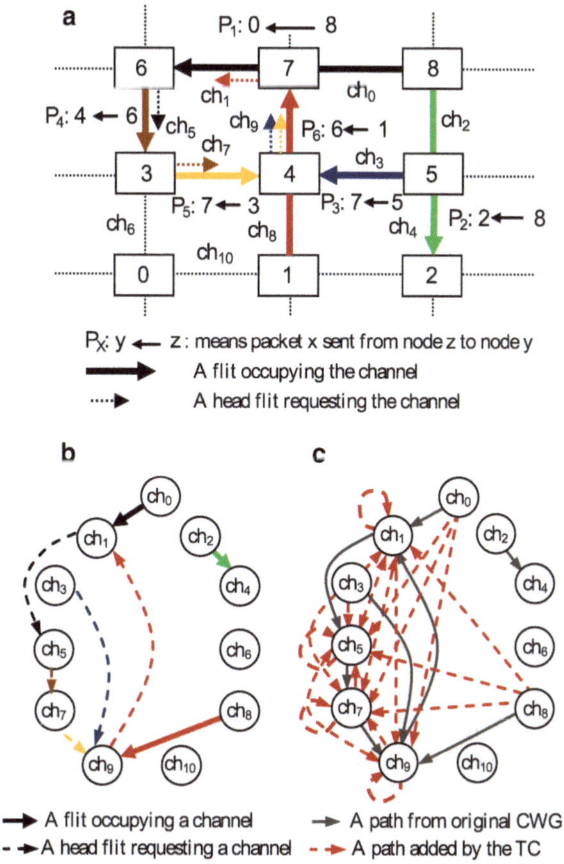

Fig. 3.3 Detecting deadlock from a 2D-NoC: (**a**) a network scenario of deadlock formation; (**b**) the CWG of the network; (**c**) the TC of the CWG, the set of channels $\{ch_1, ch_5, ch_7, ch_9\}$ satisfy the DES definition (Eq. 3.1)

The computational procedure of TC is outlined in Algorithm 1. It has three nested loops containing an $O(n)$ core. In lines 2–10, it converts the directed graph, CWG, to a Boolean Adjacency matrix to reflect the existing paths in the graph. In lines 12–18, the TC is computed. The code state T^k should show a path from vertex i to vertex j if (1) T^{k-1} already shows a path from i to j, passing through one of the vertices in $1 \ldots k-1$, or (2) T^{k-1} shows a path from i to k and a path from k to j; hence there will be a path from i to j through k. Lines 19–23 check if there exist any self-reflexive path in the computed TC, i.e., there is at least single channel i where the TC matrix $T_{i,i} = True$.

The computation of the TC is expensive. In terms of software the algorithm needs a computational complexity of $O(n^3)$, where n is the number of vertices in the CWG (i.e., number of network channels). The algorithm requires only a complexity of $O(n^2)$ as storage [11]. In hardware, however, there are many proposed previous works to speed up the TC computation time to $O(n)$ using a systolic array of $O(n^2)$ processing elements [19]. In this work, however, the NoC distributed architecture is used to simplify the computation complexity. A distributed architecture to realize

Algorithm 1 Checking the existence of a DES in software using TC computation

Definition: CWG is the channel wait-for graph for a NoC $\mathscr{I}$.
Input: $\mathscr{C}$ is a set that contains all the vertices in CWG.
$\mathscr{E}$ is a set that contains all the edges in CWG.
Output: Return *True* if there exists a deadlock-equivalent set in the CWG (self-reflexive path).

$n \Leftarrow |\mathscr{C}|$
for $i = 1$ to n **do**
  for $j = 1$ to n **do**
    if $(i \neq j \wedge (i, j) \in \mathscr{E})$ **then**
      $A_{ij} \leftarrow True$
    else
      $A_{ij} \leftarrow False$
    end if
  end for
end for
$T^0 \Leftarrow A$
for $k = 1$ to n **do**
  for $i = 1$ to n **do**
    for $j = 1$ to n **do**
      $T_{ij}^k \leftarrow T_{ij}^{k-1} \vee \left(T_{ik}^{k-1} \wedge T_{kj}^{k-1}\right)$
    end for
  end for
end for
if ($\exists i : T_{i,i}^n$) **then**
  return *True*
else
  return *False*
end if

the TC computation is presented in the next section. Moreover, the reflexive property is exploited to further simplify DES detections in NoCs to realize the proposed deadlock detection.

3.4 Transitive Closure (TC) Network Architecture

3.4.1 TC Computation with TC-Networks

Dynamic programming (DP) can yield the solution for the TC [11] with an opportunity for solving the computation using a parallel architecture. Mapping TC computation to a parallel computational platform can be achieved with the introduction of a TC-network. The network has a parallel architecture and can be used to compute the TC solution through the simultaneous propagation of successive inferences. Lam and Tong [20] introduced DP-networks to solve a set of graph optimization problems with an asynchronous and continuous-time computational context. This inference network is inherently stable in all cases and has been shown

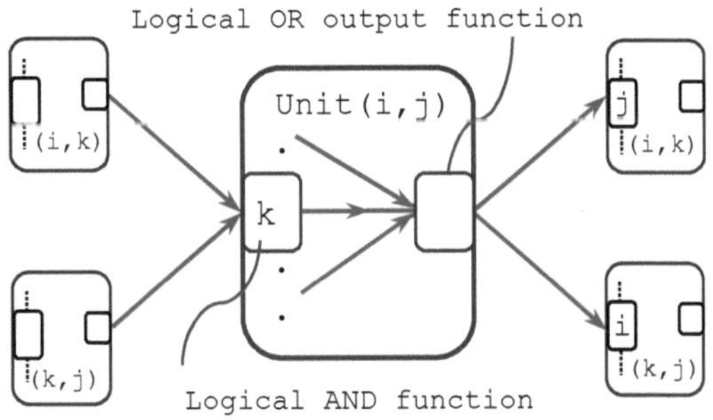

Fig. 3.4 An illustration of a general unit interconnection in the distributed network structure [20]. The output of Unit (i,j) will be the input of other units according to the problem network structure

to be robust with an arbitrarily fast convergence rate [20]. A parallel computational network for solving the dynamic routing problem for NoCs is also proposed in [23]. This work introduces a TC-network to discover the existence of DES at run-time. Similar to the DP-network [20, 23], the proposed TC-network can be realized using a distributed architecture which closely couples with the NoC infrastructure.

The TC-network is constructed by the interconnection of autonomous computational units. The structure of a unit and links in a general inference network is shown in Fig. 3.4. Each unit represents a binary relation (i, j) between two objects and there are h sites, neighboring units, to perform the inference action as defined in the site function. The value of the agreeing relation between i and j is then determined by resolving the conflict among all of the site outputs. Basically, if $S_k(i, j)$ represents the site output at the k-th site and $g(i, j)$ stands for the unit output of unit (i, j), then the TC computation can be stated in terms of network structure:

$$S_k(i,j) = g(i,k) \wedge g(k,j) \tag{3.2}$$

$$g(i,j) = \vee_{\forall k} S_k(i,j) \tag{3.3}$$

where $\wedge$ (*AND* function) is the inference for the site function and the conflict-resolution operator for the unit function. The operator $\vee$ (*OR* function) denotes the unit which resolves the binary relation (i, j). The computational units will be interconnected in the same way as the NoC structure. Each unit represents a router node and a link signifies a communication channel. A distributed network can readily be implemented using such realization.

The delay of TC-network convergence to deadlock-detection depends on the size of the DES and the network topology, which both determine the delay of information propagation within the NoC and the delay of each computational unit. As can be seen, each unit involves $O(|A|)$ AND operations and one OR operation where $|A|$ is the number of adjacent edges. Hence, the solution time is $O(k|A|)$ where k is the number of iterations evaluated by each unit. In a software computation, k equals to the number of nodes in the network which guarantees that all nodes have been updated. Nonetheless, in the hardware implementation with parallel execution, k will be determined by the network structure and $|A|$ AND operations can be

Table 3.1 DES analysis of a TC-network for different network topologies

Network topology	# of nodes	# of channels	Smallest DES	Largest DES
k-ary 1-cube	k	$2k$	k	k
k-ary 2-cube	k^2	$4k^2$	4	$3k^2-8$
k-ary 2-mesh	k^2	$4k^2-4k$	4	$3k^2-3k-2$

executed in parallel. Each computational unit can simultaneously compute the new expected output for all neighbor nodes. The delay of TC-network convergence to deadlock detection is directly proportional to the size of the DES, which determines the delay of information propagation within the TC-network and the delay of TC computational units. The worst case delay for the TC-network to converge to the deadlock detection is investigated for several popular network topologies in the NoCs literature. Some topologies are excepted because they are inherently deadlock free in conjunction with the routing function stated in Definition 4, e.g., tree, bus, star, butterfly, etc. Table 3.1 shows the largest DES for several network topologies as a function of the network size (k). The induction is used to find these equations. Consider the k-ary 1-cube (ring) topology: the maximum number of channels in such a network is $2k$, while the largest and the only possible DES is k as this will be the longest dependency cycle. Likewise for the k-ary 2-mesh network topology (2D mesh) of k^2 nodes with k rows and k columns, the smallest DES in this network will span over four nodes while the largest DES will be $3k^2-3k-2$ and thus, the detection time will depend on the network topology and on how deadlocks are distributed over the network nodes.

3.4.2 Coupling TC-Network to NoC

An on-chip communication network defines the graph vertices of its TC-network. This provides an opportunity for TC computation embedding a TC-unit at each node. Unlike general computer networks, where internode information can only be exchanged through packets, on-chip networks can take advantage of additional dedicated wires to transmit control data between routers. The TC-network shown in Fig. 3.5 comprises of distributed computational units and links between them. The topology of the network resembles the defined graph topology, which is the communication structure of a NoC. Although a 2D mesh network is used here as an example, the proposed methodology can be generalized to any network topology constrained only by Definition 2. At each node, there is computational unit which implements Eqs. 3.2 and 3.3 for each channel. The output of the unit will be propagated to neighbor units via interconnects. The TC-network tightly couples to the NoC and each computational unit locally exchanges the run-time data, such as local channels' occupation and request status, with the router.

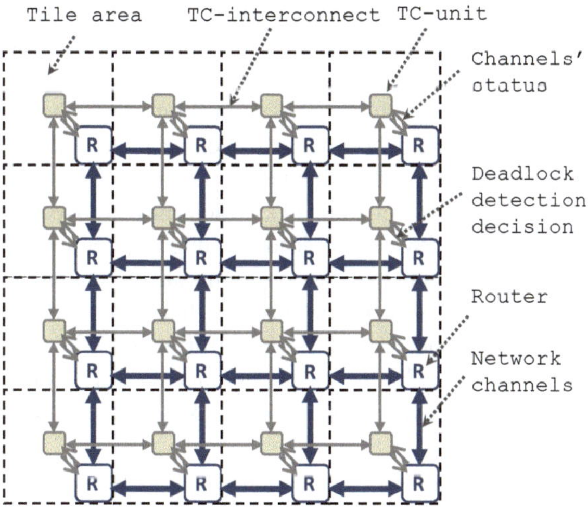

Fig. 3.5 A schematic showing the proposed TC-network (TC-unit and TC-interconnects) coupled to a NoC based SoC, drwaing not to scale

In deadlock detection/recovery strategy the ultimate goal of detecting a deadlock is to invoke a recovery mechanism. Recovering one packet from each DES will be sufficient to break the deadlock dependency cycle. However, performing exhaustive TC calculation for the entire network will discover concurrently all the channels in the network that are members of the same DES and will add recovery overhead by requesting to release all channels. Returning to Fig. 3.3c, the TC graph exhibits four self-reflexive paths around ch_1, ch_5, ch_7 and ch_9 and all of them are members of the same DES (satisfy Eq. 3.1). To reduce the hardware complexity of TC-networks the transitive-closure algorithm can be solved by finding a path in a multiple-source single destination network. In this case, it is possible to employ a light-weight network, a few logic gates and a few wires, based on mutual exclusion units to implement the TC-network to discover the existence of a self-reflexive path for each channel (i.e., there exists *i* channel in TC matrix (T) where $T_{i,i} = True$). The mutual exclusion circuit ensures that only one channel can use the TC-network at any time and channels are checked sequentially using a simple token-ring protocol (hence, finding path(s) in multiple-source multiple-destination network).

One possible token-ring path for mesh and torus networks is the one presented in [2, 3], a Hamiltonian cycle. The token-ring could be implemented as an asynchronous circuit [3] or could be clocked using router clocks. However, the synchronous implementation is not a desirable option as it requires the same clock to span the whole chip. This work implements an asynchronous token-ring protocol circuit. However, this choice requires synchronizer circuits to achieve the synchronization with the TC-unit in each router node (see Sect. 3.4.3).

Circulating the token in the network is equivalent to changing the destination and finding a path in the multiple-source single destination problem. Each TC-unit seizing the token will initiate checking of the corresponding router channels. The rest of the units will implement a transitive property that passes the test signal to neighbor units if, and only if, there is a chain of channel dependencies between its input and output. This makes the TC-network self-pruning and will keep switching power to minimum. The function performed in the TC-unit that seizes the token is described in Algorithm 2. The algorithm starts checking each channel in the router node. In a case where a self-reflexive path is detected for a particular channel, line 10, it means the channel is part of a DES and that channel corresponding deadlock flag is set (line 11), which may be used by the router node to trigger a recovery. Otherwise, the deadlock flag is reset and the next channel will be tested. Once the TC-unit finishes checking all the channels of the corresponding router, it passes the token to the next neighbor unit. The delay time of the TC-network to converge and provide useful information will mainly depend on the DES size. However, the TC-network can produce a valid output even if it does not converge in a single clock cycle, since a deadlock is a steady and persistent event [14].

Algorithm 2 Pseudo code of the TC-unit computation

```
Definitions :-
    par denotes parallel operations.
    n is the number of output channels from neighbour routers.
    m is the number of output channels to neighbour routers.
    tp[m] are temporary buffers.
Input :-
    tc_rx[n] the TC-network signals from neighbour routers,
    ch_oc[n] the channels occupation status from local router,
    ch_req[n][m] the channels request status from local router,
    ch_sel[m] the channel to check for self-reflexive loop.
Output :-
    tc_tx[m] the TC-network signals to neighbours routers,
    dd_flags[m] the deadlock detection flags to local routers.
 1: par for i = 1 to m do
 2:     if ch_sel[i] then
 3:         tc_tx[i] ← True
 4:     end if
 5:     tp[i] ← False
 6:     par for j = 1 to n do
 7:         tp[i] ← tp[i] ∨ (tc_rx[j] ∧ ch_oc[j] ∧ (ch_req[j][i] = True))
 8:     end par for
 9:     if ch_sel[i] then
10:         if tp[i] then
11:             dd_flags[i] ← True
12:         else
13:             dd_flags[i] ← False
14:         end if
15:     else
16:         dd_flags[i] ← False
17:         tc_tx[i] ← tp
18:     end if
19: end par for
```

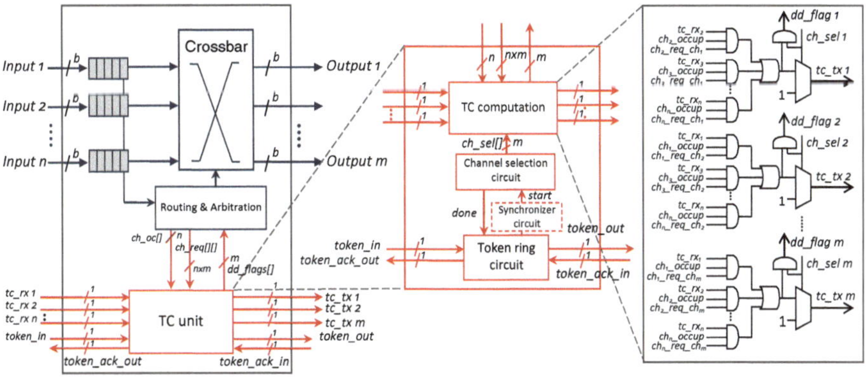

Fig. 3.6 Schematic of the router. (*Left*) Top-level view. (*Middle*) Block implementing the TC-unit. (*Right*) Block implementing the TC-computation

3.4.3 Hardware Implementation

Figure 3.6 (left-part) shows a schematic of a state-of-the-art router, with n input channels and m output channels, augmented by the TC-unit and TC-interconnects to implement the proposed deadlock detection method. The TC-unit block is shown in more details in the central part of the figure. It consists of three components: a TC-computation unit implementing Eqs. 3.2 and 3.3, a channel selection circuit implementing a channel sequence selector with a time delay based on the worst case delay scenario (presented in Sect. 3.5.2.4) and a token-ring protocol to initiate checking of the router channels attached to it.

The TC-computation block exploits the TC-interconnects, $tc\_rx[]$, provided by adjacent neighbors; the local channels' occupation, $ch\_oc[]$, and request, $ch\_req[][]$, provided by the local router and the selected channel, $ch\_sel[]$, produced by the channel selection circuit. It generates the TC-interconnects, $tc\_tx[]$, to adjacent neighbors and deadlock detection flags, $dd\_flags[]$, to the local router. The size of these signals depends on the NoCs topology; for instance, for a 2D mesh network $n, m \in \{North, East, South, West\}$. The description of the TC-computation can be found in Algorithm 2 and the circuit is shown in Fig. 3.6 (right-part). The token-ring protocol is used to change the destination node and initiate the checking of channels at that router node. A possible implementation of the asynchronous token-ring protocol is to use the David Cell circuit [32]. More details on implementing the token-ring protocol can be find in [2].

The asynchronous implementation of the token-ring requires two wires from the previous router in the Hamiltonian chain and two wires to the next router in the chain. Once a TC-unit seizes the token, it will initiate a *start* signal to the channel selection circuit. After checking all channels, a *done* signal is sent back to the token-ring circuit to pass the token to the next neighbor router. The implementation presented in Fig. 3.6 illustrates that the router requires some additional circuits and wires in order to perform the proposed deadlock detection. These requirements are evaluated and compared, in Sect. 3.5.2.3, with similar router architecture using the state-of-the-art timeout detection mechanism.

The architecture introduced in Fig. 3.6 is only one of the possible router implementations. Different optimizations could be applied, depending on the particular case under consideration. A point of interest could be the generalization of the approach in order to check more than one router channel concurrently. This can be accomplished by circulating more than one token in the token-ring protocol chain, thus activating more than one router in the network to perform the checking of its channels. It should be noted that a further TC-unit block and TC-interconnect would be required, thus leading to additional area/power increase. However, for the investigated traffic patterns and network topology the results are not substantially different.

3.5 Results and Discussion

3.5.1 Evaluation Methodology

The TC-network is evaluated in this section by comparing the percentage of detected deadlocks with the heuristic timeout mechanism [3] for different network traffic and different flits injection rate. The percentage of detected deadlocks is obtained as the ratio between the number of packets detected as deadlock over the total number of packets (received and detected). Once a packet is presumed as deadlocked, it is removed from the network. The consumed network energy caused by dropping detected packets is computed. In general, the energy dissipated by a network is divided into the following categories: (1) Routing energy, which depends on the routing type; (2) Selection energy, which refers to the type of selection if the routing algorithm returns more than one option; (3) Forwarding energy, which is used in sending a flit; (4) Receiving energy, which is used in receiving a flit and (5) Waiting energy, which is related to the time the header flit remains waiting until a successful routing takes place. The packet dropping energy is defined as:

$$\begin{aligned} E_{dd\_resolving} &= h \times \left(E_{forward} + E_{receiving}\right) + Flit_{age} \\ &\quad \times \left[E_{wait} + f \times \left(E_{routing} + E_{selection}\right)\right] \end{aligned} \tag{3.4}$$

where E denotes energy, $E_{dd\_resolving}$ is the energy consumed to resolve any detected deadlock, h is the number of hops the flit passed before being aborted, and $Flit_{age}$ is the number of clock cycles the flit existed in the NoC (either moving or waiting for resources to be freed) and f is a Boolean flag that adds the last term to the equation if the flit type is head (if the blocked flit is head it will continue consuming energy by trying to reserve an output channel in each clock cycle).

The performance evaluation was carried out using a modified version of Noxim [16]. In particular, Noxim is modified by introducing the TC-network and the timing-based deadlock detection methods [3, 24, 25]. Without any loss of generality, a NoC with the following specification is chosen for the evaluation: a mesh topology with five ports router architecture, fully adaptive routing with random selection function, no virtual channels, and a crossbar switch – these can be used in a wide range of NoC configurations. Each input channel consists of four flit buffers and one clock cycle is assumed for routing and transmission time across the crossbar and a channel. The results are captured after a warm up period of 10,000 clock cycles. The overall simulation time is set to 300,000 clock cycles. To ensure the accuracy of results captured, with a higher degree of confidence, the simulation at each injection rate is repeated five times with different seeds and their mean values taken.

3.5.2 *Evaluation Results*

3.5.2.1 Synthetic Traffic

Figure 3.7 shows the performance results for a 4×4 2D mesh NoC and a *uniform* distribution of packet destinations[1] with different injection rate. The packet lengths are randomly generated between 2 and 16 flits. Examining Fig. 3.7a, the majority of detected deadlocks using the timeout method [3] are false alarms. For instance, 22% of the packets injected in the network at higher injection rate are detected as part of deadlocks with the threshold value set to 32. The TC-network instead detected that less than 1% of packets are in true deadlocks, consistent with literature [33] which stated that deadlock is an infrequent event.

Figure 3.7b shows the network average delay versus the throughput for full load range. It shows that a smaller timeout threshold value used in the timeout mechanism improves these two important network metrics because it detects more false deadlocks and, by dropping them, will alleviate network contention that may turn into congestion. Also, resolving the detected deadlocks by merely dropping the detected packets without retransmission increases the NoC consumed energy but adds no significance in terms of the network throughput and latency.

By examining Fig. 3.7a, b, the threshold value of 256 could be selected as the best value for such a network setting as it produces a minimum detection percentage of 2.7% with good throughput and latency. The selection of the best threshold value for different network settings (packets' length, buffer size and traffic type) was the goal of several studies [22, 24, 25, 30]. However, the TC-network method detects

[1]This kind of traffic pattern is the most commonly used traffic in network evaluation [14], even though it is very gentle because it naturally balances the load all over the network.

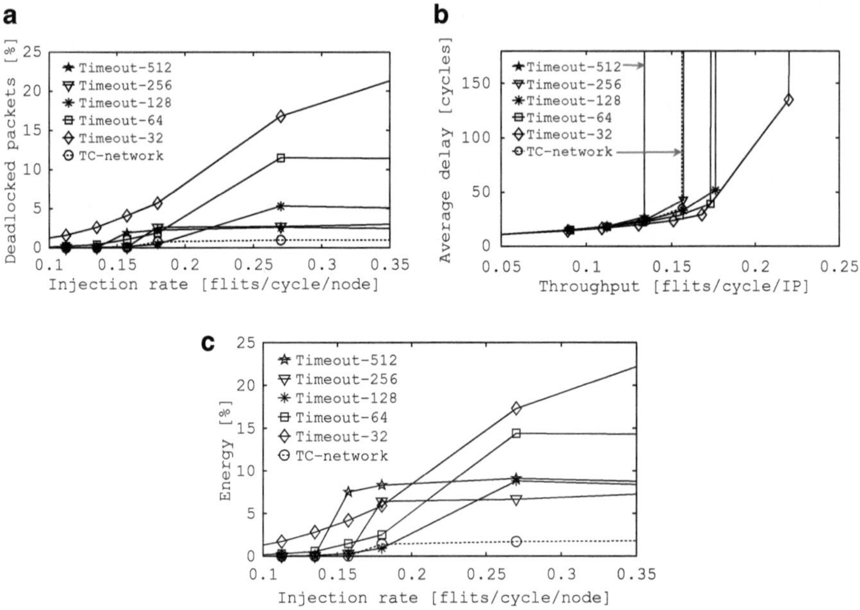

Fig. 3.7 Performance evaluation of the proposed deadlock detection method (TC-network) and the timeout mechanism under *uniform* traffic scenario. (**a**) Percentage of detected deadlocked packets to the total received packets. (**b**) NoC performance with different injection rates. (**c**) Percentage of total energy consumed to resolve detected deadlocks to the total NoC energy

true deadlocks and produces similar performance figures to the timeout mechanism with a 256 threshold value and without the need for any parameters tuning.

Figure 3.7c shows the percentage of energy consumed to resolve detected deadlocks to the total network consumed energy. The figure is similar to, but not linearly proportional to the detection percentage figure (Fig. 3.7a). This is because significant energy consumption is caused by the routing function [16] which repeatedly tries to route the head flit until the time out value has passed in the case of the timeout method (see Eq. 3.4). For instance, the timeout-128 detects 5.3% at saturation and it wastes 8.8% of the total consumed energy by aborting these packets, while the Timeout-512 detects 2.6% at saturation and wastes 9.1% energy. There are two underlying reasons for this: the first is that the NoC with a bigger threshold value is delivering fewer flits at the given simulation time (Fig. 3.7b); the second reason is that the $Flit_{age}$ in Eq. 3.4 is directly proportional to the threshold value in the case when a packet is detected as deadlocked.

To investigate different traffic scenarios and NoC sizes, Fig. 3.8 shows the performance results with the *bit-reversed* traffic.[2] The network size is 8×8 mesh

[2]Each node with binary address $\{b_{n-1}, b_{n-2}, \ldots, b_0\}$ sends a packet to the node with address $\{b_0, b_1, \ldots, b_{n-1}\}$.

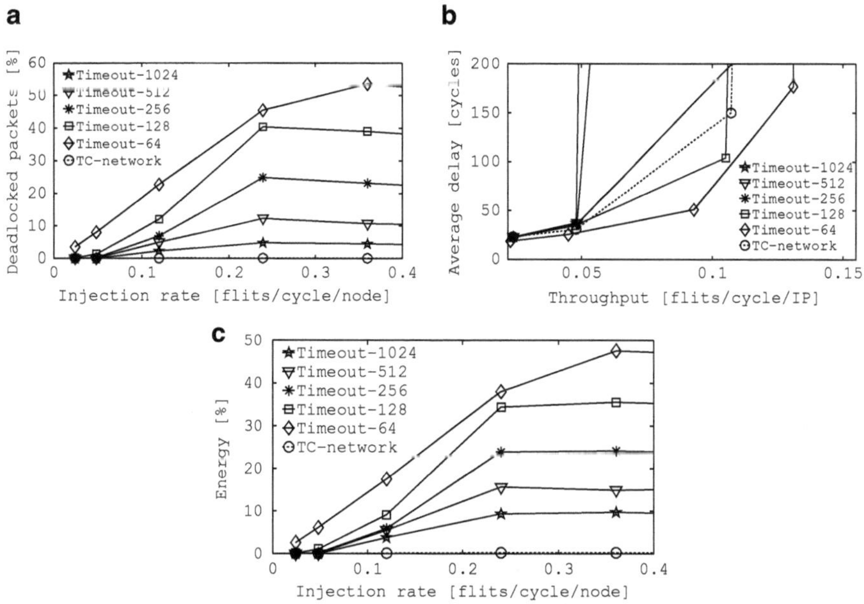

Fig. 3.8 Performance evaluation of the proposed deadlock detection method (TC-network) and the timeout mechanism under *bit-reversed* traffic scenario. (**a**) Percentage of detected deadlocked packets to the total received packets. (**b**) NoC performance with different injection rates. (**c**) Percentage of total energy consumed to resolve detected deadlocks to the total NoC energy

and the packet sizes are randomly chosen between 32 and 64 flits. The results, in general, show a similar trend to the previous example. Here, the threshold value of 1,024 could be selected as the best threshold. The TC-network method detects around 0.07% of packets as deadlocked and dropping them consumed energy of less than 0.2% compared to 4% detected using Timeout-1024 and a waste of energy of around 10%.

Moreover, this study implemented the deadlock detection techniques proposed in [24, 25], which were proposed to enhance the accuracy of deadlock detection over the crude timeout method [3], i.e., to reduce the false-positive-deadlock alarms. The threshold value for each simulated timeout technique is selected as 16, 32, 64, 128, 256, 512, and 1,024 clock cycles, but only the results for 64 and 256 cycles are presented, since the only major difference is the measures scaling up or down. In the following figures, the deadlock detection methods proposed in [3, 24] and [25] will be labeled as *crude, sw* and *fc3d* respectively, followed by the used timeout threshold value.

For the rest of the synthetic traffic scenarios, the results are summarized in Table 3.2. The results are presented for a single injection rate which is labeled as

Table 3.2 The percentage of TC-network improvement compared to timing based methods for different threshold values and traffic scenarios

		crude-64		crude-256		sw-64		sw-256		fc3d-64		fc3d-256		TC-network	
Traffic	IR	DD	Er	DD	Er	DD	Er	DD	Er	DD	Er	DD	Er	DD	Er
Shuffle	0.14	50	40	33	30	33	25	20	17	25	17	16	12	0.10	0.3
Transpose	0.10	35	28	13	11	11	7.3	1.6	1.3	3.1	1.9	0.7	0.4	0	0
Butterfly	0.24	29	34	12	17	17	17	3.7	4.6	9.6	10	2.1	2.5	0	0
Random^1	0.05	18	16	2.8	2.7	7.4	5.6	1.8	1.3	6.3	4.5	2.1	1.5	0.07	0.1
Random^2	0.05	20	16	4.9	4.6	8.6	5.5	3.9	3.0	7.1	4.9	3.2	2.2	0.69	1.2
TC Improvement		176×	83×	75×	41×	89×	70×	36×	31×	59×	45×	28×	22×	–	–

'crude' refers to the work in [3], 'sw' refers to the work in [24] and 'fc3d' refers to the work in [25]. *DD*: Percentage of detected deadlocks, *Er*: Percentage of total energy consumed to resolve detected deadlocks, *IR*: injection rate, Random^1: uniformly distributed traffic with four hot spots located at the corners, Random^2: uniformly distributed traffic with four hot spots located at the center

the saturation packet injection rate (PIR),[3] since the majority of detections occur after the network saturation point. Table 3.2 illustrates the efficacy of the proposed run-time deadlock detection method over the three different existing timing based methods [3, 24, 25]. The TC-network method significantly outperforms timing-based deadlock detection mechanisms by avoiding false detections and, as a result, reducing energy wastage to resolve all detected deadlocks for all synthetic traffic scenarios presented in the table. It should be noted that the quantitative analysis presented in the last row of Table 3.2 as "TC-improvement" would vary according to the injection rate and it is used here to summarize, on average, by what factor the TC-network reduces the detected deadlocked packets compared to the timing-based schemes [3, 24, 25]. Therefore, the TC-improvement magnitude is not necessarily the same in the case where the evaluation merged the entire load range. The results presented are for 8×8 mesh with packet sizes randomly chosen between 32 and 64.

3.5.2.2 Real Application Traffic

As a more realistic communication scenario, a generic MMS [17], which includes an *h263* video encoder, an *h263* video decoder, an *MP3* audio encoder, and an *MP3* audio decoder, is evaluated. In [17], the MMS is partitioned into 40 distinct tasks, which are then assigned and scheduled onto 25 selected IP cores. The topological mapping of IP cores into tiles of a 5×5 mesh-based NoC architecture has been obtained by using the approach presented in [5]. As can be observed from Fig. 3.9a, c, again the TC-network method surpasses the timing based deadlock detection methods with various threshold values. Interestingly the TC-network detects zero deadlocks and thus does not consume energy for retransmission for this particular traffic, while the crude timeout method [3] detects a considerable number of false deadlocks and wastes a considerable amount of energy in recovering from these false deadlocks. For instance, up to 27% of the MMS traffics are detected as deadlocks with a 64 threshold value. The other timing-based deadlock detection techniques [24, 25] reduce the false deadlock alarms, compared to the crude timeout method [3], but they do not eliminate them.

Figure 3.9b shows the network average delay versus the throughput for full MMS load range. Once more, the result suggests that a smaller threshold value used in the timeout mechanism improves these two network metrics. This is because it detects more false deadlocks and, by dropping them, alleviates network contention that may lead to congestion. However, timing-based detection methods with smaller threshold values increase the amount of wasted energy to recover from false deadlocks. Therefore, another experiment was conducted to evaluate the total energy (transmission + recovery) required to drain a fixed amount of data (2 MB MMS traffic) for different network loads. Table 3.3 shows the result of this experiment

[3] A network starts saturating when an increase in injection rate does not result in a linear increase in throughput [6].

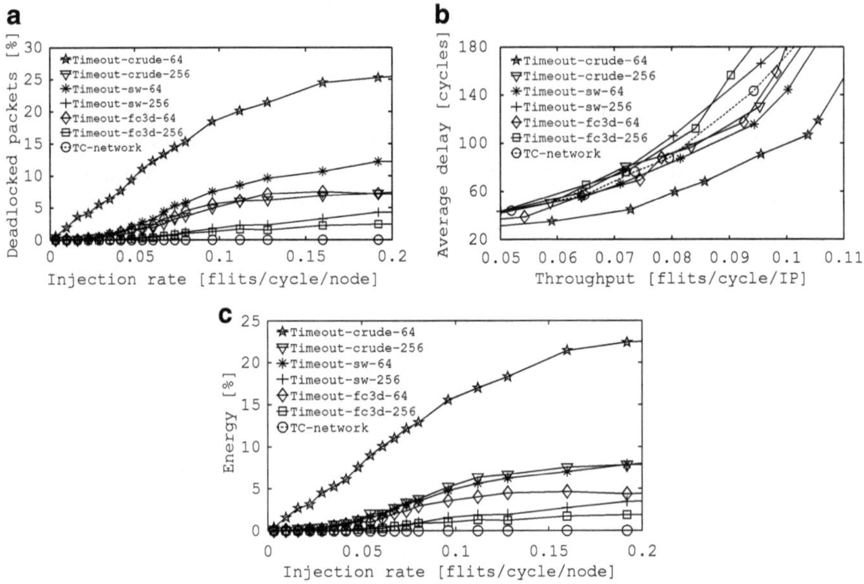

Fig. 3.9 Performance evaluation of the proposed deadlock detection method (TC-network) and the timeout mechanism under MMS traffic with different injection rates. (**a**) Percentage of detected deadlocked packets to the total received packets. (**b**) NoC performance with different injection rates. (**c**) Percentage of total energy consumed to resolve detected deadlocks to the total NoC energy

using different deadlock detection methods. The last column in the table shows the accumulated energy saving using the proposed detection method for the three IRs presented in the same table. The energy saving is different for the different detection methods utilized and for different threshold values; for instance, the TC-network can save up to 5% of energy to run the 2 MB MMS traffic compared to the timing based method [3] with a threshold value equal to 64.

3.5.2.3 Area and Power Estimation

It is crucial in designing NoCs that routers should not consume a large percentage of silicon area compared to the IP core blocks. For this study, two fully adaptive routers based on the timeout and the TC-network methods were designed in Verilog. These were then synthesized using Synopsys Design Compiler and mapped onto the UMC 90 nm technology library. Confirming well-known findings from, for instance [6], the hardware synthesis result shows that the FIFO buffer area significantly dominates the logic of the router. The buffer area mainly depends on the flit size. For a flit size of 64 bits and FIFO buffers with a capacity of four flits, it was found that the TC circuit adds only 0.71% area overhead to the total router area compared

Table 3.3 Energy consumption (mJ) to drain 2 MB of multi-media system data for different network loads and different deadlock detection methods

Method	Injection rate = 0.05			Injection rate = 0.15			Injection rate = 0.2			Accumulated
used	$E_{trans.}$	$E_{recover}$	E_{total}	$E_{trans.}$	$E_{recover}$	E_{total}	$E_{trans.}$	$E_{recover}$	E_{total}	energy saving (%)
TC-network	0.7152	0.0	0.7152	0.8040	0.0	0.8040	0.8123	0.0	0.8128	–
Timeout-crude-64	0.7002	0.0366	0.7368	0.6736	0.1696	0.8432	0.6693	0.2068	0.8761	5.05
Timeout-crude-256	0.7204	0.0035	0.7239	0.7548	0.0566	0.8114	0.7530	0.0787	0.8317	1.48
Timeout-sw-64	0.7181	0.0046	0.7226	0.7497	0.0621	0.8117	0.7653	0.0671	0.8324	1.47
Timeout-sw-256	0.7247	0.0004	0.7230	0.7911	0.0223	0.8134	0.7987	0.0301	0.8288	1.40
Timeout-fc3d-64	0.7136	0.0036	0.7171	0.7864	0.0362	0.8226	0.7873	0.0381	0.8244	1.36
Timeout-fc3d-256	0.7182	0.0002	0.7184	0.8090	0.0127	0.8218	0.7978	0.0155	0.8133	0.91

Table 3.4 Area and power contributions of the TC-unit and different timeout circuits to the total router area and power

Module name	Module area to total router area (%)	Module power to total router power (%)
TC-unit	0.71	0.44
Timeout-crude-1024	2.93	0.97
Timeout-crude-256	2.33	0.81
Timeout-crude-64	1.56	0.70
Timeout-sw-1024	3.12	1.16
Timeout-sw-256	2.54	0.98
Timeout-sw-64	1.76	0.91
Timeout-fc3d-1024	3.53	1.51
Timeout-fc3d-256	2.94	1.38
Timeout-fc3d-64	2.12	1.24

to 2.9% for the timeout [3] implementation with 10 bit threshold counter. Thus the proposed method yields an area gain of more than 2% in each network router circuit compared to the timeout implementation.

Power consumption is also an important system performance metric. The power dissipated in each router design was determined by running Synopsys Design Power on the gate-level netlist of the router with different random input data streams as test stimuli. It was found that the power dissipated by the TC-unit is 0.44% of the entire router power compared to 0.97% dissipated by the crude timeout circuit, with 10 bit threshold counter. This provides a power saving of 0.53%. In the Intel TeraFlop 80-tiles chip implementation [31] the communication power (Router + Links) estimation is 28% of the tile power profile. The Router power, however, is 83% of the communication power. Taking these numbers into consideration, and the power result from the proposed router synthesis, suggests that the TC-unit will dissipate less than 0.01% of the total tile power in similar NoC implementations.

Moreover, an investigation was carried out into the area and power contributions of different timing-based deadlock detection methods' circuits with different time-threshold values and they were compared to the TC-unit circuit implementation. The area gain and the power saving with the TC-unit implementation compared to different timeout implementations can be observed in Table 3.4. The table clearly shows that the techniques used in [24] and [25] introduce more area and power overheads compared to the crude timeout implementation and this is due to the extra hardware requirements required to implement these techniques. The TC-network implementation, however, not only improves the performance of the deadlock detection method, but also minimizes the area and power overheads, as can be seen in Table 3.4.

However, the implementation of the TC-network needs some extra control wires, as shown in Fig. 3.6. The $tc\_rx[n]$ input wires and $tc\_tx[m]$ output wires to/from each TC-unit are coming/going to TC-units in neighbor router nodes.

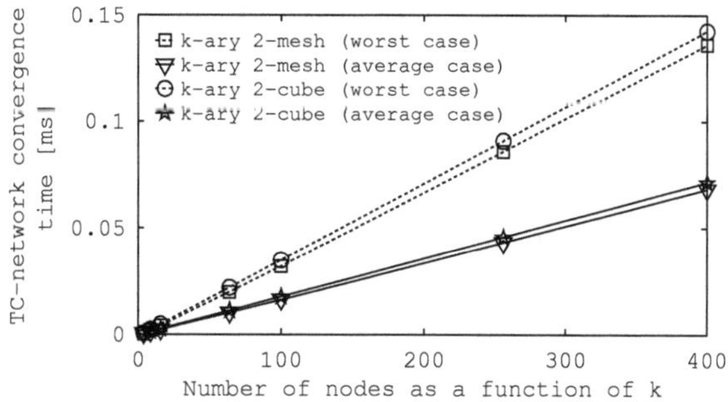

Fig. 3.10 TC-network convergence time for different network topologies and sizes

The *token_in*, *token_ack_out*, *token_out* and *token_ack_in* signals are used to propagate asynchronously the token, i.e., changing the destination node, in the token-ring protocol implemented in this work. Overall, the total number of wires required to support the deadlock detection based on TC-network is

$$TC\_Network_{wires} = n + m + 4, \tag{3.5}$$

where n is equal to the number of input channels and m is equal to the number of output channels in each of the router nodes. Considering the data used in the experiments (2D mesh network where $n, m \in \{North, East, South, West\}$ and flit size of 64 bits), the TC-Network wires are 12 in this case, which is less than 2% of the total wiring cost of the NoC. This number is small and will become smaller with a bigger flit size, as well as being independent of the capacity of the input and output buffers.

3.5.2.4 Delay Estimation

In order to study the operation delays, first the TC-unit's critical path gates delay is calculated using the *sdf* file generated after synthesizing the circuit using worst case library. Secondly, the interconnect delay calculation assumes the tiles are arranged in a regular fashion on the floorplan with 2×1.5 mm tile size, similar to the Intel TeraFLOPS chip [31]. The maximum interconnect length between routers is 2 mm. The load wire capacitance and resistance are estimated, using the Predictive Technology Model (PTM) [28], to be $0.146\,fF$ and $1.099\,\Omega/\mu\text{m}$ respectively. The wire delay between TC units, TC-interconnect, can be readily calculated based on the distributed RC model [8]. In reference to Table 3.1, the worst and average convergence times of the TC-network for different NoC topologies can be estimated. Figure 3.10 shows the worst and average times to discover a deadlock for different network topologies and sizes. It is expected that the delay required by the TC-network to converge to the desired output will depend on the network topology and the size of the DES.

3.6 Summary and Conclusions

NoCs with adaptive routing are susceptible to deadlock, which could lead to performance degradation or system failure. This chapter studies deadlock detection and recovery, as opposed to deadlock avoidance. Detecting deadlocks at run-time is challenging because of their highly distributed characteristics. This work presents a deadlock detection method that utilizes run-time transitive closure (TC) computation to discover the existence of deadlock-equivalence sets, which imply loops of requests in NoCs. This detection scheme guarantees the discovery of all true-deadlocks without false alarms, in contrast with state-of-the-art approximation and heuristic approaches. A distributed TC-network architecture, which couples with the NoC architecture, is also presented to realize the detection mechanism efficiently.

The proposed method is rigorously evaluated using a cycle-accurate simulator and synthesis tools. Experimental results confirm the merits and the effectiveness of the proposed method. It drastically outperforms timing-based deadlock detection mechanisms by eliminating false detections and, thus, reduces energy wastage to recover from false alarms for various traffic scenarios, including real-world application. The new method eliminates the need for any kind of time out mechanism and delivers true deadlock detections independent of the network load and message lengths, rather than approximating with congestion estimation, as in the existing methods. It has been observed that timing based methods may produce two orders of magnitude more deadlock alarms than the TC-network method. Moreover, the hardware overhead for the TC-network has been examined. The implementations presented in this chapter demonstrate that the hardware overhead of TC-networks is insignificant.

References

1. R. Al-Dujaily, T. Mak, F. Xia, A. Yakovlev, M. Palesi, Run-time deadlock detection in networks-on-chip using coupled transitive closure networks, in *Design, Automation Test in Europe Conference Exhibition (DATE)*, Grenoble-France, Mar 2011, pp. 1–6
2. R. Al-Dujaily, T. Mak, F. Xia, A. Yakovlev, M. Palesi, Embedded transitive closure network for runtime deadlock detection in networks-on-chip. IEEE Trans. Parallel Distrib. Syst. **23**(7), 1205–1215 (2012)
3. K.V. Anjan, T.M. Pinkston, DISHA: a deadlock recovery scheme for fully adaptive routing, in *Proceedings of the 9th International Parallel Processing Symposium*, Santa Barbara-CA, 1995, pp. 537–543
4. K.V. Anjan, T.M. Pinkston, J. Duato, Generalized theory for deadlock-free adaptive wormhole routing and its application to disha concurrent. Proc. 10th Int. Parallel Processing Symp., Honolulu-Hawaii, Apr. 1996. IEEE Computer Society
5. G. Ascia, V. Catania, M. Palesi, Multi-objective mapping for mesh-based noc architectures, in *Proceedings of the 2nd IEEE/ACM/IFIP international conference on Hardware/software codesign and system synthesis*, Stockholm, Sweden (ACM), pp. 182–187 (2004).

6. G. Ascia, V. Catania, M. Palesi, D. Patti, Implementation and analysis of a new selection strategy for adaptive routing in networks-on-chip. IEEE Trans. Comput. **57**(6), 809–820 (2008)
7. L. Benini, G. De Micheli, Networks on chips: a new SoC paradigm. IEEE Comput. **35**(1), 70–78 (2002)
8. A. Chandrakasan, J.M. Rabaey, B. Nikolic, *Digital Integrated Circuits: A Design Perspective* (Prentice Hall, London/Upper Saddle River, 2002)
9. G-M Chiu, The odd-even turn model for adaptive routing. IEEE Trans. Parallel Distrib. Syst. **11**(7), 729–738 (2000)
10. J.G. Christopher, L.M. Ni, The turn model for adaptive routing. Proc. Int. Parallel Process. Symp. **41**(5), 874–902 (1994)
11. T.H. Cormen, C.E. Leiserson, R.L. Rivest, *Introduction to Algorithms* (MIT/McGraw-Hill, Cambridge, Mass., 2001)
12. W.J. Dally, C.L. Seitz, Deadlock-free message routing in multiprocessor interconnection networks. IEEE Trans. Comput. **C-36**(5), 547–553 (1987)
13. W.J. Dally, B. Towles, Route packets, not wires: on-chip interconnection networks, DAC'2001, Las Vegas, Nevada, USA (ACM) pp. 684–689, (2001)
14. W.J. Dally, B. Towles, *Principles and Practices of Interconnection Networks* (Morgan Kaufmann, Amsterdam/San Francisco, 2004)
15. J. Duato, S. Yalamanchili, L.M. Ni, *Interconnection Networks: An Engineering Approach* (Morgan Kaufmann/San Francisco, CA, 2003)
16. F. Fazzino, M. Palesi, D. Patti, Noxim: network-on-chip simulator, 2010, http://noxim.sourceforge.net/
17. J. Hu, R. Marculescu, Energy- and performance-aware mapping for regular noc architectures. IEEE Trans. Comput. Aided Des. Integr. Circuits Syst. **24**(4), 551–562 (2005)
18. J.H. Kim, L. Ziqiang, A.A. Chien, Compressionless routing: a framework for adaptive and fault-tolerant routing. IEEE Trans. Parallel Distrib. Syst. **8**(3), 229–244 (1997)
19. S-Y. Kung, S-C. Lo, P.S. Lewis, Optimal systolic design for the transitive closure and the shortest path problems. IEEE Trans. Comput. **C-36**(5), 603–614 (1987)
20. K.P. Lam, C.W. Tong, Closed semiring connectionist network for the Bellman-Ford computation. IEE Proc. Comput. Digit. Tech. **143**(3), 189–195 (1996)
21. A. Lankes, T. Wild, A. Herkersdorf, S. Sonntag, R. Reinig, Comparison of deadlock recovery and avoidance mechanisms to approach message dependent deadlocks in on-chip networks, in *NOCS'10*, Grenoble-France, 2010, pp. 17–24
22. P. Lopez, J.M. Martinez Rubio, J. Duato, A very efficient distributed deadlock detection mechanism for wormhole networks, in *HPCA'98*, Las Vegas-Nevada (IEEE Computer Society, 1998)
23. T. Mak, P.Y.K. Cheung, W. Luk, K.P. Lam, A DP-network for optimal dynamic routing in network-on-chip, in *CODES+ISSS'09*, Grenoble-France (ACM, 2009), pp. 119–128
24. J.M. Martínez-Rubio, P. López, J. Duato, A cost-effective approach to deadlock handling in wormhole networks. IEEE Trans. Parallel Distrib. Syst. **12**(7), 716–729 (2001)
25. J.M. Martinez-Rubio, P. Lopez, J. Duato, FC3D: flow control-based distributed deadlock detection mechanism for true fully adaptive routing in wormhole networks. IEEE Trans. Parallel Distrib. Syst. **14**(8), 765–779 (2003)
26. L.M. Ni, P.K. McKinley, A survey of wormhole routing techniques in direct networks. Computer **26**, 62–76 (1993)
27. T.M. Pinkston, S. Warnakulasuriya, Characterization of deadlocks in k-ary n-cube networks. IEEE Trans. Parallel Distrib. Syst. **10**(9), 904–921 (1999)
28. PTM: Predictive technology model, (Dec. 2010) http://ptm.asu.edu/
29. C. Seiculescu, S. Murali, L. Benini, G. De Micheli, A method to remove deadlocks in networks-on-chips with wormhole flow control, in *DATE'10*, Dresden-Germany, 2010, pp. 1625–1628
30. L. Soojung, A deadlock detection mechanism for true fully adaptive routing in regular wormhole networks. Comput. Commun. **30**(8), 1826–1840 (2007)

31. S.R. Vangal, J. Howard, G. Ruhl, S. Dighe, H. Wilson, J. Tschanz, D. Finan, A. Singh, T. Jacob, S. Jain, V. Erraguntla, C. Roberts, Y. Hoskote, N. Borkar, S. Borkar, An 80-tile sub-100-w teraflops processor in 65-nm CMOS. IEEE J. Solid State Circuits **43**(1), 29–41 (2008)
32. V.I. Varshavsky (Ed.), *Self-timed Control of Concurrent Processes: The Design of Aperiodic Logical Circuits in Computers and Discrete Systems* (Kluwer Academic, Norwell, MA, USA, 1990)
33. S. Warnakulasuriya, T. Pinkston, Characterization of deadlocks in interconnection networks, in *IPPS'97*, Geneva-Switzerland (IEEE Computer Society, 1997), pp. 80–86

Chapter 4
The Abacus Turn Model

Binzhang Fu, Yinhe Han, Huawei Li, and Xiaowei Li

Abstract Applications' traffic tends to be bursty and the locations of hot-spot nodes vary with time. This will dramatically aggregate the blocking problem of wormhole-switched Network-on-Chip (NoC). Most of state-of-the-art traffic balancing solutions, which are based on fully adaptive routing algorithms, may introduce large time/space overhead to routers. On the other hand, partially adaptive routing algorithms, which are time/space efficient, are lack of even or sufficient routing adaptiveness. Due to the lack of a practical model to dynamically keep the routing algorithm deadlock free, most of reconfigurable routing algorithms, which could provide on-demand routing adaptiveness for reducing blocking, are off-line solutions. In this chapter, we will discuss the abacus turn model (AbTM). It was proposed to keep the network deadlock-free by dynamically applying forbidden turns. By this way, the AbTM-based routing algorithms are deadlock-free and on-line reconfigurable. In addition to the AbTM, this chapter also includes a brief discussion about the routing reconfiguration techniques. On detecting the congestion, the reconfigurable routing algorithm first decides the new routing rules. Afterwards, the routing reconfiguration algorithm is exploited to transform the routing function from the old one to the new one. A well-designed routing reconfiguration technique could finish the process in a very short time without blocking and dropping packets. Furthermore, this chapter also briefly discusses the possible solutions to implement reconfigurable routing algorithms, such as table-based solution and LBDR-based solution. Finally, the routing performance will be discussed at the end of this chapter.

B. Fu (✉) • Y. Han • H. Li • X. Li
Institute of Computing Technology, Chinese Academy of Sciences, Beijing, China
e-mail: fubinzhang@ict.ac.cn; yinhes@ict.ac.cn; lihuawei@ict.ac.cn; lxw@ict.ac.cn

M. Palesi and M. Daneshtalab (eds.), *Routing Algorithms in Networks-on-Chip*,
DOI 10.1007/978-1-4614-8274-1_4, © Springer Science+Business Media New York 2014

4.1 The Problem

The power, memory and instruction-level parallelism walls are forcing the processors to integrate more and more cores. For example, Tilera's TILE64 processor integrates 64 cores [1], Intel's single-chip cloud computer [2] and terascale processor [3] have 48 and 80 cores, respectively.

To efficiently interconnect such a large number of elements, Network-on-Chip (NoC) has been widely viewed as the replacement to the shared buses or dedicated wires [1,3,4]. Generally, an NoC consists of routers and links. Adjacent routers are connected by links according to the network topology, such as the ring (as shown in Fig. 4.1a) and 2D meshes (as shown in Fig. 4.1b). These kinds of topologies are popular because their planar structures facilitate the IC manufacturing. The recent advances in 3D IC technologies have motivated the 3D topologies [5], such as 3D meshes. In this chapter, most discussions are based on 2D meshes for simplicity.

The network topology determines the ideal performance of a network since it determines the network diameter and the degree of path diversity. To achieve the ideal performance, the network resources, such as buffer capacity and channel bandwidth, are assumed to be allocated without any waste. Thus, an efficient flow control technique is expected. To reduce buffer requirement and packet latency, the wormhole flow control technique has become one of the main paradigms. In wormhole switched networks, a packet is divided into fixed-size flow control digits (flits), including a head flit, several payload flits, and a tail flit. Because a packet could move across several nodes simultaneously, very short network delay is incurred. However, once the head flit is blocked, all flits must stay along the path. This, together with the fact that applications' traffic tends to be very bursty, increases the possibility of network blocking significantly.

Once a packet is blocked, its queueing delay will increase dramatically and finally become the dominant contributor to latency. Increased network latency may degrade applications' performance or could even cause QoS violations. Traffic load balancing techniques, such as traffic-aware (or congestion-aware) routing algorithms, are promising ways to reduce network blocking [6–11]. Traffic-aware routing algorithms make routing decisions based on run-time network status, so that packets could be routed evenly along all legal paths. Some paths are illegal, because routing packets along them may cause routing deadlock or livelock. Generally, most of the traffic-aware routing algorithms are fully adaptive, so there is always a possibility of making selection. However, current fully adaptive routing

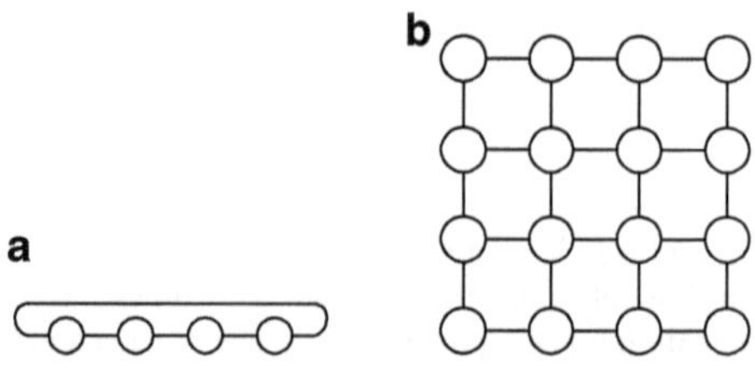

Fig. 4.1 Examples of network topologies, (**a**) ring (4-ary 1-cube), (**b**) 4-ary 2-mesh

algorithms may require a large number of virtual channels (VC) [12–14], or assume a conservative flow control technique [15–17].

The VC, which was viewed as cheap and abundant, is expensive in the NoC scenario [18]. First, increasing the number of VCs often means increasing the router latency. The reason is that VA (VC allocation) delay increases with the number of VCs and at most times VA is the critical stage of virtual-channel routers [19]. Second, increasing the number of VCs often means increasing the router area since buffers are the main contributor to the router area and adding VCs often requires more buffers. Routing algorithms following the Duato's theory [15] assume a conservative flow control technique that a queue never contains flits belonging to different packets [15–17]. This facilitates reducing the number of VCs, but usually degrades the performance of networks carrying many back-to-back short packets [19], e.g., the control packets for cache coherence. To address this problem, the whole-packet-forwarding flow control technique was recently proposed [20]. With this extension, a queue may contain multiple short packets. We should note that though Duato's theory [15] is adopted, the NoCs using state-of-the-art fully adaptive routing algorithms require at least $2\times$ VCs per physical channel. For example, for a CMP system with directory-based cache coherence protocol, MSI for L1 cache and MOSI for L2 cache, at least five virtual networks are required to avoid deadlock. Therefore, totally five VCs per physical channel are required by NoCs using xy routing, and at least 10 VCs by those using fully adaptive routing.

Unlike the fully adaptive routing algorithms, the VC-free partially adaptive routing algorithms are more cost-efficient and can be realized with aggressive flow control techniques [21,22]. Glass and Ni [21] proposed the turn model for designing partially adaptive routing algorithms without VCs. The turn model classifies all possible turns into clockwise and counter-clockwise abstract cycles. In each kind of abstract cycles, one turn is prohibited to avoid deadlock. Based on the turn model, literature [21] further proposed three partially adaptive routing algorithms, namely the west-first, north-last and negative-first. Unfortunately, the degree of adaptiveness of above algorithms is highly uneven. To address this problem, Chiu [22] proposed the odd-even turn model, where NW[1] and SW turns are prohibited in odd columns,[2] and EN and ES turns are prohibited in even columns. Odd-even routing could provide more even adaptiveness, however none of the long-distance (>2 hops) node pairs are provided with full adaptiveness.

Both insufficient and uneven routing adaptiveness, of state-of-the-art partially adaptive routing algorithms, may degrade the network performance. Applications' traffic tends to be very bursty [23], and the location of hot spots varies with time. For example, we show the traffic patterns of FFT during execution cycles $900 \sim 1{,}000$ and $1{,}500 \sim 1{,}600$ in Fig. 4.2a, b respectively. Figure 4.2a shows that node (3, 1)

[1] A packet takes an NW turn when it changes its direction from north to west [21].

[2] A column is called odd (respectively, even) column if its coordinate in dimension-x is an odd (respectively, even) number [22].

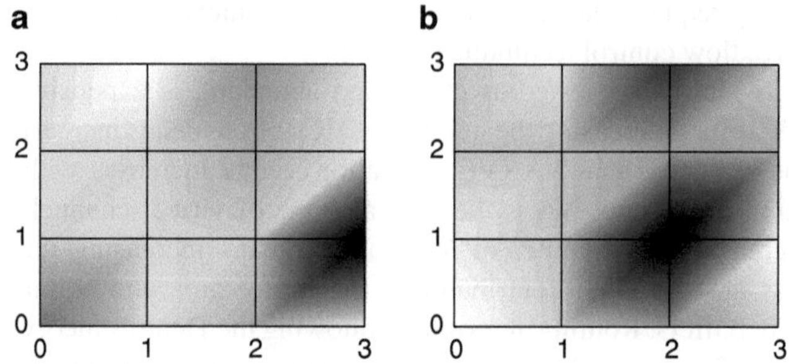

Fig. 4.2 Traffic variations in FFT with 16 cores, (**a**) time = 900 ∼ 1,000, (**b**) time = 1,500 ∼ 1,600

is the hot spot during execution cycles 900 ∼ 1,000, and Fig. 4.2b shows that nodes (2, 0) and (2, 3) become the hot spots after 500 cycles. To avoid congestions, we expect that the packets towards the hot spots are provided with full adaptiveness. Furthermore, since the positions of hot spots vary with time, the routing algorithm is expected to be able to provide full adaptiveness to all node pairs. Unfortunately, none of the state-of-the-art partially adaptive routing algorithms could meet these requirements.

According to the above observations, the expected routing algorithm should be:

1. Time/space-efficient, i.e., no VC and routing table requirement,
2. Able to provide full adaptiveness to all node pairs.

It has been proved that it is impossible to design a deadlock-free fully adaptive routing without VCs for a wormhole-switched mesh network [24]. Thus, the routing algorithm should be partially adaptive. Partially adaptive routing algorithms cannot provide full adaptiveness to all node pairs at all times. Therefore, the second requirement is reduced to "able to" not "always" provide full adaptiveness to all node pairs. In other words, the routing algorithms could provide full adaptiveness to some node pairs at some time, and to others when the traffic pattern changes.

To achieve above objectives, routing algorithms should be reconfigurable. We should distinguish the reconfigurable routing algorithm from the routing reconfiguration algorithm. The former indicates that the routing algorithm could adjust itself to adapt to network variations, such as the detection of faults or hot-spot nodes. The routing reconfiguration algorithms, on the other hand, are proposed to statically [25, 26] or dynamically [27–30] load a new routing algorithm, which was computed offline, to replace the old one without introducing deadlock. Most of the state-of-the-art reconfigurable routing algorithms are designed for tolerating faults, such as [31–35]. The common feature of them is that the reconfiguration is triggered by faults. If we simply view the nodes belonging to congestion regions as faulty (temporally), the packets could be routed without entering the congestion regions. However, according to fault-tolerant routing algorithms [31–35], faulty nodes, actually those locating in congestion regions, are not allowed to send and receive packets. Otherwise, the network may be deadlock. However, we cannot forbid sending packets to hot-spot nodes that are prone to be included into congestion regions. Therefore, fault-tolerant reconfigurable routing algorithms cannot be directly used to address the congestion problem. It is also possible to reconfigure the network topology, such as [36, 37]. However, few of them addresses the congestion problems.

Above observations encourage us to design a new reconfigurable routing algorithm. The major challenge is to address the deadlock problem. When the a hot-spot node is detected, there is no existing rule to follow to generate the new deadlock-free routing algorithm. Generally, deadlock happens when packets wait for each other in a cycle. To prevent deadlock, the VC-free routing algorithms should keep the channel dependence graph (CDG) acyclic [38]. Due to the high complexity, most of the cycle elimination algorithms, such as [39–42], are off-line solutions. Turn models [21, 22] provide a simple way to keep the CDG acyclic, but they are all static and cannot be directly used to generate reconfigurable routing algorithms. To address this problem, the Abacus Turn Model (AbTM), which will be discussed in this chapter, was recently proposed in literature [43].

4.2 The Static Turn Models

To design a deadlock-free routing algorithm, it is required to guarantee that the CDG is acyclic. Before the turn model [21] was proposed, the most popular way had been to add physical or virtual channels. The turn model, on the other hand, does it by prohibiting some turns. Since it is quite cost-efficient and very easy to implement, the turn model has become very popular, especially for networks without virtual channels. Several years later, Chiu [22] found that the routing algorithms based on turn model generate uneven routing adaptiveness. To address this problem, the odd-even turn model, which prohibits different turns in odd and even columns, was proposed. The AbTM stems from the turn models [21, 22]. Learning them is important for understanding the AbTM. Unlike the AbTM-based routing algorithms, the routing algorithms based on [21, 22] are not reconfigurable. Thus, we name them as static turn models to distinguish them from AbTM.

4.2.1 The Turn Model

The turn model is proposed for n-dimensional meshes and k-ary n-cubes. Following the turn model, there are six basic steps to design a deadlock-free routing algorithm.

1. Partition all virtual channels, except the wraparound channels, into sets according to the virtual direction. Note that the virtual channels of a physical channel are divided into separate sets since they have different virtual directions. All wraparound channels are put into one set.

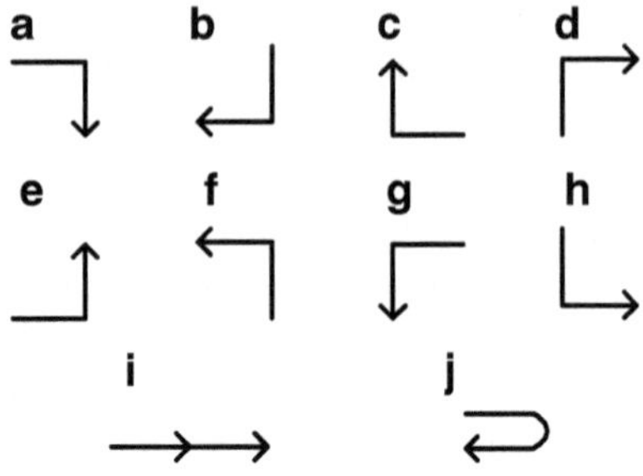

Fig. 4.3 Possible turns in 2D meshes, (**a**) ES turn, (**b**) SW turn, (**c**) WN turn, (**d**) NE turn, (**e**) EN turn, (**f**) NW turn, (**g**) WS turn, (**h**) SE turn, (**i**) 0° turn, (**j**) 180° turn

2. Identify all possible turns from one virtual direction to another, but ignore 0° and 180° turns.
3. Identify all abstract cycles, which are the simplest cycles in each plane of the topology.
4. Prohibit one turn in each abstract cycle.
5. Incorporate turns from the set of wraparound virtual channels as many as possible without introducing cycles.
6. Incorporate 0° and 180° turns as many as possible without introducing cycles.

Partitioning virtual channels into sets and using only the turns allowed by steps 4, 5 and 6 lead to that the virtual channels are accessed in a strictly increasing or decreasing order. Thus, the network is deadlock-free. Furthermore, routing algorithms based on turn model are livelock-free since a packet will definitely arrive its destination due to the finite number of channels.

In the following part, we will show how turn model is applied to a 2D mesh. Furthermore, we assume that there is no virtual channel in the network. A 2D mesh has $m \times n$ nodes, where m (resp., n) is the radix of dimension x (resp., y). Each node d has an address $d : (d_x, d_y)$, where $d_x \in \{0, 1, 2, \ldots, m-1\}$ and $d_y \in \{0, 1, 2, \ldots, n-1\}$. Two nodes $d : (d_x, d_y)$ and $e : (e_x, e_y)$ are neighbors in dimension x (resp., y) if and only if $|d_x - e_x| = 1$ and $d_y = e_y$ (resp., $|d_y - e_y| = 1$ and $d_x = e_x$). If two nodes are neighbors in dimension x (resp., y), they are connected by a bidirectional row (resp., column) channel. Each bidirectional row (resp., column) channel consists of two physical channels with opposite directions: EW and WE (resp., NS and SN) channels. Particularly, EW (resp., WE, NS, and SN) channel is used to forward packets from east to west (resp., west to east, north to south, and south to north). Each $m \times n$ mesh consists of m columns and n rows. Each row (resp., column) consists of m (resp., n) nodes with the same coordinate in dimension y (resp., x).

A packet moving towards direction X makes an XY turn if it turns to direction Y, where $X, Y \in \{E, W, N, S\}$ and E (resp., W, N, and S) refers to direction east (resp., west, north, and south). Of these 16 different combinations, 8 combinations lead to 90° turns. For example, if a packet moving towards east ($X = E$) turns to south ($Y = S$), then an ES turn is introduced (as shown in Fig. 4.3a). Four of these combinations lead to 0° turns. A 0° turn is possible as long as there are more than one channel in a direction. For example, Fig. 4.3i shows that X and Y are both

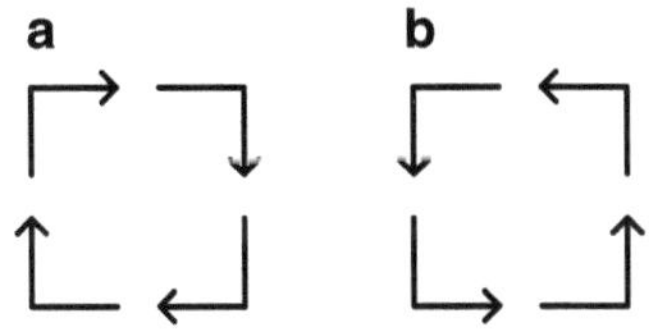

Fig. 4.4 Abstract cycles in a 2D mesh, (**a**) clockwise abstract cycle, (**b**) counter clockwise abstract cycle

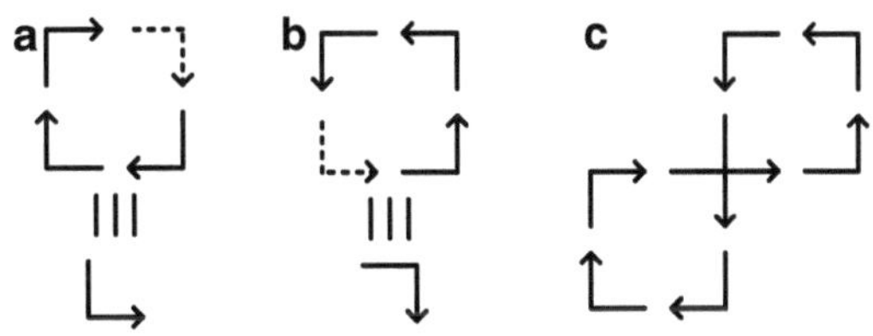

Fig. 4.5 Forbidding one turn in each abstract cycle may still cause circular channel dependence, (**a**) equivalent SE turn, (**b**) equivalent ES turn, (**c**) equivalent cycle

east (E). The last four combinations lead to 180° turns. A 180° turn is introduced if X and Y are opposite. For example, Fig. 4.3j shows that a packet makes a 180° turn by changing its direction from east to the west.

Eight 90° turns can be divided into two sets, i.e., the clockwise and counter-clockwise turns, according to the rotation direction. For example, ES, SW, WN and NE turns are clockwise turns, and EN, NW, WS and SE turns are counter-clockwise turns. As shown in Fig. 4.4a, four clockwise turns form the clockwise abstract cycle, and the other four counter-clockwise turns form the counter-clockwise abstract cycle.

Turn model avoids deadlock by prohibiting one turn in each abstract cycle. Because there are 4 different turns in each abstract cycle, there are totally 16 different combinations to prohibit the 2 turns. Of these 16 combinations, 4 combinations are illegal. The reason is that making three clockwise turns continuously is equivalent to making one counter-clockwise turn. For example, Fig. 4.5a shows that making SW, WN and NE turns continuously is equivalent to making a counter-clockwise SE turn. Similarly, making EN, NW and WS turns continuously is equivalent to making a clockwise ES turn. Therefore, the combination (ES, SE) is illegal since a cycle also can be formed by the allowed six turns as shown in Fig. 4.5c. Due to the same reason, three other combinations are illegal, i.e., the combinations (SW, WS), (WN, NW) and (NE, EN).

Of these 12 legal combinations, only 3 combinations are unique if rotation symmetry is considered. As shown in Fig. 4.6, they are named as west-first, negative-first, and north-last, respectively. In 2D meshes, there is no wraparound turn, so that the step 5 can be ignored. In step 6, we incorporate all 0° turns, and forbid all 180° turns for simplicity.

With west-first routing algorithm, the SW and NW turns are forbidden respectively in clockwise and counter-clockwise abstract cycles. West-first routing can be proved to be deadlock-free by numerating channels in such a way that channels are accessed in a strictly decreasing order. Due to the limited space, the detailed proof is omitted here. Since packets are not allowed to turn to the west during the transmission, packets should be first routed to the west if the destination is

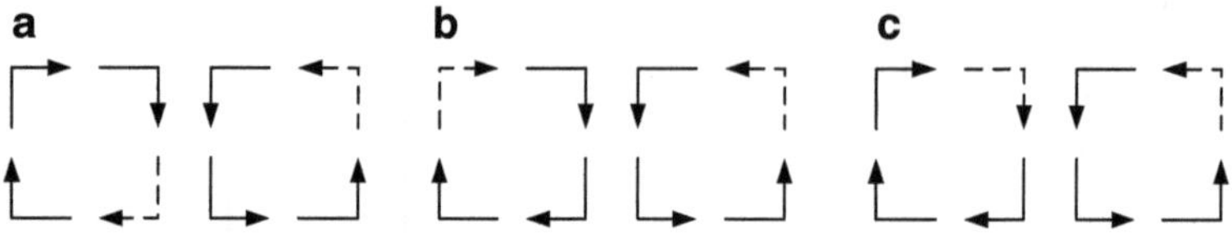

Fig. 4.6 Three unique combinations considering rotation symmetry, (**a**) west-first routing, (**b**) north-last routing, (**c**) negative-first routing

to the west of the source, i.e., $d_x < s_x$. Therefore, although the west-first routing is adaptive, there is only one path for northwestward and southwestward packets as shown in Eq. 4.1, where $S_{west-first}$ represents the degree of adaptiveness of the west-first routing. Obviously, this is unfair, and the congestions caused by northwestward and southwestward packets cannot be avoided by west-first routing.

$$S_{west-first} = \begin{cases} \dfrac{(\triangle x+\triangle y)!}{\triangle x!\triangle y!}, & if\ d_x \geq s_x \\ 1, & otherwise \end{cases} \tag{4.1}$$

North-last routing does not allow packets moving towards north to make turns by forbidding NE and NW turns, as shown Fig. 4.6b. Like the west-first routing, north-last routing is not even either as shown in Eq. 4.2. For southeastward and southwestward packets, north-last is fully adaptive. However, for packets with $d_y > s_y$, north-last routing is equivalent to XY routing.

$$S_{north-last} = \begin{cases} \dfrac{(\triangle x+\triangle y)!}{\triangle x!\triangle y!}, & if\ d_y \leq s_y \\ 1, & otherwise \end{cases} \tag{4.2}$$

As shown in Fig. 4.6c, negative-first routing forbids ES and NW turns. That means a packet cannot make a turn from a positive direction to a negative direction, i.e., from east to south and from north to west. As shown in Eq. 4.3, negative-first routing is also uneven. For northeastward and southwestward packets, negative-first routing is fully adaptive. For example, for northeastward packets, they can be routed to north or east arbitrarily since both north and east are positive directions. For northwest and southeast packets, on the other hand, negative-first routing is deterministic. For example, for northwestward packets, negative-first routing is equivalent to XY routing since west is the negative direction.

$$S_{negative-first} = \begin{cases} \dfrac{(\triangle x+\triangle y)!}{\triangle x!\triangle y!}, & if\ (d_x \leq s_x\ and\ d_y \leq s_y) \\ & or\ (d_x \geq s_x\ and\ d_y \geq s_y) \\ 1, & otherwise \end{cases} \tag{4.3}$$

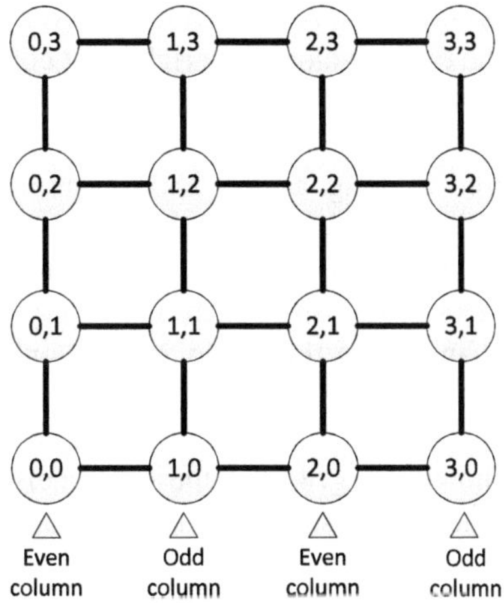

Fig. 4.7 A 4 × 4 mesh has two odd and even columns

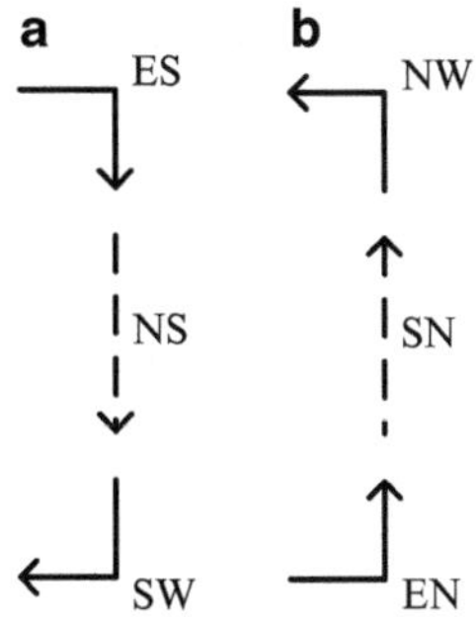

Fig. 4.8 Rightmost columns on the waiting path, (**a**) clockwise column, (**b**) counter-clockwise column

4.2.2 The Odd-Even Turn Model

The degree of adaptiveness represents the ability of a routing algorithm to reduce the chances that packets are blocked continuously. The three partially adaptive routing algorithms, i.e., west-first, north-last and negative-first routing, generate highly uneven routing adaptiveness as discussed in the above section. To address this problem, Chiu [22] proposed the odd-even turn model. The odd-even turn model first classifies network columns into odd and even columns. In a 2D mesh, a column is an odd (resp., even) column if coordinate of the column in dimension-0 is an odd (resp., even) number. For example, as shown in Fig. 4.7, nodes (0, 0), (0, 1), (0, 2) and (0, 3) form an even column since their coordinate in dimension-0 is an even number.

The main idea of odd-even turn model is to prevent the formation of rightmost column segments of any circular waiting path. As shown in Fig. 4.8, there are two kinds of rightmost column: clockwise column and counter-clockwise column. The clockwise rightmost column, as shown in Fig. 4.8a, consists of an ES turn, an SW turn, and several NS channels. To break the clockwise rightmost columns, Chiu [22] proposed to prohibit ES turn in even columns and SW turn in odd columns. To break the counter-clockwise rightmost columns, EN and NW turns are forbidden in even and odd columns, respectively.

The routing algorithms, which are based on odd-even turn model, are deadlock-free as long as the 180° turns are prohibited. This can be easily proved by contradiction. Assuming that a deadlock happens, thus there should be some channels waiting for each other make a cycle according to Dally and Seitz's theory [38]. Since there is no 180° turn, the cycle must has both row and column channels. In this cycle, there should be a rightmost column line segment, which consists of a sequence of column channels with the same direction. Let nodes S and E be the start and the end nodes of this rightmost segment, respectively. As shown in Fig. 4.8a, if these channels are NS channels, an ES and SW turn should be made at the start and the end nodes, respectively. If the rightmost column belongs to an odd column, then the SW turn at the end node is prohibited. Contradiction arises. Otherwise, if the column is an even column, then the ES turn at the start node is prohibited. Contradiction also arises. If the channels are SN channels, the proof is similar. The conclusion is that both clockwise and counter-clockwise rightmost column segments cannot be formed with odd-even turn model, thus the channel dependence cycles never form in the network. Therefore, the network is deadlock-free. This conclusion is important, because it is also the foundation of the AbTM.

The major advantage of the odd-even turn model over the original turn model is that routing algorithms following odd-even turn model provide more even routing adaptiveness to different source-destination node pairs. For simplicity, we name the minimal routing algorithm following the odd-even turn as odd-even routing. Calculating the adaptiveness of odd-even routing is more complicated than that of west-first, north-last and negative-first routing. Let's define $\triangle x = d_x - s_x$ and $\triangle y = d_y - s_y$, where (s_x, s_y) and (d_x, d_y) are the coordinates of source and destination nodes, respectively. The packet is called an NE (resp., SE, NW, and SW) packet if $\triangle x > 0$ and $\triangle y \geq 0$ (resp., $\triangle x > 0$ and $\triangle y < 0$, $\triangle x \leq 0$ and $\triangle y \geq 0$, and $\triangle x \leq 0$ and $\triangle y < 0$). The even column, which allows NW and SW turns, is an allowable column for NW and SW packets. The odd column, which allows EN and ES turns, is an allowable column for NE and SE packets. Let $h = \lceil \frac{d_x}{2} \rceil$ and $h' = \lceil \frac{d_x - 1}{2} \rceil$, the adaptiveness of odd-even routing can be calculated as Eq. 4.4 for NE and SE packets, and as Eq. 4.5 for NW and SW packets.

$$S_{odd-even-(NE+SE)} = \begin{cases} \dfrac{(|\triangle y| + h')!}{|\triangle y|!h'!}, & \textit{if column } s_x \textit{ is an allowable column} \\ & \textit{and } |\triangle x| \textit{ is an odd number} \\ \dfrac{(|\triangle y| + h)!}{|\triangle y|!h!}, & \textit{otherwise} \end{cases} \quad (4.4)$$

$$S_{odd-even-(NW+SW)} = \begin{cases} \dfrac{(|\triangle y|+h)!}{|\triangle y|!h!}, & if\ column\ s_x\ is\ an\ allowable\ column \\ & or\ |\triangle x| = 0 \\ \dfrac{(|\triangle y|+h')!}{|\triangle y|!h'!}, & otherwise \end{cases} \tag{4.5}$$

Comparing with Eqs. 4.1–4.3, we could find that the adaptiveness of odd-even routing is much more even than that of west-first, north-last and negative-first routing. This benefit stems from the fact that different turns are prohibited in odd and even columns. However, from Eqs. 4.4 and 4.5, we find that odd-even routing does not provide fully adaptiveness for any source-destination node pair, which has the distance larger than 2. Without full adaptiveness, the ability to reduce the chances that packets are blocked continuously degrades largely. The major reason that causes the limitations of turn model and odd-even turn is that routers prohibit some kinds of turns statically. For example, all routers prohibits NW and SW turns with west-first routing. In fact, turns are also resources. Allocated with more turns, there are more chances to route around congestions for a packet. The turn model and odd-even turn model allocate the turn resources statically regardless of the network status. Next section, the AbTM, which can dynamically allocate turn resources, will be discussed.

4.3 The Abacus Turn Model

The AbTM inherits the basic idea of [22] that "a network is deadlock free if both clockwise and counter-clockwise rightmost columns are removed from the network", but it uses a more flexible way to realize it. To form a clockwise rightmost column, as shown in Fig. 4.9a, the ES turn should be above an SW turn. Thus, there are three ways to avoid the formation of the rightmost column. The first way is to prohibit the SW turn at all routers as shown in Fig. 4.9b. The second way is to

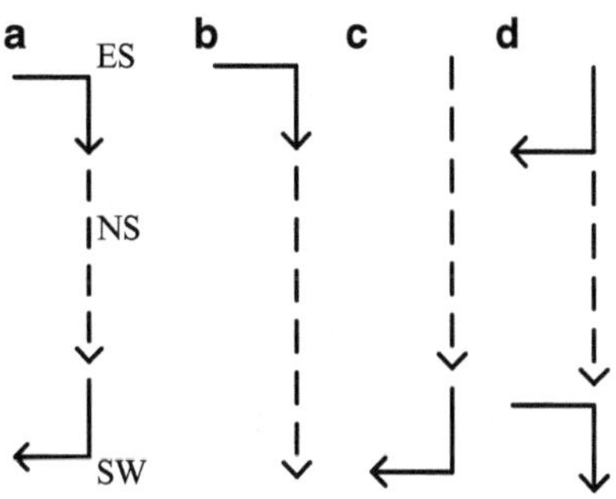

Fig. 4.9 The clockwise rightmost column, (**a**) ES above SW, (**b**) ES only, (**c**) SW only, (**d**) ES below SW

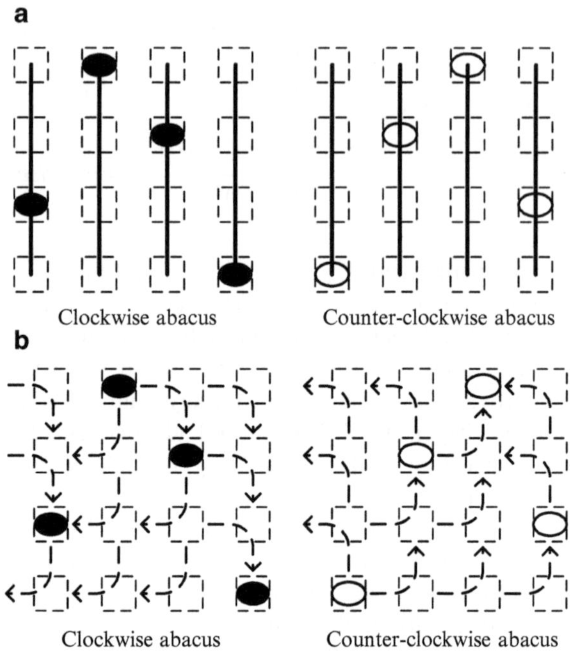

Fig. 4.10 An AbTM example, (**a**) network with beads, (**b**) network marked with prohibited turns. *Dashed blocks* represent possible position for beads, *solid ellipses* represent clockwise beads, *hollow ellipses* represent counter-clockwise beads, *dashed arrows* indicate the prohibited turns, and allowed turns are not shown for simplicity

prohibit the ES turn at all routers as shown in Fig. 4.9c. Third, we also could remove the clockwise rightmost column by forbidding all ES turns above any SW turn as shown in Fig. 4.9d. Thus, in each column, there is a node, above which all ES turns are prohibited, and below which all SW turns are prohibited. In this way, there is no ES turn above any SW turn in any column, thus no clockwise rightmost column can be formed. We name this kind of node as clockwise *bead*. Actually, the first two ways are special cases of the third one. For example, the first way can be realized by moving the *bead* to the top of the column. Note that the top node also allows the SW turn, but this turn will never be used. Similarly, in each column, there is also a counter-clockwise *bead*, above which NW turn is prohibited and below which EN turn is prohibited. Thus, the counter-clockwise rightmost column is removed from the network too.

A 4×4 mesh is compared to an abacus as shown in Fig. 4.10a. Each column is viewed as a wire with two sliding *beads*, i.e., the clockwise *bead* and counter-clockwise *bead*. Clockwise (resp., counter-clockwise) *beads*, which are separately controlled in different columns, are utilized to regulate the distribution of clockwise turns (resp., counter-clockwise turns). Generally, AbTM rules can be summarized as:

1. Nodes above clockwise (resp., counter-clockwise) bead prohibit ES (resp., NW) turn,
2. Nodes below clockwise (resp., counter-clockwise) bead prohibit SW (resp., EN) turn,

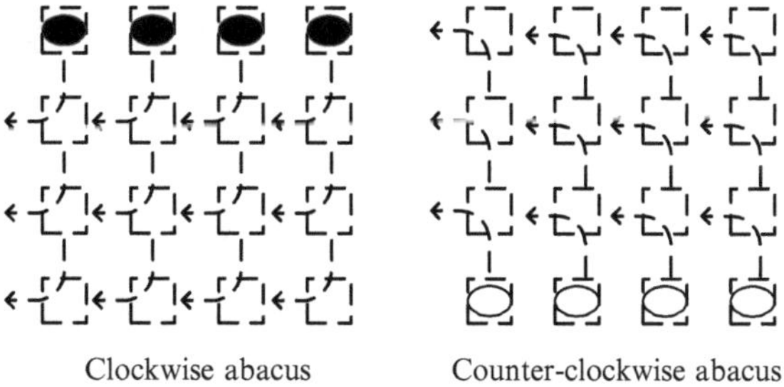

Fig. 4.11 AbTM-based west-first routing algorithm. All routers, except the clockwise and counter-clockwise bead holders, prohibit NW and SW turns

3. Clockwise (resp., counter-clockwise) bead holer does not prohibit clockwise (resp., counter-clockwise) turns.

Figure 4.10b shows the distribution of prohibited turns represented by dashed arrows. According to the AbTM, the clockwise (resp., counter-clockwise) bead holder allows all clockwise turns: i.e., ES, SW, WN and NE turns (resp., all counter-clockwise turns: i.e., EN, NW, WS and SE turns). The router above clockwise (resp., counter-clockwise) beads allows all clockwise (resp., counter-clockwise) turns except the ES (resp., NW) turn. On the other hand, the router below clockwise (resp., counter-clockwise) beads allows all clockwise (resp., counter-clockwise) turns except the SW (resp., EN) turn. Turns are distributed in such a way that none of clockwise and counter-clockwise rightmost columns can be formed. Actually, the *beads* represent the crunch points. Routers above and below *beads* are required to prohibit different turns, so that CDG is acyclic. Thus, wherever the *beads* are located, the network is deadlock-free. When the *beads* move along the column, the relative position between routers and *beads* varies. Thus, turns allowed and prohibited by routers vary accordingly. In fact, turns are resources and reflect the ability of a router to forward packets, i.e., the adaptiveness. If a turn is prohibited after the *beads* movement, corresponding routing adaptiveness is reduced. Otherwise, the adaptiveness is increased. Therefore, when the traffic pattern changes, we could adjust routing adaptiveness by moving *beads* inside each column.

West-first, negative-first and odd-even routing algorithms all can be realized with AbTM directly. For example, Fig. 4.11 shows that the west-first routing could be realized by moving all clockwise *beads* to the top of each column, and all counter-clockwise ones to the bottom. Compared with west-first routing, the routing algorithm realized by AbTM has one difference that the clockwise (resp., counter-clockwise) *bead* holders don't prohibit clockwise (resp., counter-clockwise) turns. However, the SW (resp., NW) turns, which are allowed by the clockwise (resp., counter-clockwise) *bead* holders, will never be required since they are at the top (resp., bottom) of these columns as we discussed above. Thus, it is equivalent to the case that these turns are prohibited.

Negative-first routing could be realized by moving both clockwise and counter-clockwise *beads* to the bottom of each column. Since it is simple, we omit the detailed discussion. As shown in Fig. 4.12, odd-even routing also could be realized

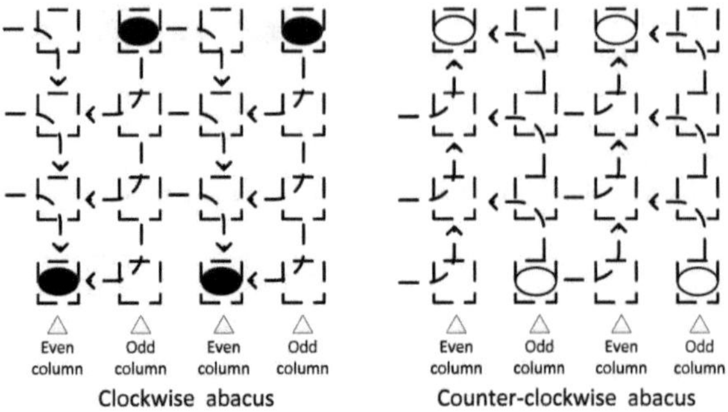

Fig. 4.12 AbTM-based odd-even routing algorithm

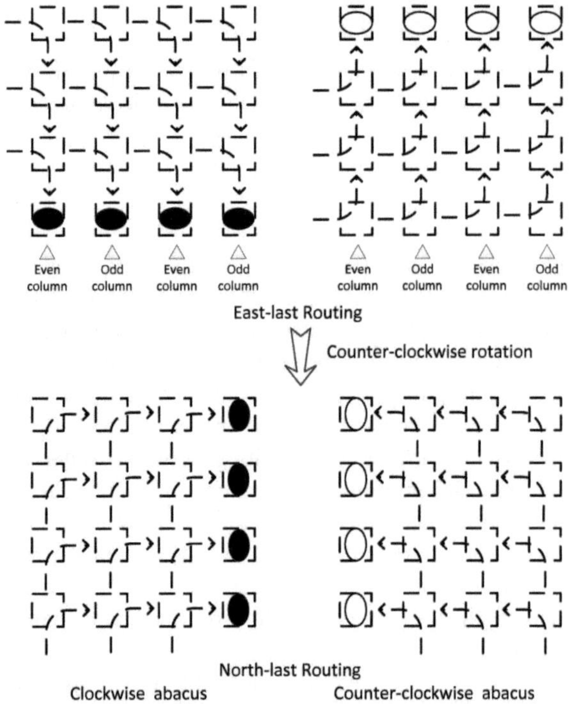

Fig. 4.13 North-last routing could be realized by rotating east-last routing 90° in counter-clockwise direction. East-last routing could be realized with AbTM

with AbTM. In even columns, we cloud move clockwise and counter-clockwise *beads* to bottom and top to prohibit ES and EN turns, respectively. In odd columns, on the other hands, clockwise and counter-clockwise *beads* should be moved to the top and the bottom to prohibit SW and NW turns, respectively.

North-last routing algorithm cannot be realized with AbTM directly since it prohibits the NE turn in clockwise abstract cycle. NE turn does not belong to the rightmost column, thus AbTM-based routing algorithms always allow the NE turn. However, if rotation symmetry is considered, then the north-last routing can be realized by rotating east-last routing 90° in the counter-clockwise direction. As shown in Fig. 4.13, the east-last routing algorithm could be realized by moving

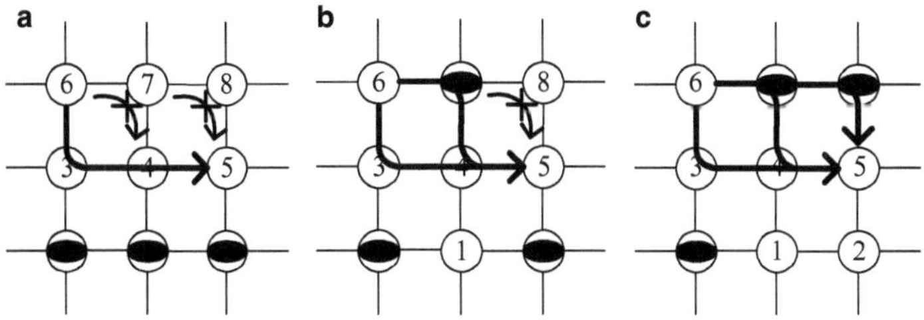

Fig. 4.14 An example of AbTM-based reconfigurable routing

clockwise and counter-clockwise *beads* to the bottom and the top in each column, respectively. After a 90° rotation in the counter-clockwise direction, the east-last routing becomes the north-last routing.

By now, designing a deadlock-free routing algorithm is reduced to assigning the positions of clockwise and counter-clockwise *beads* inside each column. As long as the positions are determined, the network is deadlock-free. We prove this as the following theorem.

Theorem 1. *The network following abacus turn model is deadlock-free.*

Proof. We prove this theorem by contradiction. We assume a network following *abacus* turn model is deadlock, so that there should be a turn dependence cycle in the network according to [21]. According to [22], there must be a clockwise or counter-clockwise rightmost column in this cycle. Without the loss of generality, we assume it is a clockwise rightmost column. Therefore, there must be a node, N_x, allowing ES turn above a node, N_y, allowing SW turn. According to the *abacus* turn model, N_x should be the clockwise *bead* or below it, and N_y should be the clockwise *bead* or above it. Therefore, N_x cannot be above N_y. Contradiction arises. ⊓

Compared with turn model [21] and odd-even turn model [22], the major advantage of AbTM is that it is dynamically reconfigurable. For each $k \times k$ mesh, there are two beads in each column and k potential positions for each *bead*, thus there are $k \times k$ combinations for each column and totally $(k \times k)^k$ combinations for a network. That means there are $(k \times k)^k$ different deadlock-free routing algorithms can be realized with AbTM in a $k \times k$ mesh. For example, for a 4×4 mesh, there are 65,536 different combinations. Once the *beads* are moved, the routing algorithm is reconfigured. Therefore, we could move the *beads* in each column up or down to optimize the network performance under different network status. The following example shows how does the AbTM-based reconfigurable routing tackle the congestion problem.

Example 1. As shown in Fig. 4.14a, clockwise *beads* are initially located on the bottom row. Thus, routers above *beads* prohibit the ES turn. Meanwhile, a new hot spot (node-5) is detected, and node-6 is expected to send packets to it in a relatively long period. Because there is only one available minimal path between them, this path is highly prone to congestions. To balance traffic, node-6 wants to have more available paths. Thus, it makes a complaint to node-7 about the prohibited ES turn at each time it tries to send packets. Meanwhile, node-7 collects the complaints, and negotiates with the bead holder, i.e., the node-1. The holder will evaluate the requirements, and determine whether to give up the *bead*. Here, node-1 will pass the *bead* up as shown in Fig. 4.14b. Receiving the *bead*, node-7 could enable the ES turn according to the AbTM. Thus, node-6 could exploit two available paths to balance the traffic. Similarly, node-7 will make complaints to node-8. Finally, node-8 also gets the ownership of the clockwise *bead* as shown in Fig. 4.14c. By now, all minimal paths could be used to forward packets from node-6 to node-5 to reduce the congestion.

4.4 The AbTM-Based Reconfigurable Routing

In this section, we will discuss two AbTM-based reconfigurable routing algorithms, i.e., the arm-wrestling and tug-war algorithms, and the related issues, such as the routing reconfiguration algorithms and the implementation issues.

4.4.1 The Algorithms

Following the AbTM, the network is always deadlock-free no matter where the clockwise and counter-clockwise *beads* locate in columns. Thus, for a reconfigurable routing algorithm, its main job is to decide the distributions of *beads* in each column to optimize the network performance. In other words, the reconfigurable routing algorithm should determine the time and direction of the bead movement.

According to the AbTM, four turns, WN, NE, WS and SE, are always enabled since they are not critical. Other four turns, ES, SW, EN, and NW, are updated with the *bead* movement. These four turns can be further classified into two groups, i.e., clockwise group (ES and SW) and counter-clockwise group (EN and NW). Generally, moving a *bead* means disabling a turn in the old *bead* holder and enabling the other turn in the same group in the new holder. Thus, the turn requirement is a nature metric to determine the direction of *bead* movement.

With the requirements of each turn, moving *bead* can be compared to moving a block (mass is 0) as shown in Fig. 4.15. The *bead* is compared to the block. For example, the north neighbor give a force, F_{up}, which reflects its eagerness for ES turn, to pull up the clockwise *bead*. On the other hand, the south neighbor represents its eagerness for SW turn by giving a force, F_{down}. The current holder gives a friction, f, to prevent the motion. The direction of f is always opposite to the direction of *bead* movement. For example, if $F_{up} > F_{down}$, the *bead* is going to be moved up. Thus, f reflects holder's eagerness for SW turn to prevent the movement. Otherwise, it reflects its eagerness for ES turn to prevent moving down the *bead*. Furthermore, a threshold (Th) is set to prevent "shaking" that a *bead* is frequently

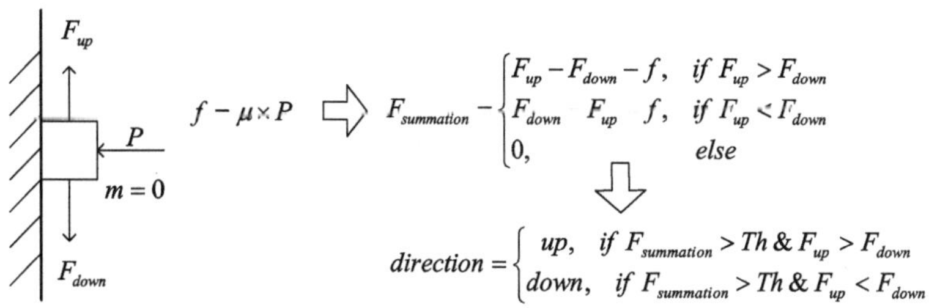

Fig. 4.15 An illustrative example of moving a block

moved up and down. Therefore, the force summation should be bigger than the Th to move the *bead*. For moving counter-clockwise *beads*, F_{up} reflects north neighbor's eagerness for NW turn, and F_{down} reflects south neighbor's eagerness for EN turn.

Mathematically, the forces can be described as Eqs. 4.6 ∼ 4.9, where CT_{xy} and CT_{xy-z} are the counters used to record the times that the xy turn was required at the *bead* holder and the z neighbor respectively. For example, the upward force on the clockwise *bead* is CT_{es-n} as shown in Eq. 4.6, where CT_{es-n} is the counter used to record the times that the ES turn was required at the north (n) neighbor. As shown in Eq. 4.7, the downward force on the clockwise *bead* is CT_{sw-s}, which represents the times that the SW turn was required at the south neighbor. Each router maintains four counters, i.e., CT_{es}, CT_{sw}, CT_{en} and CT_{nw}, each for one critical turn. These counters are updated once the corresponding turn is required. We say a turn is required if a packet makes this turn or a complaint about this turn is received. As discussed in the above example, a router makes a complaint to its neighbor once it detects that the neighbor prohibits a turn which is needed by itself. Without the "complaint" mechanism, the counters record how many times the turn was used instead of the times the turn was required or needed. Thus, the "complaint" mechanism is important for AbTM-based reconfigurable routing algorithms. In [43], complaints are only transferred between neighbors. If a router could accurately make complaints to all routers which will be used by the packets being processed, then the reconfigurable routing will be more powerful.

$$F_{up} = \begin{cases} CT_{es-n}, & clockwise; \\ CT_{nw-n}, & counter\text{-}clockwise. \end{cases} \tag{4.6}$$

$$F_{down} = \begin{cases} CT_{sw-s}, & clockwise; \\ CT_{en-s}, & counter\text{-}clockwise. \end{cases} \tag{4.7}$$

$$f = \begin{cases} CT_{sw}, & clockwise \,\&\, F_{up} > F_{down}; \\ CT_{es}, & clockwise \,\&\, F_{up} < F_{down}; \\ CT_{en}, & counter\text{-}clockwise \,\&\, F_{up} > F_{down}; \\ CT_{nw}, & counter\text{-}clockwise \,\&\, F_{up} < F_{down}; \end{cases} \tag{4.8}$$

$$F_{summation} = \begin{cases} CT_{es-n} - CT_{sw-s} - CT_{sw}, & clockwise \; \& \; CT_{es-n} > CT_{sw-s}; \\ CT_{sw-s} - CT_{es-n} - CT_{es}, & clockwise \; \& \; CT_{es-n} < CT_{sw-s}; \\ CT_{nw-n} - CT_{en-s} - CT_{en}, & counter\text{-}clockwise \; \& \; CT_{nw-n} > CT_{en-s}; \\ CT_{en-s} - CT_{nw-n} - CT_{nw}, & counter\text{-}clockwise \; \& \; CT_{nw-n} < CT_{en-s}; \\ 0, & else. \end{cases} \tag{4.9}$$

Arm-wrestling algorithm is a local mechanism, which only considers the turn requirements of current holder as well as its direct north and south neighbors. The local mechanism simplify the implementation, but the requirements of farther neighbors may be not fulfilled in time. To address this problem, the *tug-war* algorithm that differs from *Arm-wrestling* in the definition of F_{up} and F_{down} was proposed.

Basically, *tug-war* is a non-local mechanism. For the *tug-war* algorithm, the nodes above *bead* are viewed as a group, their total requirements to a corresponding turn are defined as the F_{up}. As shown in Eq. 4.10, the total requirements for ES (resp., NW) turn of all north neighbors are viewed as the F_{up} on clockwise (resp., counter-clockwise) *beads*. On the other hand, nodes below *bead* are viewed as an another group, and their total requirements to the turn are defined as the F_{down}. Equation 4.11 shows that the F_{down} on clockwise (resp., counter-clockwise) *bead* is the total requirements for SW (resp., EN) turn of all south neighbors. To get the total requirement, each router increases its local requirement by the requirement received from upstream nodes, and propagates the result to the downstream node through out-of-band dedicated wires. As shown in Eq. 4.12, to highlight the requirement of nodes closer to *bead*, the total requirements are divided by 2 at each node.

$$F_{up} = \begin{cases} T_{es-n}, & clockwise; \\ T_{nw-n}, & counter\text{-}clockwise. \end{cases} \tag{4.10}$$

$$F_{down} = \begin{cases} T_{sw-s}, & clockwise; \\ T_{en-s}, & counter\text{-}clockwise. \end{cases} \tag{4.11}$$

$$T_{xy} = CT_{xy} + T_{xy-n(s)}/2 \tag{4.12}$$

4.4.2 Reconfiguration Algorithms

Moving *beads* is a hard job due to the difficulty of keeping network deadlock-free. Taking Fig. 4.16a as an example, node-1 holds the clockwise *bead*. Thus, it allows the SW turn, and node-4 could utilize it to forward packets to node-0. As shown in Fig. 4.16b, node-1 will prohibit SW turn after the movement of *bead*. If node-4

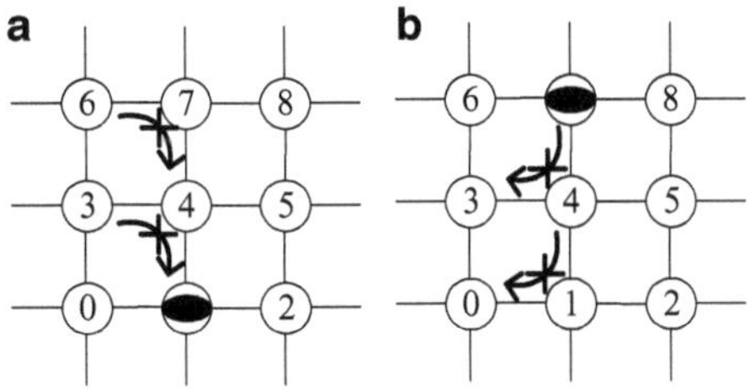

Fig. 4.16 The distribution of prohibited turns, (**a**) before bead movement, (**b**) after bead movement

does not notice that change in time, the packets sent to node-0 could be blocked at node-1. Furthermore, if node-7 enables ES turn when the SW turns, to be disabled by node-1 and node-4, are still hold by packets, then a clockwise rightmost column maybe formed. Thus, the AbTM is violated.

To solve these problems, *bead* movement should follow two rules:

1. A turn can be disabled iff no packet requires it,
2. A turn can be enabled iff there is no packet holding the other turn which is needed to form a rightmost column.

In general, h turns should be prohibited and h other turns should be enabled after a h-hops movement of a *bead*. For example, to move the *bead* from node-1 to node-7, SW turns on node-1 and node-4 should be prohibited, and ES turns on node-4 and node-7 should be enabled. Thus, it is a hard job to meet above two rules simultaneously since a large number of turns will be involved, especially in a large-scale network. To reduce the complexity of *bead* movement, it is divided into h steps, within which the *bead* is moved up/down just one hop. This basic step is named as *bead passing* and is viewed as a safe operation. Exploiting this safe operation has two advantages. First, *bead passing* has a good scalability since it can be locally realized by the cooperation of adjacent routers. Second, *bead passing* itself can guarantee the network deadlock-free during the reconfiguration. Therefore, the routing designers do not need to consider the deadlock problem when designing their own reconfigurable routing algorithms.

Within each step, only one turn should be prohibited at original *bead* holder. For example, to move *bead* from node-1 to node-4, node-1's SW turn should be prohibited. Before disabling the turn, node-1 should notify node-4 to stop injecting southwest packets[3] which may require the SW turn on node-1. On receiving the notification, node-4 labels its south output port as "southwest unaccepted". After then, all southwest packets will be routed to node-3 since the WS turn is always allowed. However, node-4 cannot send acknowledgement to node-1 right now, because there may be southwest packets that have already been routed based on the old information. Thus, node-4 should drain such kind of packets first before sending the acknowledgement.

[3]Southwest packets are those whose destination is on the southwest to current node.

a

```
1. IF MoveUp
2.   Notify north neighbor to drain
         southwest packets in local,
         east, and north input queues.
3.   Wait for acknowledgement.
4.   Drain southwest packets in north
         input queue.
5.   Disable SW and Pass bead up
6. ELSIF MoveDown
7.   Notify west neighbor to drain
         southeast packets in local,
         west, and north input queues.
8.   Wait for acknowledgement.
9.   Drain southeast packets in west
         input queue.
10.  Disable ES and Pass bead down
11. ENDIF
```

b

```
1. IF MoveUp
2.   Notify west neighbor to drain
         northeast packets in local,
         west, and south input queues.
3.   Wait for acknowledgement.
4.   Drain northeast packets in west
         input queue.
5.   Disable EN and Pass bead up
6. ELSIF MoveDown
7.   Notify south neighbor to drain
         northwest packets in local,
         east, and south input queues.
8.   Wait for acknowledgement.
9.   Drain northwest packets in south
         input queue.
10.  Disable NW and Pass bead down
11. ENDIF
```

Fig. 4.17 The pseudo code of bead passing, (**a**) clockwise bead passing, (**b**) counter-clockwise bead passing

Draining packets is an extensively-studied topic in the field of reconfiguration algorithms [25–30]. In this chapter, we borrow the idea of [26] that exploits the reconfiguration token. The main difference is that not all packets are required to be drained. In the above example, node-4 only has to drain southwest packets. Because node-4 receives southwest packets only from its local, east, and north input ports, reconfiguration tokens can only be inserted into the end of the queues belonging to these three input ports. On receiving all three tokens, node-4 can conclude that there is no southwest packets routed to node-1 existed in itself. Thus, it could send acknowledgement to node-1. We should note that the token cannot be inserted until the tail flit arrives. In other words, there is no packet will be divided by a token.

On receiving the acknowledgement, node-1 has the confidence that no new southwest packet will be received from its north neighbor. Unfortunately, there may be southwest packets still queued in the north input port. Thus, it should first drain the southwest packets in itself before disabling the SW turn. After then, it could prohibit the SW turn and pass the *bead* up. Once node-4 receives the *bead*, it enables its ES turn and notifies its west neighbor, node-3, that it could receive southeast packets from now on. By now, one iteration of *bead passing* has finished. Above operations are summarized in the line-2 to line-5 in Fig. 4.17a. As shown in Fig. 4.17, moving down the clockwise *beads* and moving up/down counter-clockwise *beads* follow the same procedure, but different packets should be drained in different neighbors.

Theorem 2. *bead passing does not introduce deadlock into network.*

Proof. We also prove this theorem by contradiction. To distinguish packets routed by the old routing algorithm from those routed by the new routing algorithm, we refer them as old packets and new packets respectively. We assume a network encounters deadlock during *bead passing*. According to [22], there should be a clockwise or counter-clockwise rightmost column. Without the loss of generality, we assume it is a clockwise rightmost column. Hence, there should be a packet, P_x, taking a ES turn is above another packet, P_y, taking a SW turn. There are four different combinations of P_x and P_y: {new, new}, {new, old}, {old, new}, and {old, old}. Since the network following the AbTM, the combination cannot be {new, new} and {old, old} as we proved above. If the combination is {new, old}, P_x must take the ES turn on the new clockwise *bead* because nodes above it do not allow the ES turn, P_y must take the SW turn on the old *bead* because nodes below it do not allow the SW turn, and the *bead* is moving up. Therefore, P_x takes a ES turn **after** the declaration of new *bead* and P_y takes a SW turn **before** the declaration. Hence, combination {new, old} is impossible. Due to the similar reason, combination {old, new} is impossible either. Contradiction arises. □

By now, we have shown that bead passing does not introduce deadlock into network. However, whether the bead passing operation is efficient is largely unknown. To keep network deadlock free, bead passing requires to drain old packets. Thus, the draining time will affect the efficiency of bead passing dramatically. To drain old packets, we insert a token behind the tail flit of each selected input FIFO. If the tail flit has not yet arrived, the draining process should be postponed. Therefore, if the packet to be drained is long, the waiting time may be unacceptable. In a real CMP, there are different types of packets. For example, the request packets without data are usually short, while response packets with data are usually long. At most times, AbTM-based routing will allocate more bandwidth to response packets, since they require more bandwidth. That means packets to be drained are short packets at most times. Therefore, the time used to drain packets is relatively short. Furthermore, if the congestion is caused by short packets, AbTM-based routing is still effective. The reason is that the congestion is inherently caused by large flows. In other words, if short packets cause the congestion, then the number of short packets should be large. It is possible that the short packet, which is injected at the beginning time of bead passing, does not get benefit of this reconfiguration. However, the following short packets can still take advantage of the increased bandwidth. We left the experiments about how does the packet length affect the reconfiguration time as future work.

4.4.3 Implementation Issues

Generally, there are two ways to implement routing algorithms, i.e., logic-based and table-based. Logic-based routing algorithms, such as xy routing, are time/space efficient. However, they are lack of flexibility for reconfiguration. Table-based solution is flexible, but it may introduce large area and timing overhead. Recently, a logic-based distributed routing (LBDR) was proposed [18]. LBDR routing provides enough flexibility to implement AbTM-based reconfigurable routing algorithms. As shown in Fig. 4.18a, LBDR routing provides an effective way to implement

a

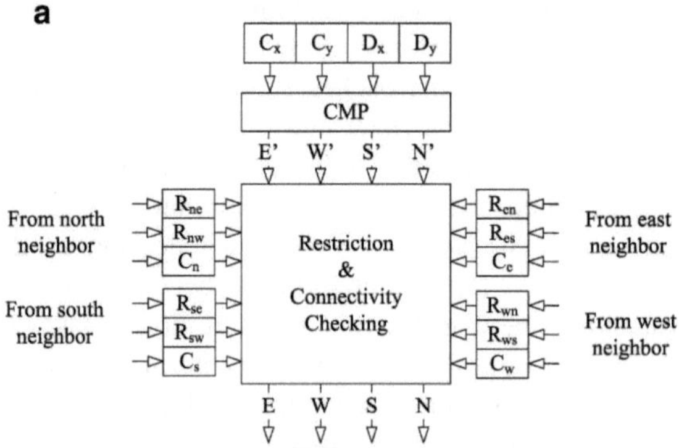

$$E = C_e \bullet (E' \bullet \overline{N'} \bullet \overline{S'} + E' \bullet N' \bullet R_{en} + E' \bullet S' \bullet R_{es})$$
$$W = C_w \bullet (W' \bullet \overline{N'} \bullet \overline{S'} + W' \bullet N' \bullet R_{wn} + E' \bullet S' \bullet R_{ws})$$
$$S = C_s \bullet (S' \bullet \overline{E'} \bullet \overline{W'} + S' \bullet E' \bullet R_{se} + S' \bullet W' \bullet R_{sw})$$
$$N = C_n \bullet (N' \bullet \overline{E'} \bullet \overline{W'} + N' \bullet E' \bullet R_{ne} + N' \bullet W' \bullet R_{nw})$$

b

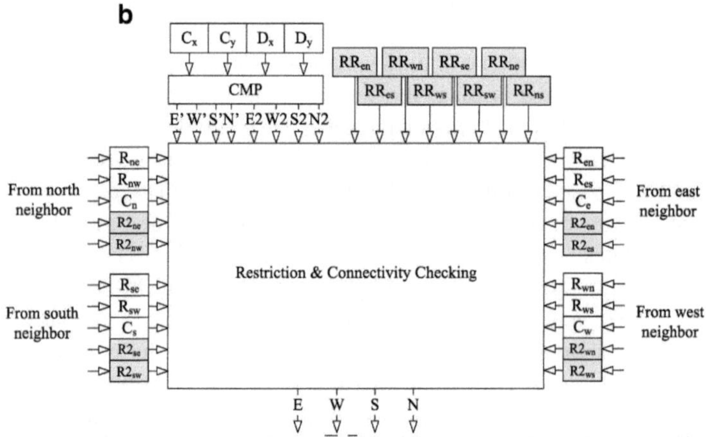

$$E = C_e \bullet (ip_l + ip_w + ip_s \bullet RR_{ne} + ip_n \bullet RR_{se}) \bullet (E' \bullet \overline{N'} \bullet \overline{S'} + E' \bullet N' \bullet R_{en} + E' \bullet S' \bullet R_{es} + E2 \bullet N' \bullet R2_{en} + E2 \bullet S' \bullet R2_{es})$$
$$W = C_w \bullet (ip_l + ip_e + ip_s \bullet RR_{nw} + ip_n \bullet RR_{sw}) \bullet (W' \bullet \overline{N'} \bullet \overline{S'} + W' \bullet N' \bullet R_{wn} + W' \bullet S' \bullet R_{ws} + W2 \bullet N' \bullet R2_{wn} + W2 \bullet S' \bullet R2_{ws})$$
$$S = C_s \bullet (ip_l + ip_n + ip_e \bullet RR_{ws} + ip_w \bullet RR_{es}) \bullet (S' \bullet \overline{E'} \bullet \overline{W'} + S' \bullet E' \bullet R_{se} + S' \bullet W' \bullet R_{sw} + S2 \bullet E' \bullet R2_{se} + S2 \bullet W' \bullet R2_{sw})$$
$$N = C_n \bullet (ip_l + ip_s + ip_e \bullet RR_{wn} + ip_w \bullet RR_{en}) \bullet (N' \bullet \overline{E'} \bullet \overline{W'} + N' \bullet E' \bullet R_{ne} + N' \bullet W' \bullet R_{nw} + N2 \bullet E' \bullet R2_{ne} + N2 \bullet W' \bullet R2_{nw})$$

Fig. 4.18 The routing logic of original (**a**) LBDR and (**b**) LBDR$_e$

distributed routing by utilizing eight routing bits (R_{en}, R_{es}, R_{se}, R_{sw}, R_{wn}, R_{ws}, R_{ne}, and R_{nw}) and 4 connectivity bits (C_e, C_s, C_w, and C_n). The routing bits represent whether the corresponding neighbor accepts such kind of packets. For example, $R_{en} = 1$ means that the east neighbor accepts the northeast packets by allowing the EN turn. The connectivity bits indicate whether the neighbor is connected.

The LBDR routing is separated into two steps. The first step is carried out by the CMP module that compares the coordinates of current router and destination router, i.e., the C_x, C_y, D_x, and D_y as shown in Fig. 4.18a. The results point out the directions which can take the packet closer to the destination. For example, $E' = 1$ indicates that the destination is on the east of the current route. The second step checks the routing and connectivity bits, and determines whether the corresponding output ports can be taken. For instance, the east output can be used if the east neighbor is connected, i.e., $C_e = 1$, and one of the following three conditions holds.

1. The destination is on the east of the current router, i.e., $E' \cdot \bar{N'} \cdot \bar{S'} = 1$,
2. The destination is on the northeast and the east neighbor allows the EN turn, i.e., $E' \cdot N' \cdot R_{en} = 1$,
3. The destination is on the southeast and the east neighbor allows the ES turn, i.e., $E' \cdot S' \cdot R_{es} = 1$.

The LBDR routing may reduce the adaptivity of implemented routing algorithms due to its local visibility. To tackle this problem, the authors proposed the LBDR_e that has two-hops visibility. The LBDR_e routing is separated into three steps. The first step is still carried out by the CMP module. The results indicate the directions where the packets should be routed to reach the destination, as well as whether the destination is two-hops away along that direction. For example, as shown in Fig. 4.18b, if the destination is on the east, then $\text{E}' = 1$; if the destination is also two-hops away, then $\text{E2} = 1$.

The second step checks whether the turn made at the current router is allowable. To this end, each router adds 8 routing restriction bits, i.e., the RR_{xy},[4] and should remember the packets' input channel represented by ip_l (local), ip_e (east), ip_w (west), ip_s (south), and ip_n (north). Furthermore, packets from local input are allowed to be routed through any output port, and the packets with 0°-turn, such as those from west input to east output, are always allowable.

The third step checks the routing restrictions of neighbors. Besides the direct neighbors, LBDR_e also checks the restrictions of two-hops neighbors. Thus, as shown in Fig. 4.18b, LBDR_e requires two more routing bits for each output port. The added routing bits, labeled as R2_{xy}, indicate whether the packet can take a XY turn at the router two hops away from the current router through the x direction. For example, R2_{en} is true means that the router, two hops away from the current router through the east direction, allows the EN turn. To sum up, an output port, such as the east (E) port, is selected at this step, one of the following conditions holds:

1. The destination is on the east of the current router, i.e., $E' \cdot \bar{N'} \cdot \bar{S'} = 1$,
2. The destination is on the northeast and the east neighbor allows the EN turn, i.e., $E' \cdot N' \cdot R_{en} = 1$,
3. The destination is on the southeast and the east neighbor allows the ES turn, i.e., $E' \cdot S' \cdot R_{es} = 1$,

[4]Note that, in [18], authors defined RR_{xy} as that if RR_{xy} is true thus the packets from input x are not allowed to be routed to output y at the current router. In this chapter, however, to keep compatible with the definition of R_{xy}, we redefine the RR_{xy} as that if RR_{xy} is true thus the packets towards direction x are allowed to change its direction to y, i.e., the current router allows the xy turn. This redefinition does not change the core principles of LBDR_e, but notably facilitates the presentation.

4. The destination is on the northeast and two-hops along the east direction as well as the two-hops east neighbor allows the EN turn, i.e., $E2 \cdot N' \cdot R2_{en} = 1$,
5. The destination is on the southeast and two-hops along the east direction as well as the two-hops east neighbor allows the ES turn, i.e., $E2 \cdot S' \cdot R2_{es} = 1$.

It is worthy to note that LBDR routing can be applied to any topology, such as mesh, +, b, d, p, and q topologies, where any node can communicate with others through one minimal path defined in the original mesh. Since AbTM assumes WN, NE, ES, and SE turns are always enabled, we could remove four routing bits, R_{se}, R_{wn}, R_{ws}, and R_{ne}. Furthermore, a reconfiguration module realizing the *bead-passing*-based *arm-wrestling* or *tug-war* algorithms should be added. The reconfiguration module is used to update these four routing bits based on the runtime network status according to *arm-wrestling* or *tug-war* algorithms. Because reconfiguration module does not add delay to the critical path, the time efficiency of LBDR routing is inherited.

4.4.4 AbTM-Based Routing Is Deadlock and Livelock Free

In the following part, we will discuss the characteristics of AbTM-based routing, i.e., deadlock-free, livelock-free, and no packets will be dropped or suspended during the reconfiguration. We should note that there is an important assumption that the network should be initialized to be deadlock-free and LBDR-connected. LBDR-connected means that the network topology is supported by LBDR, and initially there is at least one minimal LBDR-path for any node pair.

Lemma 1. *AbTM-based routing always provide at least one minimal path for any two nodes at any time*

Proof. According to AbTM, the WN, NE, WS and SE turns are always enabled at each node. Therefore, for any node pair, if AbTM-based routing could provide a minimal path consisting of these four turns only, the lemma is proved. Otherwise, we should prove that the critical turns (ES, SW, EN, and NW) in the path will never be prohibited.

First, we assume that the topology is a regular mesh, as shown in Fig. 4.19a. In this case, for any node, such as node-4, any other node can be reached by utilizing uncritical turns. For example, to reach northeast destination, such as node-8, AbTM-based routing could route the packet to north until the intermediate node on the same row as the destination, and then route it to the destination by taking a NW turn. For other directions, the critical turns are not required either. Therefore, there is always a minimal path between any node pair at any time.

Second, we assume that there are two nodes that will be disconnected by removing a critical turn. Without the loss of generality, we assume the critical turn is a ES turn. In this case, there must be an intermediate node that does not have the south neighbor ($C_s = 0$), such as the node-4 in Fig. 4.19b. Otherwise, this

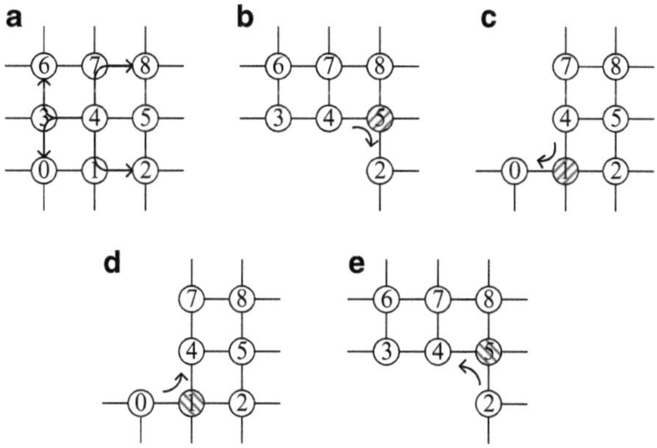

Fig. 4.19 Nodes are always connected in AbTM-based routing, (**a**) mesh, (**b**) ES-critical, (**c**) SW-critical, (**d**) EN-critical, (**e**) NW-critical

intermediate node could send the packet to its south neighbor because SE turn is always enabled. Because any two nodes are connected initially, the ES turn at node-5 should be enabled. Hence, the clockwise *bead* is or above node-5. Without the loss of generality, we assume node-5 holds clockwise *bead*. To prohibit the ES turn at node-5, the *bead* should be pulled down. To pull down the *bead*, the SW turn requirement of nodes below node-5 should be larger than the ES turn requirement of nodes above node-5 according to *arm-wrestling* and *tug-war*. However, the SW requirements of nodes below node-5 is always 0 due to the absence of west neighbors. Therefore, clockwise *bead* will never be below node-5, then the ES turn at node-5 is always enabled.

The proofs for node pairs requiring SW, EN, and NW turns are similar and omitted. We should note that, in Fig. 4.19c, node-4 does not have the west neighbor, so that the SW turn at node-1 cannot be prohibited. In this case, the ES turn requirements of node-7 is also 0 since it does not have west neighbor either. Otherwise, if the node-6 exits, the network topology is a horizontally-reversed "C" which is not supported by LBDR routing. In Fig. 4.19d, node-7 does not have the west neighbor due to the same reason. □

Theorem 3. *AbTM-based routing does not drop and suspend packets during the reconfiguration.*

Proof. Generally, a packet is dropped by an intermediate node if there is no route for that packet. According to Lemma 1, AbTM-based routing provides at least one minimal path for any node pair at any time. Therefore, dropping packet will not happen.

According to Lemma 1, there is always a minimal path, which will not be changed by the reconfiguration, between any two nodes. Therefore, any packet can proceed along the unchanged minimal path without suspension during the reconfiguration. □

Theorem 4. *AbTM-based routing is deadlock and livelock free.*

Proof. According to Theorems 1 and 2, AbTM-based routing is deadlock free.

AbTM-based routing is minimal, so that a packet will get closer to its destination at each hop. Furthermore, Theorem 3 shows that no packet is suspended, thus all packets will definitely reach their destinations. Thus, AbTM-based routing is livelock free. □

It is worthy to note that above discussions are based on an assumption that the network is fault free. In reality, faults may occur at anywhere and anytime [44]. A LBDR-connected network may be disconnected by faults. In the presence of faults, a fault-tolerant routing is needed to guarantee the connectivity of the network. Recently, there are some fault-tolerant routing algorithms proposed based on turn models, such as [31–35]. Due to the difference between the AbTM and the original turn model, whether the fault-tolerant routing algorithms [31–35] can be directly applied to AbTM-based networks is still an open question.

4.5 Evaluation

In this section, the performance and implementation overhead of AbTM-based reconfigurable routing algorithms will be discussed.

4.5.1 Methodology

AbTM provides a safe way to dynamically tune the routing algorithm, where "safe" means that the reconfiguration and reconfigured routing algorithm are both deadlock free. Therefore, we select four baseline routing algorithms also proposed to address the "safe" problem. These algorithms include a deterministic routing algorithm: the xy routing, two partially adaptive routing algorithms: west-first [21] and odd-even routing [22], and a minimal fully adaptive routing [15]. Recently proposed routing algorithms are used to improve the load balancing (or address the "port selection" problem), such as CQR [45], O1Turn [46], and RCA [10]. Thus, AbTM is orthogonal to them.

With 2D meshes, adaptive routing algorithms may produce two candidate output ports at each hop. In such cases, the port with more credits is selected. All routers, except [15], assume one VC per virtual network, and each VC is assigned with a 4-entries queue. For [15], two VCs per virtual network are provided to avoid routing deadlock. For fair comparison, its input buffer queue is equipped with two entries. Furthermore, [15] does not reallocate an VC until the tail flit leaves. Other routing algorithms, however, can reallocate an VC as soon as the tail flit is received.

We first evaluate the routing algorithms using synthetic traffic patterns, including uniform, transpose and hotspot, in a 4×4 mesh. To show the scalability, a 8×8 mesh is simulated under the same traffic patterns. These routing algorithms are

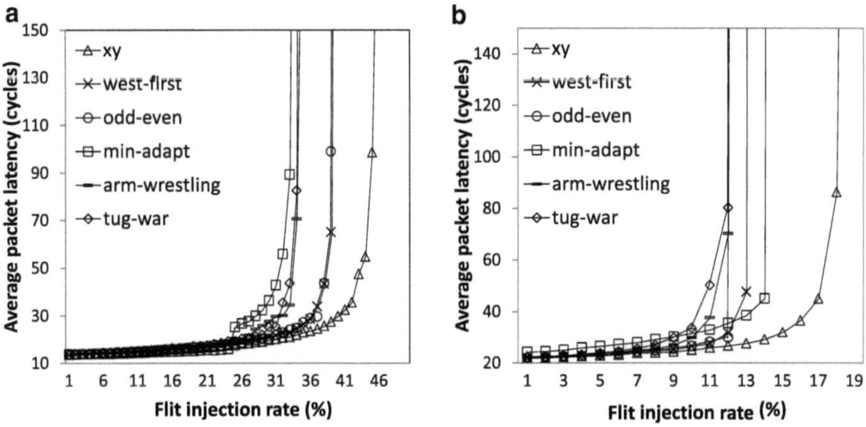

Fig. 4.20 Average packet latency under Uniform traffic patterns, (**a**) 4 × 4 mesh, (**b**) 8 × 8 mesh

also evaluated under real applications's traffic patterns through a trace-driven simulation. The GEMS simulator [47] is utilized to generate the network trace of applications. The cache coherence protocol, MSI_MOSI_CMP_directory, is adopted. This protocol requires five virtual networks to avoid protocol deadlock, and two of them are ordered. A 4 × 4 mesh, instead of the 8 × 8 mesh, is simulated to save simulation time. Benchmark suites, Splash-2 [48] and Parsec [49], are adopted in this simulation.

In the end of this section, the overhead of AbTM-based routing algorithms in timing, area and power is also discussed. To this end, six different routers with xy, west-first [21], odd-even [22], minimal fully adaptive routing [15], and AbTM-based routing algorithms are realized, respectively. The Design Compiler with the UMC 90 nm standard cell library is used to synthesize the routers. The width of input buffer queue is 64-bits, both VC and switch allocators adopt the round-robin arbiter proposed in [50], and an fully connected 5 × 5 crossbar is implemented.

4.5.2 Simulation Results and Analysis

4.5.2.1 Performance Under Synthetic Traffic Patterns

Under the uniform traffic pattern, each node sends packets to others with the same possibility. Figure 4.20a, b show the simulation results for the 4 × 4 and 8 × 8 meshes, respectively. Horizontal axis represents the injection rate in the number of flits per node per cycle, and the vertical axis indicates the average packet latency in router cycles. It has been proved that deterministic routing algorithms could get better performance under the uniform traffic pattern than adaptive routing algorithms [21, 22]. The main reason is that adaptive routing

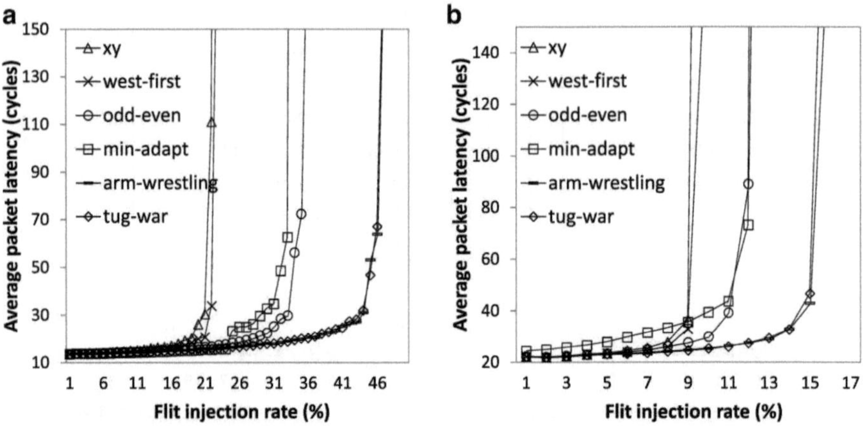

Fig. 4.21 Average packet latency under transpose traffic patterns, (**a**) 4 × 4 mesh, (**b**) 8 × 8 mesh

algorithms usually make selection based on local information, such as the number of credit of each output port. As shown in Fig. 4.20a, xy routing algorithm gets the best results as expected. Following the xy routing, west-first and odd-even, get similar results. AbTM-based routing algorithms are worse than partially adaptive routing algorithms mainly due to the reconfiguration is based on the traffic history. Unfortunately, under uniform traffic pattern, such kind of reconfiguration will be wrong at most times. For example, we assume that northeast direction carries the highest traffic load in the current interval, so that AbTM-based will allocate more adaptivity to that direction. However, northeast direction will carry the lowest traffic load in the next interval with very high possibility according to the definition of uniform traffic. The minimal fully adaptive routing algorithm [15], labeled as "min-adapt", gets the worst results due to its conservative flow control strategy that an VC cannot be reallocated until it is empty. Figure 4.20b shows the results for the 8 × 8 mesh. Most of the results are in according with that shown in Fig. 4.20a. However, the relative performance of min-adapt routing algorithm get improved in this simulation. The reason is that the contention possibility increases with the network size. When contention occurs, [15] requires packets to wait on the escape output port, i.e., the one chosen by xy routing. Thus, the min-adapt routing algorithm could exploit the long-term evenness of uniform traffic pattern just like the xy routing.

Real world applications usually generate nonuniform traffic patterns, such as transpose. Under transpose traffic pattern, source node s always sends packets to the destination d, where $d_i = s_{(i+b/2)\%b}$ and b is the number of bits used to index nodes. As shown in Fig. 4.21a, the AbTM-based routing algorithms get the best results because they could provide full adaptiveness to all packets. On the other hand, xy routing gets the worst performance since it cannot address the congestion problem. West-first gets better results than xy routing by providing full adaptiveness to eastward packets. However, the improvement is limited since westward packets

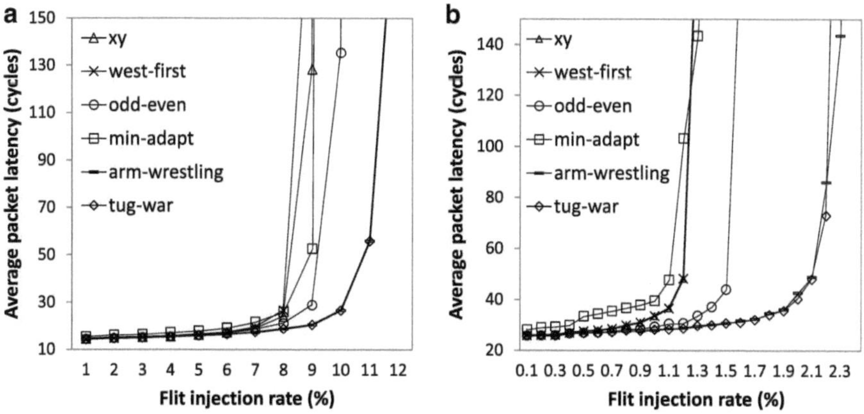

Fig. 4.22 Average packet latency under Hotspot traffic patterns, (**a**) 4 × 4 mesh, (**b**) 8 × 8 mesh

still face with the congestion problem. Odd-even routing could provide even adaptiveness to packets towards different directions, thus it could get better results than west-first and xy routing. However, it cannot get the performance as high as the AbTM-based routing algorithms due to the insufficient routing adaptiveness. Min-adapt routing algorithm also could provide full adaptiveness, but its conservative flow control technique aggregates the congestion problem. Figure 4.21b shows the simulation results in a 8 × 8 mesh. The relative performance is same, but the improvement got by the proposed algorithms is enlarged.

Applications' bursty traffic may cause hotspots, and aggregates the congestion problem. In the following two simulations, we assume four hot-spot nodes: n_0, n_4, n_8, n_{12}. Each hot-spot node gets extra 20 % traffic than others. These four nodes are selected to simulate the situation that four memory controllers are frequently accessed. In such case, westward packets are prone to be congested. As shown in Fig. 4.22a, the AbTM-based routing algorithms get the best results since they could provide full adaptiveness by allowing the NW and SW turns at each router. Min-adapt routing algorithm could provide full adaptiveness, but its conservative flow control technique leads to a low utilization of input buffers. Thus, its performance is worse than odd-even routing. Since west-first and xy routing algorithms cannot provide adaptiveness to avoid congestions, they get the worst performance. When the network is enlarged, the congestion problem is enlarged too. By successfully reducing blocking, the AbTM-based routing algorithms also get the best performance with enlarged gap.

4.5.2.2 Performance Under Applications' Traffic

According to above discussions, we could find that *arm-wrestling* and *tug-war* routing algorithms get good performance under nonuniform traffic patterns, such as transpose and hotspot. In a short range perspective, there are lots of bursts in

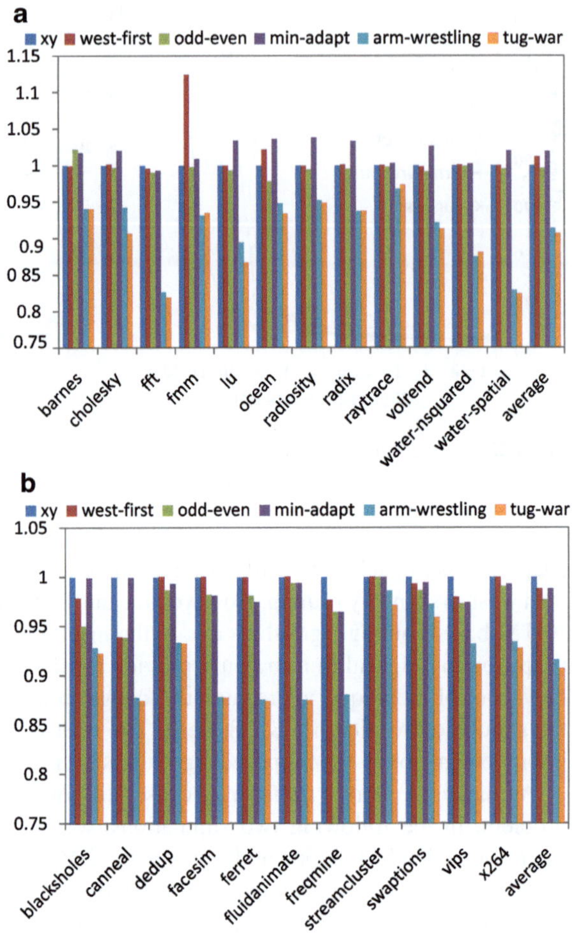

Fig. 4.23 Average packet latency across (**a**) Splash-2 and (**b**) Parsec benchmarks

application's traffic [51]. If we view the burst as hotspot traffic, then the application consists of hotspot traffic in consecutive intervals with different hot-spot nodes. If AbTM-based routing could reconfigure itself in time, we could expect the application performance improvement. The following trace-driven simulations are carried out to prove this assumption.

Figure 4.23a shows the packet latency, which is normalized to the latency of xy routing, across Splash-2 benchmarks. By dynamically allocating more adaptivity to bursty packets, the AbTM-based routing could significantly reduce the packet latency. Generally, both *arm-wrestling* and *tug-war* routing algorithms could reduce packet latency for all applications. However, due to applications' different characteristics, the improvement is different. For example, for applications generating high contented traffic, such as fft and water-spatial, the improvement is significant. For applications generating low contented traffic, such as raytrace and ocean, however,

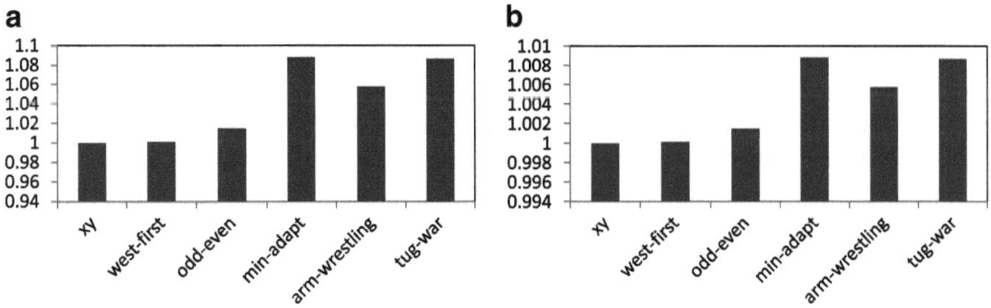

Fig. 4.24 Area overhead comparison, (**a**) router area evaluation, (**b**) tile area evaluation

the improvement will be relatively small. In general, for Splash-2 applications, AbTM-based routing algorithms could reduce the packet latency by maximally 19 and 10 % in average.

To improve the confidence, we simulate the networks again under Parsec benchmarks [49]. The simulation results are shown in Fig. 4.23b. The results lead to the same conclusion that AbTM-based routing algorithms could significantly improve the network performance. Particularly, for high-contented applications, such as canneal and freqmine, the improvement is large. For low-contented applications, however, the reduction is moderate. In general, the packet latency is reduced by 15 % maximally and 10 % in average.

4.5.2.3 Overhead

The router frequency is one of the most important performance metrics to evaluate NoCs. To achieve the high network performance, it should be as high as possible. Although AbTM-based routing algorithms dynamically update routing bits of LBDR routing, they do not modify its routing logic and add delay to the critical path. This together with the fact that LBDR routing has been proved to be timing efficient [18] prove that the AbTM-based routing algorithms are also timing efficient.

Another kind of overhead is the area overhead, which determines the chip cost. Figure 4.24a shows the router area normalized to area of xy router. According to the results, the router realizing min-adapt routing algorithm is the largest because it requires 5 VC allocators. Other routers do not have the VC allocators since they do not implement virtual channels. The *tug-war* router is the second largest, because it requires more logics to propagate traffic information. Due to the low complexity, west-first and odd-even routers are smaller than the *Arm-wrestling* router. Compared with xy router, *Arm-wrestling* and *tug-war* routers increase the area by 5 and 8 %, respectively. Taking the area of core and caches into consideration, the area increase is negligible. As shown in Fig. 4.24b, the increase of tile area will be less than 1 % with the assumption that the router area occupies 11 % area of a tile as reported by Intel [3].

Assuming the same VLSI technology and working frequency, the dynamic power of a router is largely determined by its size and activity. As discussed above, the routers do not have significant difference in area. Thus, they consume similar amount of dynamic power as the activity of them does not have significant difference either. Compared with normal routing activities, reconfigurations consume negligible power since they rarely happen. If we further take the power of cores and caches into consideration, the power overhead is negligible.

4.6 Summary

Exploiting applications' traffic information to improve NoC performance has been proved to be effective. In this chapter, we presented an on-line routing reconfiguration strategy. It dynamically reconfigures itself to provide full adaptiveness to the highest traffic load directions. The major challenge addressed is the way to keep the network deadlock-free during and after each reconfiguration. To this end, we discussed the AbTM and *bead passing*. AbTM keeps the network with a stable configuration deadlock-free, while *bead passing* keeps the network deadlock-free during the reconfiguration without suspending and dropping packets. Based on AbTM, two reconfigurable routing algorithms, i.e., the *arm-wrestling* and *tug-war*, are discussed and evaluated against both synthetic traffic and real application traffic. In this chapter, we mainly discuss the AbTM on 2D meshes. Actually, AbTM-based routing can be extended to any topology with the help of segment-based routing [39] which is proposed to design deadlock-free and fault-tolerant routing algorithms. The segment-based routing divides a network into segments, and keeps the network deadlock-free by placing bidirectional turn restrictions within each segment. So far, the segment-based routing requires off-line computation to determine how to divide the network and where to place the turn restrictions. With the AbTM-based routing, the processor allocation algorithms are required to provide the information about the segments. Thus, the AbTM-based routing could dynamically determines where to put the restrictions at runtime. AbTM-based routing is minimal since the underlying LBDR is minimal. However, non-minimal routing is expected for balancing traffic and tolerating faults. Recently, Rodrigo et al. have proposed non-minimal LBDR routing [52]. Since AbTM-based routing does not modify LBDR's routing logic, the extension can also be applied to AbTM-based routing.

References

1. S. Bell, B. Edwards, J. Amann, R. Conlin, K. Joyce, V. Leung, J. MacKay, M. Reif, L. Bao, J. Brown, M. Mattina, C.-C. Miao, C. Ramey, D. Wentzlaff, W. Anderson, E. Berger, N. Fairbanks, D. Khan, F. Montenegro, J. Stickney, J. Zook, Tile64 – processor: a 64-core soc with mesh interconnect, in *Digest of Technical Papers. IEEE International Solid-State Circuits Conference*, San Francisco, 2008, pp. 88–598

2. http://www.intel.com/content/www/us/en/research/intel-labs-single-chip-cloudcomputer.html
3. S.R. Vangal, J. Howard, G. Ruhl, S. Dighe, H. Wilson, J. Tschanz, D. Finan, A. Singh, T. Jacob, S. Jain, V. Erraguntla, C. Roberts, Y. Hoskote, N. Borkar, S. Borkar, An 80-tile sub-100-w teraflops processor in 65-nm CMOS. IEEE J. Solid-State Circuits **43**(1), 29–41 (2008)
4. W.J. Dally, B. Towles, *Principles and Practices of Interconnection Networks* (Morgan Kaufmann, San Francisco, 2004)
5. J. Kim, C. Nicopoulos, D. Park, R. Das, Y. Xie, N. Vijaykrishnan, M.S. Yousif, C.R. Das, A novel dimensionally-decomposed router for on-chip communication in 3D architectures, in *Proceedings of the 34th Annual International Symposium on Computer Architecture, ISCA'07*, San Diego (ACM, New York, 2007), pp. 138–149
6. J. Kim, D. Park, T. Theocharides, N. Vijaykrishnan, C.R. Das, A low latency router supporting adaptivity for on-chip interconnects, in *Proceedings of Design Automation Conference*, San Diego, 2005, pp. 559–564
7. A. Singh, W.J. Dally, A.K. Gupta, B. Towles, Goal: a load-balanced adaptive routing algorithm for torus networks, in *Proceedings of International Symposium on Computer Architecture*, San Diego, 2003, pp. 194–205
8. J.W. van den Brand, C. Ciordas, K. Goossens, T. Basten, Congestion-controlled best-effort communication for Networks-on-Chip, in *Design, Automation Test in Europe Conference Exhibition*, Nice, 2007, pp. 1–6
9. J. Duato, I. Johnson, J. Flich, F. Naven, P. Garcia, T. Nachiondo, A new scalable and cost-effective congestion management strategy for lossless multistage interconnection networks, in *Proceedings of International Symposium on High-Performance Computer Architecture*, San Francisco, 2005, pp. 108–119
10. P. Gratz, B. Grot, S.W. Keckler, Regional congestion awareness for load balance in Networks-on-Chip, in *Proceedings of IEEE International Symposium on High Performance Computer Architecture*, Salt Lake City, 2008, pp. 203–214
11. D. Park, R. Das, C. Nicopoulos, J. Kim, N. Vijaykrishnan, R. Iyer, C.R. Das, Design of a dynamic priority-based fast path architecture for on-chip interconnects, in *Proceedings of IEEE Symposium on High-Performance Interconnects*, Stanford, 2007, pp. 15–20
12. W.J. Dally, Virtual-channel flow control. IEEE Trans. Parallel Distrib. Syst. **3**(2), 194–205 (1992)
13. S.A. Felperin, L. Gravano, G.D. Pifarre, J.L.C. Sanz, Fully-adaptive routing: packet switching performance and wormhole algorithms, in *Proceedings of ACM/IEEE Conference on Supercomputing*, Albuquerque, 1991, pp. 654–663
14. W.J. Dally, H. Aoki, Deadlock-free adaptive routing in multicomputer networks using virtual channels. IEEE Trans. Parallel Distrib. Syst. **4**(4), 466–475 (1993)
15. J. Duato, A new theory of deadlock-free adaptive routing in wormhole networks. IEEE Trans. Parallel Distrib. Syst. **4**(12), 1320–1331 (1993)
16. Y.M. Boura, C.R. Das, Efficient fully adaptive wormhole routing in n-dimensional meshes, in *Proceedings of International Conference on Distributed Computing Systems*, Poznan, 1994, pp. 589–596
17. J.H. Upadhyay, V. Varavithya, P. Mohapatra, Efficient and balanced adaptive routing in two-dimensional meshes, in *Proceedings of IEEE Symposium on High-Performance Computer Architecture*, Raleigh, 1995, pp. 112–121
18. J. Flich, S. Rodrigo, J. Duato, An efficient implementation of distributed routing algorithms for NoCs, in *Proceedings of ACM/IEEE International Symposium on Networks-on-Chip*, Newcastle, 2008, pp. 87–96
19. L.-S. Peh, W.J. Dally, A delay model and speculative architecture for pipelined routers, in *Proceedings of International Symposium on High-Performance Computer Architecture*, Nuevo Leone, 2001, pp. 255–266
20. S. Ma, N.E. Jerger, Z. Wang, Whole packet forwarding: efficient design of fully adaptive routing algorithms for Networks-on-Chip, in *Proceedings of the 2012 IEEE 18th International Symposium on High-Performance Computer Architecture, HPCA'12*, New Orleans (IEEE Computer Society, Washington, DC, 2012), pp. 1–12

21. C.J. Glass, L.M. Ni, The turn model for adaptive routing, in *Proceedings of International Symposium on Computer Architecture*, Gold Coast, 1992, pp. 278–287
22. G.M. Chiu, The odd-even turn model for adaptive routing. IEEE Trans. Parallel Distrib. Syst. **11**(7), 729–738 (2000)
23. N. Barrow-Williams, C. Fensch, S. Moore, A communication characterisation of splash-2 and parsec, in *IEEE International Symposium on Workload Characterization*, Austin, 2009, pp. 86–97
24. L.M. Ni, P.K. McKinley, A survey of wormhole routing techniques in direct networks. Computer **26**(2), 62–76 (1993)
25. T.L. Rodeheffer, M.D. Schroeder, Automatic reconfiguration in autonet, in *Proceedings of ACM Symposium on Operating Systems Principles*, Pacific Grove, 1991, pp. 183–197
26. O. Lysne, J.M. Montanana, J. Flich, J. Duato, T.M. Pinkston, T. Skeie, An efficient and deadlock-free network reconfiguration protocol. IEEE Trans. Comput. **57**(6), 762–779 (2008)
27. O. Lysne, J. Duato, Fast dynamic reconfiguration in irregular networks, in *Proceedings of International Conference on Parallel Processing*, Toronto, 2000, pp. 449–458
28. R. Casado, A. Bermudez, J. Duato, F.J. Quiles, J.L. Sanchez, A protocol for deadlock-free dynamic reconfiguration in high-speed local area networks. IEEE Trans. Parallel Distrib. Syst. **12**(2), 115–132 (2001)
29. D. Avresky, N. Natchev, Dynamic reconfiguration in computer clusters with irregular topologies in the presence of multiple node and link failures. IEEE Trans. Comput. **54**(5), 603–615 (2005)
30. R. Casado, A. Bermudez, F.J. Quiles, J.L. Sanchez, J. Duato, Performance evaluation of dynamic reconfiguration in high-speed local area networks, in *Proceedings of International Symposium on High-Performance Computer Architecture*, Toulouse, 2000, pp. 85–96
31. J. Wu, A fault-tolerant and deadlock-free routing protocol in 2D meshes based on odd-even turn model. IEEE Trans. Comput. **52**(9), 1154–1169 (2003)
32. Z. Zhang, A. Greiner, S. Taktak, A reconfigurable routing algorithm for a fault-tolerant 2D-mesh Network-on-Chip, in *Proceedings of ACM/IEEE Design Automation Conference*, Anaheim, 2008, pp. 441–446
33. D. Fick, A. DeOrio, G. Chen, V. Bertacco, D. Sylvester, D. Blaauw, A highly resilient routing algorithm for fault-tolerant NoCs, in *Proceedings of Design, Automation Test in Europe Conference Exhibition*, Nice, 2009, pp. 21–26
34. B. Fu, Y. Han, H. Li, X. Li, A new multiple-round dimension-order routing for Networks-on-Chip. IEICE Trans. Inf. Syst. E94-D, 809–821 (2011)
35. B. Fu, Y. Han, H. Li, X. Li, Zonedefense: a fault-tolerant routing for 2-d meshes without virtual channels. IEEE Trans. Very Large Scale Integr. Syst. (2013)
36. L. Zhang, Y. Han, Q. Xu, X. Li, Defect tolerance in homogeneous manycore processors using core-level redundancy with unified topology, in *Design, Automation and Test in Europe, DATE'08*, Munich, 2008, pp. 891–896
37. L. Zhang, Y. Han, Q. Xu, X. Li, H. Li, On topology reconfiguration for defect-tolerant noc-based homogeneous manycore systems. IEEE Trans. Very Large Scale Integr. Syst. **17**(9), 1173–1186 (2009)
38. W.J. Dally, C.L. Seitz, Deadlock-free message routing in multiprocessor interconnection networks. IEEE Trans. Comput. **C-36**(5), 547–553 (1987)
39. A. Mejia, J. Flich, J. Duato, S.-A. Reinemo, T. Skeie, Segment-based routing: an efficient fault-tolerant routing algorithm for meshes and tori, in *Proceedings of International Symposium on Parallel and Distributed Processing*, Rhodes Island, 2006, p. 10
40. M. Palesi, R. Holsmark, S. Kumar, V. Catania, Application specific routing algorithms for Networks-on-Chip. IEEE Trans. Parallel Distrib. Syst. **20**(3), 316–330 (2009)
41. M.A. Kinsy, M.H. Cho, T. Wen, E. Suh, M. van Dijk, S. Devadas, Application-aware deadlock-free oblivious routing, in *Proceedings of International Symposium on Computer Architecture*, Austin (ACM, New York, 2009), pp. 208–219

42. J. Cong, C. Liu, G. Reinman, Aces: application-specific cycle elimination and splitting for deadlock-free routing on irregular Network-on-Chip, in *Proceedings of ACM/IEEE Design Automation Conference*, Anaheim, 2010, pp. 443–448
43. B. Fu, Y. Han, J. Ma, H. Li, X. Li, An abacus turn model for time/space-efficient reconfigurable routing, in *Proceedings of the 38th Annual International Symposium on Computer Architecture*, ISCA'11, San Jose (ACM, New York, 2011), pp. 259–270
44. Y. Han, Y. Hu, X. Li, H. Li, A. Chandra, Embedded test decompressor to reduce the required channels and vector memory of tester for complex processor circuit. IEEE Trans. Very Large Scale Integr. Syst. **15**(5), 531–540 (2007)
45. A. Singh, W.J. Dally, A.K. Gupta, B. Towles, Adaptive channel queue routing on k-ary n-cubes, in *Proceedings of ACM Symposium on Parallelism in Algorithms and Architectures*, Barcelona, 2004, pp. 11–19
46. D. Seo, A. Ali, W.-T. Lim, N. Rafique, M. Thottethodi, Near-optimal worst-case throughput routing for two-dimensional mesh networks, in *Proceedings of International Symposium on Computer Architecture*, Madison, 2005, pp. 432–443
47. M.M.K. Martin, D.J. Sorin, B.M. Beckmann, M.R. Marty, M. Xu, A.R. Alameldeen, K.E. Moore, M.D. Hill, D.A. Wood, Multifacet's general execution-driven multiprocessor simulator (GEMS) toolset. ACM SIGARCH Comput. Archit. News **33**(4), 92–99 (2005)
48. S.C. Woo, M. Ohara, E. Torrie, J.P. Singh, A. Gupta, The splash-2 programs: characterization and methodological considerations, in *Proceedings of International Symposium on Computer Architecture*, Santa Margherita Ligure, 1995, pp. 24–36
49. C. Bienia, S. Kumar, J.P. Singh, K. Li, The parsec benchmark suite: characterization and architectural implications, in *Proceedings of International Conference on Parallel Architectures and Compilation Techniques, PACT'08*, Toronto (ACM, New York, 2008), pp. 72–81
50. E.S. Shin, V.J. Mooney III, G.F. Riley, Round-robin arbiter design and generation, in *Proceedings of International Symposium on System Synthesis*, Kyoto, 2002, pp. 243–248
51. A.B. Kahng, B. Lin, K. Samadi, R.S. Ramanujam, Trace-driven optimization of Networks-on-Chip configurations, in *Proceedings of ACM/IEEE Design Automation Conference (DAC)*, Anaheim, 2010, pp. 437–442
52. S. Rodrigo, J. Flich, A. Roca, S. Medardoni, D. Bertozzi, J. Camacho, F. Silla, J. Duato, Addressing manufacturing challenges with cost-efficient fault tolerant routing, in *Proceedings of International Symposium on Networks-on-Chip*, Grenoble, 2010, pp. 25–32

Chapter 5
Learning-Based Routing Algorithms for On-Chip Networks

Masoumeh Ebrahimi and Masoud Daneshtalab

Abstract In this chapter, we investigate highly adaptive routing algorithms for balancing the traffic over the network based on learning approaches. The proposed methods aim to provide up-to-dated local and global congestion information at each switch. At first, the learning method is applied to a network utilizing the minimal routing. In low traffic loads, minimal methods can achieve optimal performance, while they are inefficient in avoiding hotspots when the traffic load increases. The reason of this inefficiency is that minimal methods can propagate messages through at most two directions at each switch. When the shortest paths are congested, sending more messages through them can deteriorate the congestion condition considerably. In order to address this issue, we present a non-minimal routing algorithm for on-chip networks that provides a wide range of alternative paths between each pair of source and destination switches. Initially, the algorithm determines all permitted turns in the network including 180-degree turns on a single channel without creating cycles. The implementation of the algorithm provides the best usage of all allowable turns to route messages more adaptively in the network. On top of that, for selecting a less congested path, an optimized and scalable learning method is utilized. The learning method is based on local and global congestion information and can estimate the latency from each output channel to the destination region.

5.1 Introduction

Networks-on-Chip (NoC) has emerged as a solution to address the communication demands of future multicore architectures due to its scalability, reusability, and parallelism in communication infrastructure [1]. The performance and efficiency

M. Ebrahimi (✉) • M. Daneshtalab
University of Turku, Turku, Finland
e-mail: masebr@utu.fi; masdan@utu.fi

M. Palesi and M. Daneshtalab (eds.), *Routing Algorithms in Networks-on-Chip*, DOI 10.1007/978-1-4614-8274-1_5,

of NoCs largely depend on the underlying routing model which establishes a connection between input and output channels in a switch.

Learning-based approaches have attracted considerable attention in industry and academia because they provide effective models for problems where optimal solutions are analytically unavailable or difficult to obtain. Reinforcement learning is a type of learning which is based on the common-sense idea that if an action is followed by a satisfactory state or by an improvement, then the tendency to produce that action should be strengthened, i.e., reinforced. On the other hand, if the state becomes unsatisfactory, then that particular action should be suitably punished [2, 3]. Q-Learning [4] is one of the algorithms in the reinforcement learning family of machine learning. In the Q-Learning approach, the learning agent first learns a model of the environment on-line and then utilizes this knowledge to find an effective control policy for the given task. Q-Routing [5] is a network routing method based on Q-Learning models which enables a network to learn a routing policy in order to minimize the delivery time of messages to reach their destinations [6, 7]. Q-Routing methods are implemented by allowing each switch to maintain a table of Q-values, where each value is an estimate of the time it takes for a message to be delivered to a destination, if sends via a neighboring switch. In this method, each switch updates its routing table when receives new congestion information. Congestion wires are used to transfer the congestion information between the neighboring switches. Q-Routing methods allow a network to be continuously adapted to traffic load changes.

In wormhole switching, messages are divided into small flits traversing within the network in a pipelined fashion. This approach eliminates the need for the allocation of large buffers in intermediate switches along the path. However, a message waiting to be allocated to an outgoing channel may prohibit other messages from using the channels and buffers and thereby wasting channel bandwidth and increasing latency. Adding virtual channels can alleviate this problem, but it is an expensive solution.

Non-minimal methods can partially overcome this blocking problem and reduce the waiting time of messages by delivering them via alternative paths. In contrast, performance can severely deteriorate in non-minimal methods due to the uncertainty in finding an optimal path as they may choose longer paths and meanwhile delivering messages through congested regions. Moreover, non-minimal methods can suggest minimal and non-minimal paths between a source and destination but this flexibility is at the cost of a more complex switch structure or additional virtual channels. On the other hand, an output selection function should choose a single channel from a set of predetermined channels to forward a message to the next hop. This becomes one of the main challenges involved in designing an efficient non-minimal method to select a less congested path from a set of alternative paths. An output channel should not be selected only based on local information as it may route messages through paths which are not only longer but also highly congested. On the other hand, even if a global knowledge of the network is provided, due to a large number of alternative paths, finding a less congested path is questionable which demands an intelligent method to cope with.

5.2 Learning Methods and Minimal Routing

In this section, we first investigate a minimal and adaptive routing algorithm, and then the learning-based approach is built upon this minimal routing. We called the obtained method, LM (Learning method applied to Minimal routing).

5.2.1 DyXY, the Minimal and Adaptive Routing Algorithm

A basic requirement to distribute messages over the network is to have some degree of adaptiveness when routing messages. We utilize a minimal and fully adaptive routing algorithm named Dynamic XY (DyXY) [8]. In this algorithm, which is based on the static XY algorithm, a message can be sent either to the X or Y dimension. DyXY uses one and two virtual channels along the X and Y dimensions, respectively. This is the minimum number of virtual channels that can be employed to provide fully adaptiveness.

The network can be proved to be deadlock-free as follows. The network is partitioned into two sub-networks, one covering +X and the other one covering –X. In this way, the sub-networks are disjoint along the X dimension. Moreover, the first and second sub-network use the first and second virtual channel along the Y dimension, respectively, and thus the sub-networks are disjoint along the Y dimension as well. Each switch in DyXY has seven pairs of channels, i.e. East (*E*), West (*W*), North-vc1 (*N1*), North-vc2 (*N2*), South-vc1 (*S1*), South-vc2 (*S2*), and Local (*L*).

5.2.2 LM: Learning Approach Applied to Minimal Routing

Let us explain the learning method by this assumption that the source s sends a message to the destination d through one of its neighboring switches x (see Fig. 5.1). The time it will take for a message to reach from the source s to the destination d is bounded by the sum of three quantities [6]: (1) B_x: the waiting time of the message in the input buffer of the switch x (2) δ: the transmission delay over the link from the switch s to the switch x (3) $Q_x(n,d)$: the time it would take to send the message from the switch x to the destination switch d via the least congested neighboring switch (e.g. the switch n). This value is extracted from the Q-Table of the switch x.

By sending the message over the link from the switch s to the switch x, the transmission delay δ is obtained. However, this value is considered as negligible in this work. To measure the B_x value, instead of the waiting time in the input buffer, we use the number of occupied slots in the input buffer of the switch. Finally, $Q_x(n,d)$ can be extracted from the Q-Table of the switch x as soon as the output channel of the message is determined. These quantities are summed together and form a new estimated latency from the switch s to the destination d obtained at the switch x.

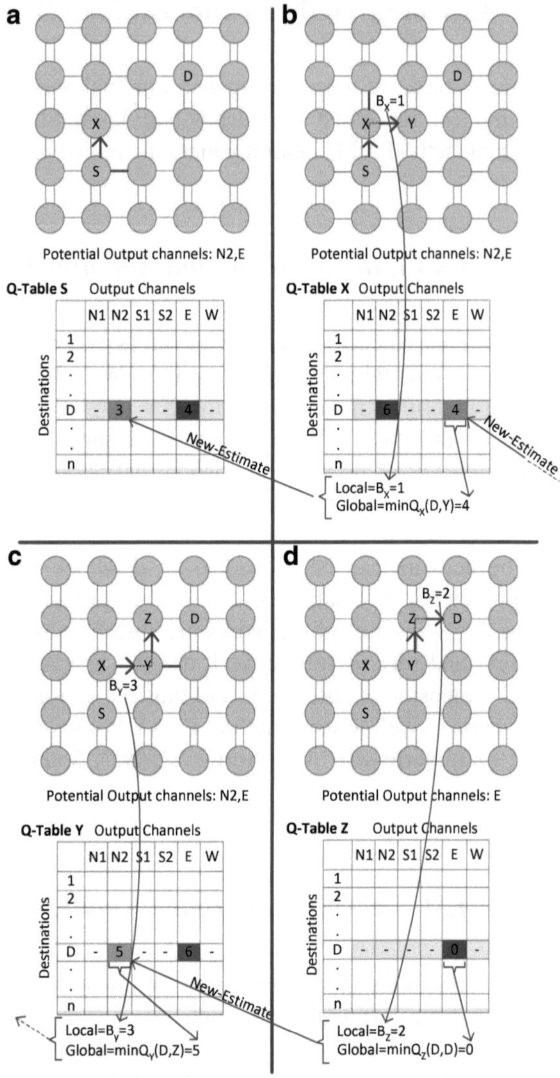

Fig. 5.1 An example of LM (a learning method build upon a minimal routing)

This information is sent back to the upstream switch *s* and updates the old estimated latency. This is performed by averaging the old and new estimated latency values.

Figure 5.1 shows an example where a message is sent from the source *s* to the destination *d*. As illustrated in Fig. 5.1a, the source switch *s* maintains a table including the estimated latency values it takes for a packet to reach from this switch to different destinations (i.e. 1 to *n* destinations in an $n \times n$ network). Each entry of this table belongs to one destination switch in the network and each column determines the output channels of the switch.

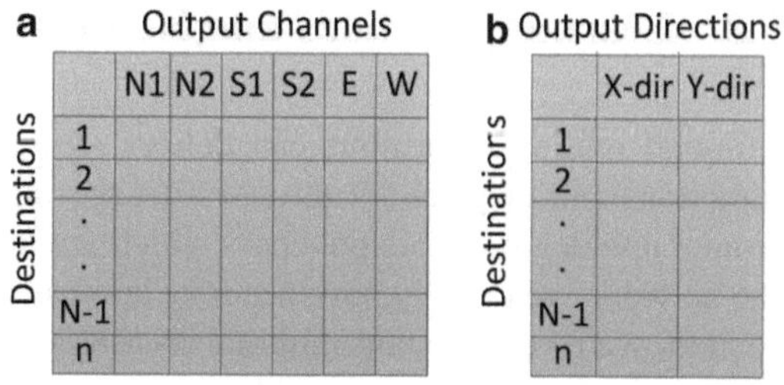

Fig. 5.2 (**a**) A typical Q-Table. (**b**) An optimized Q-Table, called Q-Routing table

Based on the minimal routing algorithm, DyXY, the message can be sent either through the output channel *E* or *N2* from the switch *s* to the switch *d* (Fig. 5.1a). According to the Q-Table values, sending the message through the output channel *N2* leads to the lowest latency, so the message is delivered through this channel. When the message arrives to the switch *x* (Fig. 5.1b), it has to be waited in the input buffer before grant access to one of output channels. This waiting time is modeled by the occupied slots in the input buffer of the switch *x* (local estimated latency: B_x). At the switch *x*, there are two output channels to deliver the message toward the destination as *E* and *N2*. Based on the Q-Table of the switch *x*, the output channel *E* has a smaller estimated latency to reach the destination switch than the output channel *N2* (global estimated latency: Q_x), and thus the message is sent through the output channel *E*. At this point, by summing up the local (B_x) and global ($Q_x(y,d)$) estimated latencies, a new value is obtained which shows the estimated latency from the switch *s* to the switch *d*. By using congestion wires, this information sends back to the switch *s* to update the corresponding entry (row: *d*; column: *N2*) of the table. The data message continues its path toward the destination by passing through the switch *y*. Similarly, after determining the output channel, local and global information is sent back to the switch *x* through congestion wires. This new information updates the corresponding entry of the switch *x* (row: *d*; column: *E*). A similar procedure is applied when the message arrives to the switch *z*. Finally the message reaches the destination and the congestion information is sent back to the previous switch. As can be obtained from this example, by delivering a single message from a source to a destination, only one entry from each table is updated. However, Q-Tables are gradually updated by propagating different messages within the network over time.

As shown in Fig. 5.2a, typically the size of a Q-Table is $n \times m \times k$ where n is the number of switches in the network, m is the number of output channels per switch, and k is the size of each entry in the Q-Table. Since a message can be delivered through at most two directions, the size of Q-Table can be decreased to $n \times 2 \times k$. We call this table, the Q-Routing table. As shown in Fig. 5.2b, the number of columns can be reduced to two; one is allocated to the X dimension and the other one to the Y dimension. Obviously, with a larger number of virtual channels, more number of columns is needed.

5.3 Learning Methods and Non-minimal Routing

Minimal routing algorithms can deliver messages through at most two minimal directions and they cannot reroute messages around congested areas. They suffer from a low degree of adaptiveness which are inefficient in distributing traffic over the network even if they have accurate knowledge of the network condition. In this section, we propose a non-minimal routing algorithm, named HARA, with a high degree of adaptiveness to provide more output options at each switch [9] and then we apply a leaning method on top of it [10]. We called the obtained method, LNM (Learning method applied to Non-Minimal routing).

5.3.1 HARA: Highly Adaptive Non-minimal Routing Algorithm

The proposed non-minimal routing algorithm, called HARA (Highly Adaptive Non-Minimal Routing Algorithm), is based on the Mad-y method which has been introduced by Glass and Ni [11]. Mad-y utilizes a double-Y network where the X and Y dimensions have one and two virtual channels, respectively (Fig. 5.3a shows a double-Y switch). In a 2D mesh network, three types of turns can be taken as: 0-degree, 90-degree, and 180-degree turn (U turns). By taking a 0-degree turn, a message transmits in a same direction with a possibility of switching between virtual channels. 0-degree turns are further divided into 0-degree-vc and 0-degree-ch. The turn is called 0-degree-vc if the virtual channel changes (Fig. 5.3b) while it is 0-degree-ch if by taking the turn neither the direction nor the virtual channel changes (Fig. 5.3c). By taking a 90-degree turn, a message transmits between the switches in perpendicular directions (Fig. 5.3d). By taking a 180-degree turn, a message is transferred to a channel in the opposite direction. The turn is called 180-degree-vc (Fig. 5.3e) when the virtual channel changes; otherwise, it is represented as 180-degree-ch (Fig. 5.3f). Note that, in all figures, the vc1 and vc2 are differentiated by "–" and "=", respectively.

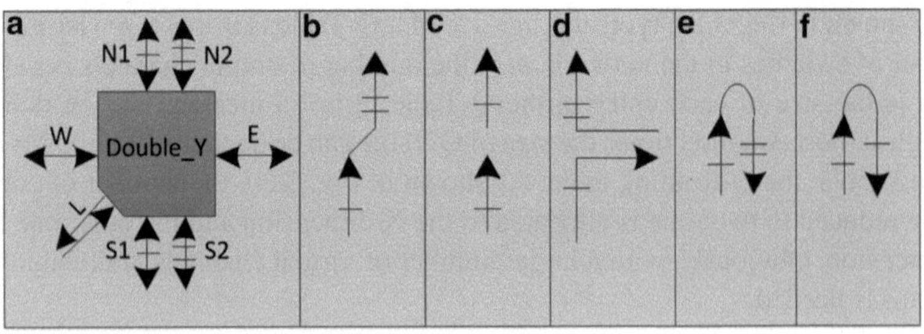

Fig. 5.3 (**a**) A switch in a double-Y network; (**b**) 0-degree-vc; (**c**) 0-degree-ch; (**d**) 90-degree; (**e**) 180-degree-vc; (**f**) 180-degree-ch

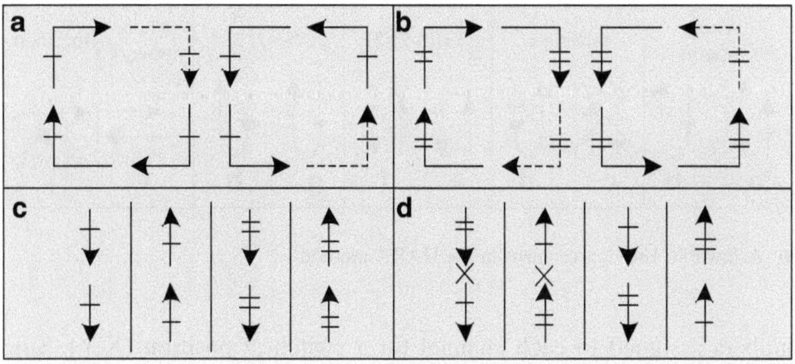

Fig. 5.4 (**a**) 90-degree turns in vc1; (**b**) 90-degree turns in vc2; (**c**) 0-degree-ch; (**d**) 0-degree-vc

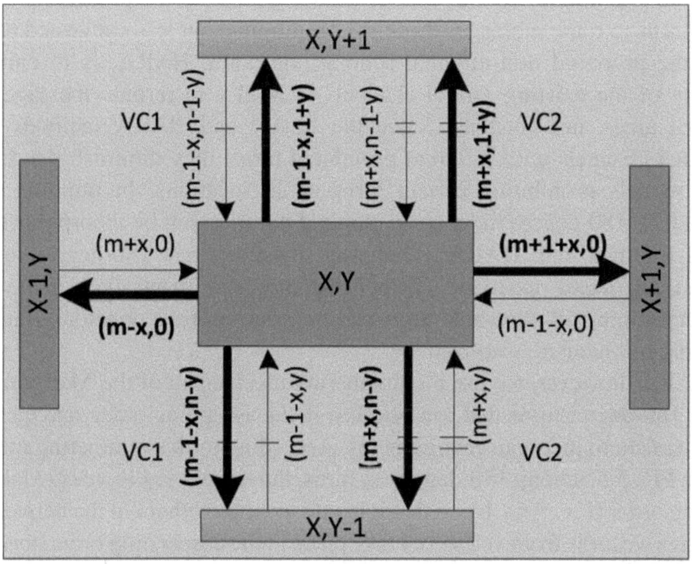

Fig. 5.5 Channel numbering in the Mad-y method

In order to avoid deadlock, the Mad-y method [11] prohibits some turns in the double-Y network. For example, as shown in Fig. 5.4d, 0-degree-vc turns from vc2 to vc1 may cause deadlock in the network and they are prohibited. The other 0-degree turns such as the 0-degree-ch turns (Fig. 5.4c) and the 0-degree-vc turns from vc1 to vc2 (Fig. 5.4d) are permitted. As illustrated in Fig. 5.4a, b, out of sixteen 90-degree turns that can be potentially taken in a network, four of them cannot be used in Mad-y. Finally, 180-degree turns are not allowed in Mad-y. To prove the freedom of deadlock, a two-digit number (a,b) is assigned to each output channel of a switch in an $n \times m$ mesh network. According to the numbering mechanism, a turn connecting the input channel (I_a,I_b) to the output channel (O_a,O_b) is called an ascending turn when $(O_a > I_a)$ or $((O_a = I_a)$ and $(O_b > I_b))$. Figure 5.5 shows

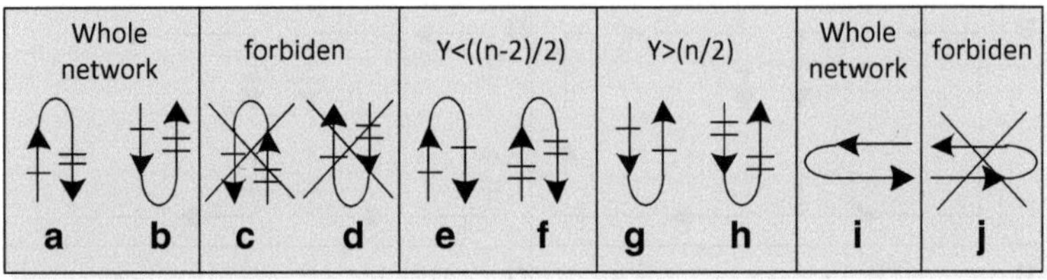

Fig. 5.6 Allowable 180-degree turns in the HARA method

the numbers assigned to each channel for a switch at position (X,Y). Since this numbering mechanism causes the messages to take the permitted turns in the strictly increasing order, so that Mad-y is deadlock-free.

As Mad-y is a minimal and adaptive routing method, it cannot fully utilize the eligible turns to route messages through non-minimal but less congested paths. The aim of the proposed non-minimal routing algorithm, HARA, is to enhance the capability of the existing virtual channels in Mad-y to reroute messages around congested areas and hotspots. Since the Mad-y and HARA methods combine two virtual channels with different prohibited turns, they diminish the drawbacks of turn models prohibiting certain turns at all locations. In minimal routings, (e.g. Mad-y), 180-degree turns are prohibited but they can be incorporated in non-minimal routings (e.g. HARA). One way to incorporate 180-degree turns is to examine them one by one to see whether each turn leads to any cycle in the network. After determining all allowable turns and in order to prove deadlock-freeness, the numbering mechanism is utilized.

In HARA, however, we use the numbering mechanism of the Mad-y method to learn all 180-degree turns that can be taken in the ascending order, and then modify the mechanism to meet our requirements. According to this numbering mechanism shown in Fig. 5.5, among 180-degree-vc turns, those from vc1 to vc2 are taken in the ascending order (Fig. 5.6a, b), so that it is safe to employ them in the network. As all 180-degree-vc turns from vc2 to vc1 take place in the descending order, thereby they cannot be used in the network (Fig. 5.6c, d). Now, let us examine the 180-degree-ch turn connecting the first virtual channel of the north output port to the same virtual channel of the north input port (Fig. 5.6e). As shown in Fig. 5.5, the label on the north output channel with vc1 is $(m-1-x, 1+y)$ and the label on the input channel of the north direction along the same virtual channel is $(m-1-x, n-1-y)$. The turn takes place in the ascending order if and only if $n-1-y$ is greater than $1+y$. Therefore, this turn can be safely added to a set of allowable turns if the Y coordinate of a switch is less than $(n-2)/2$. Similarly, in Fig. 5.6f, the 180-degree-ch turn on the vc2 of the north direction is permitted if the Y coordinate of a switch is less than $(n-2)/2$. 180-degree-ch turns on the south direction (either on vc1 or vc2) are permitted if and only if the Y coordinate of a switch is greater than $n/2$ (Fig. 5.6g, h). Finally, the 180-degree-ch turn on the west direction is always permitted (Fig. 5.6i) while the 180-degree-ch turn on the east direction is prohibited in the network (Fig. 5.6j).

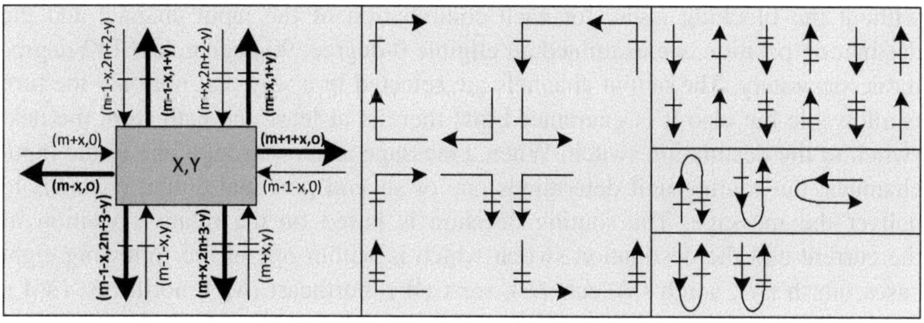

Fig. 5.7 The numbering mechanism of HARA along with all eligible turns in the network

As shown in Fig. 5.6e–h, there are four conditional 180-degree turns; two of those are allowable only in the northern part of the network and two others in the southern part of the network. This not only increases the complexity of the routing function but also imposes heterogeneous routing function to switches. To overcome this issue, we modify the numbering mechanism such that two turns are permitted in all locations of the network (Fig. 5.6g, h) and two others are prohibited in the whole network (Fig. 5.6e, f). The numbering mechanism of HARA along with all permitted turns in the network is shown in Fig. 5.7. As can be observed from this figure, all allowable turns are taken in the ascending order.

In the following we prove that HARA is deadlock-free and livelock-free.

Theorem 1 *HARA is deadlock-free*

Proof If the numbering mechanism guarantees that all eligible turns are ordered in the ascending order, no cyclic dependency can occur between channels. As can be observed from Fig. 5.7, all connections between input channels and output channels to form the eligible turns in HARA take place in the ascending order and thus HARA is deadlock-free.

Theorem 2 *HARA is livelock-free*

Proof In HARA, when a message moves to the east direction, it can never be routed back to the west direction. Therefore, in the worst case, the message may reach to the leftmost column and then moves to the east direction toward the destination column without the possibility of routing to the west direction again. Therefore, after a limited number of hops, the message reaches the destination, and Theorem 2 is proved.

In the non-minimal routing, only eligible turns can be employed at each switch but it is not sufficient to avoid blocking in the network. In fact, there is no possibility of creating cycles but messages might be blocked forever. The reason is that by utilizing the allowable turns, a message may not be able to find a path to the destination from the next hop and it is blocked. On the other hand, one of the aims of HARA is to fully utilize all eligible turns to present a low-restrictive adaptive method in the double-Y network. To achieve the maximal adaptiveness without the blocking issue, for each combination of the input channel and the destination position, we examined all eligible 0-degree, 90-degree, and 180-degree turns, separately. The output channels are selected in a way that not only the turn is allowable but also it is guaranteed that there is at least one path from the next switch to the destination switch. When a message arrives through one of the input channels, the routing unit determines one or several potential output channels to deliver the message. The routing decision is based on the relative position of the current and the destination switch which is within one of the following eight cases: north (*N*), south (*S*), east *(E*), west (*W*), northeast (*NE*), northwest (*NW*), southeast (*SE*), and southwest (*SW*). All permissible output channels of HARA, for each pair of the input channel (*inCh*) and destination position (*pos*) are shown in Table 5.1. The adaptivity provided by Mad-y is illustrated in Table 5.2. By comparing these two tables, it can be easily obtained that HARA offers a large degree of adaptiveness to route messages. One of the drawbacks of non-minimal methods is in their complexity due to considering different conditions in the routing decisions. However, as shown in Fig. 5.8 (i.e. extracted from Table 5.1), the implementation of HARA is very simple.

Figure 5.9 shows an example of the HARA method in a 5×5 mesh network in which the source switch 7 sends a message to the destination switch 14. According to Table 5.1, the message arriving from the local channel and going toward the destination in the northeast position has six alternative choices (i.e. *N1*, *N2*, *S1*, *S2*, *E*, and *W*); among them, the output channels *N1*, *N2*, and *E* introduce the minimal paths and *S1*, *S2*, and *W* indicate the non-minimal paths. Since the neighboring switches in the shortest paths are in the congested area, the message is sent to a neighboring switch located in a non-minimal path that is less congested. Again, at the switch 2, all the minimal paths are congested, so the message is sent to the switch 1 which is less congested. The same strategy is repeated until the message reaches the destination switch. This example shows the capability of the HARA method to reroute messages around the congested areas.

5.3.2 LNM: Learning Approach Applied to Non-minimal Routing

We utilize an optimized learning model for the selection function of the proposed non-minimal approach. This method is called LNM (Learning method applied to Non-Minimal routing). As already mentioned, the size of a Q-Table is $n \times m \times k$ where n is the number of switches in the network, m is the number of output channels per switch, and k is the size of each entry in the Q-Table. This size can be decreased to $n \times 2 \times k$ in the Q-Routing table. However, the required area of Q-Routing is still very large and increases as the network size enlarges.

In non-minimal approaches, the number of columns cannot be reduced to two as a message might be sent through any of output channels. To address the size and

Table 5.1 Potential output channels offered by HARA

pos / inCh	N	S	E	W	NE	NW	SE	SW
L	N1,N2, S1, W	N1, S1,S2, W	N1,N2, S1,S2, E,W	N1, S1, W	N1,N2, S1,S2, E,W	N1, S1, W	N1,N2, S1,S2, E,W	N1, S1, W
N1	N2, S1, W	S1,S2, W	N2, S1,S2, E,W	S1, W	N2, S1,S2, E,W	S1, W	N2, S1,S2, E,W	S1, W
N2	–	S2	S2,E	–	S2,E	–	S2,E	–
S1	N1,N2, S1, W	N1, S1,S2, W	N1,N2, S1,S2, E,W	N1, S1, W	N1,N2, S1,S2, E,W	N1, S1, W	N1,N2, S1,S2, E,W	N1, S1, W
S2	N2	–	N2,E	–	N2,E	–	N2,E	–
E	N1,N2, S1, W	N1, S1,S2, W	N1,N2, S1,S2, E,W	N1, S1, W	N1,N2, S1,S2, E,W	N1, S1, W	N1,N2, S1,S2, E,W	N1, S1, W
W	N2	S2	N2,S2, E	–	N2,S2, E	–	N2,S2, E	–

Table 5.2 Potential output channels offered by Mad-y

pos / inCh	N	S	E	W	NE	NW	SE	SW
L	N1,N2	S2	E	W	N1,N2,E	N1,W	S1,S2,E	S1,W
N1	–	S1,S2	E	W	–	–	S1,S2,E	S1,W
N2	–	S2	E	–	–	–	S2,E	–
S1	N1,N2	–	E	W	N1,N2,E	N1,W	–	–
S2	N2	–	E	–	N2,E	–	–	–
E	N1,N2	S2	–	W	N1,N2	N1,W	S1,S2	S1,W
W	N2	S2	E	–	N2,E	–	S2,E	–

ALGORITHM: Determining all eligible output channels
____
Definitions: inch, outCh: input and output Channel; pos: destination Position
____
if *pos={L}* **then** *outCh(L) <= '1';*
if *pos={E* or *NE* or *SE}* **then** *outCh(E) <=' 1';*
if *inCh={L* or *N1* or *S1* or *E}* **then** *outCh(W) <= '1';*
if *inCh={L* or *S1* or *E}* **then** *outCh(N1) <= '1';*
if *inCh={L* or *N1* or *E* or *S1}* **then** *outCh(S1) <= '1';*
if *(inCh/= {N2})* and *(pos={N* or *E* or *NE* or *SE})* **then** *outCh(N2) <= '1';*
if *(inCh/={S2})* and *(pos={S* or *E* or *NE* or *SE})* **then** *outCh(S2) <= '1';*
end if;

Fig. 5.8 Determining all eligible output channels of HARA

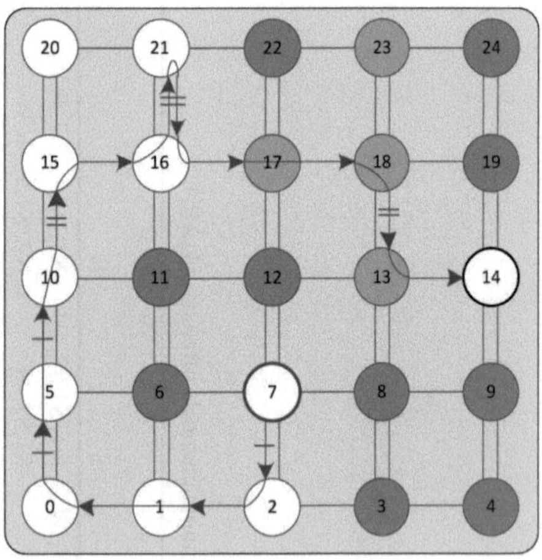

Fig. 5.9 An example of HARA

scalability problem of Q-Tables, we proposed a new table for the LNM approach which is called Region-based Routing table (R-Routing). As illustrated in Fig. 5.10, each row of this table corresponds to one of the eight different positions of the destination switch (i.e. *N*, *S*, *E*, *W*, *NE*, *NW*, *SE*, and *SW*) and each column indicates

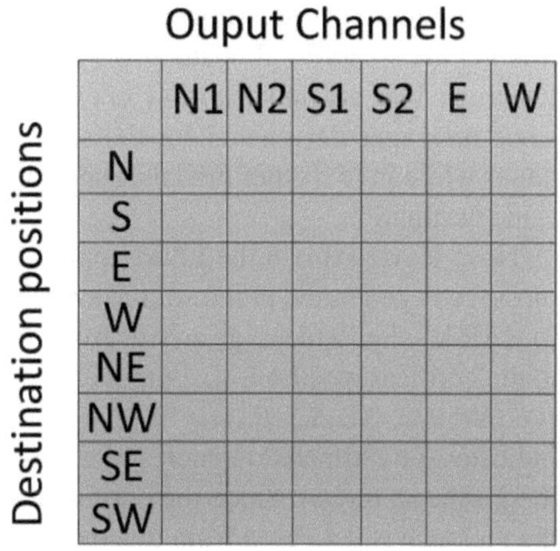

Fig. 5.10 A R-Routing table

Table 5.3 The area overhead

Size\method	Q-Routing (bytes)	C-Routing (bytes)	R-Routing (bytes)
8×8	128	40	24
16×16	512	64	24
32×32	2,048	160	24

an output channel (i.e. *N1*, *N2*, *S1*, *S2*, *E*, and *W*). Regardless of the network size, the size of R-Routing tables is $8 \times 6 \times k$ that is considerably smaller than Q-Routing and Q-Tables. There is another type of tables, called C-Routing which decreases the size of Q-Tables by taking advantages of the clustering approach [12]. The size of C-Routing tables is $(l + c) \times m \times k$, consisting of two parts: 1) the cluster part having a size of $c \times m \times k$ where c is the number of clusters 2) the local part having a size of $l \times m \times k$ where l is the number of switches in each cluster. The clustering approach suffers from the scalability issue since the size of C-Routing tables can become rather large as the network scales up. There are some other concerns regarding the clustering model such as determining the cluster size for different network sizes or partitioning the network when the network size is not a multiple of the cluster size.

The required sizes for Q-Routing, C-Routing, and R-Routing tables are reported in Table 5.3, where $k = 4$ (the size of each entry in a table) and $l = 4$ (the number of switches within each cluster for the clustering approach,). Note that the reported areas for Q-Routing and C-Routing tables are based on the assumption in which no virtual channels is used while the size of R-Routing table is measured considering one virtual channel along the Y dimension. As can be seen in this table, not only the sizes of R-Routing tables are very small but also they are independent of the network size.

While the sizes of R-Routing tables are small enough to be applicable in NoCs, someone might think that by using R-Routing tables, the accuracy of the estimated latency values toward each destination is reduced. In fact, under real traffic conditions, each entry of a R-Routing table is inherently influenced by the switches which are in more communication with the current switch at that period. Therefore, it is not necessary to allocate a row for every switch in the network. Moreover, R-Routing tables are updated more occasionally than Q-Routing and C-Routing tables since messages designated for the same region can be used to update R-Routing tables while in two other models, each entry is updated only by the messages for the same destination.

Now, let us explain the LNM approach using the example of Fig. 5.11 where a message is generated at the source switch *s* for the destination switch *d*. According to HARA, when a message arrives from the local input channel and the destination is to the northeast position, six output channels can be selected to forward the message (i.e. *N1*, *N2*, *S1*, *S2*, *E*, and *W*). In Fig. 5.11a, the colored entry of the Q-Table indicates the estimated latencies of a message from each possible output channel to the northeast region. Since the output channel *N1* has the lowest estimated latency, the message is delivered from this output channel toward the destination switch.

At the switch *x*, the message is received by the input channel *S1* (Fig. 5.11b). According to the information indicated in Table 5.1, multiple output channels can be used to forward the message (i.e. *N1*, *N2*, *S1*, *S2*, *E*, and *W*). Among eligible output channels, the output channel *E* has the lowest latency, and thus it is selected for sending the message to the switch *y*. At this time, the local and global congestion values should be returned to the switch *s*. The number of occupied buffer slots at the input buffer of the switch *x* is counted as the local information (i.e. in this example $B_x = 1$). The minimum estimated latency of routing messages from the switch *x* to the destination region via the neighboring switch *y* is considered as the global latency and it is extracted from the Q-Table of the switch *x* (i.e. in this example $minQ_x(y,d) = 4$). By summing up the local and global information, a new estimation is obtained which shows the latency from the switch *s* to the destination *d*. Finally, the corresponding entry of the Q-Table at the switch *s* (i.e. row: *NE*; column: *N1*) should be updated with the new value. This is done by taking the average of the old and new latency estimations (Fig. 5.11a).

At the switch *y*, the message is received via the west input channel (Fig. 5.11c). Among the three possible output channels (i.e. *N2*, *S2*, and *E*), the one with the lowest latency is selected. Upon connecting the input channel to the output channel of the switch *y*, local and global information are returned to the switch *x*. The local congestion shows the number of occupied slots at the input buffer of the switch *y* (i.e. $B_y = 3$) while the global congestion indicates the estimated latency from the switch *y* to the destination switch *d* via the neighboring switch *z* (i.e. $minQ_y(z,d) = 5$). The sum of the local and global values is a new latency estimation to reach the destination from the switch *x*. As shown in Fig. 5.11b, the corresponding entry of the Q-Table at the switch *x* is updated taking an average of the new estimated value (i.e. $B_y + minQ_y(z,d)$) and an existing estimation ($Q_x(y,d)$).

Finally, the message arrives at the switch *z* from the input channel *S2* (Fig. 5.11d). This message can reach the destination by sending through the output channel *N2* or *E*. The output channel *E* has the lowest latency value and it is selected for routing the message. The local latency (i.e. the number of occupied slots at the input buffer of the switch *z*) is 3 while the global latency to the destination is

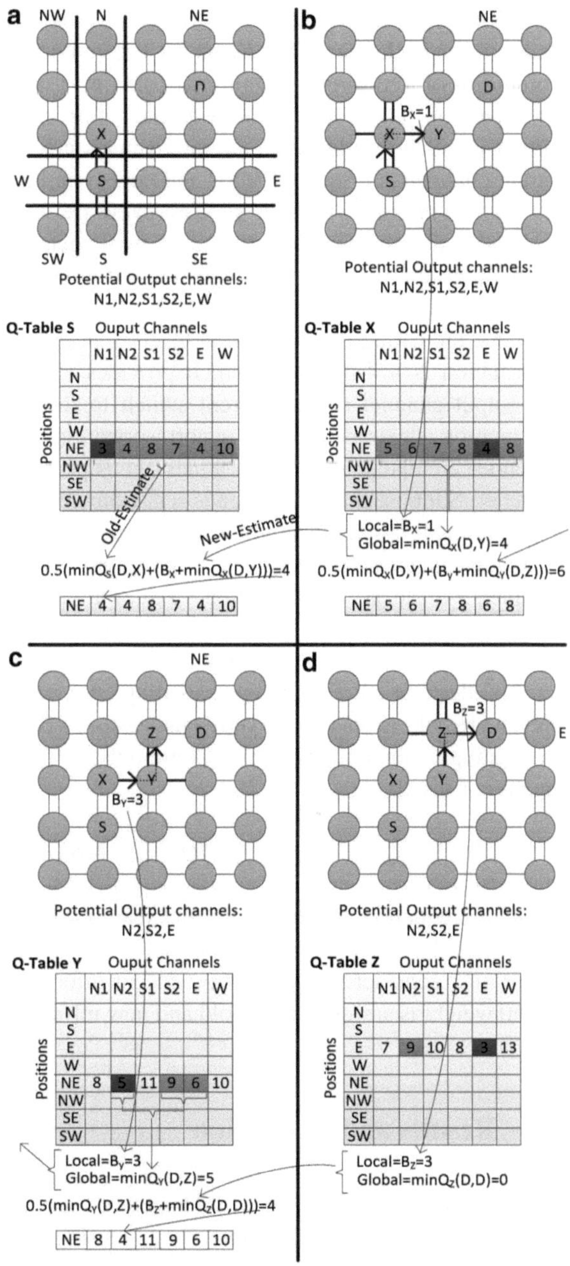

Fig. 5.11 An example of LNM (a learning method build upon a non-minimal routing)

equal to 0 as the message reaches the destination in the next hop. Similarly, the latency values are returned to the switch *y* and the Q-Table is updated with this information (Fig. 5.11c). Hence, as messages are propagated inside the network, Q-Tables gradually incorporate more global information [13].

Q-Routing models have an initial learning period during which it performs worse than minimal schemes. The reason for this temporal inefficiency is that there is a possibility of choosing non-minimal paths even if the network is not congested. To cope with this problem, in the initialization phase, all entries of Q-Tables are initialized such that minimal output channels are set to "0000" and non-minimal output channels are set to "1000" and never can be less than it. Accordingly, in a low traffic condition, only the shortest paths are selected while non-minimal paths are used to distribute traffic when the network gets congested.

In order to transfer the congestion information, LNM utilizes a 4-bit wire between each two neighboring switches. The local congestion information is a 2-bit value indicating the congestion level of an input buffer. The global congestion information is a 4-bit value which provides a global view of the latency from the output channel of the current switch to the destination region. This global information is extracted from the corresponding entry of the R-Routing table. The Q-Values are updated whenever a message is propagated between two neighboring switches. Suppose that a message is sent from the switch x to the destination switch d by passing through the neighboring switch y and then the switch z with the lowest estimated latencies. At the switch y, upon connecting the input channel to the output channel, 2-bit local and 4-bit global values are aggregated into a 4-bit value (with the maximum value of "1111") and then it is transferred to the switch x. This value is a new estimation of the latency from the selected output channel of the switch x to the destination d. The corresponding entry of the Q-Table at the switch x is updated by taking an average of the new estimated value (i.e. $B_y + minQ_y(d,z)$) and an existing estimation ($Q_x(d,y)$). Commonly, in Q-Routing models, the following equation is used:

$$Q_x(d,y) = (1-\alpha)\,Q_x(d,y) + \alpha\,(B_y + minQ_y(d,z))$$

In this equation, α represents the learning rate at which newer information overwrites the older one. With the factor of 0 no learning is performed while a factor of 1 considers only the most recent information [14]. In our simulation, a 50–50 weight is assigned to the old and new information, so that $\alpha = 0.5$.

5.4 Results and Discussion

The efficiency of the LM and LNM methods are compared with the DBAR [15] approach. DABR is an adaptive routing using local and non-local congestion information. The performance of the DBAR approach is extensively discussed in [15] and it is compared with XY, NoP [16] and RCA [17] methods. In this work, the adaptivity of DBAR is similar to the DyXY routing algorithm. A wormhole-based NoC simulator is developed with VHDL to model all major components of the on-chip network and simulations are carried out to determine the latency characteristic of each network. The message length is uniformly distributed between

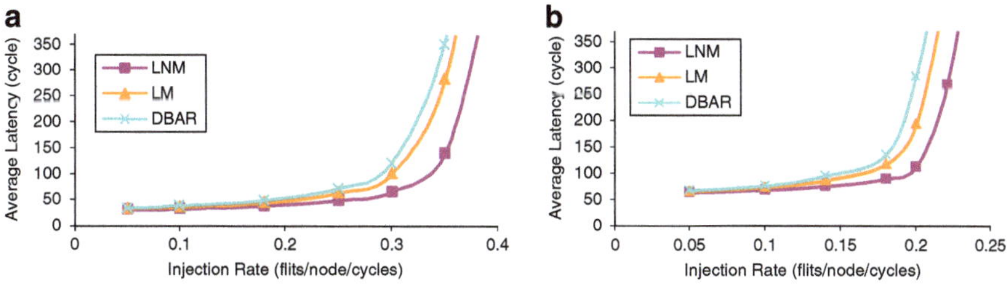

Fig. 5.12 Performance evaluation in (**a**) 8 × 8 and (**b**) 14 × 14 mesh network under the uniform traffic model

5 and 10 flits. The data width is set to 32 bits and each input channel has the buffer (FIFO) size of 8 flits. The simulator is warmed up for 12,000 cycles and then the average performance is measured over another 200,000 cycles. Two synthetic traffic profiles including uniform random and hotspot, along with SPLASH-2 [18] application benchmarks are used.

5.4.1 Performance Evaluation Under Uniform Traffic Profile

In Fig. 5.12, the average communication delay as a function of the average message injection rate is plotted for 8 × 8 and 14 × 14 mesh networks. As observed from the results, in low loads, the Q-Routing schemes (LNM and LM) behave as efficiently as DBAR. As load increases, DBAR is unable to tolerate the high load condition, while the Q-Routing schemes learn an efficient routing policy. LNM leads to the lowest latency due to the fact that it can distribute traffic over both minimal and non-minimal paths. In fact, in DBAR and LM, messages use minimal paths so that they are routed through the very center of the network which creates permanent hotspots. Correspondingly, messages traversing through the center of the network will be delayed much more than they would use non-minimal paths. Due to the fact that the LNM method can reroute messages, it alleviates the congested areas and performs considerably better than other schemes. Using minimal and non-minimal routes along with the intelligent selection policy reduces the average network latency of LNM in an 8 × 8 network (near the saturation point, 0.3) about 34 % and 45 %, compared with LM and DBAR, respectively.

5.4.2 Performance Evaluation Under Hotspot Traffic Profile

In simulations, given a hotspot percentage of *H*, a newly generated message is directed to each hotspot switch with an additional *H* percent probability. We simulate the hotspot traffic with a single hotspot switch at (4,4) and (7,7) in 8 × 8

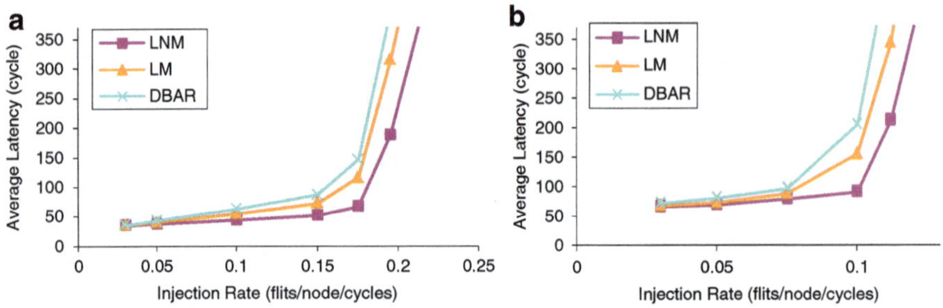

Fig. 5.13 Performance evaluation in (**a**) 8 × 8 and (**b**) 14 × 14 mesh network under hotspot traffic model with $H = 10\ \%$

and 14 × 14 mesh networks, respectively. The performance of each network with $H = 10\ \%$ is illustrated in Fig. 5.13. As can be observed from the figure, LM works better than DBAR while LNM outperforms both of them. In an 8 × 8 mesh network, the performance gain near the saturation point (0.18) is about 41 % and 53 %, compared with LM and DBAR, respectively. The results reveal that using the non-minimal scheme along with the Q-Routing policy can distribute the traffic efficiently.

5.5 Performance Analysis Under Application Traffic Profile

Application traces are obtained from the GEMS simulator [19] using some application benchmark suites selected from SPLASH-2 [18]. We use a 64-switch network configuration, including 20 processors and 44 L2-cache memory modules. For the CPU, we assume a core similar to Sun Niagara and use SPARC ISA [20]. Each L2 cache core is 512 KB, and thus, the total shared L2 cache is 22 MB. The memory hierarchy is governed by a two-level directory cache coherence protocol. Each processor has a private write-back L1 cache (split L1 I and D cache, 64 KB, 2-way, 3-cycle access). The L2 cache is shared among all processors and split into banks (44 banks, 512 KB each for a total of 22 MB, 6-cycle bank access), connected via on-chip switches. The L1/L2 block size is 64B. Our coherence model is based on a MESI-based protocol with distributed directories, with each L2 bank maintaining its own local directory. The simulated memory hierarchy mimics SNUCA [21] while the off-chip memory is a 4 GB DRAM with a 220-cycle access time. Figure 5.14 shows the average message latency across six benchmark traces, normalized to DBAR. LNM provides lower latency than other schemes and it shows the greatest performance gain in Ocean with 32 % reduction in latency (vs. LM). The average performance gain of LNM across all benchmarks is up to 27 % vs. LM and 35 % vs. DBAR.

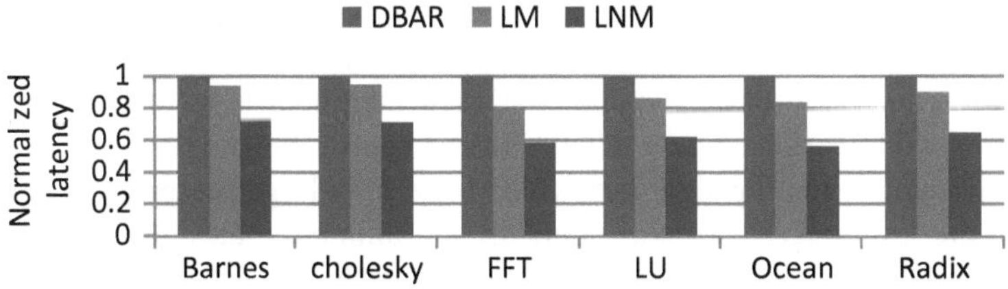

Fig. 5.14 Normalized latency under different application benchmarks normalized to DBAR

Table 5.4 Hardware implementation details

Network platforms	Area (mm^2)	Avg. power (mw) dynamic and static	Max. power (mw) dynamic and static
DBAR	6.780	2.39	3.30
LM	6.815	2.57	3.51
LNM	6.822	2.81	3.06

5.5.1 Hardware Analysis

To assess the area overhead and power consumption of LNM, the whole platform of each scheme is synthesized by Synopsys Design Compiler. Each scheme includes switches, communication channels, and congestion wires. For synthesis, we use the UMC 90 nm technology at the operating frequency of 1 GHz and supply voltage of 1 V. We perform place-and-route, using Cadence Encounter, to have precise power and area estimations. The power dissipation of each scheme is calculated under the hotspot traffic profile near the saturation point (*0.18*) using Synopsys PrimePower in an 8 × 8 mesh network. The layout area and power consumption of each platform are shown in Table 5.4. Comparing the area cost of the platform using LNM with the platforms using LM and DBAR indicates that learning approaches consumes more power and a higher area overhead than DBAR. The LNM platform consume more average power because of rerouting messages around the congestion areas which increases the hop counts. The results indicate that the maximum power of the LNM is 7–13 % less than that of the LM and DBAR platforms, respectively. This is achieved by smoothly distributing the power consumption over the network using the highly adaptive routing scheme which reduces hotspots in NoCs. The maximum power values, reported in the table, belong to the switch designated as the hotspot one, (4,4).

5.6 Conclusion

This chapter aimed to reduce congestion in the network by employing learning approaches. At first, a learning-based method is built upon a minimal and fully adaptive routing algorithm. This method is able to learn the underlying traffic condition and makes a routing decision based on this information. However, it will not lead to an optimal performance. The reason is that in minimal routing, packets are limited to at most two directions at each intermediate switch and thus they cannot be well distributed over the network. To solve this problem, we introduced a highly adaptive routing algorithm which provides a large degree of adaptiveness. To find the less congested route among all non-minimal routes, the learning-based approach is utilized. For this purpose, a table is needed at each switch. These tables are relatively small and designed in a scalable manner. The experimental results confirm the advantages of combining the non-minimal routing and learning-based approaches over traditional methods.

References

1. A. Jantsch, H. Tenhunen, *Networks on Chip* (Springer, New York, 2003)
2. X. Dai, C.-K. Li, A.B. Rad, An approach to tune fuzzy controllers based on reinforcement learning for autonomous vehicle control. IEEE Trans. Intell. Transport. Syst. **6**(3), 285–293 (2005)
3. R.S. Sutton, A.G. Barto, *Reinforcement Learning: An Introduction* (MIT Press, Cambridge, 1998)
4. C. Watkins, P. Dayan, Technical note: Q-Learning. Mach. Learn. **8**(3), 279–292 (1992)
5. J.A. Boyan, M.L. Littman, Packet routing in dynamically changing networks: A reinforcement learning approach, in *Proceedings of Advances in Neural Information Processing Systems 6* (1994), pp. 671–678
6. F. Farahnakian, M. Ebrahimi, M. Daneshtalab, P. Liljeberg, J. Plosila, Adaptive reinforcement learning method for networks-on-chip, in *Proceedings of SAMOSXIII* (2012), pp. 236–243
7. F. Farahnakian, M. Ebrahimi, M. Daneshtalab, P. Liljeberg, J. Plosila, Q-learning based congestion-aware routing algorithm for on-chip network, in *Proceedings of IEEE 2nd International Conference on Networked Embedded Systems for Enterprise Applications* (2011), pp. 1–7
8. M. Li, Q.-A. Zeng, W.-B. Jone, DyXY – a proximity congestion-aware deadlock-free dynamic routing method for network on chip, in *Proceedings of 43rd ACM/IEEE Design Automation Conference*, (2006), pp. 849–852
9. M. Ebrahimi, M. Daneshtalab, P. Liljeberg, J. Plosila, H. Tenhunen, LEAR – A low-weight and highly adaptive routing method for distributing congestions in on-chip networks, in *Proceedings of 20th Euromicro International Conference on Parallel, Distributed and Network-Based Processing (PDP)* (2012), pp. 520–524
10. M. Ebrahimi, M. Daneshtalab, F. Farahnakian, J. Plosila, P. Liljeberg, M. Palesi, H. Tenhunen, HARAQ: Congestion-aware learning model for highly adaptive routing algorithm in on-chip networks, in *Proceedings of International Symposium on Networks-on-Chip* (2012), pp. 19–26
11. C.J. Glass, C.J. Glass, L.M. Ni, L.M. Ni, Maximally fully adaptive routing in 2D meshes, in *Proceedings of International Conference on Parallel Processing* (1992), pp. 101–104
12. M.K. Puthal, V. Singh, M.S. Gaur, V. Laxmi, C-Routing: An adaptive hierarchical NoC routing methodology, in *Proceedings of IEEE/IFIP 19th International Conference on VLSI and System-on-Chip (VLSI-SoC)* (2011), pp. 392–397
13. W. Feng, K.G. Shin, Impact of selection functions on routing algorithm performance in multicomputer networks, in *Proceedings of the 11th International Conference on Supercomputing* (1997), pp. 132–139

14. C. Feng, Z. Lu, A. Jantsch, J. Li, M. Zhang, A reconfigurable fault-tolerant deflection routing algorithm based on reinforcement learning for network-on-chip, in *Proceedings of the Third International Workshop on Network on Chip Architectures* (2010), pp. 11–16
15. S. Ma, N. Enright Jerger, Z. Wang, DBAR: An efficient routing algorithm to support multiple concurrent applications in networks-on-chip, in *Proceedings of the 38th Annual International Symposium on Computer Architecture* (2011), pp. 413–424
16. G. Ascia, V. Catania, M. Palesi, D. Patti, Implementation and analysis of a new selection strategy for adaptive routing in networks-on-chip. IEEE Trans. Comput. **57**(6), 809–820 (2008)
17. P. Gratz, B. Grot, S.W. Keckler, Regional congestion awareness for load balance in networks-on-chip, in *Proceedings of IEEE 14th International Symposium on High Performance Computer Architecture, HPCA* (2008), pp. 203–214
18. S.C. Woo, M. Ohara, E. Torrie, J.P. Singh, A. Gupta, The SPLASH-2 programs: Characterization and methodological considerations, in *Proceedings of 22nd Annual International Symposium on Computer Architecture* (1995), pp. 24–36
19. M.M.K. Martin, D.J. Sorin, B.M. Beckmann, M.R. Marty, M. Xu, A.R. Alameldeen, K.E. Moore, M.D. Hill, D.A. Wood, Multifacet's general execution-driven multiprocessor simulator (GEMS) toolset. SIGARCH Comp. Archit. News **33**(4), 92–99 (2005)
20. P. Kongetira, K. Aingaran, K. Olukotun, Niagara: A 32-way multithreaded Sparc processor. IEEE. Micro. **25**(2), 21–29 (2005)
21. B.M. Beckmann, D.A. Wood, Managing wire delay in large chip-multiprocessor caches, in *Proceedings of the 37th annual IEEE/ACM International Symposium on Microarchitecture* (2004), pp. 319–330

Part II
Multicast Communication

Chapter 6
Efficient and Deadlock-Free Tree-Based Multicast Routing Methods for Networks-on-Chip (NoC)

Faizal Arya Samman and Thomas Hollstein

Abstract This chapter presents a new efficient and deadlock free tree-based multicast routing method and concept. The presented deadlock-free multicast routing algorithm can be implemented on a network-on-chip (NoC) router microarchitecture, realizing a mesh planar network topology. The NoC microarchitecture supports both deadlock-free static and efficient adaptive tree-based multicast routing. Multicast packets are routed and scheduled in the NoC by using a flexible multiplexing/interleaving technique with wormhole switching. The flexibility of the proposed multicast routing method is based on a locally managed packet identity (ID-tag) attached to every flit. This concept allows different packets to be interleaved at flit-level in a single buffer pool on the same link. Furthermore, a pheromone tracking strategy presented in this chapter, which is used to reduce communication energy in the adaptive tree-based multicast routing method. The strategy is used to perform efficient spanning trees for the adaptive tree-based multicast routing which are generated at runtime.

6.1 Introduction

Recent multicomputers and large-scale multiprocessor systems have been developed towards collective communication services to reduce communication overheads of parallel computations running on these systems. These collective communication

F.A. Samman (✉)
Universitas Hasanuddin, Fakultas Teknik, Jurusan Teknik Elektro,
Jl. Perintis Kemerdekaan Km. 10, Makassar 90245, Indonesia
e-mail: faizalas@unhas.ac.id

T. Hollstein
Department of Computer Engineering, Tallinn University of Technology,
Akadeemia tee 15A, 12618 Tallinn, Estonia
e-mail: thomas@ati.ttu.ee

M. Palesi and M. Daneshtalab (eds.), *Routing Algorithms in Networks-on-Chip*,
DOI 10.1007/978-1-4614-8274-1_6,

services include *one-to-many* communication such as *multicast* (the same message is sent from one source node to an arbitrary number of destination nodes), *one-to-all* communication such as *broadcast* (the same message is sent from one source node to all nodes (entries) in the network) and *scatter* (different messages are sent from one source node to all entries in the network), *many-to-one* communication (one destination node receives different messages from an arbitrary number of source nodes), and *all-to-one* communication such as *reduce* (one destination node combines different messages from an arbitrary number of source nodes by performing a certain operation such addition, multiplication, maximum, minimum, or a logical operation).

Among the aforementioned collective communications, multicast and broadcast communications are the most interesting communication modes. Both communication modes are needed in many parallel algorithms and applications in multiprocessor systems. They also require special attention and handling in the network communication protocol layers. A processing node in a multiprocessor system can inject a multicast message into a network by sending separate individual copies of the message from the source to every destination node. However, this unicast-based multicast delivery is energy- and time-consuming [24].

In large-scale off-chip multiprocessor systems, collective communication services, such as multicast communication, have been a fundamental prerequisite for some data parallel computer languages. In a distributed shared-memory (DSM) parallel programming model, the multicast data communication service is applied to efficiently support shared-data invalidation and updating on distributed nodes. This is being used in numerous parallel algorithms, e.g. parallel search and parallel graph algorithms. In a single-program multiple-data (SPMD) and in a data parallel programming models, a variety of process control operations and global data movements such as reduction, replication, permutation, segmented scan and barrier synchronization [16] can benefit from multicast services.

In the Multiprocessor System-on-Chip (MPSoC) or multicores systems domains, parallel computing problems can potentially be solved based on parallel programming models. Therefore, multicast communication services, which should be implemented in upper protocol layers (typically on software level) and bottom protocol layers (typically on hardware level), will also be an important issue in a NoC-based multiprocessor context. Modern 3D chip stacking technologies enable on-Chip DSM systems with local DRAM memory being available at every processing node. Multicast communication services are in this context essential for efficient implementation of memory coherency protocols. Furthermore, in the upper protocol layers, multicast services are implemented as Application Programming Interface (API) routines (programming model) that can be used by users to develop parallel computing programs.

Further information on multicast routing issues can be found in [10]. This chapter is addressing a router hardware architecture supporting an efficient tree-based adaptive multicast routing method for NoCs that is implemented in bottom communication network protocol layers, i.e. on network, data link and physical layers according to the Open Systems Interconnection (OSI) model [30].

6.2 Occurrence of a Deadlock Problem Due to Multicast Dependency

This chapter describes the use of an efficient multicast routing method for NoCs. However, the use of the tree-based multicast routing method may lead to a multicast dependency problem ending up in a deadlock, as presented in Fig. 6.1. This multicast dependency can cause a multicast deadlock configuration (as described in Duato's Book [10]). In this case, multicast packets block each other and cannot move further.

The deadlock problem occurs especially if packets switched with the wormhole method or virtual cut-through switching are not short enough, there is not enough buffer space to store the contenting wormhole packet, and/or arbitration rules are not well organized to handle the multicast contention. In node (2,2) as presented in the figure, packet A blocks the flow of two multicast branches of packet B (to west and east), while in node (2,1), packet B blocks the flow of two multicast branches of packet A (to west and to east). Due to the "wait and hold" situation in both network switch nodes, both message A and B cannot move further.

In this chapter a NoC architecture called XHiNoC (eXtendable Hierarchical Network-on-Chip) is presented, which proposes a *novel wormhole cut-through switching* concept [26] based on a local identity-based (ID-based) interleaved routing organization, in which the ID-tag of each packet is locally updated on each communication link [23]. The XHiNoC's concept has also exhibited a novel multicast routing method [22, 24, 25] for NoCs based on the interleaved routing organization with local Identity (ID) management. In the XHiNoC routers, the just described deadlock configuration problem (due to multicast dependencies) is solved efficiently by using a so-called "*hold and release tagging mechanism*".

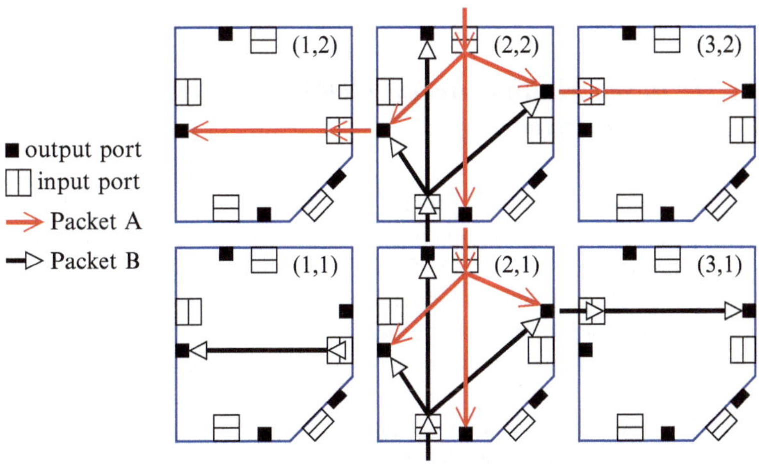

Fig. 6.1 Deadlock configuration in tree-based multicast routing

The hold-release tagging mechanism is based on an asynchronous data switching approach. A multicast flit will not be switched to an outgoing port, before its multicast request has been granted by the arbiter unit at the considered outgoing port. Each individual request can be granted asynchronously. Once the individual request is granted, it will be switched out towards the next link. The Hold-Release Tagging Mechanism applies the following principle to escape from flit contention due to multicast dependency. "*If a multicast flit from an input port n has an number N of requests at time t_s, then each single request to an output port m can be forwarded from the input port n to this output port m in the next time stage if and only if it receives a grant by an arbitration unit at the input port n, while the other requests must be held in the input port if they did not receive a grant by their requested output ports. In each next time stage, a single request, which has been granted before, must be reset to prevent improper flit replication. If all requests have been granted, then the multicast flit can be released from the queue in input port n* [25]".

Most of this chapter contents are adopted from our paper published by Elsevier Science [27]. Therefore, we would like to thank Elsevier Science for the Courtesy/Permission to republish some parts of the issue.

6.3 ID-Based Routing Organization

In the XHiNoC routing concept, an ID-tag is attached on each flit of a data packet. Flits belonging to the same data packet will have the same local ID-tag on each communication link. Based on this concept, multicast communications can be handled dynamically and pre-processing algorithms are not required before injecting multicast messages.

6.3.1 ID-Based Multiple Access (IDMA) Packet Stream Interleaving

Figure 6.2 illustrates, how five packets or data streams (*a*,*b*,*c*,*d* and *e*) can be interleaved on each communication link. For instance, the ID-tag allocation of the stream *a* sent from *Sw*4 to *Sw*3 via *Sw*5 and *Sw*2 can be seen. Its ID-tags are mapped and allocated to ID slot 2 on Local input link of *Sw*4, ID slot 1 on West input link of *Sw*5, ID slot 1 on North1 input link of *Sw*2, ID slot 1 on West input link of *Sw*3 and at last, to ID slot 0 on Local output link of *Sw*3. Flits belonging to the same packet or stream will have the same ID-tag on a specific link. Therefore, the ID-tag attached on every flit enables each packet or streaming data flit to be switched to the correct routed direction. In other words, the ID-tag represents the compressed form of the routing direction made by the header flit when reserving communication resources.

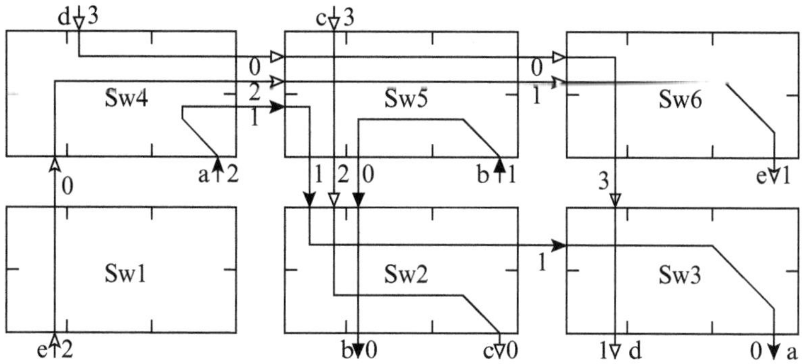

Fig. 6.2 ID-tag-based routing/switching organization and connection scheduling

Each streaming data flit can extract the required routing direction from the routing table that has been indexed in accordance with the local ID-tag of the packet stream.

6.3.2 Routing Table and ID Management

The local ID-tag attached on each flit of the packet (see Fig. 6.5) is updated and dynamically changed once a flit is switched to a new outgoing port. The ID update is made by an ID management unit located at every output port. Figure 6.2 shows the detailed view of the link sharing between stream *a*, *b* and *c* as presented in Fig. 6.2. The communication resource (link) connecting South1 output port of Switch 5 (*Sw*5) and North1 input port of Switch 2 (*Sw*2) assigns packet stream *a*, *b* and *c* with ID-tag 1, 0 and 2, respectively. Tables presented on the right side of the switches are the ID-slot table of the South1 output port at *Sw*5 and the routing table (*LUT*) of the North1 input port at *Sw*2. The content of the ID-slot table represents the ID-tag mapping function of each packet stream. The content of the routing table represents the routing directions to keep the correct routing tracks for each flit of the interleaved packet streams.

Figure 6.2 shows how the ID-tag of a stream header coming from NORTH port with ID-tag 3 is updated after being switched. The ID update process working as follows: When the IDM detects a new incoming stream or packet header, then it searches for a free ID slot on the output link by checking the ID-slot state in the corresponding table. In the example case, the ID-tag 2 is free. The ID is then assigned as the new ID-tag on the next link segment. The ID-slot 2 is indexed based on the associated incoming local ID-tag 3 and the incoming direction (NORTH). Hence, a data flit following the packet/stream at any instant time coming from NORTH port with ID-tag 3 will be assigned also with the new ID-tag 2 in the outgoing SOUTH port. In the same phase, the ID-tag 2 flag is set from "free" to "used" state, and the number of used IDs (UID) is incremented. When the UID has

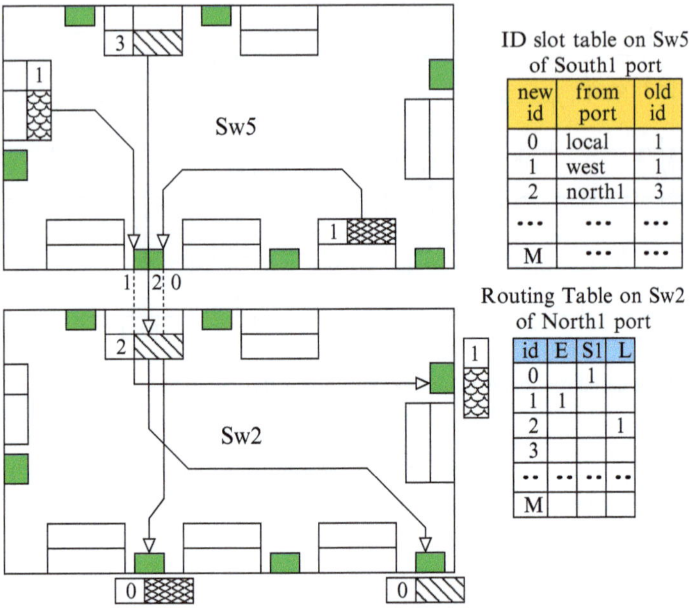

Fig. 6.3 IDMA: packet flit (wire-through-share) interleaving

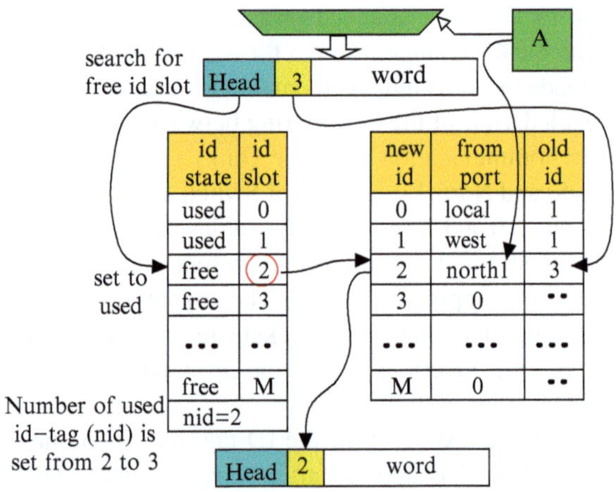

Fig. 6.4 IDMA: ID-tag updating

reached the number N of available ID slots, then "empty free ID flag" is set. When a tail flit (the end of stream data) is passing through the outgoing port, then the state of the related ID-tag 2 state is set from "used" to "free", the UID is decremented and the information related to this ID-number is then deleted from the ID Slot Table concurrently (Figs. 6.3 and 6.4).

More information about the dynamic ID management can be found in detail in [22–24, 26].

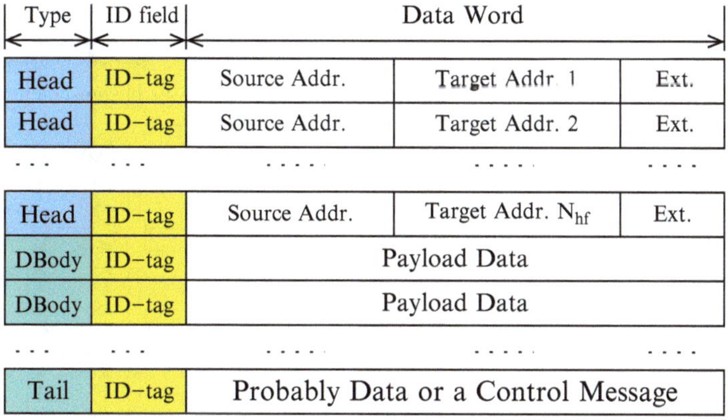

Fig. 6.5 The packet format for a multicast message

6.3.3 *Multicast Packet Format*

The packet format used in the XHiNoC, supporting the ID-based routing organization, is depicted in Fig. 6.5. A multicast packet consists of a number N_{hf} of header flits (which is equal to the number of multicast destinations N_{dest}), any amount of payload flits and a tail flit. The flit flow control field of every data flit consists of 3-bit flit *type* header and a 4-bit packet ID (*Identity*). Therefore in a 32-Bit instance of the system, each flit of the packet has 39-bit width, i.e. 32-bit data word plus 7-bit control field. The *type* can be *header*, *data body*, and the *end of databody* (*last/tail flit*) as shown in the Fig. 6.5. Flits belonging to the same message will always have the same local ID label on one and the same communication link. The ID number attached in each flit will vary over different communication links to support a *wire-sharing concept* with *flit-level message interleaving*. This concept scales well with increasing mesh network sizes.

Definition 1. A multicast message (even if the size is very large) is not divided into several sub-packets. Therefore, when N_{pf} number of payload flits that will be sent to N_{dest} number of destination nodes, the size of the multicast message will be $N_{hf} + N_{pf} + 1\ tailflit$.

6.4 Microarchitecture of the XHiNoC Router

6.4.1 *2D Mesh Planar Topology*

By for instance using the West-First routing algorithm, the adaptivity of the multicast-tree in the West direction is limited by the South–West and North–West prohibited turns, where multicast adaptive tree-based routing cannot be made if the

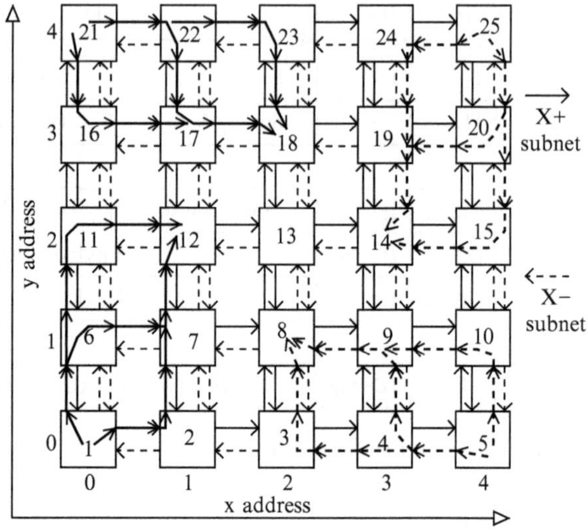

Fig. 6.6 2D planar mesh network

destination addresses are located in the South-West and North-West quadrant area [24]. These prohibited turns must be implemented in the routing algorithms to avoid the occurrence of a deadlock configuration. In order to cover such problems, a planar 2D NoC architecture with mesh topology is also presented in this chapter. The NoC is divided into two sub-networks in order to increase the degree of adaptivity of the routing functionality. A planar adaptive routing algorithm has been firstly introduced in [6], in which virtual channels (VCs) are introduced to support adaptive routing and to couple the sub networks. The main difference of the presented approach is that, instead of using VCs, we replace them with a double physical communication link to increase the link and switch bandwidth capacity.

Figure 6.6 shows an example of the 2-D mesh 5×5 network. The Network-on-Chip is physically divided into two subnetworks i.e., $X+$(depicted in solid line arrows) and $X-$subnetworks (depicted in dashed line arrows). If the x-distance between source and target nodes ($x_{offs} = x_{target} - x_{source}$) is zero or positive, then packets will be routed using the $X+$subnetwork. If x_{offs} is zero or negative, then the packets will be routed through the physical channels of the $X-$subnetwork. The ports connected with vertical y-direction links of $X+$ and $X-$subnetworks are denoted by (North1, South1) and (North2, South2), respectively. The packets being routed through the $X+$subnetwork will have adaptivity to make West–North1, West–South1, North1–East and South1–East turns as well as West–East, North1–South1 and South1–North1 straightforward (non-turn) routing direction. The packets being routed through the $X-$subnetwork will have adaptivity to make East–North2, East–South2, North2–West and South2–West turns as well as East–West, North2–South2 and South2–North2 straightforward routing directions.

The planar adaptive routing technique on a mesh topology has been firstly introduced in [6] and is deadlock-free by principle. Instead of using virtual channels to implement the interconnects between NORTH and SOUTH port as made in [6], we prefer to implement two physical channels to separate the NORTH–SOUTH link interconnects for the $X+$ and $X-$subnetworks. The objectives for this modified topology are the preservation of the router performance and the increase of the network bandwidth. In case that virtual channels would have been implemented between the NORTH and SOUTH ports, then we needed to add two virtual queues at both incoming and outgoing ports. This would degrade the router performance characteristic or increase the data transfer latency, therefore we substitute VCs by adding two additional ports (NORTH2 and SOUTH2 ports) to the existing mesh router.

In a typical 5-port mesh switch, a 2D $N \times M$ mesh network will have $N \times (M-1) + M \times (N-1)$ full-duplex directional communication links. By using 2D planar-based mesh router, the $N \times M$ mesh network will provide more communication resources, i.e. $2N \times (M-1) + M \times (N-1)$ full-duplex directional links (and additionally the local core connections).

6.4.2 Static and 2D Planar Adaptive Routing Algorithms

In this chapter, we will evaluate different models of NoCs with mesh topology by using four mesh prototypes with different routing algorithms. The first model ('plnr' prototype) uses a 2D multicast planar adaptive routing algorithm that is presented in Algorithm 4 and is implemented based on the planar mesh topology. The adaptive routing decision is made based on the number of used ID slots (or the number of free ID slots) on each possible alternative routing direction and the record of the routing path made by other headers of the same message, which is later explained in Sect. 6.5.3 and presented in Algorithm 2. The routing algorithm is divided into two subalgorithms for $X+$ and $X-$subnetworks.

The second prototype ('xy' prototype) uses a static XY routing algorithm in the standard mesh topology, where messages are first routed in horizontal X-direction and then to vertical Y-direction. Hence, North–East turn and North–West as well as South–East and South–West are prohibited in the static XY routing algorithm. The remaining two models ('wf-v1' and 'wf-v2' prototypes) use the minimal West-First routing algorithms. In the West-First routing algorithm, packets will always be firstly routed to West direction, if its destination nodes are located in western area relative from its source node or current position. When the destination nodes are located in eastern area of its source node or current position, then the packet can be routed adaptively to East, North or South directions [10].

6.4.3 Generic Modules and Modular Design

The XHiNoC router microarchitecture is developed based on a modular concept and units are grouped into incoming block and outgoing block components as presented

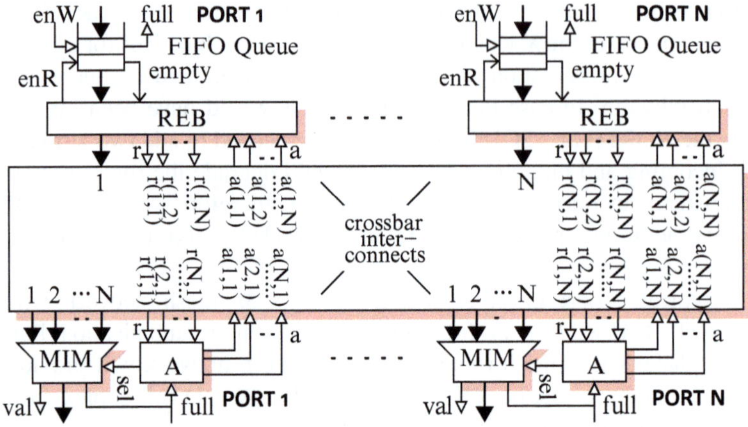

Fig. 6.7 General XHiNoC router architecture

in Fig. 6.7. Each module is modeled based on generic code, which is strongly related to the number of input-output connectivities of each port. The components located at input ports are *FIFO buffers* and a *Routing Engine with Data Buffering (REB)*. The components located at output ports are an *Arbiter (A)* and a *Crossbar Multiplexer with ID-Management Unit (MIM)*. The working principles and mechanisms of the Arbiter, REB and MIM units are explained in detail in [22] and [24].

Figure 6.8 shows the components in an incoming port of the XHiNoC router. In the REB module, there are a *Grant-Multicasting Controller* (GMC), a *Routing Engine* (RE), a *Route Buffer* and a *Read-Logic Unit* (RLU). The GMC consists of combinatorial logic and is used to control the acceptance of the multicast routing acknowledge (grant) signals from output ports. The RLU is also a combinatorial logic, which is used to control the read-operation of a flit from the FIFO buffer into the Route Buffer. When a routing direction for a flit is being decided by the RE unit, then this flit will be concurrently buffered in the Route Buffer. This concurrent step is introduced in order to reduce the number of internal pipeline stages in the XHiNoC router, and it can improve the router performance accordingly. The RE unit consists of the *Routing State Machine (RSM)* and the *Routing Reservation Table (RRT)*, which consists of a number of H preservable routing slots.

The design of the XHiNoC routers can be fully parametrized and customized on demand. Each VHDL entity contains generic code, which enables the derivation of new VHDL modules with a specific architecture and a number of input/output pins according to the specification. The custom-generic modular-based design approach also enables us to easily generate irregular NoC topologies.

Before running a real experiment, the different routing paths that will be performed by the aforementioned tree-based multicast routing methods (except for the wf-v1 multicast method) are shown in Fig. 6.9. A multicast message is injected from node (2,2) to 10 multicast destinations. The ‘xy’ multicast router performs 24 link usages in the NoC. While the ‘plnr’ and ‘wf-v2’ multicast routers perform

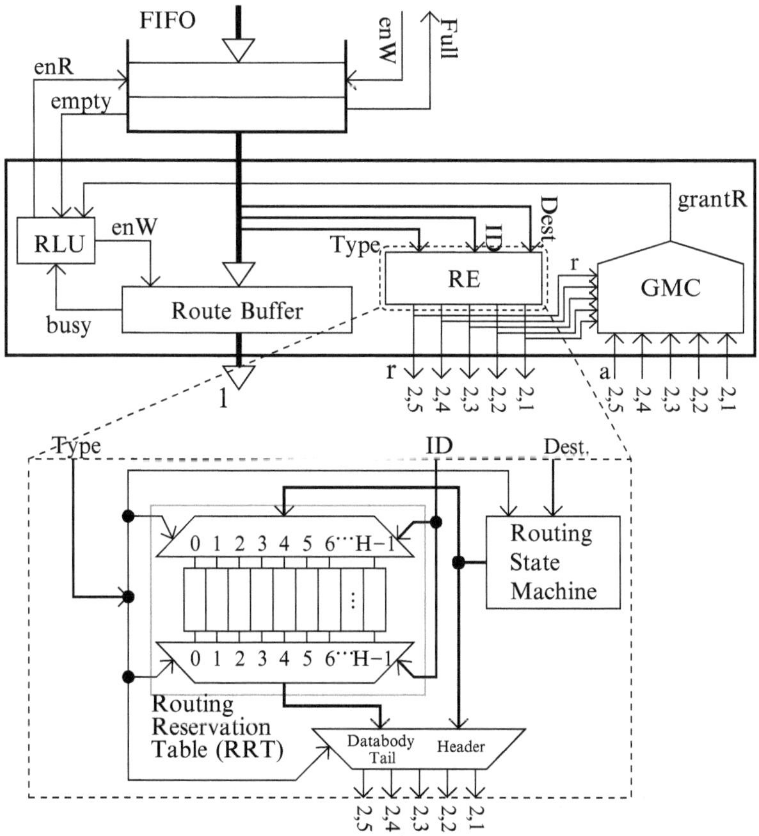

Fig. 6.8 Input port of a 5-port router (for the static XY routing, the signal paths of the number of used ID slots are removed)

with only 19 and 21 link segment usages in the NoC respectively. The traffic metric is defined as the number communication links used by a multicast packet to travel from a source node to destination nodes. From this example it can be concluded, that the planar adaptive multicast router can potentially reduces the communication energy of the multicast data transmission. The following experiment will show us the result of a more complex data distribution scenario.

6.5 Efficient Runtime Spanning Tree Configurations

The tree-based multicast routing algorithm implemented in the XHiNoC architecture is deadlock-free. The multicast NoC router architecture is designed and implemented based on the novel theory for deadlock-free multicast routing presented in [25].

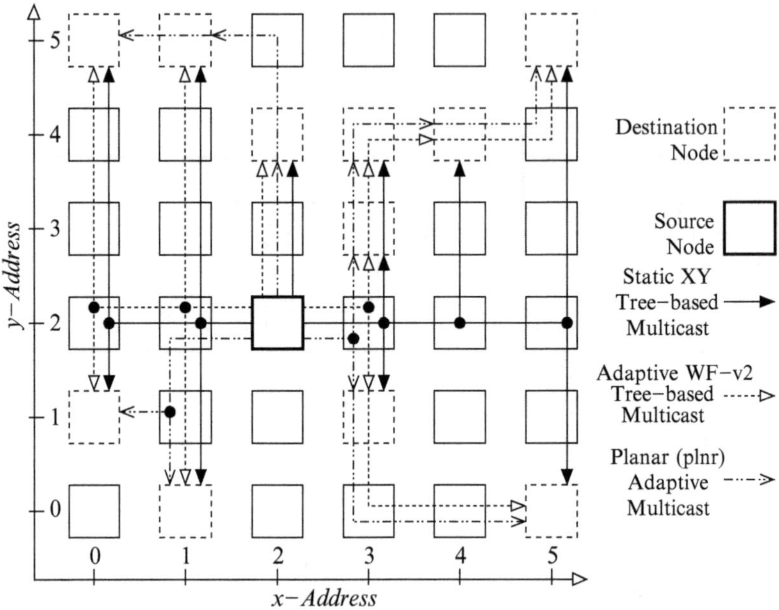

Fig. 6.9 The traffic patterns by using static tree-based, minimal adaptive west-first and minimal planar adaptive multicast routing methods

The problem of inefficient runtime spanning tree configuration that can affect the overall throughput of the generated multicast trees [24]. Because of this uncovered issue, in any circumstance, the adaptive routing cannot show better performance, since the data rate of the multicast tree depends on the slowest data rate in all spanning trees or branches of the multicast tree. Therefore, based on the presented local ID management concept, an *efficient method for runtime multicast spanning tree configurations* by using a minimal adaptive routing algorithms based on a so-called *pheromone tracking strategy* is proposed in this contribution. Minimizing the size of spanning trees (total multicast communication traffic) will not only reduce the communication energy but also decrease the probability of forming spanning trees having slower data rates.

6.5.1 Routing and Multicasting Procedure

The routing engine (RE) units in XHiNoC consist of combination of a *Routing State Machine* (*RSM*) unit and a *Routing Reservation Table* (*RRT*) unit. The combination is aimed at supporting a *runtime link interconnect configuration.*

Definition 2. A Multicast Routing Slot within a Routing Reservation Table is defined as

$$T_{mcs}(k, r_{dir}) \in \{0, 1\} \quad (6.1)$$

where $k \in \Omega = \{0, 1, 2, \cdots, N_{slot} - 1\}$, N_{slot} is the number of ID slots on a link. r_{dir} is a routing direction, where $r_{dir} \in D = \{1, 2, 3, \cdots, N_{outp}\}$, and N_{outp} is the number of I/O ports in a router.

Definition 3. A Routing Reservation Table (RRT) of the RE unit at an input port of a multicast router is defined as

$$T(k) = [T_{mcs}(k,1) \;\; T_{mcs}(k,2) \;\; \cdots \;\; T_{mcs}(k, N_{outp})] \quad (6.2)$$

$T(k)$ contains a binary-element vector. Hence, T has 2D (matrix) size of $row \times column = N_{slot} \times N_{outp}$.

Based on Definitions 2 and 3, a binary-encoded multicast routing direction $r_{dir}^{bin} = enc(r_{dir})$ is introduced and has a size of N_{outp} number of binary elements. For example, if $N_{outp} = 5$, then $enc(1) = [1\ 0\ 0\ 0\ 0]$, $enc(2) = [0\ 1\ 0\ 0\ 0]$, $enc(3) = [0\ 0\ 1\ 0\ 0]$, $enc(4) = [0\ 0\ 0\ 1\ 0]$ and $enc(5) = [0\ 0\ 0\ 0\ 1]$. Algorithm 1 describes the ID-based Routing Organization between the *RSM* and the *RRT* unit. If a *RE* unit identifies a flit $F(type, I)$ as a header flit ($type = header$) from the output of a FIFO buffer with local ID-tag $I \in \Gamma$, where $\Gamma = \Omega$ then the routing function of *RSM* unit $f_{RSM}(A_{dest})$ will determine a routing direction r_{dir}. This means that the r_{dir} is calculated logically based on destination address A_{dest} attached in the header flit and current address of the router, and the routing direction is written in the slot number $k = I$ of the *RRT* unit. In the next time periods, when the *RE* units identify payload flits with the same ID-tag number (ID-tag number I) with the previously forwarded header flit, then their routing direction will be taken up directly from the slot number $k = I$ in the *RRT* unit.

Algorithm 1 Runtime IDMA-based multicast routing mechanism

Read Data Flit from Queue : $F_n(type, I)$
1: $k \Leftarrow I$; A_{dest} is obtained from Header flits
2: BEGIN *Multicast routing* (r_{dir}^{bin})
3: **if** $type$ is $Header$ **then**
4: $\quad r_{dir} \Leftarrow f_{RSM}(A_{dest})$; $T(k, r_{dir}) \Leftarrow 1$
5: $\quad r_{dir}^{bin} = enc(r_{dir})$
6: **else if** $type$ is $Response$ **then**
7: $\quad r_{dir} \Leftarrow f_{RSM}(A_{dest})$
8: $\quad r_{dir}^{bin} = enc(r_{dir})$
9: **else if** $type$ is $Databody$ **then**
10: $\quad r_{dir}^{bin} \Leftarrow T(k)$
11: **else if** $type$ is $Tail$ **then**
12: $\quad r_{dir}^{bin} \Leftarrow T(k)$; $T(k) \Leftarrow \emptyset$
13: **end if**
14: END *Multicast routing*

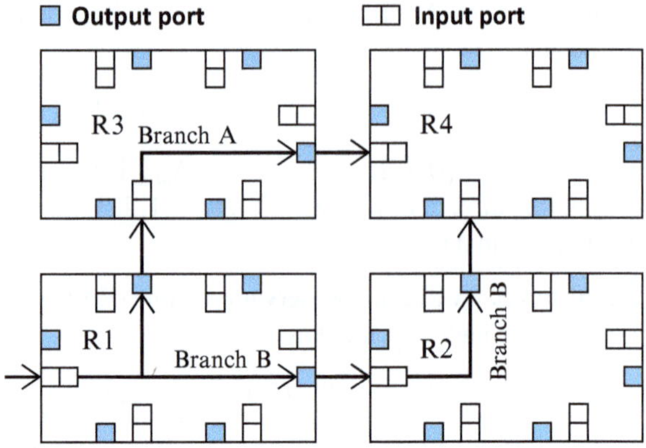

Fig. 6.10 Example of an inefficient spanning tree in adaptive tree-based multicasting

There are three main steps to send a multicast message towards multiple destinations. The first step is to forward all header flits for the multicast tree routing setup and ID-slot reservation. The second step is to multicast (replicate) the payload flits to follow the path set up previously by the header flits. The last step is to set free the reserved local ID-slot by the tail flit. The detail procedure can be found in [24] and is formally described in Algorithm 1.

Definition 4. A runtime tree-based multicast routing configuration of a message that will be sent to a number of N_{dest} multicast destinations is established by sending N_{hf} number of header flits, where $N_{dest} = N_{hf}$. The multicast header flits $H_j(I), j \in \{1,2,\cdots,N_{hf}\}$ can be ordered arbitrarily, where I is the ID-tag of the headers at a certain (input) link n. Thus, we can further define that $F_n(header, I) = H_j(I)$.

6.5.2 *Inefficient Spanning Tree Problem*

The aim of an efficient adaptive multicast routing is to minimize the communication energy. Figure 6.10 shows an example of an inefficient adaptive routing. A tree-based multicast message coming from the WEST input port of the router $R1$ forms two branches in different routing directions i.e., a branch to NORTH (branch A) and a branch to EAST (branch B) direction. It can be assumed, that the branches A and B are established by header flit 1 and header flit 2, belonging to the same multicast message. The header flits will be routed to multicast destinations (x_{t1}, y_{t1}) and (x_{t2}, y_{t2}), respectively, by using a *minimal adaptive routing algorithm*.

Postulate 1. *In a regular 2D mesh network, if a header flit $H_j(I)$ in a current mesh node (x,y) will be routed to a destination node (x_j, y_j) by using a minimal adaptive routing algorithm, then there will be at most two alternative output port directions (otherwise the routing path would not be minimal). When $|x_{offs,j}| = |x_j - x| > 0$ and $|y_{offs,j}| = |y_j - y| > 0$, then there will be two alternative output directions, i.e. m_1 and m_2, where if m_1 is an output direction that can reduce $|x_{offs,j}|$, then m_2 is an output direction that can reduce $|y_{offs,j}|$, or vice versa.*

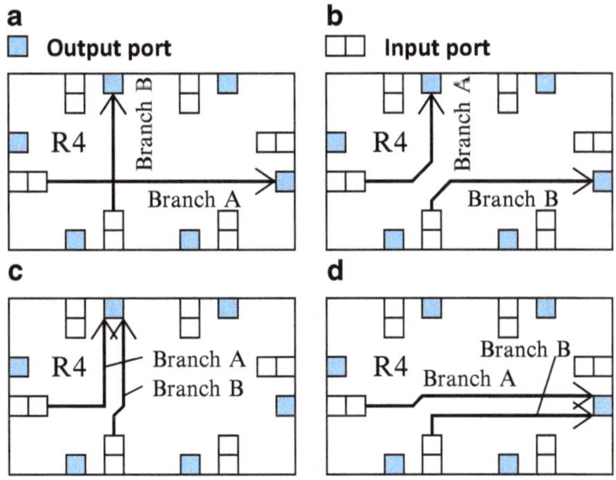

Fig. 6.11 Consequences of the inefficient spanning trees

In the router $R3$ in Fig. 6.10, the multicast message is routed from SOUTH to EAST direction (branch A), while in the router $R2$, the multicast message is routed from WEST to NORTH direction (branch B). Finally these two branches are then routed to the same router (router $R4$). In this case, the multicast tree branches (spanning trees) are inefficient in term of communication energy. The communication energy can be reduced if the router $R1$ performs only the multicast tree branch A or branch B.

Postulate 2. *If two header flits ($H_j(I)$ and $H_k(I)$) having the same ID-tag I (hence, belonging to the same multicast message) are routed from the same input port n in a router node (x,y) at two consecutive times t_{H_j} and t_{H_k} where $t_{H_j} < t_{H_k}$ (which means that H_j is routed firstly before H_k), then an inefficient runtime spanning tree configuration can happen when H_k (which will be routed to destination node (x_k,y_k), where $|x_{offs,k}| = |x_k - x| > 0$ and $|y_{offs,k}| = |y_k - y| > 0$) does not follow to an output port m which has also been selected previously for routing of H_j (having destination node (x_j,y_j)), where $\langle x_j - x = x_{offs,j} = x_{offs,k}$ and $y_j - y = y_{offs,j} = y_{offs,k}\rangle$ or $\langle x_{offs,j} = 0$ and $y_{offs,j} = y_{offs,k}\rangle$ or $\langle x_{offs,j} = x_{offs,k}$ and $y_{offs,j} = 0\rangle$.*

Figure 6.11 depicts four possible situations that can occur in the router $R4$ as the further disadvantageous consequences of the inefficient multicast spanning trees configured from Fig. 6.10. These situations can happen since the number of free ID slots on each communication link as the parameter of the adaptive routing algorithm may change dynamically. Figure 6.11a, b show a *tree-branch crossover problem*, in which the inefficient spanning trees are propagated through different outgoing ports. If we assume that the current address of router $R4$ is (x_{curr},y_{curr}) and the target nodes of the tree branches A and B are (x_{t1},y_{t1}) and (x_{t2},y_{t2}) such that $x_{offset1} = x_{t1} - x_{curr} > 0$ and $y_{offset1} = y_{t1} - y_{curr} > 0$ as well as $x_{offset2} = x_{t2} - x_{curr} > 0$ and $y_{offset2} = y_{t2} - y_{curr} > 0$, then in any circumstance, the inefficient situation might happen again in the next intermediate nodes.

Figure 6.11c, d illustrate a *tree-branch interference problem*, where the inefficient spanning trees interfere into the same outgoing port. This situation will lead to inefficient multicast communication time (increase of communication latency and data workloads) because of the self-contention problem. Two multicast messages will be forwarded from different input ports to the same output port, carrying identical data payload (double workloading).

6.5.3 Solution: Efficient Adaptive Routing Selection with Pheromone Tracking Strategy

The problematic configurations presented in Figs. 6.10 and 6.11 are not only inefficient in terms of communication energy (because the inefficient traffic will overburden the NoC), but also in term of communication latency, since the inefficient traffic can degrade the data rate of the multicast traffic. The problems reduce the NoC performance while increasing power consumption.

We solve the aforementioned problem not by designing a specific multicast path optimization algorithm that should be run at compile time or before injecting multicast messages (*pre-processing algorithm*). Compile-time path optimization algorithms such as the optimal spanning tree algorithm are suitable for *source routing approach*, where the routing paths for the overall pathes of a multicast message from source to destination node are determined at source node before the message is injected to the network. In contrast, in XHiNoC the routing algorithm used to route unicast and multicast messages is the same, the routing decisions are made at runtime and locally executed hop-by-hop on every port of each router. Thus, we do not follow the approach of a path pre-processing optimization algorithm for the sake of initiation-time-efficiency.

In order to avoid the previously described problems, each time a routing engine has two alternative output ports for making a routing decision, then a new adaptive output selection strategy between two alternative output ports will be applied. A simple abstract view of the adaptive selection strategy is outlined in Algorithm 2. The basic concept of the proposed algorithm is the identification of the track records (*pheromone* trails) of other previously-routed header flits that belong to the same multicast message. This concept is designed in order to avoid inefficient spanning trees branches of the multicast tree.

Definition 5. A pheromone trail checking is an operation to check the binary state of the multicast routing slots $T_{mcs}(k,m_1)$ and $T_{mcs}(k,m_2)$. The operation is made by a header flit $H_j(I)$ having ID-tag number $I = k$ that will be alternatively routed to output directions m_1 and m_2, where $m_1, m_2 \in D$.

The hardware implementation of the efficient adaptive routing selection function with pheromone tracking strategy will not result in a more complex operation in the XHiNoC microarchitecture. The router complexity can be reduced naturally

Algorithm 2 Multicast: adaptive routing selection strategy

Begin Function **Select**(*port* m_1, *port* m_2)
if *Routing* can be made to *port* m_1 and *port* m_2 **then**
  if *Routing* for the same *Multicast Packet* has been made to *port* m_1 and not yet to *port* m_2 **then**
    Return $Routing = port\ m_1$
  else if *Routing* for the same *Multicast Packet* has been made to *port* m_2 and not yet to *port* m_1 **then**
    Return $Routing = port\ m_2$
  else if *Routing* for the same *Multicast Packet* has not been made to *port* m_1 and *port* m_2, or has been made both to *port* m_1 and *port* m_2 **then**
    if $UsedID(port\ m_1) < UsedID(port\ m_2)$ **then**
      Return $Routing = port\ m_1$
    else
      Return $Routing = port\ m_2$
    end if
  end if
end if
End Function

Algorithm 3 Logical view of Algorithm 2

Incoming Data Flit : $F_n\,(type, ID)$
$T(k,m)$: The slot of RRT for a flit with ID=k to direction m
$UsedID(m)$: Number of used/reserved ID slots in direction m

Begin Function **Select**(m_1, m_2)
$k \Leftarrow ID$
if $\neg T(k,m_1)\ \&\ UsedID(m_1) \leq \neg T(k,m_2)\ \&\ UsedID(m_2)$ **then**
  Return m_1
else if $\neg T(k,m_1)\ \&\ UsedID(m_1) > \neg T(k,m_2)\ \&\ UsedID(m_2)$ **then**
  Return m_2
end if
End Function

by using the advantageous feature of our dynamic runtime table-based routing management and control. In order to achieve an efficient operation, the logical view of the multicast adaptive routing selection strategy is proposed in Algorithm 3.

The operation of Algorithm 3 (logical view of the efficient adaptive multicast routing algorithm) in accordance with Definition 5 (pheromone trail checking strategy) will be illustrated and explained in the following example. Let us assume that a multicast header having ID-tag $k = 3$ can be alternatively routed to output port m_1 and m_2, where the number of used (already reserved) ID-tags at the output ports is $UsedID(m_1) = 0101(5)$ and $UsedID(m_2) = 0010(2)$, respectively. Let us also assume that a previously routed header, which belongs to the same multicast group as the current header (also with ID-tag $k = 3$), has been routed to port m_1 such that $T(3,m_1) = 1$, and there is no header with ID-tag $k = 3$ that has been routed to port m_2 such that $T(3,m_2) = 0$. Based on

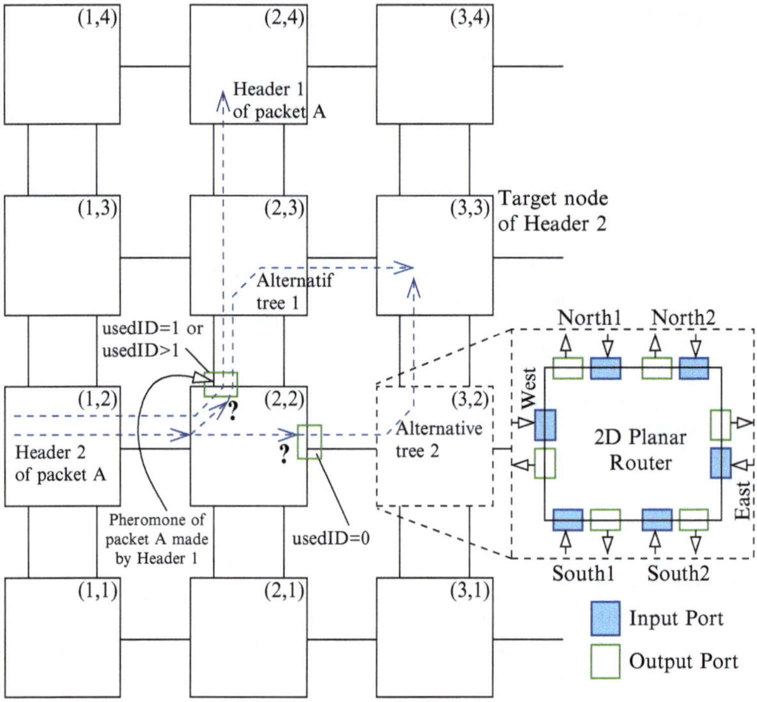

Fig. 6.12 Example illustration of the pheromone tracking strategy

Algorithm 3, we evaluate $\neg T(3,m_1)\ \&\ UsedID(m_1) < \neg T(3,m_2)\ \&\ UsedID(m_2)$, or $0\ \&\ 0101\ (00101) < 1\ \&\ 0010\ (10010)$. The operator "$\neg$" is a negation operator. The operator "&" represents a concatenation operator, such that the negation of $T(k,m)$ is inserted as the most-significant-bit of the $UsedID(m)$ logical value. Therefore, according to Algorithm 3. the header is then routed to port m_1, because it has lower cost value than the port m_2.

For the sake of clarity, Fig. 6.12 illustrates how a packet header will track a pheromone made by other header flit that belong to the same packet, in order to set up an efficient spanning tree. As shown in this figure, Header 1 of packet A has arrived router node (2,4). The path traversed by Header 1 is shown in the figure. At the same time, Header 2 of packet A is now at router node (2,2), targeting the destination router node (3,3). So Header 2 can select either output port East or North 1 to configure one of two alternative spanning tree branches as depicted in the figure. The number of used IDs at output port North 1 is 1 or probably more than 1 if any other packets have also reserved IDs. Although the number of used IDs at East direction outport is 0 or less than the number of used IDs at output port North 1, Header 2 will finally select the North 1 port to configure the alternative spanning tree branch 1, because Header 2 has detected a pheromone trail made by Header 1, which has been previously routed at the router node (2,2).

According to Definition 5 and Algorithm 3, Header 2 can detect the pheromone of Header 1 and track/follow the same spanning tree made by the Header 1, because both header flits belong to the same packet, i.e. both have the same local ID-tag at each communication link. From Fig. 6.12, we can see that by following the pheromone tracking strategy, Header 2 will finally add only one communication link (link between nodes (2,3)–(3,3), through alternative tree 1), instead of two communication links (links between nodes (2,2)–(3,2) and between node (3,2)–(3,3), through alternative tree 2). Therefore, the communication energy of the multicast communication performed by the packet A will be less and the overall communication more efficient.

It it is obvious, that the pheromone trail tracking can be efficiently and logically implemented by detecting the bit-value of the routing slot $T(k,m)$. This detected bit is used as the most-significant bit (MSB) in the concatenation operation with the $UsedID$ bit-signal, i.e. $T(k,m)\&UsedID(m)$. In general, the router will route a header flit into an output port having more free ID-tags. However, the pheromone trail, which is represented by the routing slot $T(k,m)$, has a higher priority over the number of already reserved ID slots. This pheromone trail checking method is used to avoid inefficient multicast spanning tree as presented visually in Fig. 6.10.

In this section four algorithms that are used to route packets in NoCs have been presented. Algorithm 1 is an ID-based routing algorithm that is generally used in the XHiNoC router by default. Algorithm 4 represents a planar adaptive routing scheme, that routes packets in the 2D planar NoC using two sub-networks with dual physical channels in the north and south direction. The most important contributing algorithms are Algorithms 2 and 3. These algorithms are solving the problem of the inefficient adaptive tree-based multicast. The algorithms can also be simply derived and implemented based on the existing physical state information of the router microarchitecture.

6.6 Experimental Results

In this section, four XHiNoC multicast router prototypes with different multicast routing algorithms are compared. The first prototype is a multicast router working with a planar adaptive routing algorithm on a NoC mesh architecture, which is presented with **'plnr'** acronym in the figures. The second prototype applies a static XY multicast routing algorithm on the mesh standard architecture (**'xy'**). The third and fourth prototype are multicast routers in a standard mesh architecture applying an adaptive West-First (WF) routing algorithm (**'wf-v1'** and **'wf-v2'**). The adaptive WF multicast router version 1 (**'wf-v1'**) is a multicast router without implementation of an adaptive selection strategy to avoid an inefficient spanning tree (branches of the multicast tree) as presented in [24]. Thus in this prototype, the multicast trees are formed freely without considering the track records of the other previously-routed header flits belonging to the same multicast group. The adaptive WF multicast router version 2 (**'wf-v2'**) implements the adaptive selection strategy presented in the Algorithm 2 to avoid an inefficient spanning tree problem.

Algorithm 4 2D Planar adaptive routing algorithm

Network is partitioned into two subnets: X^+ and X^- Subnet.
Set of output ports in X^+ Subnet: $\{EAST, SOUTH\ 1, NORTH\ 1, LOCAL\}$.
Set of output ports in X^- Subnet: $\{WEST, SOUTH\ 2, NORTH\ 2, LOCAL\}$.
Select(m_1, m_2) is selection function between output port m_1 or m_2.

$X_{offs} = X_{target} - X_{source}$
$Y_{offs} = Y_{target} - Y_{source}$
while *Packet* is in *Subnet* X^+ i.e.$(X_{offs} \geq 0)$ **do**
  if $X_{offs} = 0$ and $Y_{offs} = 0$ **then**
    $Routing = LOCAL$
  else if $X_{offs} = 0$ and $Y_{offs} > 0$ **then**
    $Routing = NORTH\ 1$
  else if $X_{offs} = 0$ and $Y_{offs} < 0$ **then**
    $Routing = SOUTH\ 1$
  else if $X_{offs} > 0$ and $Y_{offs} = 0$ **then**
    $Routing = EAST$
  else if $X_{offs} > 0$ and $Y_{offs} > 0$ **then**
    Routing=**Select**$(NORTH\ 1, EAST)$
  else if $X_{offs} > 0$ and $Y_{offs} < 0$ **then**
    Routing=**Select**$(SOUTH\ 1, EAST)$
  end if
end while
while *Packet* is in *Subnet* X^- i.e.$(X_{offs} \leq 0)$ **do**
  if $X_{offs} = 0$ and $Y_{offs} = 0$ **then**
    $Routing = LOCAL$
  else if $X_{offs} = 0$ and $Y_{offs} > 0$ **then**
    $Routing = NORTH\ 2$
  else if $X_{offs} = 0$ and $Y_{offs} < 0$ **then**
    $Routing = SOUTH\ 2$
  else if $X_{offs} < 0$ and $Y_{offs} = 0$ **then**
    $Routing = WEST$
  else if $X_{offs} < 0$ and $Y_{offs} > 0$ **then**
    Routing=**Select**$(NORTH\ 2, WEST)$
  else if $X_{offs} < 0$ and $Y_{offs} < 0$ **then**
    Routing=**Select**$(SOUTH\ 2, WEST)$
  end if
end while

The experiment is setup by applying a multicast random data distribution (traffic) scenario to validate the theorem and methodology of the proposed deadlock-free multicast routing. Figure 6.13 depicts the distribution of the source-destination unicast-multicast communication partners in the 2D 8×8 mesh architecture (64 network nodes). The numerical symbol at the bottom and the left side of the mesh network represents the 2D node x-address and y-address. From the figure it can be seen, that 9 multicast communication partners (**M6**, **M8**) and 4 unicast communication pairs (**U**) are placed. Three of 9 multicast communication sources have 8 multicast targets (**M8**), while the remaining 6 multicast sources have 6 multicast destinations (**M6**).

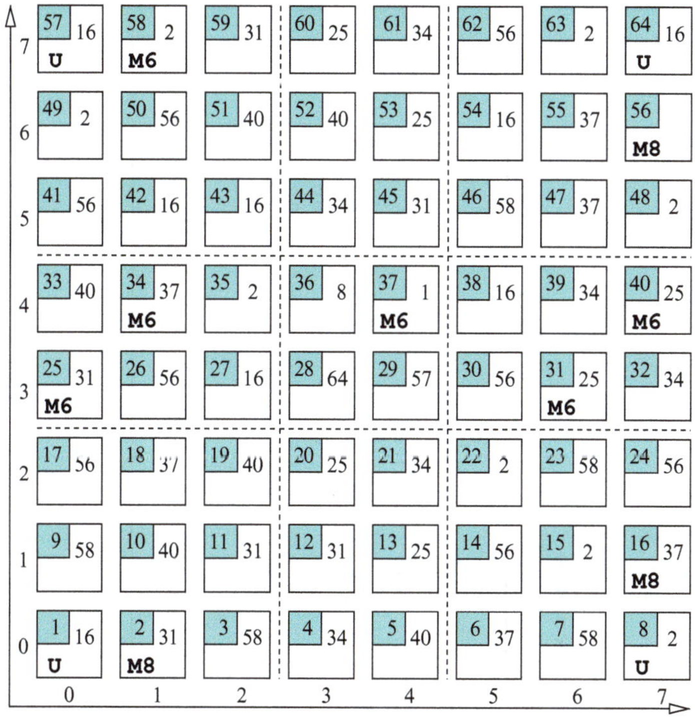

Fig. 6.13 Distribution of the source-destination communication partners

Every NoC router in Fig. 6.13 is depicted with a square block with numerical attributes. A numerical symbol in the small square block at the top-left side of a NoC router node represents the node number. The numerical symbol at the top-right side in the NoC router node represents the communication partner of the node, from which the NoC router node will receive a message. For example, the network node at node address $(2,1)$ (2D node address, 2 is the x-horizontal address and 1 is the y-vertical address) has node number 11. At the top-right side in this node, we see the numerical value 31 (this means that the node will receive a packet from mesh node 31 located in the node address $(6,3)$).

The boldface symbols (**U**, **M6** and **M8**) at the bottom-left corner of the router box represent that the network node will send a unicast message (**U**) or a multicast message with a number of 6 target nodes (**M6**) or 8 target nodes (**M8**). For example, the mesh node at address $(7,1)$ (mesh node number 16) is attributed with (**M8**). This means that the node will send a multicast message into 8 destination nodes. We can find the target nodes of the multicast message sent from the mesh node number 16 by identifying mesh nodes having numerical symbol 16 at the right-side in the router box. In order to find easily the partners of each unicast and multicast communications, Table 6.1 gives an overview on the unicast and multicast communication partners/groups of the source-destination distribution presented in Fig. 6.13.

Table 6.1 Unicast and multicast communications for the random multicast test traffic scenario

Comm. group	Type	Source	Targ. 1	Targ. 2	Targ. 3	Targ. 4	Targ. 5	Targ. 6	Targ. 7	Targ. 8
Comm. 1	M6	(1,4)	(6,4)	(7,3)	(4,2)	(3,0)	(3,5)	(4,7)	–	–
Comm. 2	M6	(7,4)	(0,4)	(3,6)	(2,6)	(2,2)	(1,1)	(4,0)	–	–
Comm. 3	M6	(0,3)	(6,3)	(3,2)	(4,1)	(7,4)	(4,6)	(3,7)	–	–
Comm. 4	M6	(6,3)	(0,3)	(4,5)	(3,1)	(2,1)	(1,0)	(2,7)	–	–
Comm. 5	M6	(1,7)	(7,6)	(5,5)	(6,2)	(6,0)	(2,0)	(0,1)	–	–
Comm. 6	M6	(4,4)	(1,4)	(1,2)	(7,1)	(5,0)	(6,5)	(6,6)	–	–
Comm. 7	M8	(1,0)	(7,0)	(6,1)	(5,2)	(2,4)	(7,5)	(6,7)	(0,6)	(1,7)
Comm. 8	M8	(7,1)	(2,3)	(5,4)	(2,5)	(1,5)	(5,6)	(0,7)	(7,7)	(0,0)
Comm. 9	M8	(7,6)	(1,6)	(0,5)	(5,3)	(1,3)	(0,2)	(5,1)	(5,7)	(7,2)
Comm. 10	U	(0,0)	(4,4)	–	–	–	–	–	–	–
Comm. 11	U	(0,7)	(4,3)	–	–	–	–	–	–	–
Comm. 12	U	(7,0)	(3,4)	–	–	–	–	–	–	–
Comm. 13	U	(7,7)	(3,3)	–	–	–	–	–	–	–

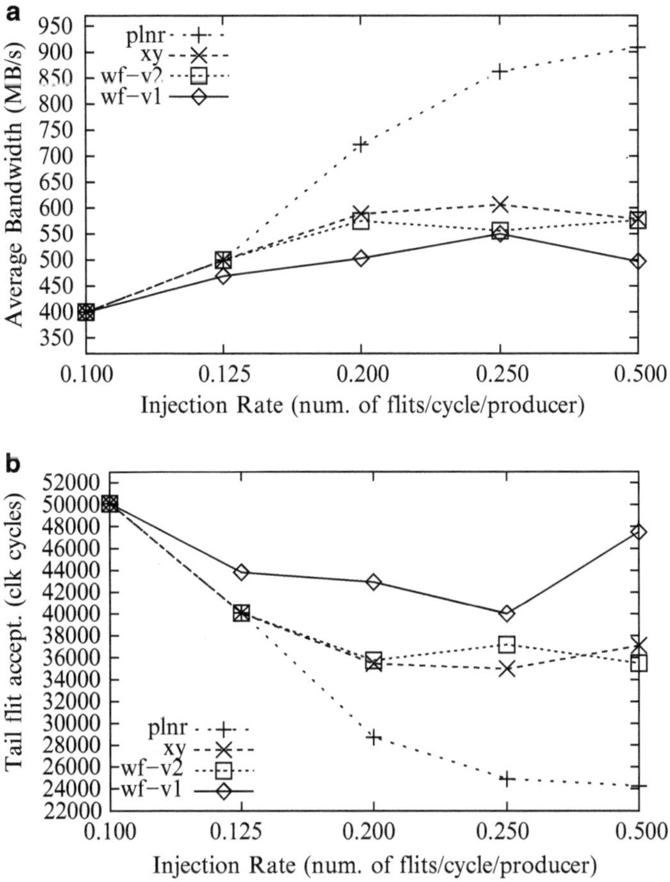

Fig. 6.14 Average bandwidth and tail flit arrival latency measurement versus expected data injection rates for multicast random test scenario. (**a**) Average bandwidth. (**b**) Average tail flit latency

6.6.1 Performance Measurements

The measurements of the average bandwidth and tail flit acceptance latency with various expected data injection rates are depicted in Fig. 6.14. The measurements are made for five different expected data injection rates, i.e. 0.1, 0.125, 0.2, 0.25 and 0.5 flits/cycle (fpc) where each source node injects a 5000-flit packet (equivalent with $4 \times 5{,}000 = 20$ kB data words). It seems that for all multicast routing algorithms, the average BW increases as the injection rate is increased. However, the average BW will tend to saturate, when the injection rate is set nearly to maximum injection rate (the maximum injection rate is 1 flit/cycle).

Table 6.2 Total performed traffic on each link direction for different tree-based static and adaptive multicast routing methods

Routers	South2	North2	South	West	North	East	TOT.
plnr	30	28	34	65	27	46	230
xy	–	–	83	40	89	36	248
wf-v2	–	–	79	40	80	46	245
wf-v1	–	–	85	40	81	86	292

Based on measurements with 1 GHz data frequency, each link has a maximum bandwidth (BW) capacity of 2,000 MB/s. The number of clock cycles required to transfer the 20 kB data words to a target node j is measured by counting the number of clock cycles until receiving the tail flit of the packet (i.e. the 5,000th or the last flit), which is defined as N^j_{TC}. Since there are 64 target nodes in the traffic scenario shown in Fig. 6.13, the average tail flit acceptance values plotted in Fig. 6.14b are $\frac{1}{64}\sum_{j=1}^{64} N^j_{TC}$. The BW measurement on a target node j is calculated as $B_j = (N^j_{TC})^{-1} \times 20{,}000$ B. Consequently, the average actual BW values plotted in Fig. 6.14a are $\frac{1}{64}\sum_{j=1}^{64} B_j$. It can be seen, that the tree-based multicast router with planar adaptive routing algorithm shows the best performance both in terms of the average bandwidth (Fig. 6.14a) and the average latency of the tail flits acceptance (Fig. 6.14b) for all expected data rates. If the data injection rates are very low (e.g. 0.1 fpc), then the performance of the multicast routers will be the same. The performance of the planar adaptive multicast router will be significantly better compared to the other multicast routers, when the data is expected to be transmitted with higher data rates.

6.6.2 Communication Energy Evaluation

Table 6.2 shows the comparisons of the total performed communication traffic for the four tree-based multicast router prototype scenarios. The number of the traffic represents the number of communication resources (communication links) being used to route the unicast/multicast message from source to destination nodes. The metric of the traffic number is measured by counting the number of used/reserved ID slots at all router output ports at peak performance. This performance metric (the number of traffic) can also be used as measurement unit for the communication energy of the evaluated multicast routers. This metric is also interesting for data comparison with other works in the future, since it is technology-independent.

As shown in Table 6.2, it seems that the planar adaptive multicast router ('plnr') consumes less communication resources compared to the other multicast routers, i.e. about 230 communication links followed by the west-first adaptive multicast routing with the efficient spanning tree method ('wf-v2'), and then the static tree-based multicast router that uses XY routing algorithm ('xy').

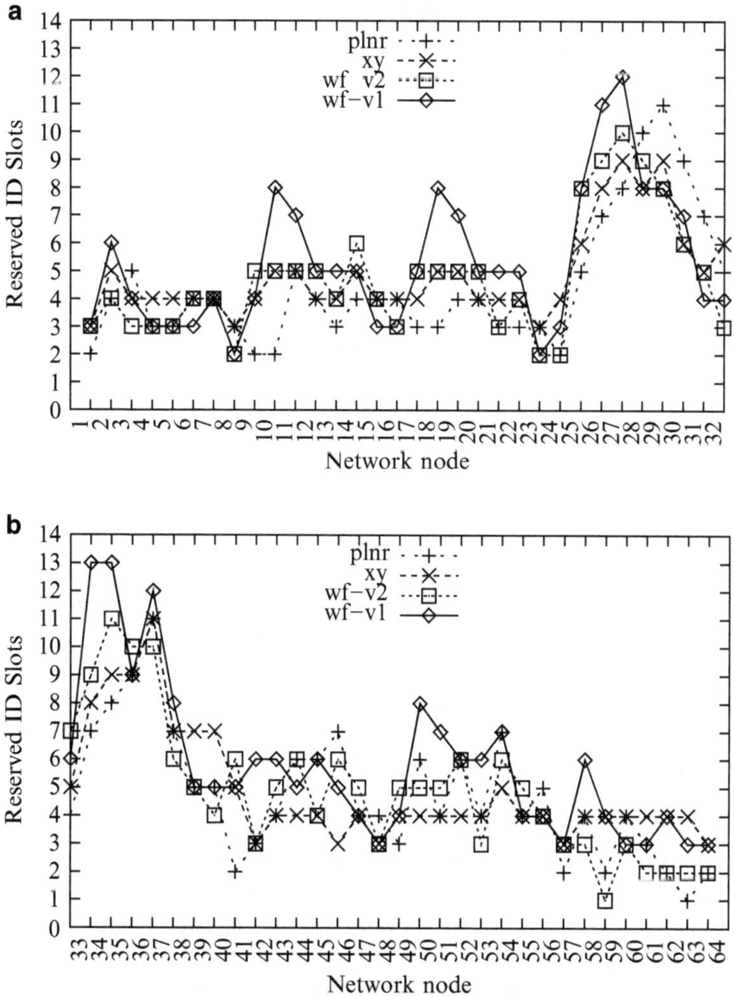

Fig. 6.15 Reserved (used) overall ID slots for multicast random test scenarios. (**a**) Node 1–32. (**b**) Node 32–64

In order to see the traffic configurations in the network in detail, Fig. 6.15 shows the 2D view of the total ID slot reservations in every NoC router node. Figure 6.15a depicts the total reserved ID slots for the NoC router node 1 until node 32, while Fig. 6.15a shows the total ID slots reservation for the NoC router node 33 until node 64. As depicted in Fig. 6.15, the west-first adaptive routing algorithm without the efficient spanning tree method ('wf-v1') reserves more ID-slots than the other routing algorithms at several router nodes.

6.7 State-of-the-Art of Multicast Routing Techniques for NoCs

One basic class of methods for routing multicast messages in mesh-based NoCs are *tree-based multicast routing* techniques. In tree-based multicast routing, a header ordering before submission of the packet within the source node is not required (the order of the destination addresses can be freely determined). The multicast routing will form communication paths like branches of trees connecting the source node with the destination nodes at the end points of the tree branches. The work in [4, 19] and [14] have presented the concept and methodology to route multicast messages by using tree-based methods, which has been utilized in general internetworking context. The work in [25] has presented a new theory for deadlock free tree-based multicast routing for networks-on-chip area (mesh topology). The theory is developed based on a dynamic local ID-tag routing organization and the concept of a *hold-release tagging mechanism*.

Alternatively, multicast messages can also be routed by applying path-based multicast routing methods. Here, as a prerequisite, a multiple target ordering is required before the multicast packets are sent to the network. The path-based multicast routing requires a full implementation of an adaptive routing algorithm allowing all turns in the mesh-based network topology. Therefore, virtual channels are usually needed to make a deadlock free multicast routing function. Virtual channels in the context of on-chip interconnection network will consume not only larger logic gate area but also larger power dissipation. The works in [9, 15, 16] and [5] have presented the *path-based multicast routing* methods for a mesh-based network topology.

In the Network-on-Chip (NoC) research area, some multicast NoC architectures have been introduced. Most of them use the tree-based multicast routing method [1, 11, 12, 28], and path-based multicast routing method [8, 13, 18, 21]. The virtual circuit tree multicast (VCTM) NoC [12] for example has presented a NoC that uses virtual circuit tree numbers to configure routing paths. However, compared to the presented approach, which uses runtime dynamic local ID configuration, the VCTM NoC applies a static method, where virtual circuit tables are statically partitioned among nodes.

A few NoC environments also proposed specific multicast routing methodologies such as a closed-loop path routing method [17] and a region-based routing method [21]. The recursive partitioning multicast (RPM) NoC [28], as another example, applies a recursive hop-by-hop network partitioning method to multicast packets at each intermediate node. The packets in the RPM NoC make replication at a certain node to multicast packets. The replicated packets will update the destination list attached on their header flit and make a new network partitioning recursively based on their current position. By using such scheme, the RPM method will increase the complexity of the routing computational logic.

Table 6.3 presents several NoCs that propose and provide multicast routing services for packet routing. Most of the NoC approaches route the network packets

Table 6.3 State-of-the-arts of multicast routing techniques for NoCs

	Multicast method	Switching method	Routing adaptivity	VC buffers, (buffer depth)	Logic area (technology)	Specific features
VCTM [12]	Tree-based	Circuit switching	Adaptive, static	ja. 4,8 (d.n.a)	0.0240[d] mm^2 (70 nm)	Virtual circuit tree, static VC-table partitioning
RPM [28]	Tree-based	Wormhole	d.o. target distribution	ja. 4 (4)	4.0[e] mm^2 (65 nm)	Recursive partitioning, priority-based replication
LDPM [8]	Path-based	Wormhole	Odd-even adaptive	ja[a]. 4 (8)	21,050 gates (0.25 μm)	Path-based with optimized destination ordering
bLBDR [21]	Region-based	Wormhole	Static, adaptive ext.	d.n.a (d.n.a)	0.0499 mm^2 (90 nm)	Traffic isolations, network domain partitioning
MRR [1]	Tree-based	Virtual cut-through	Adaptive	no. (20,20,10)[b]	d.n.a	Extra internal ring buffers (rotary router)
OPT, LXYROPT [11]	Tree-based	Circuit	Adaptive west-first	ja. 4 (3)	d.n.a	Pre-processing algorithm for tree generation
VC-A/D-FD [13]	Path-based	Wormhole, VCT	Static, adaptive	ja. 4 (8/10)[c]	1,172.03[f] $M\lambda^2$ (65 nm)	FIFO with address/data decoupling
Custom Mcast [29]	Tree-based	Packet	Static	no. (4)	0.18–3.06 mm^2 (70 nm)	Multicast routing at design-time (static)
COMC [18]	Path-based	Wormhole	Static	ja. 4, 6 (2)	d.n.a	Connection-oriented path-based multicast
TDM-VCC [17]	Closed-loop path	Circuit	Static	no. (d.n.a)	d.n.a	Pre-processing algorithm for TDM circuit configuration
XHiNoC	Tree-based	Wormhole cut-through	Adaptive 2-net planar	no. (2)	0.1378 mm^2 (130 nm)	Runtime dynamic local ID management

d.n.a. detail not available, *d.o.* depend on, *ext.* extendable, *VC* virtual channel, *VCT* virtual cut-through
[a]Implemented as delivery channel buffers to avoid multicast deadlock
[b]20 phits in buffering segment stage, 20 in output stage, 10 in input stage
[c]VC length is 8 for address, 10 for data
[d]Only the table (512 entries), not the overall logic area of the router
[e]One tile area (inclusive tile processor)
[f]The area is for adaptive router with wormhole switching (no further explanation about the λ unit)

by using wormhole switching method. The table compares some aspects regarding router implementation and their specific features. Some information cannot be provided in the table, because the data is not available from the considered publications in the bibliography. As shown in the table, compared to the other NoCs, the XHiNoC can be implemented with a very small size buffer (single buffer with only two data slots per input port). The FIFO Buffer is a NoC component that can consume relatively large logic area compared to other NoC components. It can also have a large power dissipation due to intensive switching activities of data that occurs in the buffer.

In Multiprocessor Systems-on-Chip (MPSoC) and chip-level multiprocessor system (CMP) applications, multicast communication services are an important and essential issue. Recent works related to NoC-based multicast communication are presented in [29] and [17]. The work in [29] presents the problem of synthesizing custom NoC architectures that are optimized for a given application (which is critical, when dependability aspects are considered as well). The there presented multicast method considers both unicast and multicast traffic flow in the input specification. But the work proposes a static solution for deadlock-free multicast routing that is fixed to specific NoC applications, i.e. the applications must be known before chip fabrication. The work presented in [17] proposes a TDM (Time Division Multiplex)-based virtual circuit configuration (TDM-VCC) where a pre-processing algorithm for time slot allocations is made before injecting multicast messages into the NoC. In some specific embedded system applications, the inter-core communication patterns are known. Therefore, a pre-processing static routing for congestion avoiding techniques can be used [20, 29], expectedly resulting in a much simpler router architecture (pre-manufacturing routing technique). Runtime dynamic adaptive routing methods [2, 3, 22] are however an interesting approach in the NoC-based multicore embedded systems, where applications may not be known in advance and dependability and reconfiguration plays a role. Indeed, some embedded IC vendors in the multicore era could potentially not only market IP cores but also system architectures [7], where many applications can be mapped onto the system architectures product (IP+NoC cores). Therefore, the implementation of the runtime dynamic adaptive tree-based multicast routing will simplify an embedded system design flow because the routing information configuration is not needed anymore on the post-manufacture (on-chip) router. In this context however, the runtime techniques will need extra area cost and complexity.

6.8 Summary

In general, NoC routers applying planar adaptive routing schemes can achieve an improved performance because of the higher bandwidth capacity of the NoC in double vertical links connecting NORTH and SOUTH ports. Nevertheless, this performance gain must be paid by logic and routing area overhead to implement the mesh planar router architecture.

The tree-based multicast routing presented in this chapter belongs to the class of runtime distributed routing techniques, in which routing decisions are made locally during application execution time (runtime) on every router/switch node based on a header's destination address. The advantage of this method is, that it scales well with increasing NoC sizes. Hence, the presented technique to prevent the inefficient spanning tree problems has been proven to be feasible and deliver good results. Compared to static methods it will probably result in a *suboptimal* or *near-optimal* multicast spanning tree, but in certain cases, a *global optimal* spanning tree may be attained.

When a static tree-based multicast routing would be used, then the configured multicast spanning trees will always be the similar, although the order of the header probes is changed. For a fixed traffic scenario, the global optimal multicast spanning tree could be attained by finding an optimum ordering of the header flits in one multicast message. The optimum ordering is strongly dependent on the multicast traffic patterns. Such a procedure would require the computation of an optimum ordering algorithm before injecting multicast packets at source nodes. However, the supplement effort will lead to extra computational power and delay, due to the extra pre-processing at the source nodes. Furthermore, the result can also be sub-optimal, since the traffic situation in the network can vary.

This issue has exactly been addressed within the presented approach, which considers local traffic situations dynamically and minimizes the communication resource and energy usage. When the presented adaptive routing algorithms are used, then the spanning tree is formed independently at runtime by header probing and additional consideration of the local traffic situation. The spanning tree topology can vary with the order of header flits in the multicast message and with indeterministic dynamically varying traffic loads in the network.

References

1. P. Abad, V. Puente, J.-A. Gregorio, MRR: enabling fully adaptive multicast routing for CMP interconnection networks, in *Proceedings of the 15th IEEE International Symposium on High Performance Computer Architecture (HPCA 2009)*, Shanghai, 2009, pp. 355–366
2. M.A. Al Faruque, T. Ebi, J. Henkel, Run-time adaptive on-chip communication scheme, in *Proceedings of the 2007 IEEE/ACM International Conference on Computer-Aided Design (ICCAD'07)*, San Jose (IEEE Press, Piscataway, 2007), pp. 26–31
3. G. Ascia, V. Catania, M. Palesi, D. Patti, Implementation and analysis of a new selection strategy for adaptive routing in networks-on-chip. IEEE Trans. Comput. **57**(6), 809–820 (2008)
4. M. Barnett, D.G. Payne, R.A. van de Geijn, J. Watts, Broadcasting on meshes with worm-hole routing. J. Parallel Distrib. Comput. **35**(2), 111–122 (1996)
5. R.V. Boppana, S. Chalasani, C.S. Raghavendra, Resource deadlocks and performance of wormhole multicast routing algorithms. IEEE Trans. Parallel Distrib. Syst. **9**(6), 535–549 (1998)
6. A.A. Chien, J.H. Kim, Planar adaptive routing: low-cost adaptive networks for multiprocessors, in *Proceedings of the 19th International Symposium on Computer Architecture*, Gold Coast, May 1992, pp. 268–277

7. M. Coppola, M.D. Grammatikakis, R. Locatelli, G. Maruccia, L. Pieralisi, *Design of Cost-Efficient Interconnect Processing Units: Spidergon STNoC* (CRC/Taylor & Francis, Boca Raton, 2009)
8. M. Daneshtalab, M. Ebrahimi, S. Mohammadi, A. Afzali-Kusha, Low-distance path-based multicast routing algorithm for network-on-chips. IET Comput. Digit. Tech. **3**(5), 430–442 (2009)
9. J. Duato, A theory of deadlock-free adaptive multicast routing in wormhole networks. IEEE Trans. Parallel Distrib. Syst. **6**(9), 976–987 (1995)
10. J. Duato, S. Yalamanchili, L. Ni, *Interconnection Networks: An Engineering Approach*, Revised Printing (Morgan Kaufmann, San Francisco, 2003)
11. W. Hu, Z. Lu, A. Jantsch, H. Liu, Power-Efficient tree-based multicast support for networks-on-chip, in *Proceedings of the 16th Asia and South Pacific Design Automation Conference (ASP-DAC'11)*, Yokohama, 2011, pp. 363–368
12. N.E. Jerger, L.S. Peh, M. Lipasti, Virtual circuit tree multicasting: a case for on-chip hardware multicast support, in *Proceedings of the 35th Annual Internationl Symposium on Computer Architecture (ISCA'08)*, Beijing, 2008, pp. 229–240
13. K.-M. Keung, A. Tyagi, Breaking adaptive multicast deadlock by virtual channel address/data FIFO decoupling, in *Proceedings of the 2nd International Workshop on Network-on-Chip Architecture (NoCArch'09)*, New York, 2009, pp. 11–16
14. D.R. Kumar, W.A. Najjar, P.K. Srimani, A new adaptive hardware tree-based multicast routing in K-Ary N-cubes. IEEE Trans. Comput. **50**(7), 647–659 (2001)
15. X. Lin, L.M. Ni, Multicast communication in multicomputer networks. IEEE Trans. Parallel Distrib. Syst. **4**(10), 1105–1117 (1993)
16. X. Lin, P.K. McKinley, L.M. Ni, Deadlock-free multicast wormhole routing in 2-D mesh multicomputers. IEEE Trans. Parallel Distrib. Syst. **5**(8), 793–804 (1994)
17. Z. Lu, A. Jantsch, TDM virtual-circuit configuration for network-on-chip. Trans. Very Larg. Scale Integr. Syst. **16**(8), 1021–1034 (2008)
18. Z. Lu, B. Yi, A. Jantsch, Connection-oriented multicasting in wormhole-switched network-on-chip, in *Proceedings of the IEEE Computer Society Annual Symposium on VLSI (ISVLSI'06)*, Karlsruhe, 2006, pp. 205–210
19. M.P. Malumbres, J. Duato, J. Torrelas, An efficient implementation of tree-based multicast routing for distributed shared-memory multiprocessors, in *Proceedings of the 8th IEEE Symposium on Parallel and Distributed Processing*, New Orleans, 1996, pp. 186–189
20. M. Palesi, R. Holsmark, S. Kumar, V. Catania, Application specific routing algorithms for networks on chip. IEEE Trans. Parallel Distrib. Syst. **20**(3), 316–330 (2009)
21. S. Rodrigo, J. Flich, J. Duato, M. Hummel, Efficient unicast and multicast support for CMPs, in *Proceedings of the 41st IEEE/ACM International Symposium on Microarchitecture (MICRO-41)*, 2008, pp. 364–375
22. F.A. Samman, T. Hollstein, M. Glesner, Multicast parallel pipeline router architecture for network-on-chip, in *Proceedings of the Design, Automation and Test in Europe Conference and Exhibition (DATE'08)*, Munich, Mar 2008, pp. 1396–1401
23. F.A. Samman, T. Hollstein, M. Glesner, Networks-on-Chip based on dynamic wormhole packet identity management. VLSI Des. J. Hindawi Publ. Corp. **2009**, 1–15 (2009)
24. F.A. Samman, T. Hollstein, M. Glesner, Adaptive and deadlock-free tree-based multicast routing for networks-on-chip. IEEE Trans. Very Larg. Scale Integr. Syst. **18**(7), 1067–1080 (2010)
25. F.A. Samman, T. Hollstein, M. Glesner, New theory for deadlock-free multicast routing in wormhole-switched virtual-channelless networks-on-chip. IEEE Trans. Parallel Distrib. Syst. **22**(4), 544–557 (2011)
26. F.A. Samman, T. Hollstein, M. Glesner, Wormhole cut-through switching: flit-level messages interleaving for virtual-channelless network-on-chip. Elsevier Sci. J. Microprocess. Microsyst. Embed. Hardw. Des. **35**(3), 343–358 (2011)

27. F.A. Samman, T. Hollstein, M. Glesner, Planar adaptive network-on-chip supporting deadlock-free and efficient tree-based multicast routing method. Elsevier Sci. J. Microprocess. Microsyst. Embed. Hardw. Des. **36**(6), 449–461 (2012)
28. L. Wang, Y. Jin, H. Kim, E.J. Kim, Recursive partitioning multicast: a bandwidth-efficient routing for networks-on-chip, in *Proceedings of the 3rd ACM/IEEE International Symposium on Networks-on-Chip (NOCS'09)*, San Diego, 2009, pp. 64–73
29. S. Yan, B. Lin, Custom networks-on-chip architectures with multicast routing. Trans. Very Larg. Scale Integr. Syst. **17**(3), 342–355 (2009)
30. H. Zimmermann, OSI reference model – the ISO model of architecture for open systems interconnection. IEEE Trans. Commun. **8**(4), 425–432 (1980)

Chapter 7
Path-Based Multicast Routing for 2D and 3D Mesh Networks

Masoumeh Ebrahimi, Masoud Daneshtalab, Pasi Liljeberg, Juha Plosila, and Hannu Tenhunen

Abstract In this chapter, we address how to implement unicast and multicast routing algorithms efficiently in 2D and 3D mesh networks. To do this, we present several partitioning methods for the path-based multicast approach with different levels of efficiency. In path-based methods, a multicast message is routed along a path and the message is transferred to the destinations along this path. Partitioning methods divide the network into several logical partitions and assign destinations to different sets; one set for each partition covering destinations that belong to that partition. Smart partitioning methods must balance the sets and reduce the path length within each partition. All of the partitioning methods can be supported by a deterministic routing algorithm. However, in order to increase the performance, we design a general minimal and adaptive routing algorithm which is based on the Hamiltonian path and can be applied to all partitioning methods. The algorithm is simple and does not require any virtual channel for neither unicast nor multicast messages.

7.1 Introduction

Communication between different cores may be in the form of unicast or multicast messages. In the unicast communication, a message is sent from a source node to a single destination node, while in the multicast communication, a message is delivered from one source node to an arbitrary number of destination nodes. Multicast can be easily implemented with almost no hardware overhead by assuming a multicast message is replicated and every instance is sent to a particular destination (this is termed unicast-based multicast, see Fig. 7.1a). However, this implementation

---

M. Ebrahimi (✉) • M. Daneshtalab • P. Liljeberg • J. Plosila • H. Tenhunen
University of Turku, Turku, Finland
e-mail: masebr@utu.fi; masdan@utu.fi; pakrli@utu.fi; juplos@utu.fi; hanten@utu.fi

M. Palesi and M. Daneshtalab (eds.), *Routing Algorithms in Networks-on-Chip*,
DOI 10.1007/978-1-4614-8274-1_7,

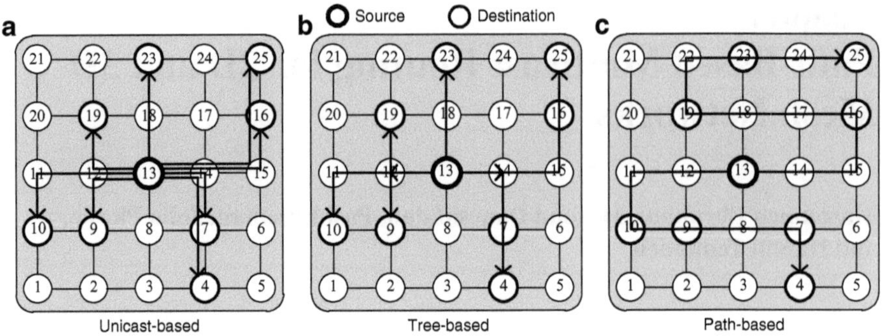

Fig. 7.1 Different multicast approaches

is inefficient. This inefficiency arises because of sending multiple copies of the same message into the network. It not only results in a significant amount of traffic but also introduces a large serialization delay at the injection point.

The vast majority of traffic in Multi-Processor Systems-on-Chip (MPSoCs) consists of unicast traffic and most studies have assumed that the traffic is only unicast. Based on this assumption, the concept of unicast communication has been extensively studied [1–4]. In these approaches, to support a multicast message a single unicast message is delivered per destination. The unicast protocols are efficient when all injected messages are unicast. However, if only a small percentage of total traffic is multicast, the efficiency of the overall system is considerably reduced. For example, let us assume that only 5 % of traffic consists of multicast messages, meaning that per 100 messages, 5 messages are multicast and 95 messages are unicast. For each unicast message, a single message is delivered into the network (i.e. 95 messages). However, if each multicast message has 10 destinations, 50 unicast messages are sent to the network in order to support the multicast operation. With this simple calculation, we notice that more than 50 % of the whole traffic is because of multicast messages (50 messages for multicast operation vs. 95 messages for unicast communication). Thereby, efficient multicast support has a large impact on the performance of chip multi-processor systems. The multicast communication is frequently present in many cache coherency protocols (e.g. directory-based protocols, token-based protocols, and Intel QPI protocol [5, 6]). For an instance, around 5 % of total traffic in a SGI-Origin protocol (i.e. a directory based protocol) consists of multicast messages. It should be taken into account that some cache coherence protocols are heavily multicast (e.g. the token-based MOESI).

Hardware-based multicast schemes can be broadly classified into path-based and tree-based methods. In the tree-based method, a spanning tree is built at the source node and a single multicast message is sent down the tree (Fig. 7.1b). The source node is considered as the root and destinations are the leaves of this tree. The message is replicated along its route at intermediate nodes and forwarded along multiple outgoing channels reaching disjoint subsets of destinations.

This replication at intermediate nodes may result in the blockage of messages [7, 8]. In the path-based method, a source node prepares a message for delivery to a set of destinations. The message carries the destination addresses in the header. The message is routed along the path until it reaches the first destination. The message is delivered both to the local core and to the corresponding output channel for continuing the path toward the next destination in the header. In this way, the message is eventually delivered to all specified destinations (Fig. 7.1c). Notice that the path-based approach does not replicate the multicast message along the path and thus it does not have such blocking issues which exist in the tree-based methods. However, the path visiting all nodes can become relatively long.

7.2 Multicast Routing in a 2D Mesh Network

In this section, we investigate the implementation of the path-based multicast approaches in a 2D mesh network. Three partitioning methods are introduced as Dual-Path (DP), Multi-Path (MP), and Column-Path (CP) along with the minimal and adaptive routing algorithm (HAMUM) [9].

7.2.1 Hamiltonian Path in a 2D Mesh Network

The path-based routing algorithms are commonly based on the Hamiltonian path where an undirected Hamiltonian path is constructed in the network [10]. A Hamiltonian path visits every node in a graph exactly once. For each node in an $a \times b$ mesh network, a label $L(x, y)$ is assigned as:

$$L(x,y) = \left\{ \begin{array}{ll} (a \times y) + (x + 1) & \textit{if } y \textit{ is even} \\ (a \times y) + (a - x) & \textit{if } y \textit{ is odd} \end{array} \right\}$$

where x and y are the coordinates of the node. Figure 7.2a shows the labeling assignment based on this equation in a 3×4 mesh network.

As exhibited in Fig. 7.2b, c, two directed Hamiltonian paths (or two sub-networks) are constructed by labeling the nodes. The high channel sub-network (G_H) starts at 1 (Fig. 7.2b) and the low channel sub-network (G_L) ends at 1 (Fig. 7.2c). If the label of the destination node is greater than the label of the source node, the routing always takes place in the high channel sub-network; otherwise, it takes place in the low channel sub-network. The destinations are divided into two groups. One group contains all the destinations that could be reached using G_H and the other contains the remaining destinations that could be reached using G_L. To reduce the path length, the vertical channels that are not part of the Hamiltonian path (the dashed lines in Fig. 7.2) could be used in appropriate directions. In fact, all messages in the high channel (low channel) sub-network follow paths in the strictly

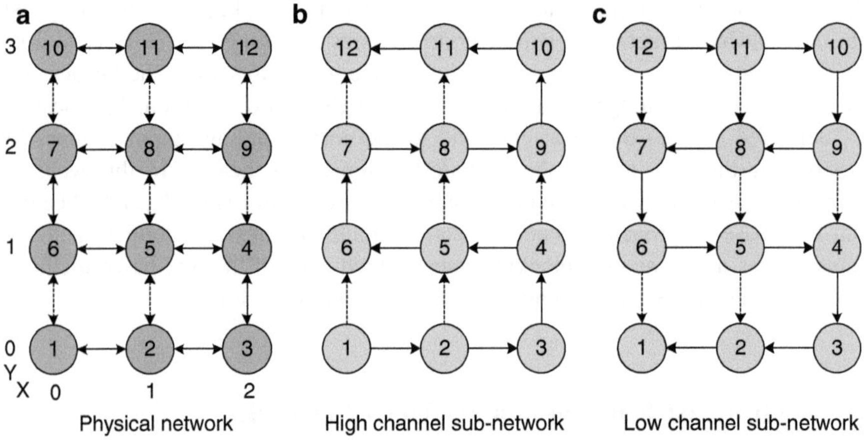

Fig. 7.2 (**a**) A 3 × 4 mesh physical network with the label assignment (**b**) high channel sub-network and (**c**) low channel sub-network [11]. *Solid lines* indicate the Hamiltonian path and *dashed lines* indicate the links that could be used to reduce the path length

ascending (descending) order (using either solid lines or dashed lines), no cyclic dependency can be formed among channels; thus the routing algorithm based on the Hamiltonian path is deadlock-free.

7.2.2 Dual-Path Partitioning

Dual-Path (DP) partitioning is a naïve method based on the Hamiltonian path with no effort toward reducing the path length. In DP, destinations are partitioned into two subsets [10]: G_H and G_L. Every node in G_H has a higher label than that of the source node and every node in G_L has a lower label than that of the source node. G_H and G_L are then sorted in the ascending order and descending order, respectively, as the label of each node is used as the key for sorting. Therefore, a multicast message from the source node will be sent to the destinations in G_H using the high channel sub-network and to the destinations in G_L using the low channel sub-network. Let us consider the example shown in Fig. 7.3 where the node 13 should send a multicast message to ten destinations as $m=(13,\{2,5,6,8,9,11,17,20,23,24\})$. Destinations are organized into two subsets. The first subset (G_H) contains all destinations that could be reached from the source node using the high sub-network as: $G_H=\{17,20,23,24\}$. The multicast message always traverses in the ascending order in the high channel sub-network or the descending order in the low channel sub-network. In this example, the traversed path by the message is $P_H=\{13,14,\underline{\mathbf{17}},18,19,\underline{\mathbf{20}},21,22,\underline{\mathbf{23}},\underline{\mathbf{24}}\}$. The second group has the remaining destinations that could be reached using the low channel sub-network $P_L=\{13,12,\underline{\mathbf{11}},10,\underline{\mathbf{9}},\underline{\mathbf{8}},7,\underline{\mathbf{6}},\underline{\mathbf{5}},4,3,\underline{\mathbf{2}}\}$.

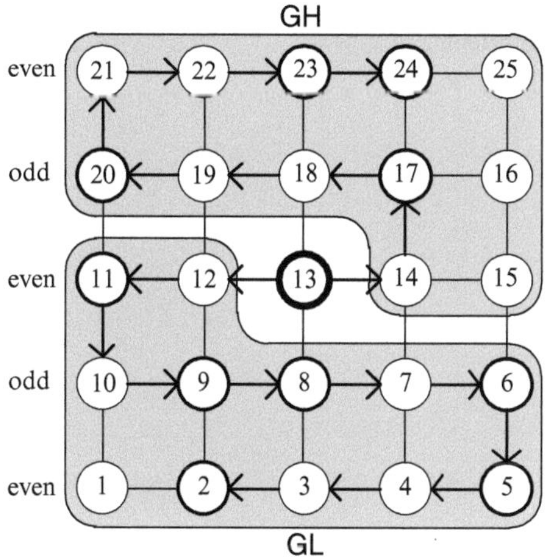

Fig. 7.3 Dual-path partitioning in a 2D mesh NoC

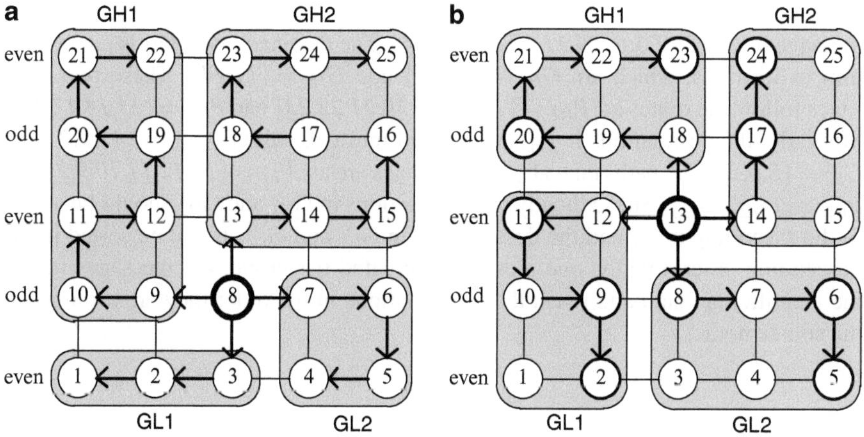

Fig. 7.4 Multi-path partitioning in a 2D mesh NoC

7.2.3 Multi-path Partitioning

To reduce the path length, in Multi-Path (MP) partitioning, G_H and G_L are also partitioned [10]. G_H is divided into two subsets (G_H: G_{H1},G_{H2}). If the source node is located in an odd row, G_{H1} covers the nodes whose x coordinates are smaller than that of the source node and the other subset, G_{H2}, contains the remaining nodes in G_H (Fig. 7.4a). If the source node is located in an even row, G_{H1} covers the nodes whose x coordinates are smaller than or equal to the source node and G_{H2} covers the rest of the destinations (Fig. 7.4b). G_L is partitioned in a similar way into two subsets (G_L: G_{L1},G_{L2}). Hence, all destinations of a multicast message are grouped into four disjointed sub-networks. An example of MP is illustrated in Fig. 7.4b for a multicast

Fig. 7.5 Column-path partitioning in a 2D mesh NoC

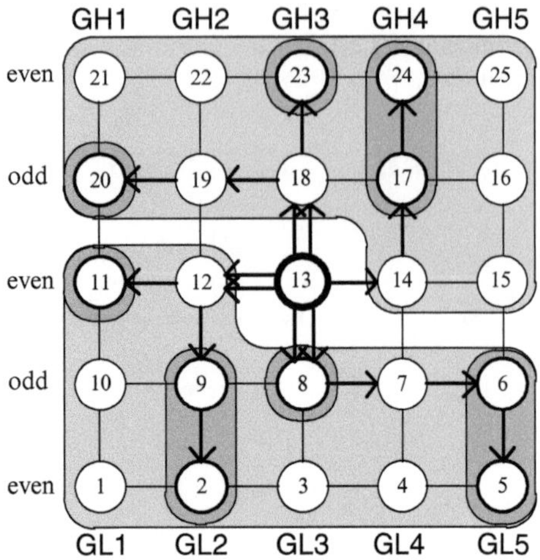

message $m = (13, \{2,5,6,8,9,11,17,20,23,24\})$. The destinations in G_H is partitioned into two groups, which are $G_{H1} = \{20,23\}$ and $G_{H2} = \{17,24\}$. These two messages follow the routes as: $P_{H1} = \{13,18,19,\underline{\mathbf{20}},21,22,\underline{\mathbf{23}}\}$ and $P_{H2} = \{13,14,\underline{\mathbf{17}},\underline{\mathbf{24}}\}$. Similarly, the destinations in G_L are divided into two subsets, $G_{L1} = \{11,9,2\}$ and $G_{L2} = \{8,6,5\}$. The paths taken by the messages are as: $P_{L1} = \{13,12,\underline{\mathbf{11}},10,\underline{\mathbf{9}},\underline{\mathbf{2}}\}$ and $P_{L2} = \{13,\underline{\mathbf{8}},7,\underline{\mathbf{6}},\underline{\mathbf{5}}\}$. Subsequently, all destinations in G_{H1} and G_{H2} should be sorted in the ascending order and the destinations in G_{L1} and G_{L2} should be sorted in the descending order. Finally, one message is created per group. All messages follow the Hamiltonian path and reach their destinations in the order they are arranged at the source node.

7.2.4 Column-Path Partitioning

In Column-Path (CP) partitioning, the destination nodes are partitioned into $2k$ subsets where k is the number of columns in the mesh. In this method, at most two messages will be sent to each column. Figure 7.5 shows an example where a multicast message is sent from the source node 13 to 10 destinations as $m = (13, \{2,5,6,8,9,11,17,20,23,24\})$. Seven copies of the message are used to achieve the desired multicast operation; the path taken by these messages are as follows: $P_{H1} = \{13,18,19,\underline{\mathbf{20}}\}$, $P_{H3} = \{13,18,\underline{\mathbf{23}}\}$, $P_{H4} = \{13,14,\underline{\mathbf{17}},\underline{\mathbf{24}}\}$, $P_{L1} = \{13,12,\underline{\mathbf{11}}\}$, $P_{L2} = \{13,12,\underline{\mathbf{9}},\underline{\mathbf{2}}\}$, $P_{L3} = \{13,\underline{\mathbf{8}}\}$, and $P_{L5} = \{13,8,7,\underline{\mathbf{6}},\underline{\mathbf{5}}\}$. Using the column-path partitioning, the path length can be reduced compared with dual-path and multi-path approaches. However, more messages should be delivered into the network (i.e. at most $2\,k$ copies of the message).

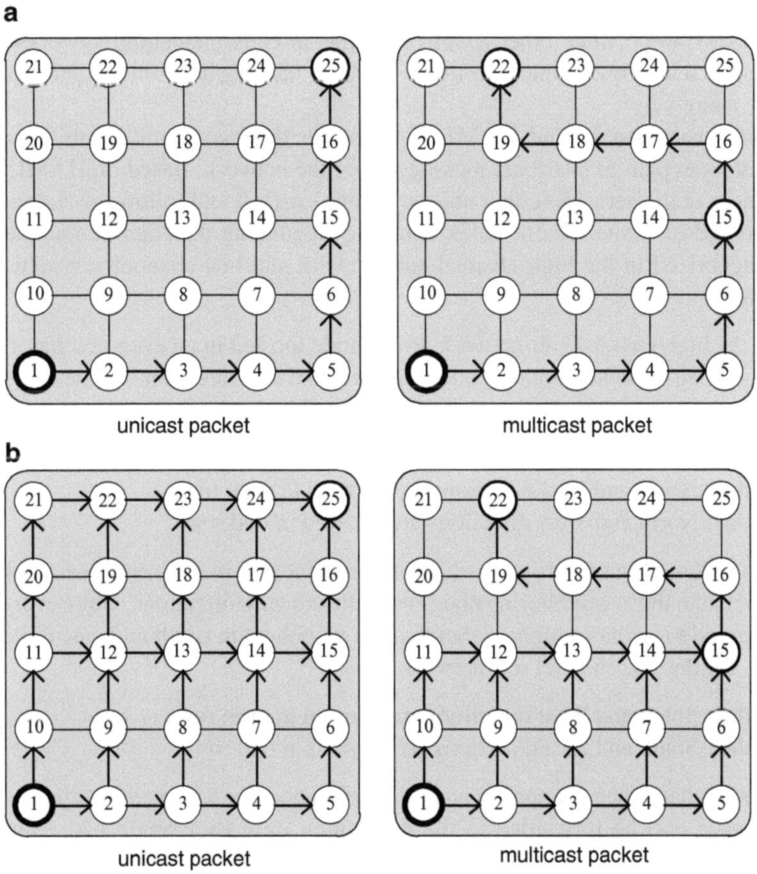

Fig. 7.6 Unicast and multicast routing (**a**) before applying HAMUM (**b**) after applying HAMUM

7.2.5 *Minimal Adaptive Routing*

All the mentioned partitioning methods are supported by a deterministic routing for both unicast and multicast traffic (Fig. 7.6a). Taking into account that the vast majority of traffic in many applications consists of unicast messages, the performance can be significantly affected by routing messages deterministically inside the network. In general, the purpose of supporting multicast traffic is to improve the performance. However, by using a deterministic algorithm, the performance is degraded due to the adaptivity restriction imposed to all unicast messages.

To improve the efficiency of the path-based method, an adaptive and deadlock-free method is presented to bring adaptivity for all partitioning methods (Fig. 7.6b).

We call this method, Hamiltonian Adaptive Multicast and Unicast Model (HAMUM) [9]. Unlike other adaptive models in communication networks which are applicable only for unicast traffic, HAMUM handles both unicast and multicast traffic adaptively.

The basic idea behind HAMUM is to use the maximum capability of the Hamiltonian path to distribute messages over the network. Based on HAMUM, all messages in the network (either unicast or multicast) should follow the Hamiltonian path based on some specific rules. The rules regulating the Hamiltonian path can be categorized in the high channel sub-network and low channel sub-network as follows:

- In the high channel sub-network, every node located in an even row has a lower label than those neighboring nodes in north and east directions; while every node in an odd row has a lower label than its neighbors in north and west directions. Therefore, for the high channel sub-network:

  Rule1: North and East directions are allowed in even rows.
  Rule2: North and West directions are allowed in odd rows.

- In the low channel sub-network, every node located in an even row has a higher label than those neighboring nodes in south and west directions; while every node in an odd row has a higher label than its neighbors in south and east directions. So, for the low channel sub-network:

  Rule1: South and West directions are allowed in even rows.
  Rule2: South and East directions are allowed in odd rows.

Notice that, when the current and destination nodes are located in adjacent rows, a message will be forwarded to the destination through a single path. Since the rules keep the messages traversing in the strictly ascending order (high channel sub-network) or descending order (low channel sub-network), it prevents the occurrence of deadlock. Figure 7.7 shows the pseudo code of the HAMUM algorithm which is executed in each node when a new message arrives. It is worth mentioning that the partitioning methods and the HAMUM algorithm are performed at run time.

As already mentioned, using HAMUM, both unicast and multicast messages can be routed adaptively inside the network. As shown in Fig. 7.8a, two unicast messages generated at the source nodes 2 and 3 can reach the destination nodes 21 and 25 through different routes. Figure 7.8b shows two examples of multicast messages as $m = (2,\{20,22\})$ and $m = (3,\{15,25\})$. As can be seen in these examples, messages can be routed adaptively between each two successive destinations. As shown in Fig. 7.8a, b all paths are compatible with Rule1 and Rule2.

In HAMUM, when there is more than one option to deliver a message, the stress values of the input buffers in the corresponding directions are examined and the message is sent to a less congested direction.

```
ALGORITHM: The HAMUM routing algorithm
                    **------------------------------------------**
Definitions:     (Xc,Yc): X and Y coordinates of the current node
                 (Xd,Yd): X and Y coordinates of the destination node
                    **------------------------------------------**
If (Yd = Yc) then
      If (Xd = Xc) then
            direction <= Local;
      Elsif (Xd > Xc) then
            direction <= East;
      Else  direction <= West;
      End if;
Elsif (Yd > Yc) then        – High Channel
      If (Yc mod 2 = 0) then     -- even rows
            If (Xd>Xc) and (Yd - Yc > 1) then
                  direction <= North or East;
            Elsif (Xd>Xc) and (Yd - Yc = 1) then
                  direction <= East;
            Else  direction <= North;
            End if;
      Elsif (Yc mod 2 /= 0 ) then  -- odd rows
            If (Xd<Xc) and (Yd - Yc > 1) then
                  direction <= North or West;
            Elsif (Xd<Xc) and (Yd - Yc = 1) then
                  direction <= West;
            Else  direction <= North;
            End if;
      End if;
Elsif (Yd < Yc) then        – Low Channel
      If (Yc mod 2 = 0) then      -- even rows
            If (Xd < Xc) and (Yc - Yd > 1) then
                  direction <= South or West;
            Elsif (Xd < Xc) and (Yc - Yd = 1) then
                  direction <= West;
            Else  direction <= South;
            End If;
      Elsif (Yc mod 2 /= 0) then  -- odd rows
            If (Xd > Xc) and (Yc - Yd > 1) then
                  direction <= South or East;
            Elsif (Xd > Xc) and (Yc - Yd = 1) then
                  direction <= East;
            Else direction <= South;
            End if;
      End if;
End if;
```

Fig. 7.7 The pseudo code of HAMUM

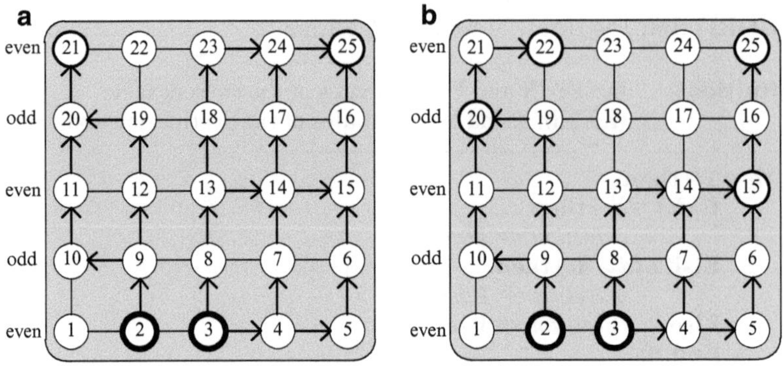

Fig. 7.8 An example of HAMUM for (**a**) unicast messages (**b**) multicast messages

Fig. 7.9 All possible shortest paths from the source nodes 1, 8, 57, and 64 to the destination node 28 using HAMUM

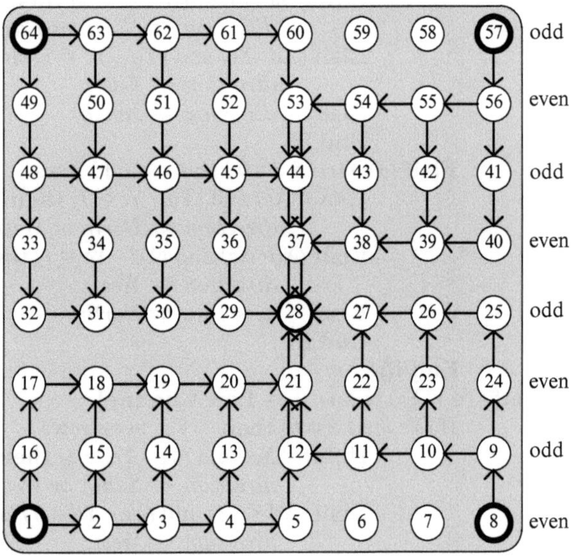

7.2.6 *Adaptiveness of HAMUM*

Since the Odd-Even model [1] is one of the most popular adaptive routing algorithms for unicast communication, we compare the adaptivity of HAMUM with the Odd-Even model. Figure 7.9 shows all possible shortest paths based on HAMUM taken by four messages in an 8×8 mesh network. All of the possible routing paths for the Odd-Even model are indicated in Fig. 7.10.

In order to compare these two algorithms with each other, we use the Degree of Adaptiveness (DoA) factor [1, 11], which is the number of minimal paths that can be taken by a message to traverse from a source node (X_s,Y_s) to a destination node (X_d,Y_d). Assuming that $\Delta x = X_d - X_s$ and $\Delta y = Y_d - Y_s$, the number of hops between

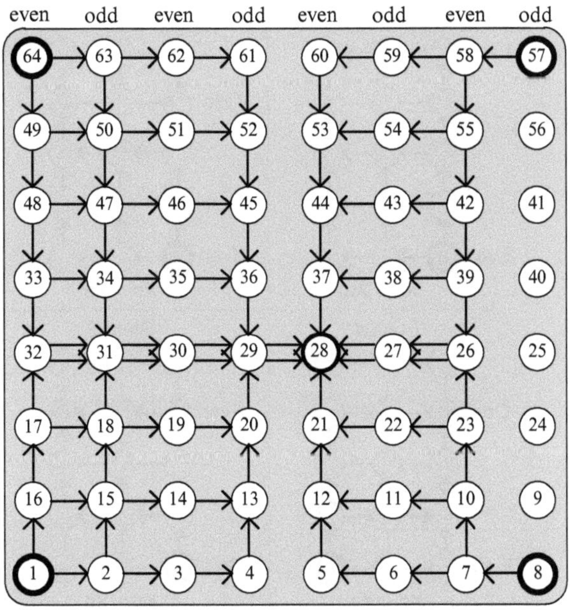

Fig. 7.10 All possible shortest paths from the source nodes 1, 8, 57, and 64 to the destination node 28 using odd-even

Table 7.1 Eight different states regarding the source and destination positions

State	Source row (odd/even)	Destination row (odd/even row)	Destination position (left/right)
1	Even	Even	Right
2	Even	Odd	Right
3	Even	Even	Left
4	Even	Odd	Left
5	Odd	Even	Right
6	Odd	Odd	Right
7	Odd	Even	Left
8	Odd	Odd	Left

the source and destination nodes are as: $d_x = |\Delta x|$, and $d_y = |\Delta y|$. The degree of adaptiveness for a fully adaptive algorithm is given by:

$$DoA\ (fully\ adaptive\ routing)_{s,d} = \frac{(d_x + d_y)!}{d_x!d_y!}$$

Based on the Hamiltonian Path, there can be eight different states considering the source row (which can be even or odd), the destination row (that can be odd or even), and the location of the destination node regarding the source node (left or right side of the source node). These states are summarized in Table 7.1. Note that when the source and destination nodes are located in the same dimension, only a single path exists.

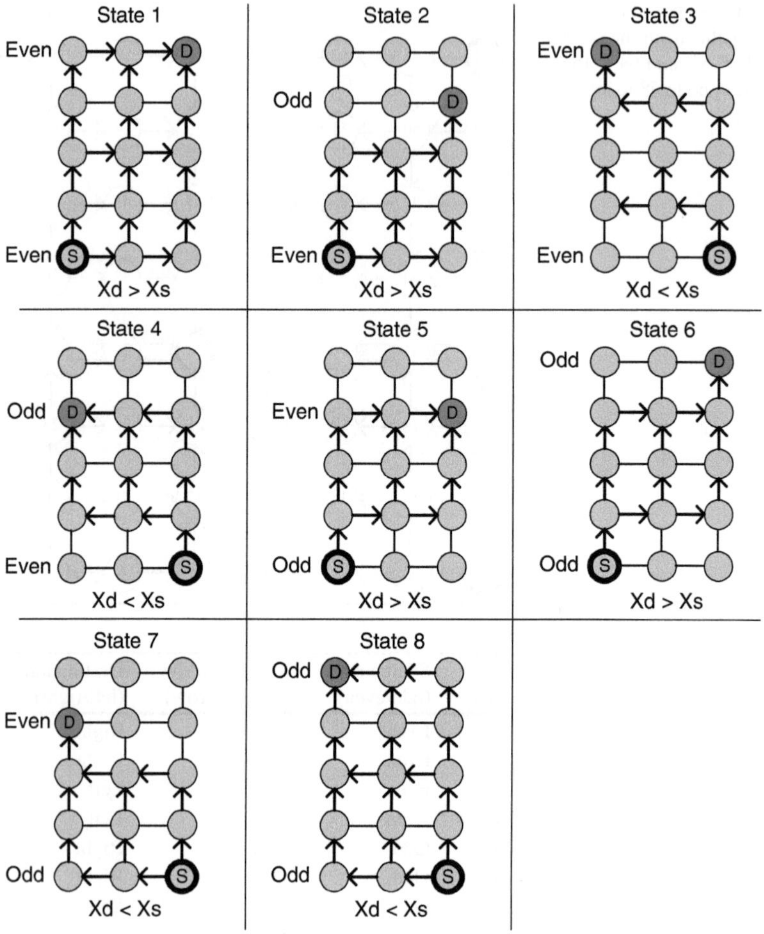

Fig. 7.11 Eight different states in the high channel sub-network

We compute the degree of adaptiveness for HAMUM in the high channel sub-network. Similar method can be applied to the low channel sub-network. As can be seen in Fig. 7.11, the degree of adaptiveness for the state 1 and 8 is equal and can be computed as:

$$DoA\ (1)_{s,d} = \frac{(d_x + D)!}{d_x!D!} \quad \text{Where } D = \left\lceil \left| \frac{d_y}{2} \right| \right\rceil$$

DoA of the other states can be calculated by:

$$DoA\ (2)_{s,d} = \frac{(d_x + D')!}{d_x!D'!} \quad \text{Where } D' = \left\lfloor \left| \frac{d_y - 1}{2} \right| \right\rfloor$$

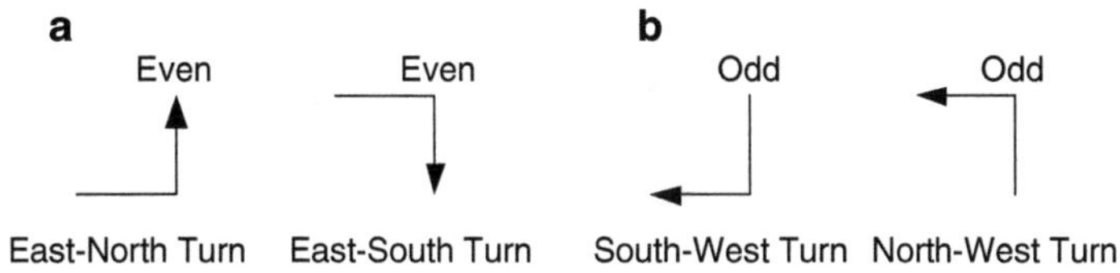

Fig. 7.12 The rules of odd-even turn model; prohibited turns in (**a**) even columns (**b**) odd columns

These equations can be summarized as:

$$DoA(HighChannel)_{s,d} = \begin{cases} DoA(1)_{s,d} & for\ states\ 1\ and\ 8 \\ DoA(2)_{s,d} & otherwise \end{cases}$$

Similarly, in the low channel sub-network, the degree of adaptiveness in two states is DoA(1) while in the other states is DoA(2).

The Odd-Even turn model restricts the locations where some types of turns can be taken. The rules of Odd-Even are applied to columns rather than rows as in HAMUM. Odd-Even rules can be described as follow:

Rule 1: East-North and East-South turns cannot be taken in even columns (Fig. 7.12a).

Rule 2: North-West and South-West turns cannot be taken in odd columns (Fig. 7.12b).

When the destination node is in the right side of the source node ($\Delta x > 0$), the degree of adaptiveness for the Odd-Even turn model is computed by [1]:

$$DoA(\Delta x > 0)_{s,d} = \begin{cases} DoA(2)_{s,d} & source : even\ and\ destination : odd \\ DoA(1)_{s,d} & otherwise \end{cases}$$

When the destination node is to the left side of the source node ($\Delta x < 0$) the following equation is obtained:

$$DoA(\Delta x < 0)_{s,d} = \begin{cases} DoA(2)_{s,d} & source : odd\ and\ destination : odd \\ DoA(1)_{s,d} & otherwise \end{cases}$$

Considering the above analysis, the degree of adaptiveness of HAMUM and Odd-Even is close to each other. However, the Odd-Even model is designed for unicast communication and cannot be utilized for multicast traffic. On the other hand, HAMUM not only is compatible with multicast traffic but also provides adaptivity for both unicast and multicast messages.

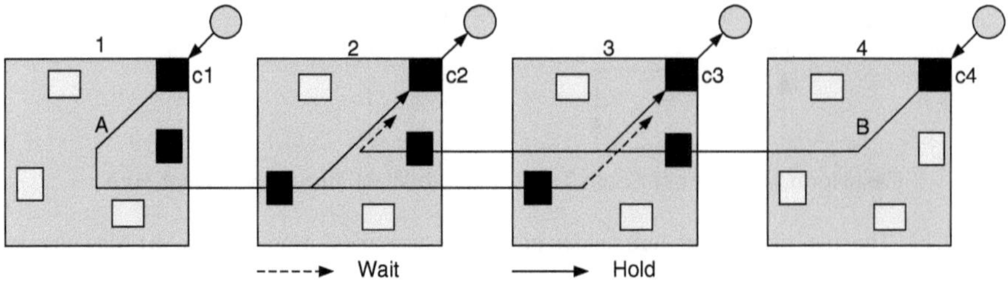

Fig. 7.13 Deadlock due to the delivery channel dependency [10]

7.2.7 Delivery Channels

In the path-based multicast mechanism, when multiple delivery channels (consumption channels) are occupied by one message along the multicast path, cyclic dependencies on the delivery channels may occur [8, 11]. In fact, in the proposed path-based multicast approaches, a message cannot be delivered to two different output channels simultaneously, unless the message should be sent to a consumption channel and an output channel. As illustrated in Fig. 7.13, the multicast message *A* is generated at the node 1 destined to nodes 2 and 3. Simultaneously, the node 4 generates the message *B* destined to the same destinations. The message *A* receives a local channel at the node 2 and waits to receive a channel at the node 3. On the other hand, the message B receives the local channel at the node 3 and waits to receive a channel at the node 2. As a result, due to the delivery channel dependency, this cycle cannot be removed and thus creates a deadlock. To prevent deadlock in delivery channels, the required number of delivery channels to avoid such deadlock is equal to $2nv$ where n is the network dimension and v is the number of virtual channels per input port [8]. As a result, at least two delivery channels are necessary and sufficient for DP, MP, and CP partitioning methods.

7.3 Multicast Routing in a 3D Mesh Network

In this section, we extend the idea of partitioning methods and the adaptive routing algorithm to a 3D mesh network. At first, different partitioning schemes are introduced, called Dual-Path (DP), Vertical-Path (VP), and Recursive partitioning (RP) methods, and then the minimal and adaptive routing algorithm (MAR) is proposed [12].

7.3.1 Hamiltonian Path in a 3D Mesh Network

Several Hamiltonian paths can be considered in a mesh topology. In an $a \times b \times c$ mesh network, each node is labeled by an integer value according to the x, y, and z

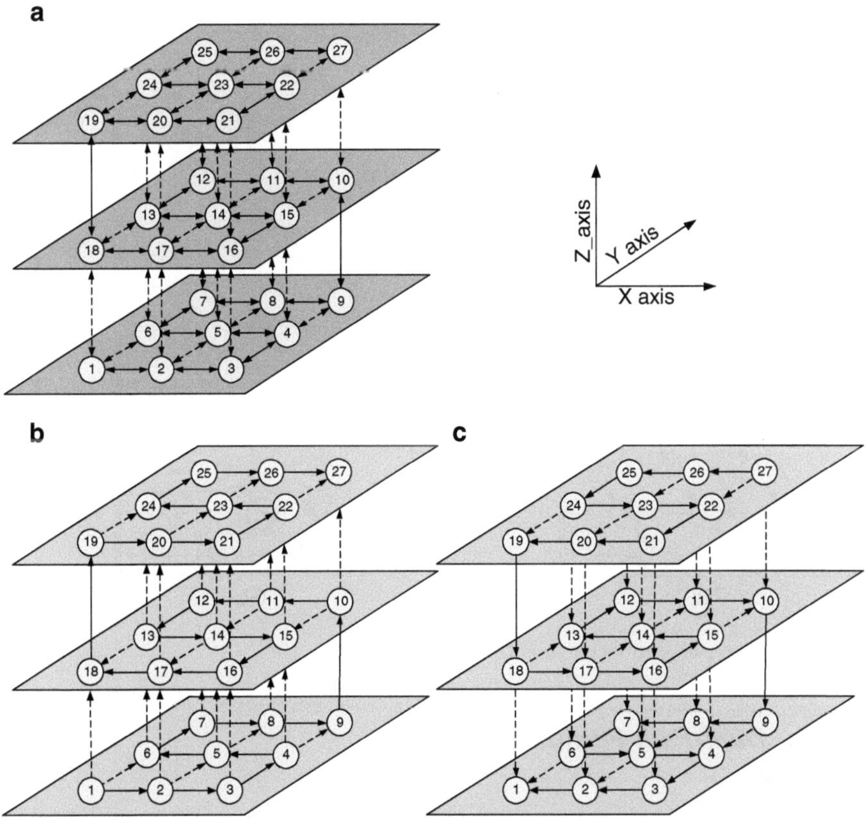

Fig. 7.14 (**a**) A $3 \times 3 \times 3$ mesh network (**b**) high channel sub-network (**c**) low channel sub-network

coordinates. The following equations show one possibility of assigning the labels, which we utilize:

$$L(x,y,z) = \{(a \times b \times z) + (a \times y) + (x+1)\} \qquad \text{where } z: even, y: even$$
$$L(x,y,z) = \{(a \times b \times z) + (a \times y) + (a-x)\} \qquad \text{where } z: even, y: odd$$
$$L(x,y,z) = \{(a \times b \times z) + (a \times (b-y-1)) + (a-x)\} \qquad \text{where } z: odd, y: even$$
$$L(x,y,z) = \{(a \times b \times z) + (a \times (b-y-1)) + (x+1)\} \qquad \text{where } z: odd, y: odd$$

The labels of nodes based on this equation are shown in Fig. 7.14a. As exhibited in Fig. 7.14b, c, two directed Hamiltonian paths (or two sub-networks) are constructed by this labeling, similar to a 2D mesh network. In a case that the label of the destination node is greater than the label of the source node, the routing always takes place in the high channel sub-network; otherwise in the low channel sub-network.

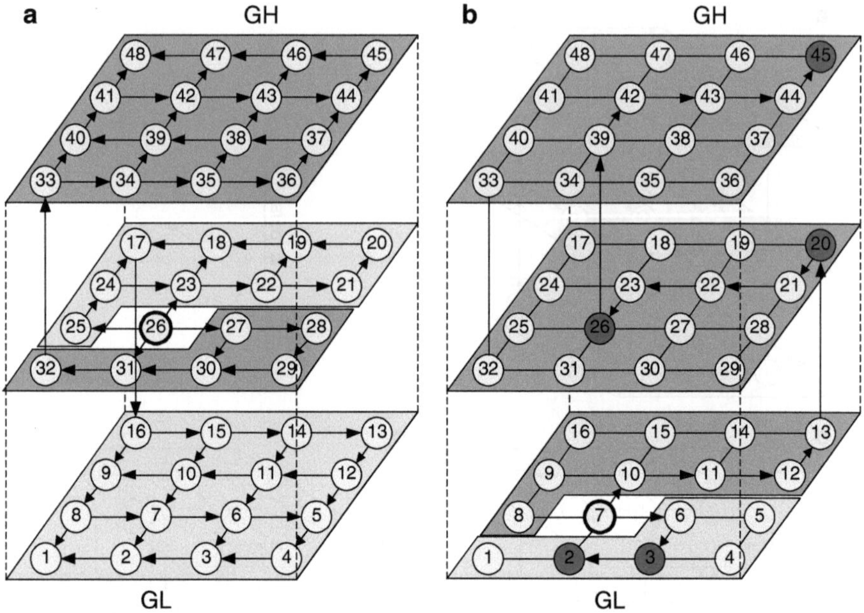

Fig. 7.15 Dual-path partitioning in a 3D stacked mesh NoC

7.3.2 Dual-Path Partitioning

It is a straight forward extension of Dual-Path partitioning (DP) from 2D to 3D NoCs and it can be seen as a naïve method since no effort is made to balance the two sets. This method is used as a reference method for the comparison purposes. Figure 7.15 shows an example of the DP partitioning approach and the portion of each partition that depends on the source node position. As illustrated in Fig. 7.15a, if the source node is located at the middle layer, two partitions cover comparable number of nodes but still with a large number of nodes in both partitions. However, in Fig. 7.15b, one partition contains considerably more nodes than the other one. Now, suppose that the multicast message $m = (7,\{2,3,20,26,45\})$ is generated at the node *7*. Destination nodes are split into two sets and should be visited accordingly to their labels: $G_H = \{20,26,45\}$ and $G_L = \{3,2\}$. The message created for the high channel sub-network uses the Hamiltonian path as follows: $P_H = \{7,10,11,12,13,\underline{\mathbf{20}},21,22,23,\underline{\mathbf{26}},39,42,43,44,\underline{\mathbf{45}}\}$ where fourteen hops are needed to reach the last destination. The message path for G_L is: $P_L = \{7,6,\underline{\mathbf{3}},\underline{\mathbf{2}}\}$ where three hops are required for delivering the message to all destinations in the low channel sub-network. In the DP approach, the number of startup messages is low and never gets larger than two. However, it suffers from high network latency due to unbalanced partitions and high probability of long paths within the network.

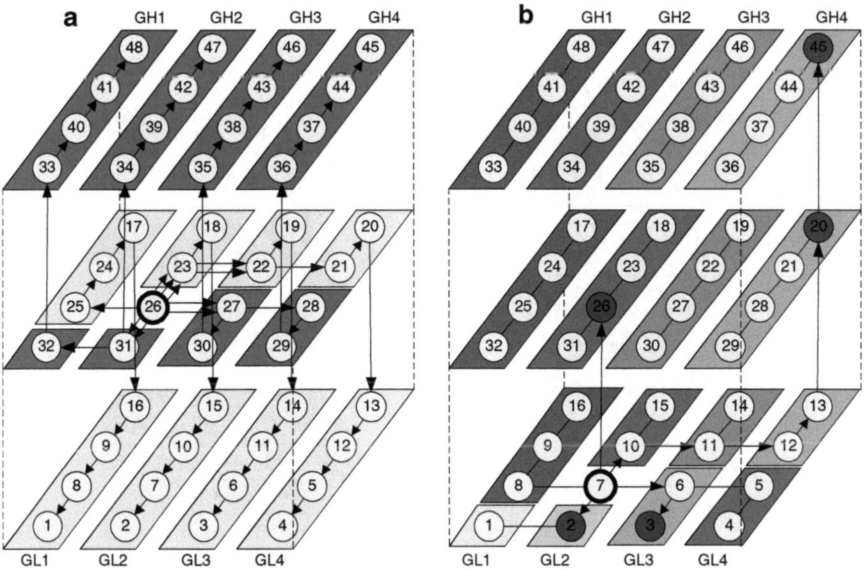

Fig. 7.16 Vertical-path partitioning in a 3D stacked mesh NoC

7.3.3 *Vertical-Path Partitioning*

In Vertical-Path partitioning (VP), similar to Dual-Path partitioning, the network is divided into high and low channel sub-networks. In an additional step, each sub-network is vertically partitioned such that each column (nodes with the same *a* value in an $a \times b \times c$ network) form a new set. As illustrated in Fig. 7.16, this scheme does not guarantee balanced partitions. For the node located at *26*, partitions are balanced, but they are not balanced when the source node is located at *7* (i.e. four sub-networks cover more nodes than the others). Moreover, the time required to prepare at most *8* messages is considered as the number of startup messages. For the multicast message $m = (7, \{2,3,20,26,45\})$, four sets are formed: $G_{H2} = \{26\}$, $G_{H4} = \{20,45\}$, $G_{L2} = \{2\}$, and $G_{L3} = \{3\}$. One message is generated for each set and paths are $P_{H2} = \{7,\underline{\mathbf{26}}\}$, $P_{H4} = \{7,10,11,12,13,\underline{\mathbf{20}},\underline{\mathbf{45}}\}$, $P_{L2} = \{7,\underline{\mathbf{2}}\}$, and $P_{L3} = \{7,6,\underline{\mathbf{3}}\}$ where the maximum hop count is six.

This scheme has several advantages over the DP method as it achieves a high level of parallelism; avoids the creation of long paths and reduces the network latency. The VP approach increases, however, the number of startup messages as it requires up to *2a* messages in an $a \times b \times c$ network. In addition, this scheme does not guarantee balanced partitions as it is balanced when the source node is located in middle layers while some partitions may cover considerably more nodes than the others when the source node is located at the top or bottom layer.

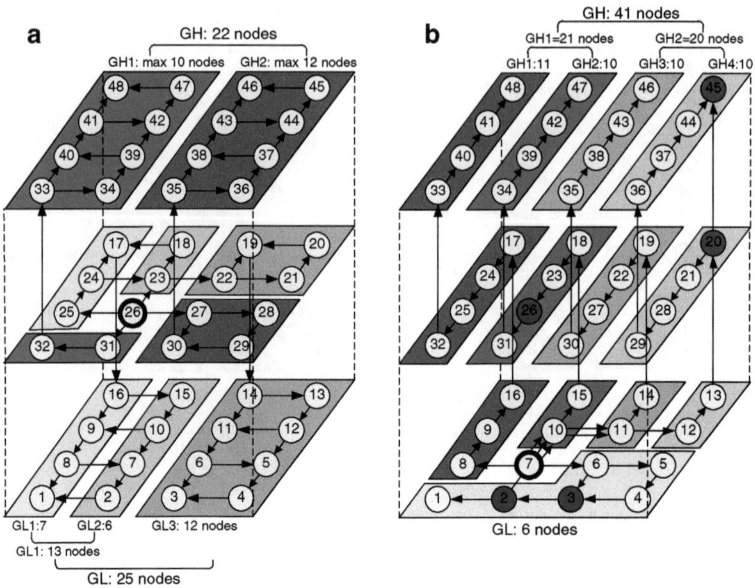

Fig. 7.17 Vertical-path partitioning in a 3D stacked mesh NoC

7.3.4 Recursive Partitioning (RP)

The objective of the Recursive Partitioning (RP) method is to optimize the number of nodes that can be included in a partition and thus to achieve a better parallelism. In this method, the network is recursively partitioned until each partition contains k nodes. In the worst case, the network is partitioned into $2a$ vertical partitions like in the VP method. We have considered the value k as a reference value indicating the number of nodes in each partition of the VP method, i.e. ($k = bc$) in an $a \times b \times c$ network. An example of the RP approach is illustrated in Fig. 7.17a where a multicast message is generated at the source node 26. The required steps of the RP method can be expressed as follows:

Step1: The value k is set to 12 in a $4 \times 4 \times 3$ network.

Step2: The network is divided into two partitions using the DP method. Figure 7.15a shows two formed partitions when the source node is located at 26.

Step3: If the number of nodes in a partition exceeds the reference value k, the partition is divided into two new partitions. This step is repeated until all partitions cover at most k nodes. Following the example of Fig. 7.15a, 22 nodes are covered by the high channel sub-network which is greater than $k = 12$. The high channel sub-network needs to be further divided into two new partitions (G_{H1} and G_{H2} as shown in Fig. 7.17a). The G_{H1} and G_{H2} partitions contain 10 and 12 nodes, respectively. Since both numbers are less than or equal to $k = 12$, no further partitioning is required for the high channel sub-network. The same partitioning technique is applied to the low channel sub-network.

Figure 7.17b shows another example of the RP method where the multicast message is $m = (7, \{2,3,20,26,45\})$. In this example, three messages are formed and their paths are $P_{H2} = \{7,10,11,12,13,\underline{20},\underline{45}\}$, $P_{H4} = \{7,\underline{26}\}$, and $P_L = \{7,6,\underline{3},\underline{2}\}$, and the maximum latency is six hops. In brief, this scheme has a similar performance in avoiding long paths as the VP method while it provides better parallelism since the number of nodes is comparable among partitions. By considering the RP approach, the creation of balanced partitions is less dependent of the source node position. Therefore, RP is able to avoid long paths in the network and increases the parallelism while keeping the number of startup messages relatively low.

7.3.5 Minimal Adaptive Routing

We present a minimal and adaptive routing algorithm (MAR) based on the Hamiltonian path. Using MAR, unicast and multicast messages can be adaptively routed inside the network. All routes used by the unicast messages are the shortest paths. Although the overall path of a multicast message might be non-minimal, the paths between each two destinations in the overall multicast path are minimal. Each node in the graph has a label (L) determined by the Hamiltonian path labeling mechanism. The MAR algorithm is implemented at the routing units and can be described in three steps as follows:

Step1: it determines the neighbors of the node u that can be used to move a message closer to its destination d. The pseudo code for Step1 is shown in Fig. 7.18.

Step2: due to the fact that in the Hamiltonian path all nodes are visited in the ascending order (high channel sub-network) or descending order (low channel sub-network), all of the selected neighbors in Step1 do not necessarily satisfy the ordering constraint. Therefore, if the labels of the selected neighbors (in Step1) are between the label of the node u and destination d, it/they can be selected as the next hop. The pseudo code for Step2 is shown in Fig. 7.18.

Step3: since the MAR algorithm provides several choices at each node, the goal of Step3 is to route a message through the less congested neighboring node. So, in the case when the message can be forwarded through multiple neighboring nodes, the congestion values of the corresponding input buffers of the candidate neighbors are checked and then the message is sent to the neighbor with the smallest stress value.

An example of the MAR algorithm is illustrated in Fig. 7.19a where the source and destination are located at the nodes 6 and 48, respectively. According to the algorithm, in the first step the neighbors are chosen in a manner that gets the message closer to its destination, i.e. $n = \{7,11,27\}$. In the second step, the selected neighbors (in Step1) are checked to determine whether they are in the Hamiltonian path or not. Since the labels of the three selected neighbors are between the labels of the current node $(u = 6)$ and the destination node $(d = 48)$, the message can be routed via each of them. Suppose that the neighbor $p = 11$ has a smallest congestion value, so the algorithm selects this neighbor to forward the message. If we continue with the node

```
ALGORITHM:  Minimal Adaptive Routing (MAR_3D)
                    **-------------------------------------------**
Definitions:  (Xc,Yc,Zc): X,Y, and Z coordinates of the current node
              (Xd,Yd,Zd): X,Y, and Z coordinates of the destination node
                    **-------------------------------------------**
------ STEP 1 ------
X_dir = East when (Xc<Xd) else West;
Y_dir = North when (Yc<Yd) else South;
Z_dir = High when (Zc<Zd) else Low;
------ STEP 2 ------
if ((Label(CurrentNode) = Label(DestNode)) then Choice<=Local;

                    ----------High Channel --------
elsif ((Label(CurrentNode) < Label(DestNode)) then
        if (Label(CurrentNode) < Label(Neighbor(X_dir))) and
           (Label(Neighbor(X_dir)) < Label(DestNode)) then
             First Choice <= Neighbor(X_dir)  end if;
        if (Label(CurrentNode) < Label(Neighbor(Y_dir))) and
           (Label(Neighbor(Y_dir)) < Label(DestNode)) then
             Second Choice <= Neighbor(Y_dir) end if;
        if (Label(CurrentNode) < Label(Neighbor(Z_dir))) and
           (Label(Neighbor(Z_dir)) < Label(DestNode)) then
             Third Choice <= Neighbor(Z_dir)  end if;

                    ----------Low Channel ---------
elsif ((Label(CurrentNode) >Label(DestNode)) then
        if (Label(CurrentNode) > Label(Neighbor(X_dir))) and
           (Label(Neighbor(X_dir)) > Label(DestNode)) then
             First Choice <= Neighbor(X_dir) end if;
        if (Label(CurrentNode) > Label(Neighbor(Y_dir))) and
           (Label(Neighbor(Y_dir)) > Label(DestNode)) then
             Second Choice <= Neighbor(Y_dir) end if;
        if (Label(CurrentNode) > Label(Neighbor(Z_dir))) and
           (Label(Neighbor(Z_dir)) > Label(DestNode)) then
             Third Choice <= Neighbor(Z_dir) end if;
end if;
```

Fig. 7.18 The pseudo code of the MAR algorithm

$u = 11$, this node has three neighboring nodes belonging to the minimal paths, i.e. $n = \{10,14,22\}$. However, only two of them $(n = \{14,22\})$ have the labels greater than the label of the current node $(u = 11)$ and lower than the label of the destination node $(d = 48)$.

Finally, according to the stress values of the input buffers, one of them is selected as the next hop. The algorithm is repeated for the rest of the nodes until the message reaches the final destination. It is worth noting that the stress value is updated whenever a new flit enters or leaves the buffer (flit events: flit_tx or flit_rx). That is, in each flit event, if the number of occupied cells of the input buffer is larger (smaller) than a threshold value, the threshold signal is assigned to one (zero).

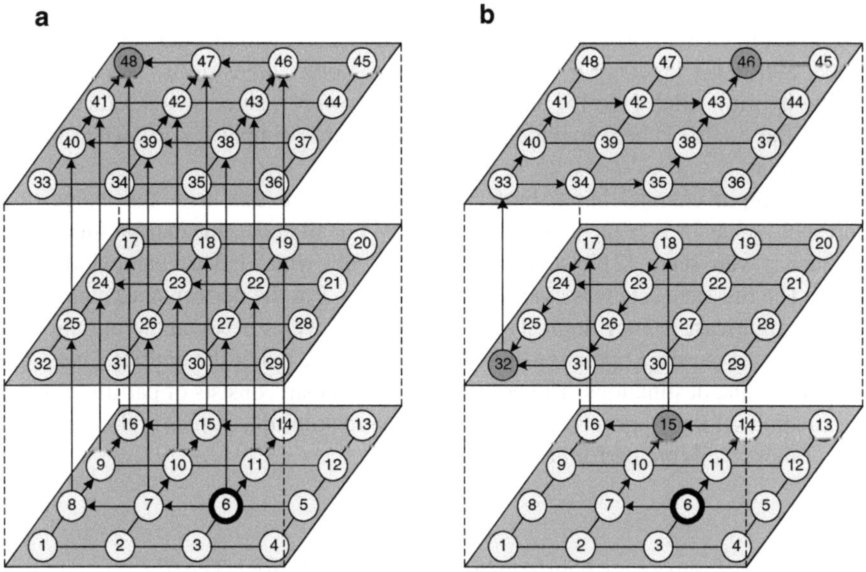

Fig. 7.19 The MAR algorithm (**a**) for a unicast message (**b**) for a multicast message

The MAR algorithm can be adapted for multicast messages such that alternative paths are used to route a message between the source node and the first destination and also between successive destinations.

An example of MAR is shown in Fig. 7.19b where the source node ($u = 6$) forwards a multicast message towards three destinations ($D = \{15,32,46\}$). The MAR algorithm provides a set of alternative paths to send a message from the source node to the first destination ($d1 = 15$). Similarly, the message can be adaptively routed between each two destinations. For example, at the node *15*, the message can make progress towards the destination *32* either by selecting the node *18* in the next layer or the node *16* in the current layer. MAR is compatible with all methods supporting the Hamiltonian path. Therefore, all partitioning methods can utilize the MAR algorithm for both unicast and multicast messages.

7.4 Results and Discussion

Dual-Path (DP), Multi-Path (MP), and Column-Path (CP) employing the HAMUM routing algorithm are implemented in a 2D mesh network. In a 3D mesh network, Dual-Path (DP), Vertical-Path (VP), and Recursive Partitioning (RP) methods along with the MAR routing algorithm are implemented. We have developed a cycle accurate wormhole-based 2D and 3D NoC simulator. The simulator inputs include the array size, the operating frequency, the routing algorithm, the link width, and the traffic type. Each switch in a 3D mesh network has 7 input/output ports, a natural extension from a 5-port 2D switch by adding two ports to make connections to the upper and lower layers [13, 14]. There are some other types of 3D switches such as the hybrid switch [13, 15] and MIRA [16], however, since switch efficiency is out of the goals of these methods, we have chosen a simple 7-port switch in our simulation.

Each input channel has a buffer (FIFO) size of 8 flits with the congestion threshold set at 75 % of the total buffer capacity. For all nodes, the data width and the frequency were set to 32 bits and 1GHz, respectively, which led to a bandwidth of 32 Gb/s. The message size was assumed to be 16 flits. For the performance metric, we use the multicast latency defined as the number of cycles between the initiation of the multicast message operation and the time when the tail of the multicast message reaches all the destinations. The preparation mechanism consists of partitioning the destination set into appropriate subsets and creating multiple copies of the message. All of these steps are performed at runtime.

7.4.1 Performance Evaluation Under Multicast Traffic

The first sets of simulations were done for a random traffic profile. In these simulations, each processing element generates data messages and injects them into the network according to the time intervals which are obtained using the exponential distribution. The mesh sizes are considered 8×8 and $4 \times 4 \times 4$ in a 2D and 3D network, respectively. In the multicast traffic profile, each processing element sends a message to a set of destinations. A uniform distribution was used to construct the destination set of each multicast message. The number of destinations was set to 10.

The average communication delay as a function of the average flit injection rate in a 2D network is shown in Fig. 7.20a. According to the results, the proposed CP multicast routing algorithm leads to the lowest latency among the two other multicast approaches (DP and MP) and a simple unicast-based multicast support (UB). Figure 7.21a shows the performance gain of using HAMUM. As observed from the results, the performance of the MP and CP schemes increases when applying HAMUM (AMP and ACP). This is due to the fact that HAMUM brings adaptivity to all unicast and multicast messages.

As can be seen from the results in Fig. 7.20b, the RP method meets lower delay than the DP and VP methods. The foremost reason for this performance gain is due to the efficiency of the RP method which not only reduces the number of hops for multicast messages but also the number of startup messages. In fact, the DP approach suffers from long paths while the performance of the VP method degrades due to a large number of startup messages. Adaptive routing algorithms obtain better performance in congested networks due to using alternative paths. In Fig. 7.21b, ARP (Adaptive RP), utilizing MAR in RP, and AVP (Adaptive VP), utilizing MAR in VP, are the adaptive models of the RP and VP, respectively. Results show that adaptive routings become more advantageous when the injection rate increases.

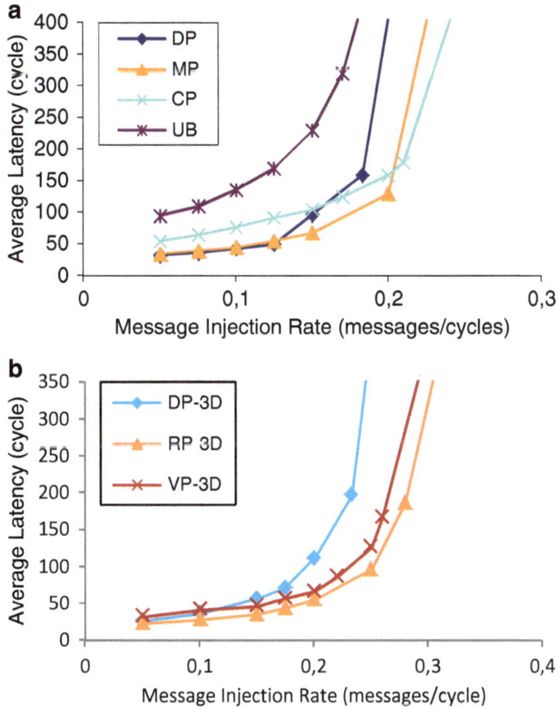

Fig. 7.20 Performance evaluation under multicast traffic model with ten destinations using deterministic routing (**a**) 8 × 8 mesh network (**b**) 4 × 4 × 4 mesh network

7.4.2 *Performance Evaluation Under Unicast and Multicast (Mixed) Traffic Profiles*

In these simulations, we employed a mixture of unicast and multicast traffic where 80 % of the injected messages are unicast messages and the remaining 20 % are multicast messages. This pattern may represent the traffic in a distributed shared-memory multiprocessor where updates and invalidation produce multicast messages and cache misses are served by unicast messages. In this experiment, the simulation parameters were similar to the previous simulations in terms of the number of destinations and array sizes. The unicast messages are also routed using the deterministic routing algorithm following the Hamiltonian path. Uniform traffic profile is considered for the unicast messages. In the uniform traffic profile, each processing element sends a message to any other processing element with an equal probability. Figure 7.22a shows the average communication latencies versus the message injection rate under the uniform traffic model for unicast messages. As these figures reveal, under this traffic profile, CP outperforms the other algorithms. As depicted in Fig. 7.23a, using HAMUM to deliver messages in MP and CP partitioning methods leads to a better performance. Figure 7.22b shows the performance gain of the RP approach over the other partitioning methods. The performance of RP can be improved by employing the MAR routing algorithm as shown in Fig. 7.23b.

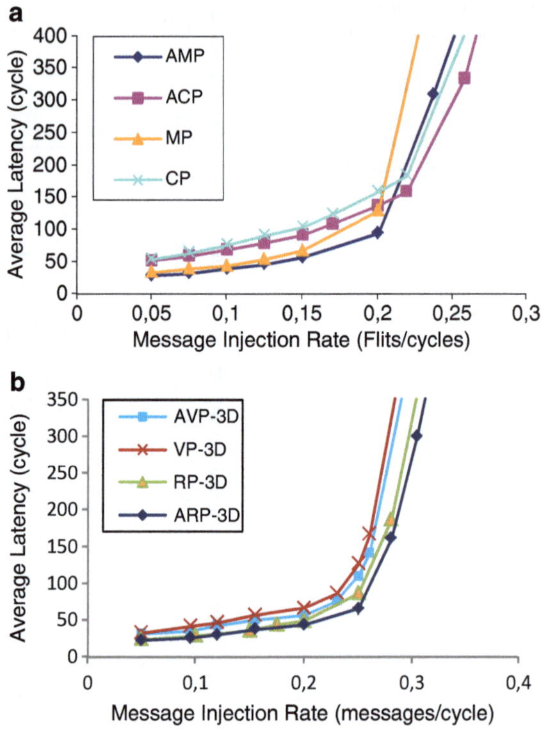

Fig. 7.21 Performance evaluation under multicast traffic model with ten destinations (**a**) 8 × 8 mesh network utilizing HAMUM (**b**) 4 × 4 × 4 mesh network utilizing MAR

7.4.3 *Performance Evaluation Under Application Traffic Profile*

To show the efficiency of the proposed model under the application traffic profiles, traces from some benchmark suites selected from SPLASH-2 [17] and PARSEC [18] are used. Traces are generated from SPLASH-2 and PARSEC using the GEMS simulator [19]. We used the X264 application of PARSEC and the Radix, Ocean, and fft applications from SPALSH-2 for our simulation.

Table 7.2 summarizes the full system configuration where the cache coherence protocol is token-based MOESI and access latency to the L2 cache is derived from the CACTI [20]. It is noteworthy that the token-based MOESI protocol is heavily based on multicast traffic. On account of our analysis on average 80 % of traffic in token-based MOESI is multicast. We form a 64-node on-chip network (i.e. 8 × 8 and 4 × 4 × 4 mesh networks). Out of the 64 nodes, 16 nodes are processors and other 48 nodes are L2 caches. In a 2D network, processors are located in the first and last rows. In a 3D network, L2 caches are distributed in the bottom three layers, while all the processors are placed in the top layer close to a heat sink so that the

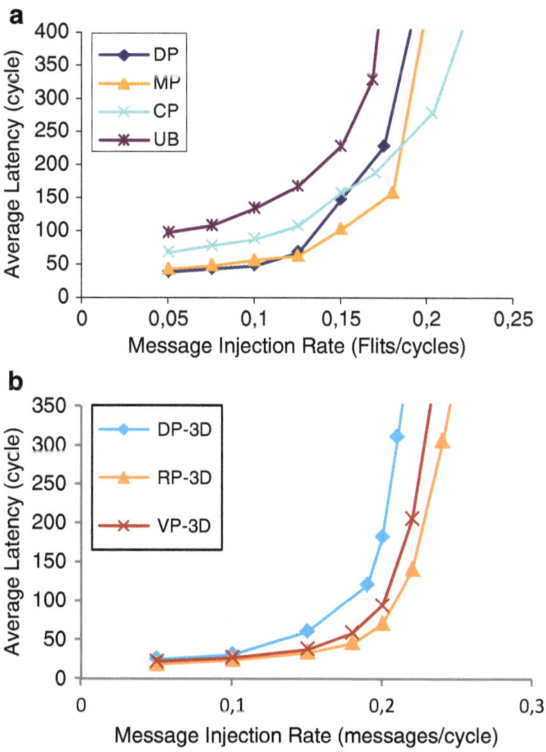

Fig. 7.22 Performance evaluation under mixed traffic (20 % multicast and 80 % unicast) with ten destinations using deterministic routing (**a**) 8 × 8 mesh network (**b**) 4 × 4 × 4 mesh network

best heat dissipation capability is achieved [16, 21]. For the processors, we assume a core similar to Sun Niagara and use SPARC ISA [22]. Each L2 cache core is 1 MB, and thus, the total shared L2 cache is 48 MB. The memory hierarchy implemented is governed by a 2-level directory cache coherence protocol. Each processor has a private write-back L1 cache (split L1 I and D cache, 64 KB, 2-way, 3-cycle access). The L2 cache is shared among all processors and split into banks (48 banks, 1 MB each for a total of 48 MB, 6-cycle bank access). The L1/L2 block size is 64B. The simulated memory hierarchy mimics SNUCA [23] while the off-chip memory is a 4GB DRAM with a 260-cycle access time.

Figure 7.24a shows the average network latency of the real workload traces collected from the aforementioned system configurations, normalized to MP in a 2D network. As can be seen from this figure, the proposed adaptive model, HAMUM, diminishes the average delay of MP and CP significantly under all benchmarks. That is, adaptive routing has an opportunity to improve performance.

As can be seen from Fig. 7.24b in a 3D network, the recursive partitioning method using MAR consistently reduces the average network latency across all tested benchmarks.

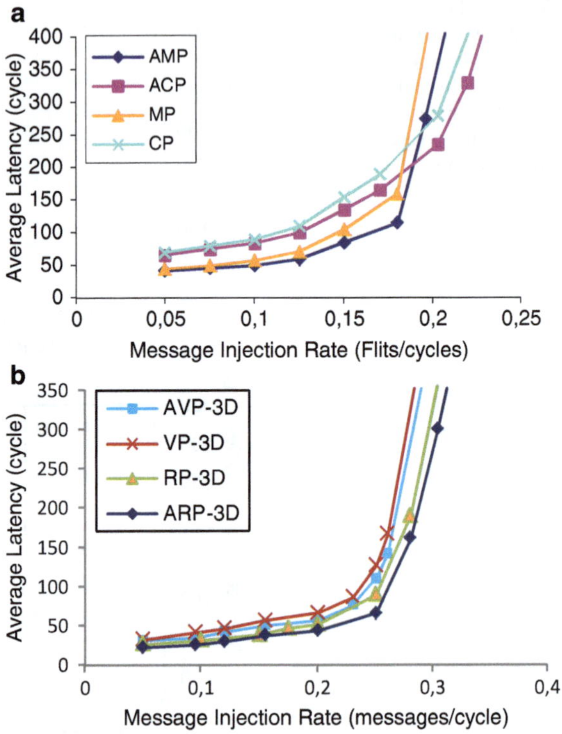

Fig. 7.23 Performance evaluation under mixed traffic (20 % multicast and 80 % unicast) with ten destinations (**a**) 8 × 8 mesh network utilizing HAMUM (**b**) 4 × 4 × 4 mesh network utilizing MAR

Table 7.2 System configuration parameters

Processor configuration	
Instruction set	SPARC, 16 processors
L1 cache	16 KB. 4-way associative, 64-bit line, 3-cycle access time
L2 cache	Shared, distributed in 3 layers, unified, 48 MB (48 banks, each 1 MB), 64-bitline, 6-Clock
Cache coherence protocol	MESI, Token-based MOESI
Cache hierarchy	SNUCA
Benchmarks	SPLASH-2, PARSEC
Access latency	260 cycles
Requests per processor	16 outstanding
Size	4GB DRAM
Workloads	
SPLASH-2	FFT, Ocean, radix
PARSEC	X264

7.4.4 Hardware Overhead

Different partitioning methods have been implemented in network interfaces while the same routing algorithm is implemented in routing units. All the partitioning methods in a 2D network utilize the HAMUM routing algorithm and those in

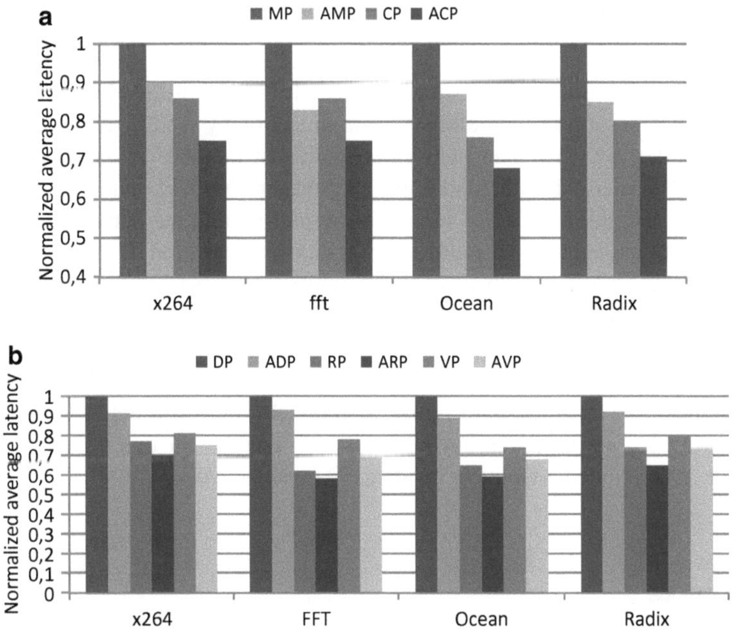

Fig. 7.24 Performance evaluation under different application benchmarks (**a**) 2D network (**b**) 3D network

a 3D network use the MAR routing algorithm. Therefore, the differences in the hardware overhead of different methods are because of the partitioning methods but not the routing units. While the same routing algorithm was used for the CP, MP, and DP multicasting schemes, various numbers of registers were employed in implementing their sorting mechanisms leading to different area overheads. To estimate the hardware cost of the proposed methods, the network area of each partitioning scheme, including switches and network interfaces, with the aforementioned configuration were synthesized by Synopsys D.C. using the UMC *90 nm* technology. The frequency and the supply voltage are *1GHz* and *1 V*, respectively. We performed place-and-route, using Cadence Encounter, to have precise power and area estimations. Based on our analysis, the area overheads of MP and CP are 3 % and 5 % higher than that of the baseline method, DP.

In a 3D network, depending on the technology and manufacturing process, the pitches of TSVs can range from *1* to *10* μm square [24]. In this work, the pad size for TSVs is assumed to be *5* μm^2 with pitch of around *8* μm^2. VP and RP schemes indicate *5 %* and *6 %* additional overhead over the area cost of DP, respectively. The area overhead of the routing algorithms, HAMUM in a 2D network and MAR in a 3D network are negligible.

7.5 Conclusion

Hardware-level multicast implementation results in a considerably better performance gain than supporting multicast communication by sending unicast messages. In this chapter, we described path-based multicast methods in 2D and 3D mesh networks. The main ideas behind these methods are twofold. One is to reduce the overall path length of multicast messages by partitioning the network into several sub-networks and sending a multicast message per partition. The second idea is to provide adaptivity to both unicast and multicast messages. We proposed three partitioning methods in a 2D mesh network, called DP, MP, and CP along with the minimal and adaptive routing algorithm HAMUM. We also proposed three partitioning methods well suited to a 3D mesh network, called DP, VP, and RP. Among them, the recursive partitioning method (RP) performs the best. In addition, MAR routing algorithm was introduced which enables all unicast and multicast messages to be routed adaptively inside a 3D mesh network.

References

1. G.-M. Chiu, The odd-even turn model for adaptive routing. Ieee Trans. Parall. Distrib. Syst. **11**(7), 729–738 (2000)
2. P. Lotfi-Kamran, A.M. Rahmani, M. Daneshtalab, A. Afzali-Kusha, Z. Navabi, EDXY – a low cost congestion-aware routing algorithm for network-on-chips. J. Syst. Arch. **56**(7), 256–264 (2010)
3. M. Ebrahimi, M. Daneshtalab, F. Farahnakian, J. Plosila, P. Liljeberg, M. Palesi, H. Tenhunen, HARAQ: Congestion-aware learning model for highly adaptive routing algorithm in on-chip networks, in *Proceedings of International Symposium on Networks-on-Chip* (Denmark, 2012), pp. 19–26
4. M. Ebrahimi, M. Daneshtalab, P. Liljeberg, J. Plosila, H. Tenhunen, CATRA- congestion aware trapezoid-based routing algorithm for on-chip networks, in *Proceedings of Design, Automation Test in Europe Conference Exhibition (DATE)* (Germany, 2012), pp. 320–325
5. N. E. Jerger, L.-S. Peh, M. Lipasti, Virtual circuit tree multicasting: A case for on-chip hardware multicast support, in *Proceedings of the 35th Annual International Symposium on Computer Architecture (ISCA)*, vol. 36 (China, 2008), pp. 229–240
6. P. Abad, V. Puente, J. Gregorio, MRR: Enabling fully adaptive multicast routing for CMP interconnection networks, in *Proceedings of IEEE 15th International Symposium on High Performance Computer Architecture (HPCA)* (USA, 2009), pp. 355–366.
7. J. Duato, S. Yalamanchili, L. Ni, *Interconnection Networks* (Morgan Kaufmann, San Francisco, 2003)
8. R.V. Boppana, S. Chalasani, C.S. Raghavendra, Resource deadlocks and performance of wormhole multicast routing algorithms. IEEE Trans. Parall. Distrib. Syst. **9**(6), 535–549 (1998)
9. M. Ebrahimi, M. Daneshtalab, P. Liljeberg, H. Tenhunen, HAMUM – A novel routing protocol for unicast and multicast traffic in MPSoCs, in *Proceedings of 18th Euromicro International Conference on Parallel, Distributed and Network-Based Processing (PDP)* (Italy, 2010), pp. 525–532
10. X. Lin, P.K. McKinley, L.M. Ni, Deadlock-free multicast wormhole routing in 2D mesh multicomputers. IEEE Trans. Parall. Distrib. Syst. **5**(8), 793–804 (1994)

11. W. Dally, B. Towles, *Principles and Practices of Interconnection Networks* (Morgan Kaufmann, San Francisco, 2003)
12. M. Ebrahimi, M. Daneshtalab, P. Liljeberg, J. Plosila, J. Flich, H. Tenhunen, Path-based partitioning methods for 3D networks-on-chip with minimal adaptive routing, *IEEE Transactions on Computer* (2012)
13. B.S. Feero, P.P. Pande, Networks-on-chip in a three-dimensional environment: a performance evaluation. IEEE Trans. Comput. **58**(1), 32–45 (2009)
14. V.F. Pavlidis, E.G. Friedman, 3-D topologies for networks-on-chip. IEEE Trans. Very Large Scale Integr. (VlSI) Syst. **15**(10), 1081–1090 (2007)
15. F. Li, C. Nicopoulos, T. Richardson, Y. Xie, V. Narayanan, M. Kandemir, Design and management of 3D chip multiprocessors using network-in-memory, in *Proceedings of ISCA-33* (USA, 2006), pp. 130–141
16. D. Park, S. Eachempati, R. Das, A.K. Mishra, Y. Xie, N. Vijaykrishnan, C.R. Das, MIRA: A multi-layered on-chip interconnect router architecture, in *Proceedings of the 35th Annual International Symposium on Computer Architecture (ISCA)* (China, 2008), pp. 251–261
17. S.C. Woo, M. Ohara, E. Torrie, J.P. Singh, A. Gupta, The SPLASH-2 programs: characterization and methodological considerations, in *Proceedings of 22nd Annual International Symposium on Computer Architecture* (Italy, 1995), pp. 24–36
18. C. Bienia, S. Kumar, J.P. Singh, K. Li, The PARSEC benchmark suite: Characterization and architectural implications, in *Proceedings of the 17th International Conference on Parallel Architectures and Compilation Techniques* (Canada, 2008), pp. 72–81
19. M.M.K. Martin, D.J. Sorin, B.M. Beckmann, M.R. Marty, M. Xu, A.R. Alameldeen, K.E. Moore, M.D. Hill, D.A. Wood, Multifacet's general execution-driven multiprocessor simulator (GEMS) toolset. Sigarch. Comput. Arch. News. **33**(4), 92–99 (2005)
20. N. Muralimanohar, R. Balasubramonian, N. Jouppi, Optimizing NUCA organizations and wiring alternatives for large caches with CACTI 6.0, in *Proceedings of the 40th Annual IEEE/ACM International Symposium on Microarchitecture* (USA, 2007), pp. 3–14
21. I. Loi, L. Benini, An efficient distributed memory interface for many-core platform with 3D stacked DRAM, in *Proceedings of Design, Automation Test in Europe Conference Exhibition (DATE)* (Germany, 2010), pp. 99–104
22. P. Kongetira, K. Aingaran, K. Olukotun, Niagara: A 32-way multithreaded sparc processor. IEEE Micro. **25**(2), 21–29 (2005)
23. B.M. Beckmann, D.A. Wood, Managing wire delay in large chip-multiprocessor caches, in *Proceedings of the 37th annual IEEE/ACM International Symposium on Microarchitecture* (Oregon, 2004), pp. 319–330
24. J. Hu, L. Wang, L. Jin, H. Z. JiangNan, Electrical modeling and characterization of through silicon vias (TSV), in *Proceedings of International Conference on Microwave and Millimeter Wave Technology (ICMMT)*, China, 2012, vol. 2, pp. 1–4

Part III
Fault Tolerance and Reliability

Chapter 8
Fault-Tolerant Routing Algorithms in Networks On-Chip

Reyhaneh Jabbarvand Behrouz, Mehdi Modarressi, and Hamid Sarbazi-Azad

Abstract As the semiconductor industry advances to the deep sub-micron and nano technology points, the on-chip components are more prone to the defects during manufacturing and faults during system life time. These components include the Networks-on-Chip (NoCs) which are expected to be an important part of the future complex multi-core and many-core chips. As a result, fault tolerant techniques are essential to improve the yield of modern complex chips. In this chapter, we propose a fault-tolerant routing algorithm that keeps the negative effect of faulty components on the NoC power and performance as low as possible. Targeting intermittent faults, we achieve fault tolerance by employing a simple and fast mechanism composed of two processes: NoC monitoring and route adaption. The former keeps the track of the on-chip traffic pattern and faulty links, whereas the latter adapts the packet paths to the current set of faulty components. This mechanism exploits the global information of the state of the NoC components and on-chip traffic pattern and aims to minimize the performance loss and the power overhead imposed by the faulty NoC links and nodes. Experimental results show the effectiveness of the proposed

R.J. Behrouz (✉)
Department of Computer Science, George Mason University, United States
e-mail: rjabbarv@gmu.edu

M. Modarressi
Department of Electrical and Computer Engineering, College of Engineering,
University of Tehran, Tehran, Iran

School of Computer Science, Institute for Research in Fundamental Sciences (IPM),
Tehran, Iran
e-mail: modarressi@ut.ac.ir

H. Sarbazi-Azad
Department of Computer Engineering, Sharif University of Technology, Tehran, Iran

School of Computer Science, Institute for Research in Fundamental Sciences (IPM),
Tehran, Iran
e-mail: azad@sharif.ir

M. Palesi and M. Daneshtalab (eds.), *Routing Algorithms in Networks-on-Chip*,
DOI 10.1007/978-1-4614-8274-1_8,

technique, in that it offers lower average message latency and power consumption and a higher reliability, compared to some state-of-the art related work.

8.1 Introduction

NoC is a scalable communication framework, proposed to overcome the traditional shared-bus communication shortcomings in multi-processors chips (CMPs) [1, 2]. Due to the capabilities of deep sub-micron technologies, the number of cores on a single chip has increased rapidly, caused thousands (or millions) of transistors being tight in a new layout consequently. Technology scaling not only arise the likelihood of permanent and transient faults, but also cause process variation in manufacturing lifecycle which results in intermittent faults. Therefore, fault management is a vital issue and should be confronted in NoCs.

During the past two decades, many approaches have been proposed to make interconnection networks fault tolerant. These approaches can be broadly categorized in two categories: fault-tolerant network architectures and fault-tolerant routing algorithms. The former category addresses fault-tolerant design of on-chip components to have them operate normally in the presence of faults, whereas the latter bypasses the faulty components.

Fault tolerant routings make local and global decisions about the path of packets, based on the position of the current faulty links and nodes. In this case, finding new paths for the packets can be done either centralized or distributed [3, 4]. Distributed algorithms spread the computation overhead of finding new paths among all nodes of the network, while in a centralized algorithm, one or more nodes are responsible to re-route the packets. Thus, the architecture of the routers which a distributed algorithm needs could be simpler than the routers which a centralized algorithm uses. However, the amount of computation to be performed in a centralized algorithm is by far more than the computations done in a distributed algorithm each node. In addition, centralized algorithms use global fault information, but distributed algorithms are acquainted about fault occurrence only based on local information from neighbors. This is mainly due to the fact that as sending the global fault information to all nodes is very costly in terms of communication overhead. Hence, the path selection decision in centralized algorithms can result in higher quality results.

In this chapter, we present a fault-tolerant routing mechanism (FaulToleReR) to cope with intermittent faults in NoCs. Intermittent faults arise under some conditions; mainly the process variations combined with hard operating conditions (such as high temperature [6]) and affect the system as long as the condition is met [5]. The faulty part will resume its normal operation when the working conditions (temperature, for example) backs to normal. Temperature increases the delay of interconnects and logic and if it exceeds a threshold, the hot component may not complete the computation and communication within the single clock cycle, leading to generate incorrect outputs [7, 8].

This routing mechanism aims to manage the paths taken by packets in an NoC which some parts become temporarily non-functional due to some working conditions, in particular high temperature. The temperature of the links and other NoC components is directly related to the traffic load on them. Consequently, the intermittent faults caused by high temperature will disappear when there is no load on the faulty link for a period of time. We use a simple model for the temperature of the components in which the temperature is only affected by the power consumption of the component. The power consumption is in turn caused by the load (activity) of the component. In this model, the second order effects, like the contribution of the temperature of the neighboring components in the temperature of each component, is ignored.

The topology of the NoC considered in this chapter changes dynamically during the system lifetime due to the links affected by intermittent faults. FaulToleReR targets such condition and tries to dynamically change the routing algorithm at run-time, in response to a change in the NoC topology. It routes packets in such a way that all of them travel over the possible shortest path between their source and destination nodes, the traffic is balanced across the chip, the bandwidth constraint of each link is satisfied, and deadlock freedom is guaranteed.

By considering the permanent faults as intermittent faults with infinite recovery time, our method can also tolerate design time and run-time permanent faults. We assume that faults have been detected by some fault detection algorithm and our focus is then on addressing different fault tolerance issues. FaulToleReR routing algorithm is orthogonal to the existing fault detection approaches and can be accompanied with any of them.

Like most application-specific NoCs, we use a table-based routing mechanism to exploit the information about the application traffic characteristics and NoC faulty links to appropriately allocate paths to traffic flows. In [9], authors compare the advantages and drawbacks of the table-based routing schemes and show that it is suitable for the NoC-based systems due to its lower cost, ease of implementation, and power efficiency.

We also take the important problem of deadlock freedom into account in the path finding algorithm. Most fault-tolerant routing algorithms use virtual channels (VC) to apply the well-known escape channel method to avoid deadlock [10], whereas some other methods achieve deadlock avoidance by placing some restrictions on the paths that can be taken by the packets. Our method can be classified as the latter class; it involves less power consumption, complexity, and cost, compared VC-based deadlock avoidance methods.

Intermittent faults occur frequently and last for several cycles. Hence, a routing algorithm should adjust itself to such frequently occurring faults. Deterministic fault-tolerant routings are not resilient to intermittent faults, while adaptive routings could adjust themselves according to topology changes which are caused by faults. However, existing adaptive routing methods use local information to choose a new route. In this case, the limited local information may lead to a decision that misroutes packets and make the packet latencies higher.

Benefitting from the global information about the fault patterns and the communication demand of each node with low overhead is the key advantage of our work over the few works that handle intermittent faults [6].

Although FaulToleReR targets intermittent faults, it can be used to route packets in any network with dynamic topology (such as wireless networks), as well as irregular mesh networks with oversized IPs (OIPs) [11].

The rest of the chapter is organized as follows. Section 8.2 gives an overview of the entire problem and the proposed solution. Section 8.3 presents the implementation details of the proposed fault-tolerant routing algorithm. The experimental results are presented in Sect. 8.4, and finally Sect. 8.5 concludes the chapter.

8.2 Problem Definition

8.2.1 NoC Architecture

The NoC under consideration is composed of $m \times n$ nodes arranged as a 2D mesh. Although the proposed NoC is not restricted to a specific switching mechanism, we use the wormhole switching [12]. The proposed fault-tolerant routing approach relies on monitoring the on-chip traffic pattern and status of each NoC link and making global decisions on packet routes based on the observed NoC status. The route of each packet is determined by a root processor in a centralized manner. Most multi-core processors and SoCs select one of the processors as a root or configuration node that manages the system [12], we assign the task of setting the packet routes in case of a change in NoC topology to this node.

Traffic monitoring is simply done at each node by keeping track of the number and destination address of the packets it sends. The weight of each flow between two nodes is an n-bit value proportional to the communication volume between them (the n-most-significant bits of the register holding the number of sent packets at the source). In this work, we set n to 6 bits to support 32 different levels of flow weight. By reducing the value of n, we can significantly reduce the power consumption of arithmetic operations, but at the price of losing the accuracy.

Data transmission between the root and other NoC nodes can be either done through the main data network by some specific kind of packets or via a separate control network (like the control network in [12]). In this work, we adopt a control network that provides a light-weight and efficient infrastructure for the root node to collect the information about NoC (traffic statistics and the position of faulty links) and send configuration information back to NoC nodes. The data transmission required for the route calculation does not impose considerable overhead to the NoC due to its infrequent invocation and small volume of transmitted data.

Here, we use the control network proposed in our previous work [12]. Collecting the required data from the NoC by the root node involves broadcasting a request from the root node to the network nodes, followed by sending the required information from the network nodes to the root on the control network. The former, broadcasting a request or a data from the root to the network, is done by the root node by sending the request on its row along both the E and W directions. Each control network node at the same row of the root node, when detecting the message, sends it to the nodes in its column along both the N and S directions. For the second transactions that carry information to the root, the control network is divided into four quadrants with respect to the root node position and the nodes inside each quadrant direct their messages to the node at the row or column in which the root node is located and then the node at that row/column forwards the message to the root node. Each node accepts the message of its previous node only when it has finished sending its message to the next node along the assigned path. This scheduling removes the need for buffering and complex arbitration on the control network. The proposed scheduling schemes work correctly for any given location of the root node, but it is more beneficial to map the root node into one of the central nodes, since it allows more parallelization in data transmission.

8.2.2 Problem Formulation

As mentioned before, temporal failure of links causes dynamic changes in NoC topology. The traffic is also assumed to be dynamic, so the root node uses the collected information about the traffic (end-point nodes and weight of each communication flow) at each time window to construct a communication task graph which will be used during the route calculation procedure. The information about unavailable links is also collected by the root node through the control network. We aim to change the routing algorithm in order to efficiently route the packets based on the new topology in a centralized manner.

It is assumed that the faults do not affect the network connectivity, i.e. there is at least one path between any two given NoC nodes that need to communicate. Otherwise, it is impossible to find a path for some packets which in turn, leads to system failure.

We model all faults as link fault, whether occurred on a router or on a link itself. The set of faulty links (represented by *FL*) includes all of the NoC links that are currently unavailable. In order to formulate the problem, we first need to provide the following definitions:

Definition 1 Communication Task Graph: The tasks and the corresponding relations are represented as a communication task graph, *CTG* which is a directed graph *G(V,E)*where each $v_i \in V$ represents a task, and a directed edge, $e_{i,j} \in E$ characterizes a communication flow from v_i to v_j. The communication volume (bits per second) corresponding to every edge $e_{i,j}$ is represented by $t(e_{i,j})$.

Definition 2 Link Bandwidth Constraint: The bandwidth constraint of each NoC link l_k should be satisfied as:

$$\forall l_k, BW(l_k) \geq \sum_{\forall e_{i,j} \in E} X^k(i,j) \tag{8.1}$$

Where $BW(l_k)$ characterizes the bandwidth of link l_k and $X^k(i,j)$ is calculated as:

$$X^k(i,j) = \begin{cases} t(e_{i,j}) & if\ l_k \in path(e_{i,j}) \\ 0 & Otherwise \end{cases} \tag{8.2}$$

The parameter $path(e_{i,j})$ is the set of links onto which *CTG* edge $e_{i,j}$ is mapped. This can be stated more simply as:

$$\forall l_k, BW(l_k) \geq BR(l_k) \tag{8.3}$$

Where $BR(l_k)$ is the total amount of traffic that travels on NoC link l_k.

Formally, the problem of dynamic route calculation can be described as follows.

Given an NoC, which is obtained by removing faulty links (*FL* members) form the total NoC links, and the NoC traffic pattern described by a communication task graph *CTG*, find a new route for each flow (each edge $e_{i,j}$ in *CTG*) that minimizes the following expression, where *route(i,j)* is the distance (in terms of hop count) between nodes *i* and *j*:

$$\sum\nolimits_{\forall e_{i,j} \in E} t(e_{i,j}) \times route(i,j) \tag{8.4}$$

The following expression can be postulated as the average message latency of the network as the latency of a packet is the sum of its zero-load latency and blocking latency. Zero load latency is the time (in terms of cycles) it will take for a packet to traverse from the source to the destination node when there is no contention and all resources are available. However, in reality, a packet should compete with other packets for the NoC bandwidth. This competition causes blocking latency in switches. If the network is not congested, the number of hops traversed by a packet to reach the destination node is roughly proportional to packet's latency. Therefore, minimizing Eq. 8.4 will lead to minimizing the average message latency in the NoC. The power consumed to transmit packets is also directly proportional to packet path length. Consequently, by specifying the packet routes in such a way that each packet is routed through a minimal path (based on the constraints of the new topology), whereas the bandwidth constraint of each link is not violated, we can guarantee that the power consumption and packet latency of the NoC is kept as low as possible. In the next section, we show how FaulToleReR algorithm can achieve this goal.

8.3 The Proposed Fault-Tolerant NoC

8.3.1 The FaulToleReR Algorithm

The FaulToleReR algorithm is capable of rerouting packets upon both topology and traffic changes by a two–step routing procedure. The first step is the initial step which finds all available paths for each flow of the *CTG*. These preliminary paths are used in the second step to find final possible paths for each source-destination pair,

The **FaulToleReR** Reconfigurable Fault-Tolerant Routing Algorithms

```
boolean finish = false;
int dis_tsh, l_tsh;
function Initionation (flows, dis_tsh) {
    Sort (flows); //According to a priority criteria
    all_paths = Find_All_Paths (flows, dis_tsh);
}// end function
function Main (flows, all_paths, faults) {
    if (faults) {
        Prune_All_Paths (all_paths, faults);
    }//enf if
    while (!finish) {
        minimals = Get_Minimal_Paths (flows, all_paths);
        finish = Path_Finder (flows, all_paths, minimals, l_tsh);
    }//end while
}// end function
function Path_Finder (flows, all_paths, minimals, l_tsh) {
    for all possible paths from all_paths {
        if (Link_Load (paths, l_tsh)) {
            if (DeadLock_Free (paths)) {
                return true;
            }//end if
        }//end if
    }//end for
}//end function
```

Fig. 8.1 Pseudo code for FaulToleReR algorithm

in order to minimize Eq. 8.4. FaulToleReR needs no virtual channel for deadlock handling, but it can use virtual channels to improve performance. The algorithm is outlined in Fig. 8.1.

The *Initiation* function launches when the system starts up and whenever the on-chip traffic (represented by the *CTG*) changes. This function consists of two sub-functions, *Sort* and *Find_All_Paths*. The *Sort* function is a modified version of *bubble sort* which sorts the flows according to some predefined priority. As we target to optimize the power and latency, the priority criteria is the communication volume of flows. Hence, heavier flows have higher priority. These sorted flows and *dis_tsh,* a path length threshold, are passed to the *Find_All_Paths* function as input, to find all possible paths for each flow that are shorter than *dis_tsh* (in terms of hop count). This function is a customized version of *Dijkstra*'s algorithm with the modifications proposed in [12]. The customized algorithm significantly reduces the computations of the algorithm and now is suitable for online execution. We refer interested readers to [12] for more details of the algorithm. The output of this sub-function is a hash set of the paths, indexed by the links they include. Once this step is done, the basic step of the algorithm begins.

After the initial process, the algorithm enters the *main* function, to find an appropriate route based on the current network configuration. This function is triggered by either the system start-up or a fault occurrence. Firstly, it checks for a topology change and fault occurrence. If it is not the first time this function launches, the set of indexed paths should be pruned by removing the faulty links.

Before explaining the rest of the algorithm, it is worth to bearing in mind that we require finding a path for each flow that satisfies the following conditions:

1. The set of all found paths should minimize Eq. 8.4 in order to reach the minimum average latency across all communication flows. The minimum value of Eq. 8.4 arises when each packet reach its destination via a shortest path. However, this may not be possible due to faulty links, deadlock occurrence, or the constraints imposed by Eq. 8.3. Therefore, some of the flows should detour from a shortest path.

   If the shortest path of a flow and the actual path assigned to it are represented by *route***(i,j)* and *route(i,j)*, respectively, we define the detour overhead parameter of a flow as:

$$d_{f_{i,j}} = route\,(i,j) - route^*\,(i,j) \tag{8.5}$$

   Since the amount of overall detour overhead is the difference between minimum average path length and average path length gained by actual routing, minimizing Eq. 8.4 is equivalent to minimizing the cumulative detour overhead of the flows defined as:

$$D = \sum_{\forall e_{i,j} \in E} d_{f_{i,j}} \tag{8.6}$$

2. The overall traffic passing over a link should not exceed the maximum available link bandwidth to avoid congestion (Eq. 8.3). We will later see that this constraint leads to a better NoC performance and increases the network tolerance against faults. If the total communication volume of two or more flows mapped onto the same link violates this constraint, an unacceptable situation called overlapping occurs.
3. Routing configuration should be deadlock and livelock free. Many algorithms do not engage with deadlock problem and use virtual channels to escape deadlock scenarios. However, FaulToleReR avoids deadlocks during path selection to relax the VC count constraint experienced by many routing algorithms.

Back to the algorithm, the *Get_Minimal_Paths* function, which is the heuristic part of FaulToleReR algorithm, aims at meeting the first condition mentioned above. This function starts from the minimum possible value for D, the overall detour overhead, and returns the set of paths by which this value is achieved (The minimum value is 0 when each flow is directed through the shortest path). If the set of selected paths is approved by *Link_Load* and *Deadlock_Free* functions which check if the candidate set meets the second and third conditions, respectively, the goal is achieved. Otherwise, *Get_Minimal_Paths* increments D by one and passes it to the *Path_Finder* function for further checks. This procedure repeats until FaulToleReR finds a set of routes that meets the aforementioned conditions.

Get_Minimal_Paths function uses two pre-defined data structures, the *memory* and the *visited.* At first, the function supposes that each flow will have its shortest path from source to destination making the total detour overhead of the network zero. Then, it sends the n-tuple of path lengths to *Path_Finder* function to check if they meet the link capacity and deadlock freedom criteria. If they the checks are passed, the algorithm updates related routing tables and terminates. Otherwise, the algorithm returns to the *Get_Minimal_Paths* function to find other paths.

When the algorithm returns to *Get_Minimal_Paths* for the second time, it chooses the last flow in a sorted list (which would be the flow with lowest communication volume), and assigns it a path length that is one step longer than its previous path length. For example, if the shortest path from source to destination for that flow is 5, it can reach the destination by traversing paths with length of 5, 7, 9, 11, and etc. So, the length of the next shorter path of this flow is 7. The length of other flows remains the same. The new set of paths has the minimum detour overhead greater than 0. Again, FaulToleReR stores the n-tuple in *visited* and feed them to the *Path_Finder* function to check if they are the final solution. In many applications, where there are many flows and it is very likely that path overlapping brings about exceeding the link capacity limit, this function may be invoked more than twice.

At the third call, the algorithm assigns the next longer shortest path to the second flow in the increasing order of communication volume, while all other flows have their original shortest path. After calculating the total detour overhead of this set and storing it in the Visited data structure, the algorithm combines this set with all other sets with different values for the same flow, saved in Visited to generate a new set. The outcome is a new n-tuple which its detour overhead is the sum of the old sets detour overhead. FaulToleReR stores this n-tuple set in memory data structure. This procedure repeats in further iteration of the algorithm and if there is a set in the Memory data structure that has lower detour overhead than the newly generated one, the algorithm selects it.

For example, consider a NoC shown in Fig. 8.2 with four faulty links and maximum link bandwidth of 100 Mbps. There are three flows with different communication volumes. Flow1 connects node 7 to node 13 with communication volume of 50 Mbps. Flow2 connects node 16 to node 18 with communication volume of 80 Mbps and Flow3 connects node 10 to node 14 with communication volume of 90 Mbps. The algorithm iterates six times to find the best solution of the given configuration. The sets and detour overheads for each iteration is shown in the figure.

We claim that the final found routes give lower average latency in comparison with other related methods, because the FaulToleReR algorithm targets minimizing Eq. 8.6. In addition, our proposed algorithm guarantees freedom from deadlock without utilizing virtual channels, but by means of some deterrent rules confining the routes from source to destination. This is accomplished by constructing a channel

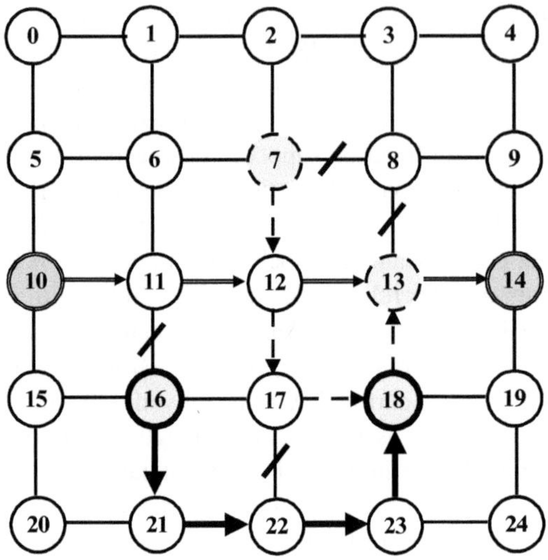

The *Visited* data structure

Path Length (f1,f2,f3)	Detour Over-head
(4,2,2)	0
(4,2,4)	100
(4,4,2)	160
(6,2,2)	180
(4,2,6)	200
(4,6,2)*	320

The *Memory* data structure

Path Length (f1,f2,f3)	Detour Over-head
(4,4,4)*	260
(6,2,4)	280
(6,4,2)	340

Fig. 8.2 An example run of the *Get_Minimal_Paths* function

dependency graph, *CDG*, of the selected paths and check if it is acyclic or not. In case of not finding any cycles, the routing configuration is certainly deadlock free.

FaulToleReR is also livelock free. In other words, it definitely finds a route for each flow to its destination as long as the network is not disconnected. As mentioned before, we place a bandwidth occupancy constraint on each link to avoid network saturation. Consequently, we extended the meaning of disconnection, in which two nodes are disconnected not only when there is no physical paths between them, but also when using the possible paths between them may lead to violating the bandwidth constraint of some links.

Our proposed algorithm applies kind of fairness in routing, so that flows with lower priority are not always sacrificed by those with higher priority. Intuitively, FaulToleReR seems to choose to detour a flow with lower communication volume in favor of an overlapping flow with higher communication volume. However, it is not

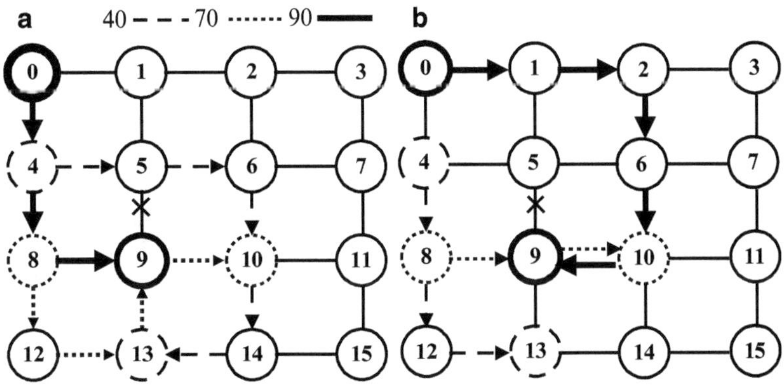

Fig. 8.3 The impact of fairness on detour overhead in the proposed routing. Routing three flows (**a**) without fairness and (**b**) with fairness

the best strategy in all cases. For example, suppose a network configuration shown in Fig. 8.3. There are three flows with communication volume of 90, 70, and 40 Mbps, and the maximum link bandwidth is 100 Mbps. As there are common links between shortest paths of the flows and the total link load cannot exceed the bandwidth limitation, some flows should detour their shortest paths. Figure 8.3a illustrates a case without fairness where two lower flows make a detour from their shortest paths to secure a shortest path for the flow with higher volume. Consequently, the total detour overhead is 200 (140 for the 70 Mbps flow and 80 for the 40 Mbps flow and according to Eq. 8.6). On the other hand, the routing configuration shown in Fig. 8.3b, which has been obtained by misrouting the largest flow, has a total detour overhead of 180, that is obviously preferable to the total detour overhead of Fig. 8.3a.

8.4 Experimental Results

8.4.1 Simulation Environment

In this section, the proposed fault-tolerant routing mechanism is evaluated under some realistic and synthetic benchmarks. The benchmark set includes some synthetic task graphs along with some existing applications that have been widely used in the literature, namely Multi Media System (MMS) [13] with 43 tasks, and Global System for Mobile Communications (GSM) decoder [14] with 80 tasks.

We evaluate the FaulToleReR algorithm by BookSim, a cycle-accurate interconnection network simulator [15]. We have augmented BookSim by Orion2.0 power library [16] in order to evaluate the power consumption of NoCs under different routing algorithms. The power results are based on a 128-bit NoC implemented in 65 nm technology and the frequency varies for different network configuration. Fault injection is done by randomly disabling some links based on a given fault rate. A Router becomes non-functional when all of the links connected to it become faulty. FaulToleReR, as mentioned before, can tolerate as many faults as the case when the aforementioned conditions in Sect. 8.3 are standing.

The network size for the synthetic benchmarks is 4×4, 6×6, and 8×8. The NoC topology is the conventional 2D mesh and the NoC adopts the wormhole switching with 4-flit packets and16-flit buffers for each virtual channel. Simulation lifetime is 6,000,000 cycles for the realistic benchmarks and 10,000,000 cycles for synthetic benchmarks.

We have selected two routing algorithms for the comparison purpose: planar routing [17] and Highly Resilient Routing Algorithm, or HRA for short, presented in [18]. Planar is a well-known adaptive routing algorithm that uses local information to adaptively route packets. It needs some virtual channels to operate correctly and places some restrictions on the virtual channel allocation in order to guarantee deadlock freedom. HRA is a state-of-the-art distributed fault-tolerant routing algorithm that reconfigures NoC to avoid faulty components. HRA prohibits some links and turns in order to prevent deadlock.

8.4.2 Average Message Latency Analysis

In the first experiment, we compare the average message latency (AML) of FaulToleReR with other considered routing algorithms under different number of randomly injected faults in the links. Figure 8.4 displays the average message latencies given by the three routing schemes under different benchmarks. We have repeated simulations for each number of faults for 10 times to obtain accurate results. The simulation results reported in this figure reveals that:

- FaulToleReR and planar have the same behavior when there are no faults. However, HRA's AML is a bit higher, due to its specific routing rules that disallow using some links and turns to avoid deadlock. Therefore, the number of available paths between two given source and destination nodes is less than that of planar and FaulToleReR which do not have these limitations.
- By increasing the number of faulty links, the AML of all routing algorithms increases. The reason is that faults reduce the network resources and the load of the faulty links is distributed among other links, leading to an increase in their load, and hence, their latency. In all three cases depicted in Fig. 8.4, however, our proposed algorithm has less AML compared to other routing schemes for the full range of the considered fault count (number of faulty links). This is due to the fact that our algorithm always finds the best path for the packets based on the current topology and communication pattern. In addition, FaulToleReR algorithm shifts the network saturation point to tolerate more faults (when the traffic is fixed and the latency is increased due to faulty links) and network saturation occurs at higher fault injection rates, compared to other routing schemes.

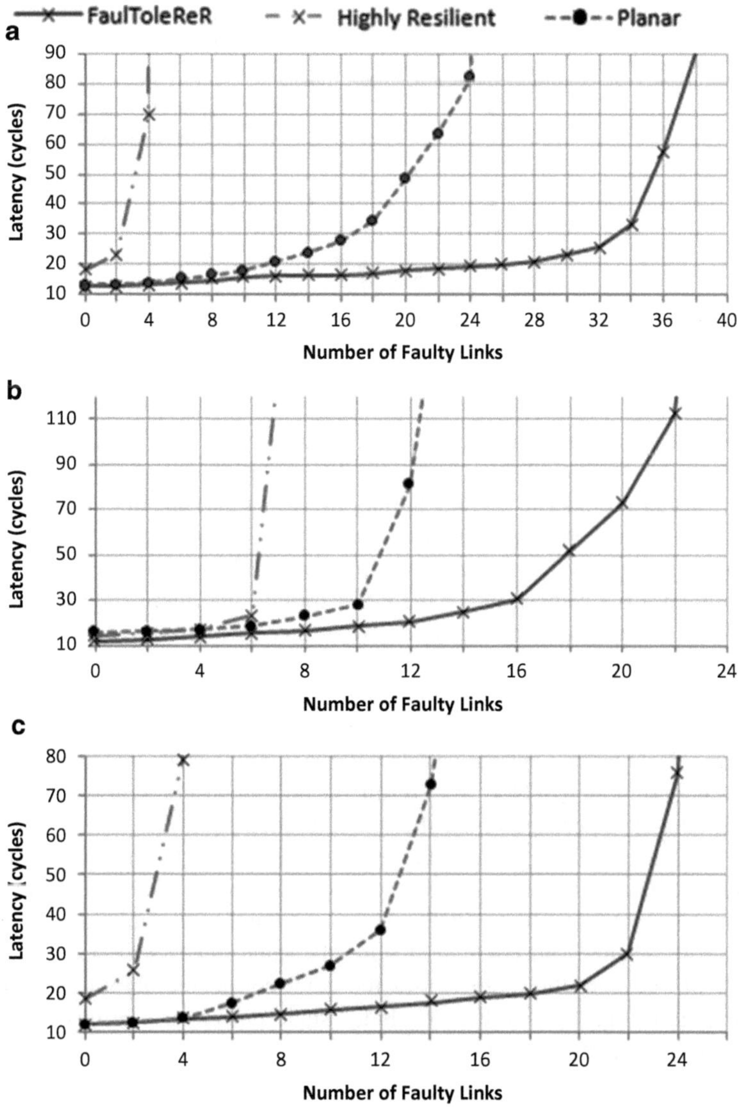

Fig. 8.4 Average message latency comparison between FaulToleReR, Planar, and a highly resilient algorithm for a given number of faults with (**a**) GSM, (**b**) MMS, and (**c**) a uniform input traffic

8.4.3 Power Analysis

We further compare FaulToleReR with the two other mentioned routing algorithms in terms of power consumption. As FaulToleReR forwards packets via shorter paths than the other algorithms, packets pass fewer intermediate routers. Hence we expect

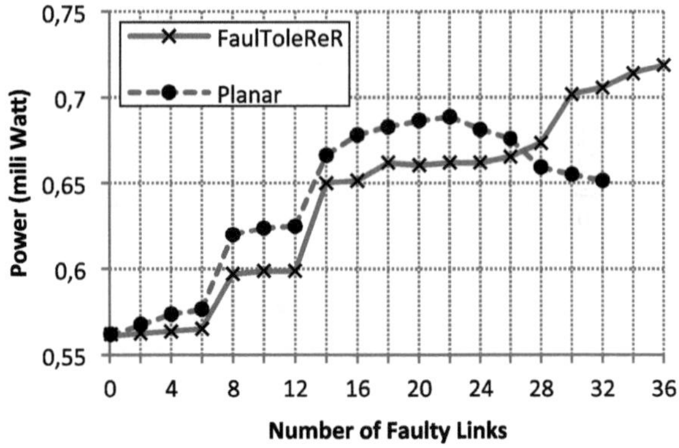

Fig. 8.5 Power consumption comparison between FaulToleReR and planar

our proposed algorithm to consume less average power than the others. The total power consumed by a packet is the sum of the power it consumes in each node by: writing/reading the packet to/from buffer, crossbar traversal, link traversal, route computation, virtual channel arbitration, and switch arbitration.

Figure 8.5 displays the power consumption of FaulToleReR and Planar routing algorithms in a 6×6 mesh under the GSM traffic. The figure shows that FaulToleReR offers less power consumption when there are less than 27 faulty links. As Fig. 8.4.a shows, the network enters the saturation state by the planar routing algorithm in presence of 22 and more faults. From this point on, FaulToleReR consumes more power as it is still working normally, while under the other algorithms, NoC traffic acceptance rate decreases significantly, leading to limiting the power consumption of NoC. As a result, the power consumption of FaulToleReR is higher than the other algorithms after the saturation point.

We then calculate the power-delay product (PDP) of the three NoCs to give a better understanding about the behavior of the routing algorithms in dealing with faults. Figure 8.6 shows PDP for a given number of faults and traffic for FaulToleReR, Planar, and HRA. Following the same trend as the latency results, the PDP of our algorithm outperforms the other considered algorithms under all benchmarks. In Fig. 8.6, we have excluded static power form results as it is not affected by the routing algorithm.

8.4.4 Reliability Analysis

In the third experiment, we investigated the relationship between the number of faulty links and the reliability of the NoC. We Consider that a routing algorithm is reliable if:

- There is no deadlock in packet routes
- When there is a physical route between nodes A and B, the routing algorithm can direct packets from A to B.
- The routing algorithm does not saturate the network

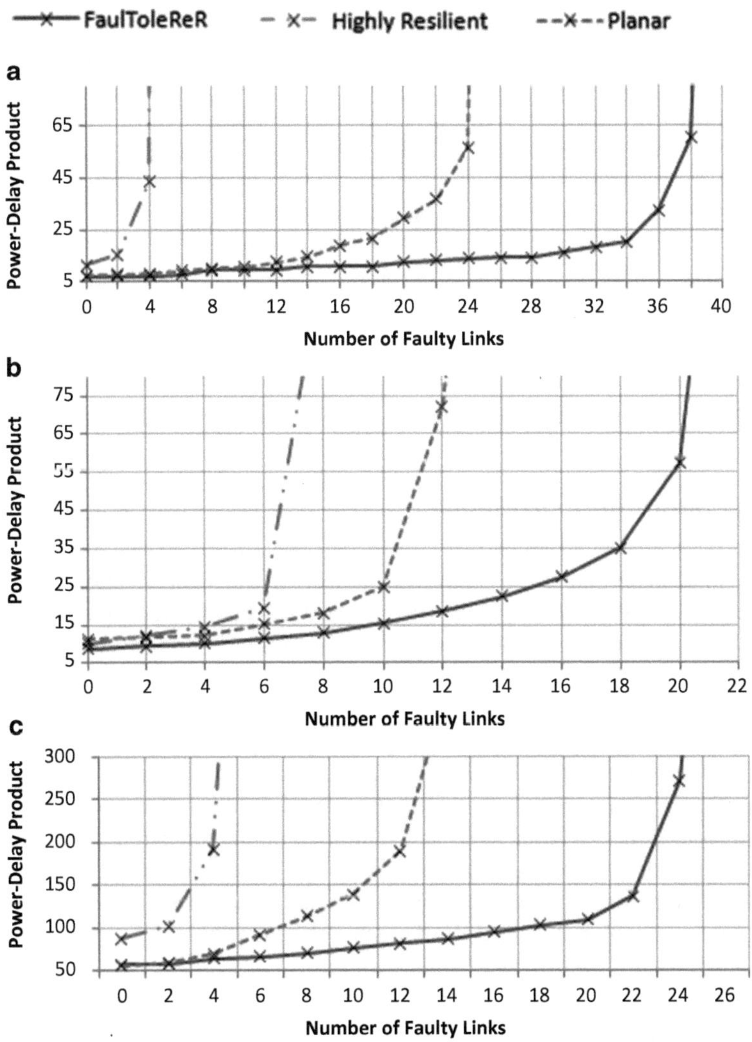

Fig. 8.6 Power-delay product comparison between FaulToleReR, Planar, and HRA for a given number of faults with (**a**) GSM, (**b**) MMS, and (**c**) a uniform input traffic

Figure 8.7 shows reliability of FaulToleReR, planar, and HRA for different NoC sizes, for different number of faults. HRA is 90% reliable when 10% of the links are broken, while this reliability is obtained by planar and FaulToleReR when 37% and 54% of links are broken, respectively.

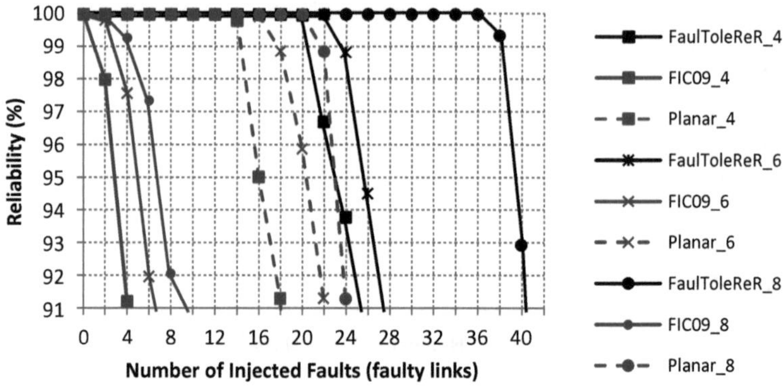

Fig. 8.7 Reliability of FaulToleReR, planar, and HRA versus number of faults

8.4.5 Saturation Threshold

In the final experiment, we study the impact of the fault location on latency. We compare FaulToleReR with the reconfigurable routing algorithm presented in [19] which tolerates single-node failure. The authors in [19] have considered nine contours (four of them at the corners, four between each two corners, and one at the middle) in a 2D mesh network where a single-node failure can occur.

Figure 8.8 depicts the packet latency of a 6×6 mesh (for a given traffic load packet/node/cycle) of four contours. According to the results shown in Fig. 8.8b, the routing scheme proposed in [19] uses an X-Y-like routing with local fault information. Hence, it leads to sooner saturation point, compared to FaulToleReR whose behavior is shown in Fig. 8.8a. Since the middle routers of the mesh (not placed at the network boundaries) play a more significant role in packet routing, a node failure in contour5 leads to earlier network saturation.

8.5 Conclusion

In this chapter, we first provided a comprehensive study on previous works related to fault-tolerant routing in NoCs. Then, we proposed a centralized routing scheme, aiming to cope with the dynamic topology changes in NoCs caused by permanent and intermittent faults. This is an important issue in NoCs in the nanotechnology era, as the on-chip components are becoming more and more susceptible to faults. The entire procedure relies on monitoring the state of NoC components (to verify whether they are faulty or not), as well as the on-chip traffic pattern, and changing the route of packets in response to a change in NoC topology (caused by removing faulty links) or on-chip traffic. Using a centralize management approach, we showed that our proposal can effectively reduce the impact on fault on NoC power and performance over previously proposed methods.

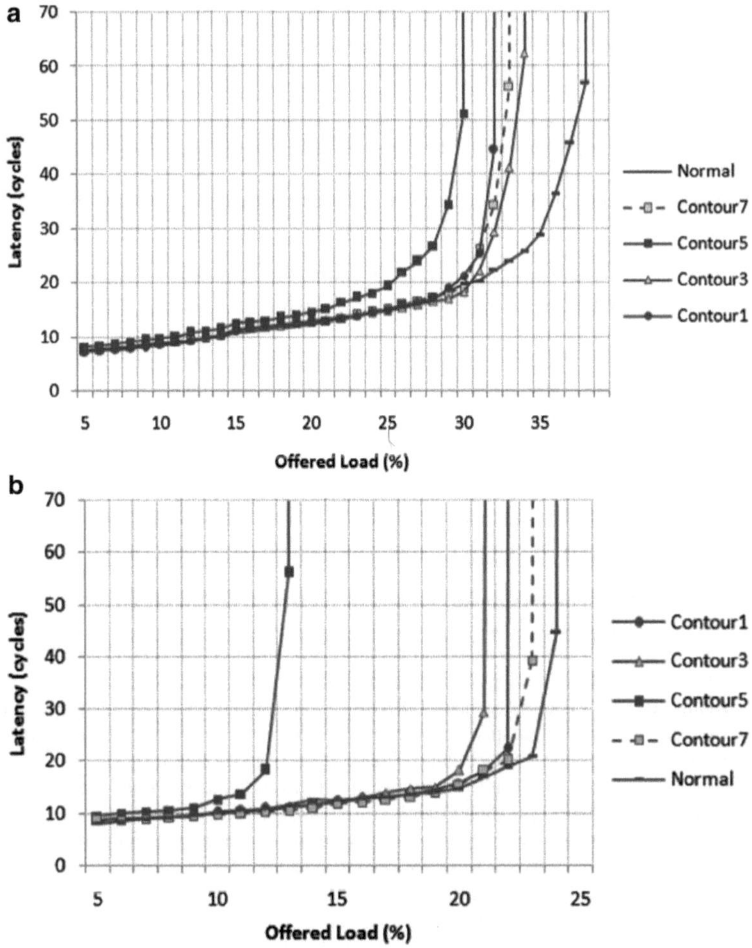

Fig. 8.8 The impact of single-node fault location and traffic load (packet/node/cycle) on (**a**) FaulT-oleReR and (**b**) the algorithm proposed in [19]

References

1. ITRS, International technology roadmap for semiconductors (2007 edition), http://public.itrs.net
2. D. Sylvester, K. Keutzer, A global wiring paradigm for deep submicron design. IEEE Trans. Comp. Aid. Des. Integ. Circuit. Syst. **19**(2), 242–252 (2000)
3. A. Alaghi, M. Sedghi, N. Karimi, M. Fathy, Z. Navabi, Reliable NoC architecture utilizing a robust rerouting algorithm, in *Proceedings of the 6th IEEE East–West Design and Test Symp. (EWDTS)* (Kharkov, Poland, 2008), pp. 101–105
4. A. Patooghy, S.G. Miremadi, XYX: A power & performance efficient fault-tolerant routing algorithm for network on chip, in *Proceedings of the 17th EuroMicro International Conference on Parallel, Distributed and Networks-Based Processing* (Weimar, Germany, 2009), pp. 245–251

5. C. Constantinescu, Intermittent faults in VLSI circuits, *in Proceedings of IEEE Workshop on System Effects of Logic Sot Errors*, Urbana-Champaign, USA, 2006
6. S. Pasricha, N. Dutt, *On-Chip Communication Architectures* (Morgan Kauffman, Boston, 2008)
7. A.H. Ajami, K. Banjerjee, M. Pedram, Modeling and analysis of nonuniform substrate temperature effects on global VLSI interconnects. IEEE Tans. Comp. Aid. Des. Integ. Circuit. Syst. **24**(6), 849–861 (2005)
8. K. Banerjee, A. Mehrotra, A.S. Vincentelli, C. Hu, On thermal effects in deep sub-micron VLSI interconnects, in *Proceedings of the Design Automation Conference (DAC)* (New Orleans, LA, USA, 1999), pp. 885–891
9. J. Hu, R. Marculescu, Exploiting the routing flexibility for energy/performance aware mapping of regular NoC architectures, in *Proceedings of Design, Automation, and Test in Europe Conference* (Munich, Germany, 2003), pp. 668–693
10. J. Duato, A necessary and sufficient condition for deadlock-free adaptive routing in wormhole networks. IEEE Trans. Parall. Distrib. Syst. **6**(10), 1055–1067 (1995)
11. S. Lin, C. Huang, C. Chao, K. Huang, A. Wu, Traffic-balanced routing algorithm for irregular mesh-based on-chip networks. IEEE. Trans. Comp. **57**(9), 1156–1168 (2008)
12. M. Modarressi, A. Tavakkol, H. Sarbazi-Azad, VIP: Virtual point-to-point connections in NoCs. IEEE Trans. Comp. Aid. Desi. Integ. Circuit. Syst. (TCAD) **29**(6) (June, 2010)
13. J. Hu, U.Y. Ogras, R. Marculscu, System-level buffer allocation for application-specific networks-on-chip router design. IEEE Tans. Comp. Aid. Des. Integ. Circuit. Syst. **25**(12), 2919–2933 (2006)
14. M. Schmitz, Energy minimization techniques for distributed embedded systems, Ph.D. thesis, University of Southampton, 2003
15. BookSim: A cycle-accurate interconnection network simulator, Version 2.0, (2013), https://nocs.stanford.edu/cgi-bin/trac.cgi/wiki/Resources/BookSim
16. A.B. Kahng, B. Li, L. Peh, K. Samadi, ORION 2.0: A fast and accurate NoC power and area model for early-stage design space exploration, in *Proceedings of Design, Automation, and Test in Europe Conference (DATE)* (Nice, France, 2009), pp. 423–428
17. R.V. Boppana, S. Chalasani, Fault-tolerant wormwhole routing algorithms for mesh networks. IEEE Trans. Comp. **44**(7), 848–864 (1995)
18. D. Fick, A. DeOrio, G. Chen, V. Bertacco, D. Sylvester, D. Blaauw, A highly resilient routing algorithm for fault-tolerant NoCs, in *Proceedings of the Design, Automation, and Test in Europe (DATE)* (Nice, France, 2009), pp. 21–26
19. Z. Zhang, A. Greiner, S. Taktak, A reconfigurable routing algorithm for a fault-tolerant 2D-mesh network-on-chip, in *Proceedings of the 45th Annual Design Automation Conference* (Anaheim, CA, USA, 2008), pp. 441–446

Chapter 9
Reliable and Adaptive Routing Algorithms for 2D and 3D Networks-on-Chip

Masoumeh Ebrahimi

Abstract Faults may have undesirable effects on the correct operation of the system or at least the performance. NoC inherently has a potential of being a more reliable infrastructure than buses by providing alternative paths between each pair of source and destination routers. However, this potential cannot be utilized without the support of fault-tolerant routing algorithms. In this chapter, we take a detail view of implementing high-performance fault-tolerant routing algorithms in 2D and 3D mesh networks. The required turn models are discussed and all fault conditions are investigated. As faults may occur links and routers, this chapter investigates both types of faults. Unlike traditional methods in which the performance is degraded significantly in faulty situations, the proposed fault-tolerant routing algorithms perform well in maintaining the performance of a faulty network. This is achieved by using the shortest paths as long as possible to bypass faults while non-minimal paths are used when necessary. The proposed methods can be adjusted to balance between reliability and performance. This chapter gives an extensive knowledge to develop a fault-tolerant routing algorithm based on the characteristics of an NoC.

9.1 Introduction

Traditionally, System-on-Chip (SoC) designers employ buses or hierarchical bus structures to interconnect Intellectual Property (IP) blocks. The advances in semiconductor technologies make it possible to integrate billions of gates and hundreds of processing units into a single chip [1]. However, bus architectures are not scalable enough to provide a high-performance communication between the processing units. On the other hand, reliability is another important issue in buses as a fault may disable the whole system or a major part of it. Network-on-Chip (NoC)

M. Ebrahimi (✉)
University of Turku, Turku, Finland
e-mail: masebr@utu.fi

M. Palesi and M. Daneshtalab (eds.), *Routing Algorithms in Networks-on-Chip*,
DOI 10.1007/978-1-4614-8274-1_9, © Springer Science+Business Media New York 2014

has emerged as a solution to address the communication demands of future SoC designs. Scalability, reusability, and reliability characteristics of NoC make it a potential candidate to be used instead of buses. NoC consists of an interconnection of routers to enable a large number of cores to communicate with each other. In an NoC, each core is connected to a router by a local network interface. Cores can communicate with each other by propagating packets through routers in the network. Each router is connected to its neighbors through bidirectional links. NoC allows the communication between the processing units via different routes, so that if one route is faulty, the other routes can be used. This characteristic enables an NoC to be a more reliable infrastructure than a bus.

As the size of a 2D network scales up, the transmission delay between distant routers is significantly increased which results in lower performance and higher power consumption. In addition, a 2D IC design imposes a very large chip area as the number of cores increases. Considering 2D design bottlenecks, the technology is moving toward the concept of 3D integrated circuits where multiple active silicon layers are vertically stacked. Combining the benefits of 3D IC and NoC schemes provides a significant performance gain for SoCs. Layers, stacked on top of each other, are connected via vertical interconnects tunneling through them. Wire bonding, contactless, micro-bump (capacitive or inductive), and through-silicon-via (TSV) are some of the vertical interconnect technologies that have been used in stacked structures [2, 3]. The TSV interconnection approaches have the potential to offer the greatest vertical interconnect density and therefore they are the most promising ones among the other vertical interconnect techniques [2, 4, 5]. The major advantages of 3D ICs are the considerable reduction in the average wire length and wire delay, resulting in lower power consumption and better performance [6–9]. Nevertheless, if the number of IP cores and memories increases in each layer, more TSVs are necessitated to handle the inter-layer communication. Since TSV employs a pad for bonding, the area footprint is augmented significantly [10].

Routing algorithms can be classified as deterministic and adaptive algorithms. The simplest deterministic routing method is dimension-order routing which is known as XY and XYZ in 2D and 3D networks, respectively. Implementations of deterministic routing algorithms are simple but they are unable to distribute packets and tolerate faults efficiently. In adaptive algorithms, a packet can traverse from a source router to a destination router through multiple paths. Adaptive routing has been used in interconnection networks to improve network performance and to tolerate faults. Adaptive routing algorithms can be either partially or fully adaptive. In partially adaptive routing algorithms, packets are limited to choose among some minimal paths, while in fully adaptive methods, packets are allowed to take any shortest paths available between the source and destination pair [4, 11, 12].

In this chapter, we discuss about high-performance fault-tolerant routing algorithms in 2D and 3D mesh networks. For this purpose we build a fault-tolerant routing algorithm on top of fully adaptive routing algorithms in both 2D and 3D network. We use this adaptive characteristic to avoid taking unnecessary non-minimal paths. In the proposed algorithms, the shortest paths are used when the source and destination are not located in the same row or column. Packets have to be rerouted around a fault when the source and destination are located along a same dimension and there is a fault between them. We will prove that minimal and non-minimal routing can be used by packets without creating any cycle in the network and thus the routing algorithms are deadlock-free.

9.2 Related Works

Several methods are presented in the realm of 2D NoCs in order to balance the traffic load over the network. DyXY [13] is a fully adaptive routing algorithm using one and two virtual channels along the X and Y dimensions, respectively. There are few partially and fully adaptive algorithms in a 3D mesh network. MAR [14] is a partially adaptive routing algorithm for 3D NoCs which is based on the Hamiltonian path. It is a simple approach providing the adaptivity without using virtual channels. A fully adaptive routing algorithm in a 3D mesh network is presented in [15], called DyXYZ. Using this algorithm, packets are able to take any shortest paths between the source and destination routers. DyXYZ requires two, four, four virtual channels along the X, Y, and Z dimensions, respectively, to provide fully adaptiveness.

A number of studies presented solutions to tolerate faulty links or routers in a 2D mesh network. The presented method in [16] can tolerate a large number of faults without using virtual channels. However, this approach takes advantage of a routing table at each router and an offline process to fill out the tables. The presented algorithm in [17] does not require any routing tables, but packets take unnecessary non-minimal paths. In this algorithm, an output hierarchy is defined for each position in the network. According to the positions of the current and destination routers, the routing algorithm scans the hierarchy in the descending order and selects the highest priority direction which is not faulty. BFT-NoC [18] presents a different perspective to tolerate faulty links. This method tries to maintain the connectivity between the routers through a dynamic sharing of surviving channels. Zhen Zhang et al. present an approach [19] to tolerate a single faulty router in the network without using virtual channels. The main idea of this algorithm is to reroute packets through a cycle free contour surrounding a faulty router. Each router should be informed about the faulty statuses of eight direct and indirect neighboring routers. DBP [20] method uses a lightweight approach to maintain the network connectivity among non-faulty routers. In this method, besides the underlying interconnection infrastructure, all routers are connected to each other via an embedded unidirectional cycle (e.g. a Hamiltonian cycle or a ring along a spanning tree). A default back-up path is used at each router in order to connect the upstream to the downstream router. All of the mentioned algorithms may take unnecessary non-minimal routes to tolerate faults, which increases the latency of packets significantly. Four high-performance fault-tolerant approaches have been presented in a 2D mesh network, which are based on using the shortest paths. Two of these algorithms tolerate faulty links (MD [21] and MAFA [22]) and two others tolerate faulty routers (HiPFaR [23] and MiCoF [24]). This chapter is mainly based on these sets of works.

Implementing deadlock-free fault-tolerant algorithms are usually complicated in a 2D mesh network while the complexity increases in a 3D mesh network. This is due to the fact that in a 2D mesh network, the algorithms are concerned with two abstract cycles in a XY plane while in a 3D network the cycles should be prevented within each layer (i.e. XY, XZ, and YZ) and between layers. The 4NP-First method [25] utilizes two separate virtual channels, one dedicated to 4N-First routing algorithm and the other to the 4P-First routing algorithm. The 4N-First and 4P-First routing algorithms are resulted from the straight forward extension of the negative-first turn model from a 2D to 3D NoC. The planar-adaptive routing algorithm [26] is a well-known method presented in the realm of interconnection networks. This algorithm requires one, three, and two virtual channels along the X, Y, and Z dimensions, respectively. The adaptivity of this method is limited to a fully adaptive routing algorithm inside a sequence of 2D planes. HamFA [27] is a Hamiltonian-based fault-tolerant algorithm presented for a 3D mesh network. This algorithm tolerates almost all situations of single faults on unidirectional links either in horizontal or vertical links without using virtual channels. The 3D-FT method is presented in [30] which is able to tolerate faults by using the shortest paths and it is the base of 3D section of this chapter.

9.3 Fault-Tolerant Routing Algorithms in a 2D Mesh Network

In this section, at first a conventional fully adaptive routing algorithm is explained. Then we propose a Reliable Routing algorithm, called RR-2D, which enables the network to be fault-tolerant by applying some rules on the fully adaptive routing. For tolerating faults, the statues of surrounding routers and links should be provided at each router. For this purpose, we introduce the fault distribution and monitoring mechanism. Finally, the reliability of RR-2D is checked considering all locations of a single faulty router and link.

9.3.1 Fully Adaptive Routing in a 2D Mesh Network

There are two types of complete cycles that can be formed in the network, known as clockwise and counter-clockwise (Fig. 9.1a). The creation of cycles may lead to

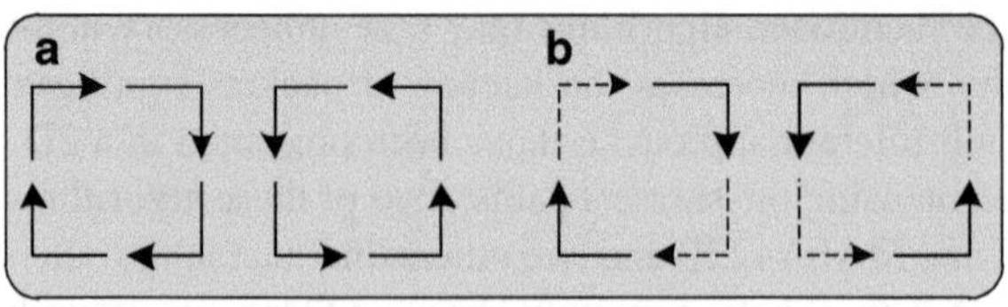

Fig. 9.1 (**a**) Clockwise and counter-clockwise turns; (**b**) permitted and prohibited turns in the XY routing algorithm (Note that *dash lines* indicate prohibited turns)

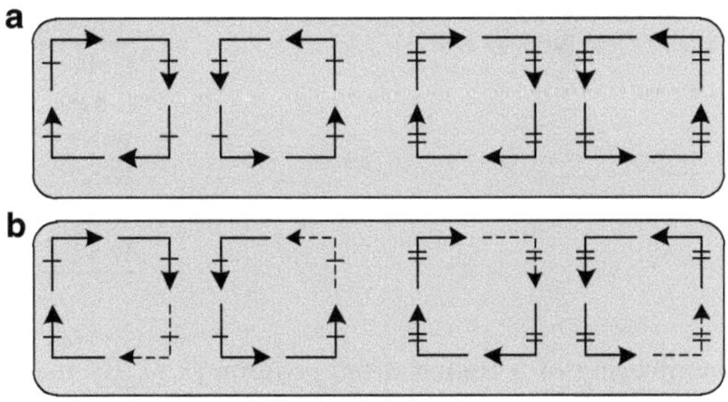

Fig. 9.2 (**a**) Four abstract cycles when using one and two virtual channels along the X and Y dimensions; (**b**) permitted and prohibited turns of RR-2D, similar to Mad-y [10]

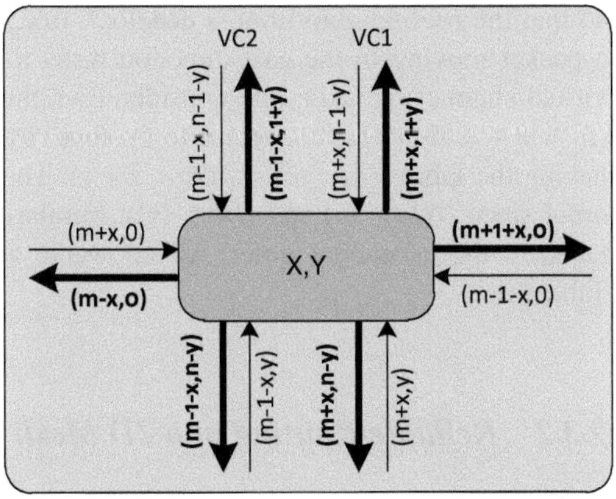

Fig. 9.3 The numbering mechanism of RR-2D similar to Mad-y [10]

deadlock in the network and thus they should be avoided. In turn models, certain turns are prohibited from each cycle in order to break all cyclic dependencies and thus avoiding deadlock. In the XY routing algorithm, for example, packets are routed along the X dimension before proceeding along the Y dimension. As shown in Fig. 9.1b, in this algorithm, two turns are avoided from each abstract cycle, and thus there is no possibility of forming a complete cycle among the remaining turns.

We utilize one and two virtual channels along the X and Y dimensions, respectively, in which four cycles might be formed in the network (Fig. 9.2a). In order to avoid deadlock, one turn is prohibited from each cycle which is shown in Fig. 9.2b. The prohibited turns in each virtual channel are taken from the Mad-y method [28]. Based on this turn model, the turns E-N1, E-S1, N1-E, N2-E, S1-E, and S2-E are permitted for eastward packets while the turns W-S1, W-S2, W-N1, W-N2, N2-W, and S2-W are allowable for westward packets.

To prove deadlock-freeness, we use a numbering mechanism similar to the Mad-y method [28]. This numbering mechanism shows that all the turns have occurred only in the ascending order, and thus no cycle can be formed in the network. A two-digit number *(a,b)* is assigned to each output channel of a router in an $n \times m$ mesh network. According to the numbering mechanism, a turn connecting the input channel *(Ia,Ib)* to the output channel *(Oa,Ob)* is called an ascending turn when $(Oa > Ia)$ or $((Oa = Ia)$ and $(Ob > Ib))$. Figure 9.3 shows the channels'

Table 9.1 All allowable and unallowable turns by RR-2D

Allowable turns		Unallowable turns
E-N1	N1-E	E-N2
E-S1	N2-E	E-S2
W-N1	N2-W	N1-W
W-N2	S1-E	S1-W
W-S1	S2-E	
W-S2	S2-W	

numbering of a router at the position *(x,y)*. By using this numbering mechanism, it is guaranteed that all permitted turns are taken strictly in the increasing order, so that the routing algorithm is deadlock-free. For instance, if the turn E-N1 (i.e. a packet moving to the east direction turns to the north direction using the first virtual channel) is taken into consideration, the west input channel with the label ($Ia = m + x$, $Ib = 0$) is connected to the first virtual channel of the north output port having the label ($Oa = m + x$, $Ob = 1 + y$). This turn takes place in the ascending order since (($Oa = Ia$) and ($Ob > Ib$)). Similarly, all the other permitted turns are taken in the ascending order. All allowable and unallowable turns are listed in Table 9.1.

9.3.2 Reliable Routing in a 2D Mesh Network

The fault-tolerant routing algorithms are usually very complex. In contrast, the proposed algorithms in this chapter are very simple and can be easily implemented. The basic idea behind these algorithms is to keep the adaptivity of packets as long as possible. It means that for example if they are two minimal directions to deliver a packet, a direction should be preferably selected in which from the next router, the packet has still some alternative paths to reach the destination router. This simple idea avoids taking unnecessary non-minimal paths and reduces the complexity of the fault-tolerant algorithms.

Let us consider the examples of Fig. 9.4 where a packet is sent from the current router C to the destination router D. In Fig. 9.4a, the packet can be sent through two minimal directions (i.e. E and N) to reach the destination router D. By sending the packet to either of directions, the packet will have two minimal choices from the next neighboring routers to reach the destination router. Therefore, if one path is faulty, there is an opportunity to send a packet through the other path. In Fig. 9.4b, however, if the packet is sent to the north neighboring router, it will have only one option to reach the destination router. So it is better to send the packet to the east direction from where there are two minimal paths toward the destination. In Fig. 9.4c, there are one and two minimal paths from the east and north neighboring routers, respectively, to reach the destination router. Thereby, the north direction is a better choice to deliver the packet. In Fig. 9.4d, there is only one possible choice to reach the destination router from either the east or north neighboring router (as

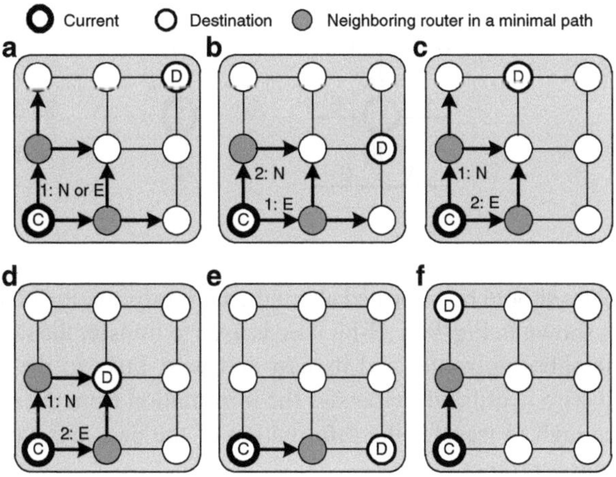

Fig. 9.4 The basic idea of the RR-2D method (Note that numbers determine the priority of selecting among different routes)

we will see in Sect. 9.3.3, to distribute less fault information, the possibility of the north direction is checked earlier than the east direction). In Fig. 9.4e, f, only one minimal route exists to reach the destination and the packet has to be routed through this single path. In this case if there is a fault in the path, the packet has to take a non-minimal route to bypass the fault. In fact, by making a decision similar to Fig. 9.4b, c, packets do not lose their adaptivity and thus they will not face similar situations as in Fig. 9.4e, f.

In general, packets are routed inside the network using the permitted turns offered by a fully adaptive routing algorithm. The adaptivity is limited when the distance between the current and the destination router reaches one in at least a dimension. RR-2D avoids reducing the distance into zero in one direction when the distance along the other direction is greater than one.

9.3.3 Fault Monitoring and Management Technique

We assume that faults are detected by a fault-detection technique and we monitor this information at each router. To be able to tolerate faults, the routing unit should be assisted by the fault information of some routers and links.

RR-2D needs to know at most the statuses of eight surrounding links to be able to make its routing decision (Fig. 9.5a). Moreover, to tolerate a faulty router, it is enough that each router is informed about the fault statues of its four neighboring routers (Fig. 9.5b). The whole information that is needed by RR-2D is shown in Fig. 9.5c. This information should be transferred to the given router using a monitoring and management mechanism. Each router is aware about the fault in its

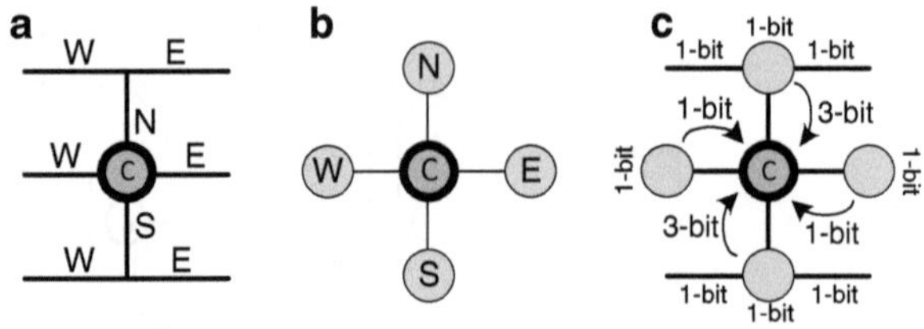

Fig. 9.5 The statuses of eight links and four routers are needed by RR-2D

instant links, but it should be informed about the fault information on the other links and routers. As shown in Fig. 9.5c, 3-bit wire is used to transfer the fault information of the north neighboring router and its east and west links to the current router. Similarly, 3-bit wire is utilized to transfer the information from the south direction. 1-bit wire is enough to transfer the information of the east and west neighboring routers to the current router.

9.3.4 Tolerating Faulty Links and Routers by RR-2D

In this section, we investigate how to tolerate faults using RR-2D. The simple rule is to check the possibility of sending a packet through the greater-distance dimension when the distance to the destination router reaches one along a dimension. The packet is sent to the greater-distance dimension if the instant link and router along this direction are non-faulty; otherwise the smaller-distance dimension is examined. In other situations (i.e. distance has not reached a along a dimension), packets have no routing limitation. Different positions of a faulty link and router for a northeastward packet are illustrated in Fig. 9.6.

As shown in Fig. 9.6a, when the distance along both X and Y dimensions reaches one (X-dir = 1 and Y-dir = 1), there will be six different positions of faults (two routers and four links). According to RR-2D, east and north directions have the same priority to be selected as there are no alternative choices from the next router to the destination router. However, according to the fault distribution mechanism (Fig. 9.5c), the current router knows about the faulty links in the NE path and also the N neighboring router while it does not know about the fault status in the whole EN path. Therefore, the availabilities of the NE path and N router are checked and if they are non-faulty the packet is sent to the north direction, otherwise the packet is possibly sent to the E direction. In the patterns A1, A2, A3, and A6 of Fig. 9.6a, the NE path and the N router are non-faulty and the packet is sent to the north direction. In the patterns A4, A5, and A7, the packet is delivered to the east direction as either the NE path or the N router is faulty.

In Fig. 9.6b, when X-dir = 1 and Y-dir = 2, a fault might occur in seven different locations of links and four locations of routers. In all patterns, the availabilities of the N link and the N router are examined before those of the east direction. In the patterns B1, B2, B3, B4, B5, B6, B7, B9, B10, and B11, the packet is sent to the

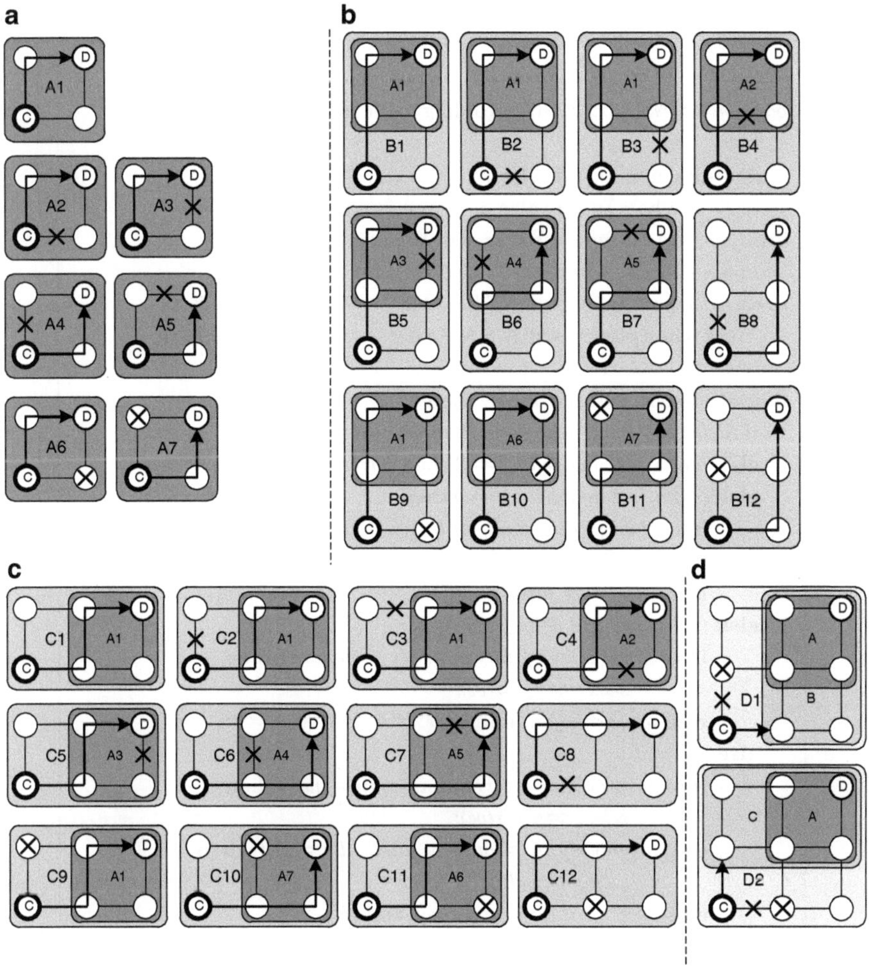

Fig. 9.6 Tolerating faulty routers and links by RR-2D

north direction as both the link and router are non-faulty, while in the next hop, one of the patterns of Fig. 9.6a arises (i.e. patterns A1 to A7). In the patterns B8 and B12 of Fig. 9.6b, the packet has to be routed to the east direction to reach the destination router. In Fig. 9.6c, when X-dir = 2 and Y-dir = 1, the availability of the E link and the E router should be checked before the N link and the N router. Therefore, in the patterns C1, C2, C3, C4, C5, C6, C7, C9, C10, and C11, the packet is sent to the east direction as both the E link and the E router are non-faulty. In the patterns C8 and C12, the packet is delivered to the north direction so that the fault is bypassed. In all the other cases (when X-dir > =2 and Y-dir > =2 in Fig. 9.6d), the packet is sent to the non-faulty direction. In the next hop, the patterns are similar to Fig. 9.6b, c. Based on this discussion, to support all single

```
Definitions:  Xc,Yc,Xd,Yd: X and Y coordinates of current and destination
routers
                    **______________________________**
position <= {NE,NW,SE,SW} according to the source and destination pos.

x_dir <= E when Xd > Xc else W;
y_dir <= N when Yd > Yc else S;
delta_x <= Xd - Xc when Xd>Xc else Xc-Xd;
delta_y <= Yd - Yc when Yd>Yc else Yc-Yd;
vc <= vc1 when position={E,NE,SE} else       --eastward packets
      vc2 when position={W,N,S,NW,SW};       --westward packets

if position={NE, NW, SE, or SW} then
   if (delta_x>=1 and delta_y=0) then select <= x_dir;
   elsif (delta_x=0 and delta_y>=1) then select <= y_dir(vc);
   elsif (delta_x=1 and delta_y>=1) then
         if neighbor(y_dir)=faulty or link(y_dir)=faulty then
                  select <= x_dir;
         else select <= y_dir(vc); end if;
   elsif (delta_x>1 and delta_y=1) then
         if neighbor(x_dir)=faulty or link(x_dir)=faulty then
                  select <= y_dir(vc);
         else select <= x_dir; end if;
   else
         if neighbor(x_dir)=faulty or link(x_dir)=faulty then
                  select <= y_dir(vc);
         elsif neighbor(y_dir)=faulty or link(x_dir)=faulty then
                  select <= x_dir;
         else select <= x_dir or y_dir(vc); end if;
   end if;
end if;
```

Fig. 9.7 RR-2D for northeast, northwest, southeast, and southwest packets

faults, only the shortest paths are used. The same perspective can be applied to the northwest-, southeast-, and southwest-ward packets. Figure 9.7 shows the pseudo code of the RR-2D routing algorithm covering all these positions.

When the packet is east-, west-, north-, or south- bounded and there is a faulty link or router in the path, the packet must be rerouted through a non-minimal path around the fault. As illustrated in Fig. 9.8a, for the eastward packet, at first the east link and router are checked and if they are non-faulty, the packet is sent to this direction. However, if either the link or the router is faulty, the packet is delivered to the north or south direction according to the congestion values. Westward packets do the same behavior (Fig. 9.8b). If a northward packet faces a fault in the north link

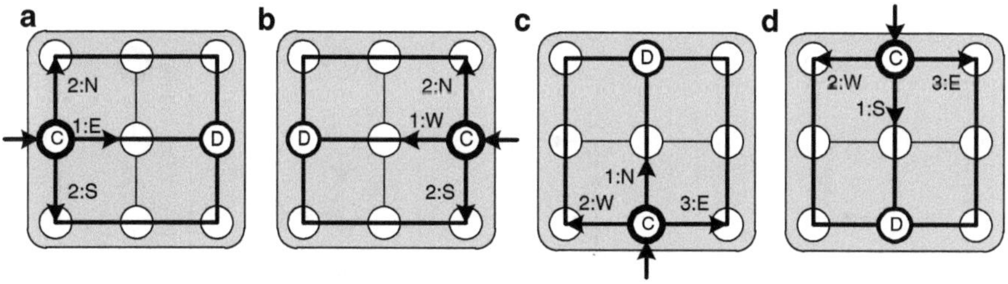

Fig. 9.8 Bypassing faults when the destination is located in the (**a**) east, (**b**) west, (**c**) north, (**d**) south positions of the source router (Note that numbers determine the priority of selecting among different routes)

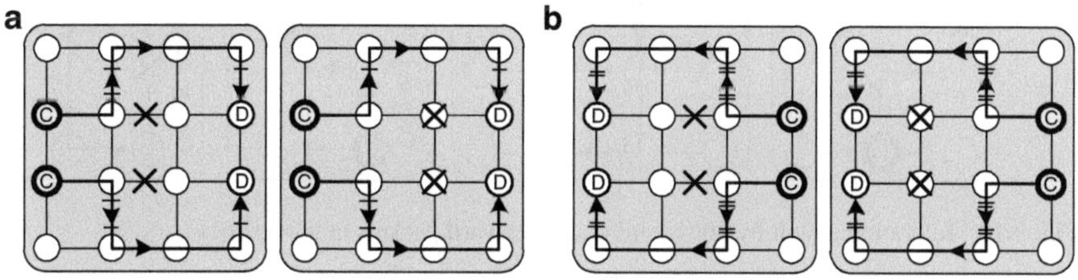

Fig. 9.9 Tolerating a fault by (**a**) eastward packets (**b**) westward packets

or router (Fig. 9.8c), the west direction is checked earlier than the east direction. It means that rerouting through the east direction is done only when the fault is located in the left borderline. A similar perspective is applied to the southward packets (Fig. 9.8d).

Now, we need to show that all the required turns for bypassing faults are in the set of allowable turns. By investigating the required turns, we notice that eastward packets use the E-N1, N1-E, E-S1, and S1-E turns to bypass a faulty link or router (Fig. 9.9a) in which all turns are in the set of allowable turns. Similarly, as shown in Fig. 9.9b, all the required turns by the westward packets are permitted (i.e. W-N2, N2-W, W-S2, and S2-W).

We should also prove that the northward and southward packets are routed in the network without creating deadlock. As illustrated in Fig. 9.10a, b, normally the northward and southward packets use the permitted turns as N2-W, W-N2, N2-E, S2-W, W-S2, S2-E to bypass faults. However, when the source and destination routers are located in the left borderline and there is a faulty link or router in the path, the required turns are N2-E, E-N2, N2-W, S2-E, E-S2, and S2-W. Among them, the E-N2 and E-S2 turns are unallowable according to our turn models (Table 9.1), but a complete cycle cannot be formed in borderline cases (as indicated in [19]) and these unallowable turns can be safely taken. The remaining part of the RR-2D routing algorithm to tolerate faulty links and routers for east-, west-, north-, and south- ward packets is shown in Fig. 9.11 (i.e. continuation of Fig. 9.7).

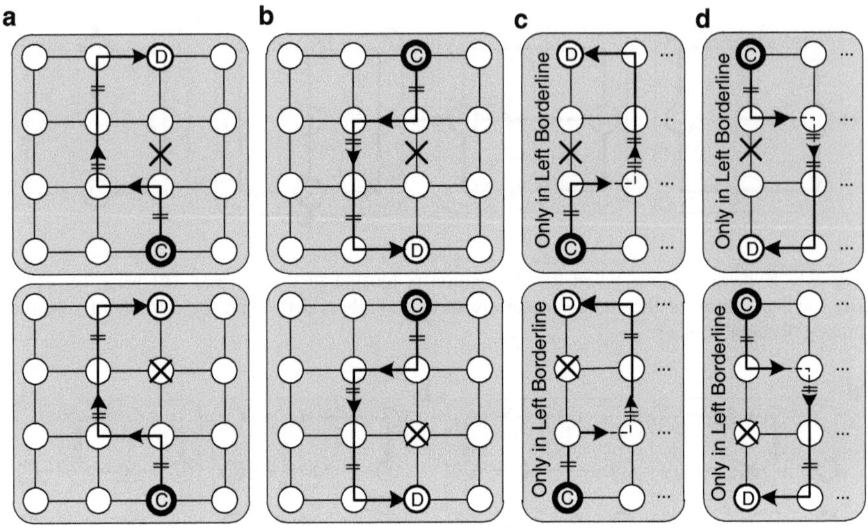

Fig. 9.10 Tolerating a fault by northward and southward packets in borderlines

```
if position={E or W} then
   if delta_y=0 then
         if neighbor(x_dir)=faulty or link(x_dir)=faulty then
                  select <= N(vc) or S(vc) based on their availability;
         else select <= x_dir; end if;
   else
         if neighbor(y_dir)=dest or link(y_dir)=faulty then
                  select <= y_dir(vc);
         else select <= x_dir; end if;
   end if;
elsif position={N or S} then
   if delta_x=0 then
         if neighbor(y_dir)=faulty or link(y_dir)=faulty then
                  if Xc/=0 then select <= west; else select <= east; end if;
         else select <= y_dir(vc); end if;
   else
         if inPort/= {E,W} and
            (neighbor(x_dir)=non-faulty or link(x_dir)=faulty) then
                  select <= x_dir;
         else select <= y_dir(vc); end if;
   end if;
end if;
```

Fig. 9.11 RR-2D for east-, west-, north-, and south- ward packets

9.4 Fault-Tolerant Routing Algorithms in a 3D Mesh Network

In this section, we present a Reliable Routing algorithm for 3D stacked mesh, called RR-3D. We investigate this method for tolerating faulty routers and links. At first, we present a fully adaptive routing algorithm in a 3D mesh network using two, two, and four virtual channels. Then we show how packets can switch between virtual channels to achieve a better fault-tolerant capability. We explain the fault monitoring and management technique and finally we investigate the proposed algorithm regarding all positions of a single faulty router and link.

9.4.1 Fully Adaptive Routing in a 3D Mesh Network

A 3D network can be divided into eight subnetworks: ((X+)(Y+)(Z+), (X+)(Y+)(Z-), (X+)(Y−)(Z+), (X+)(Y−)(Z−), (X−)(Y+)(Z+), (X−)(Y+)(Z−), (X−)(Y−)(Z+), and (X−)(Y−)(Z−)) where "+", "−" represent the channels along the positive and negative directions, respectively. A simple way to implement a fully adaptive routing algorithm is to assign a separate set of virtual channels to each subnetwork such as: ((X1+)(Y1+)(Z1+), (X2+)(Y2+)(Z1−), (X3+)(Y1−)(Z2+), (X4+)(Y2−)(Z2−), (X1−)(Y3+)(Z3+), (X2−)(Y4+)(Z3−), (X3−)(Y3−)(Z4+), and (X4−)(Y4−)(Z4−)) where the numbers indicate the virtual channel number along each dimension. With this channel assignment, four virtual channels are needed along each dimension and the network is deadlock-free as subnetworks are disjoint from each other.

In DyXYZ [15], the number of virtual channels are reduced to two along one dimension (i.e. thereby, it requires four, four, and two virtual channels along the X, Y, and Z dimensions). In this method, the whole network is further split into two main subnetworks, each having four subnetworks. Dividing the network into two parts can be done along one dimension, reducing the number of virtual channels from four to two for that dimension. Since two main subnetworks are separated, the network remains deadlock-free.

As we show, the number of virtual channels can be further reduced to two, two, and four virtual channels along the X, Y, and Z dimensions (i.e. two virtual channels less than DyXYZ). In this method, the network is partitioned into four subnetworks (Fig. 9.12) as: ((X+)(Y+)(Z*), (X−)(Y+)(Z*), (X+)(Y−)(Z*), and (X−)(Y−)(Z*)) where "+", "−" represent the channels along the positive and negative directions, respectively, and "*" stands for both positive and negative directions (i.e. a bidirectional channel). The deadlock freeness can be proved by guaranteeing that subnetworks are disjoint and cycles cannot be formed within each subnetwork. The virtual channel assignment of this algorithm is as follow: ((X1+)(Y1+)(Z1*), (X2+)(Y1−)(Z2*), (X1−)(Y2+)(Z3*), and (X2−)(Y2−)(Z4*)) where the numbers indicate the virtual channel number assigned to a dimension.

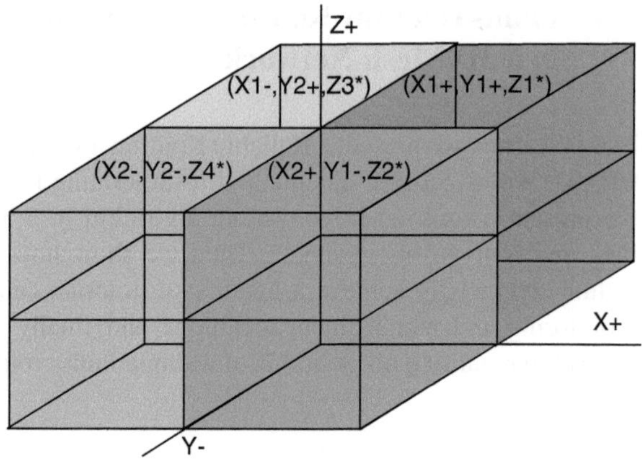

Fig. 9.12 Four disjoint subnetworks

Theorem 1 The network is deadlock-free within each subnetwork

A cycle can be formed if packets are able to take both positive and negative directions along at least two dimensions. For example, to form a cycle in a XY plane, it is necessary to take X+, X−, Y+, and Y− directions. Similarly, to form a cycle in a YZ plane, there should be a possibility of taking Y+, Y−, Z+, and Z− directions. Finally, to form a cycle in a XZ plane, X+, X−, Z+, and Z− directions should be taken by packets. As can be obtained from four different subnetworks, only one pair along the Z dimension (Z*: Z−,Z+) is completed in each of four subnetworks, and thus there is no possibility of forming deadlock within each subnetwork.

Theorem 2 The network is deadlock-free between subnetworks

To prove that the network is deadlock-free between subnetworks, it is enough to show that different subnetworks are disjoint from each other along each dimension. By a pairwise comparison among each two subnetworks, it can be easily obtained that either the direction or the virtual channel number differs along each dimension. For example, the subnetwork 1 (X1+)(Y1+)(Z1*) and the subnetwork 3 (X1−)(Y2+)(Z3*) are disjoint along the X dimension since the subnetwork 1 covers the positive direction of the virtual channel 1 (X1+) and the subnetwork 3 covers the negative direction of virtual channel 1 (X1−). The two subnetworks are also disjoint along the Y dimension since the virtual channel 1 is employed in the subnetwork 1 (Y1+) and the virtual channel 2 is used in the subnetwork 3 (Y2+). Similarly, two subnetworks use different virtual channel number along the Z dimension (Z1* and Z3*). This comparison can be done for any other pairs of subnetworks (Table 9.2).

Considering Theorem 1 and Theorem 2, we prove that the network is deadlock-free within each subnetwork and on the other hand, different subnetworks are disjoint from each other, therefore the whole network is deadlock-free.

We proved that four subnetworks are disjoint from each other. Based on this proof, all packets belong to one subnetwork and they cannot switch to the other ones. However, it limits the fault-tolerant capability as packets cannot use the channels from the other subnetworks in the case of faults.

Table 9.2 Assigning virtual channels to subnetworks

Subnetwork A	Subnetwork B	X dimension	Y dimension	Z dimension
$(X1^+)(Y1^+)(Z1^*)$	$(X1^-)(Y2^+)(Z3^*)$	Differs in vc direction	Differs in vc number	Differs in vc number
$(X1^+)(Y1^+)(Z1^*)$	$(X2^+)(Y1^-)(Z2^*)$	Differs in vc number	Differs in vc direction	Differs in vc number
$(X1^+)(Y1^+)(Z1^*)$	$(X2^-)(Y2^-)(Z4^*)$	Differs in vc number	Differs in vc number	Differs in vc number
$(X1^-)(Y2^+)(Z3^*)$	$(X2^+)(Y1^-)(Z2^*)$	Differs in vc number	Differs in vc number	Differs in vc number
$(X1^-)(Y2^+)(Z3^*)$	$(X2^-)(Y2^-)(Z4^*)$	Differs in vc number	Differs in vc direction	Differs in vc number
$(X2^+)(Y1^-)(Z2^*)$	$(X2^-)(Y2^-)(Z4^*)$	Differs in vc direction	Differs in vc number	Differs in vc number

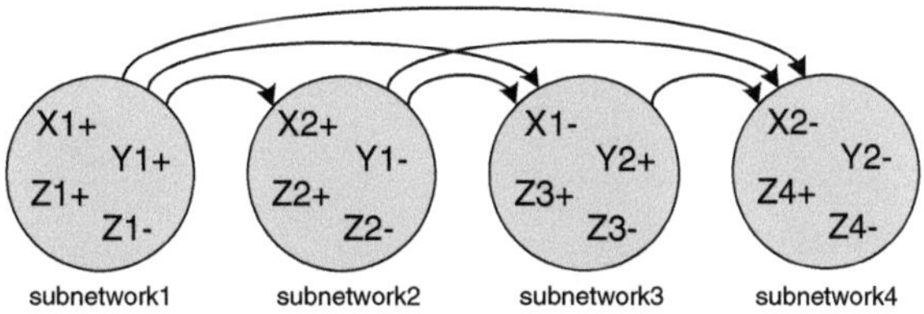

Fig. 9.13 Packets can switch between subnetworks in the strictly ascending order

To improve the fault-tolerant capability, we enable the switching between subnetworks without forming cycles. As shown in Fig. 9.13, packets can traverse between subnetworks in the strictly ascending order so that a cycle cannot be formed. It means that the packets moving in the subnetwork 1 can switch to higher subnetworks (2, 3 or 4), but after switching, they are no longer allowed to use any channels from the subnetwork 1. Similarly, packets in the subnetwork 1 or the subnetwork 2 can switch to the subnetwork 3 or 4, losing the possibility of using the channels of the subnetwork 1 or the subnetwork 2. Finally, the packets moving in the subnetwork 1, subnetwork 2, or subnetwork 3 can switch to the subnetwork 4, but they cannot use any channels than the subnetwork 4. In general, after using a channel in a higher subnetwork, switching to the lower subnetwork is not allowed.

9.4.2 Reliable Routing in a 3D Mesh Network

Similar to a 2D mesh network, in order to tolerate faults in a 3D mesh network, the fully adaptive routing algorithm is slightly modified. As shown in Fig. 9.14a, when a source is located at the router 0 and a destination is located at the router 8, 13, 14, 16, 17, 20, 22, 23, 24, 25, or 26, there are at least two minimal candidates to forward a packet. By sending a packet to any of the neighboring routers located in minimal paths, there will be at least two alternative routes from the next router to the destination. In Fig. 9.14b, when a destination is located at the router 5, 7, 11, 15, 19, or 21, there are two directions to forward a packet. However, by sending the

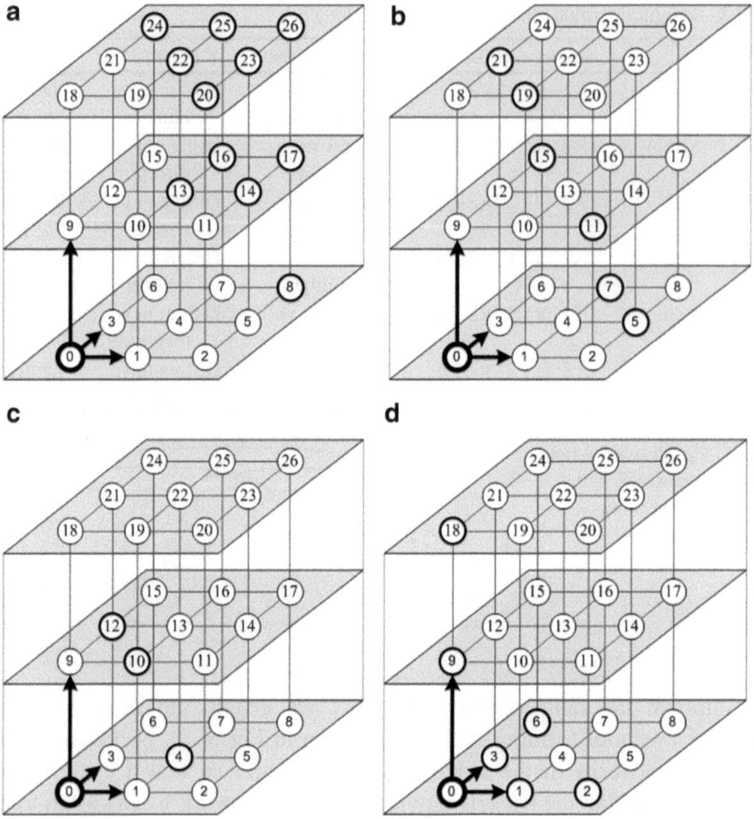

Fig. 9.14 Reliable routing for a 3D mesh network

packet to one of these directions, the adaptivity will be limited to one path while by delivering the packet to the other direction there will be at least two shortest paths to reach the destination. Thereby, it is better to select a path which maintains the adaptivity of packets. In Fig. 9.14c, when a destination is at the router 4, 10, or 12, there are again two options to forward the packet while by selecting either option the adaptivity will be lost, and thus there is not any preference to select between them. Finally, in Fig. 9.14d, packets have no alternative paths to reach the destination router 1, 2, 3, 6, 9, or 18.

9.4.3 Fault Monitoring and Management Technique

To be able to tolerate faults, some fault information should be provided at the routing unit. We assume that faults are detected by a fault-detection technique and we monitor this information at each router. RR-3D needs to monitor the faulty statuses of its six neighboring routers and eighteen surrounding links as shown in

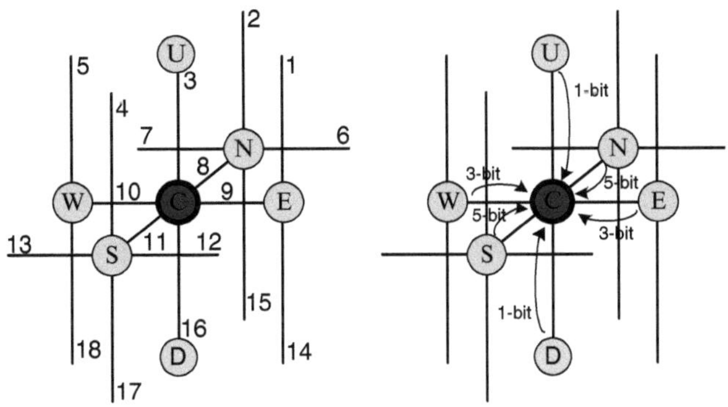

Fig. 9.15 Fault monitoring and management of RR-3D

Fig. 9.15 at each given router (i.e. router C). The links include the instant links connected to each neighboring routers (i.e. 3, 8, 9, 10, 11, and 16). The information of these instant links is already provided for the given router using a fault-detection technique. However, the fault information of the other links should be sent to the given router. These links are as: four links connected to the north neighboring router (i.e. 2, 6, 7, and 15), four links connected to the south neighboring router (i.e. 4, 12, 13, and 17), two links connected to the east neighboring router (i.e. 1 and 14), and two links connected to the west neighboring router (i.e. 5 and 18). Moreover, using a fault-detection technique, a router is informed about the fault status of itself while the information about the neighboring routers (i.e. north, south, east, west, up, and down) should be transferred to the given router. The fault management mechanism is responsible to combine the fault information at each router and transfer it to the neighboring routers. In this example, either of the north and south neighboring routers combines the fault information of the router itself and the connected four links and transfers a 5-bit information to the given router. East and west neighboring routers transfer 3-bit to the given router; 1-bit for the fault status of the router and 2-bit for the fault statuses of the links. From the up and down directions, 1-bit information is transferred to the given router which indicates the fault status of the connected router.

Now, let us investigate which information should be transferred from the current router to each of the neighboring routers. The fault status of the router C and its four links 3, 9, 10, and 16 are transferred to the north and south neighboring routers. The fault status of the router C and its two links 3 and 16 are transferred to the east and west neighboring routers. Finally, only the fault status of the router C is sent to the up and down neighboring routers.

9.4.4 *Tolerating Faulty Links and Routers by RR-3D*

In this section, we show how faulty links can be tolerated using RR-3D. When packets have at least two minimal choices (similar to Fig. 9.14a), they are sent

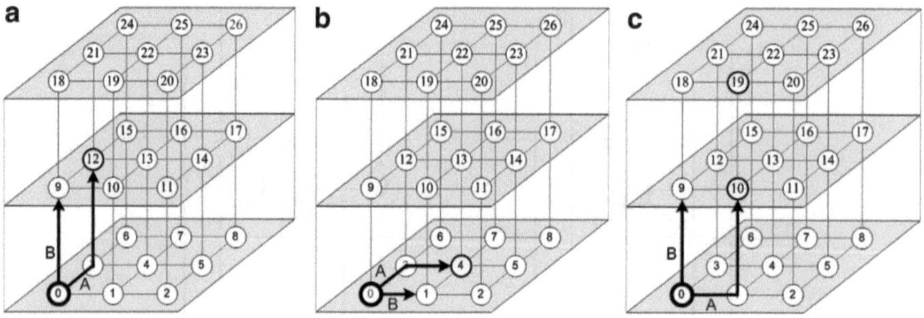

Fig. 9.16 RR-3D when the distance reaches one along two directions

to a direction with a non-faulty link and router. When the distance reaches one along two directions, it is important to send a packet through a non-faulty path as the wrong decision may result in dropping the packet or taking non-minimal routes. In Fig. 9.16a, the packet can be sent to the north or up direction. If only the neighboring links and routers are checked, the packet might be sent to the up direction. However, if the link between the router 9 and the destination router 12 is faulty, a non-minimal path should be taken. In order to avoid this situation, at first, the fault status of the NU path and the N router are checked (which are available by the management mechanism). If they are non-faulty, the packet is sent to the north direction; otherwise the up direction is selected. Similarly, in Fig. 9.16b, the statuses of the NE path and the N router are checked and if they are non-faulty, the packet is sent to the north direction; otherwise the east direction is selected. Finally, in Fig. 9.16c, the packet is sent to the east direction if the EU path and the E router are non-faulty; otherwise the up direction is selected.

When the current and destination routers are located in the same dimension and a link or router between them is faulty, the non-minimal route is necessitated. In order to avoid taking unnecessary longer paths, RR-3D always tries not to reduce the distance to zero along two dimensions when the distance along the third dimension is greater than one. For example, in Fig. 9.16c, if the source and destination are located at the router 0 and 19, respectively, the availability of the up direction is checked earlier than the east direction. The reason is that if the packet is sent to the east dimension and the link 10–19 is faulty, the packet has to take a non-minimal path. On the other hand, by sending the packet to the router 9, the packet has two alternative routes to reach the destination, and if one of the routes is faulty, the packet is sent through the other one.

As already mentioned, when packets are east-, west-, north-, or south-bounded and there is a fault in the path, they have to take a non-minimal path. RR-3D should be able to tolerate these faults as well. (It is worth mentioning that if the source and destination routers are not located in the same dimension, packets never face these conditions as faults are bypassed prior to reaching them). We take advantage of the capability of switching between subnetworks. Therefore, when the source and destination routers are located in a same dimension, packets are started routing in

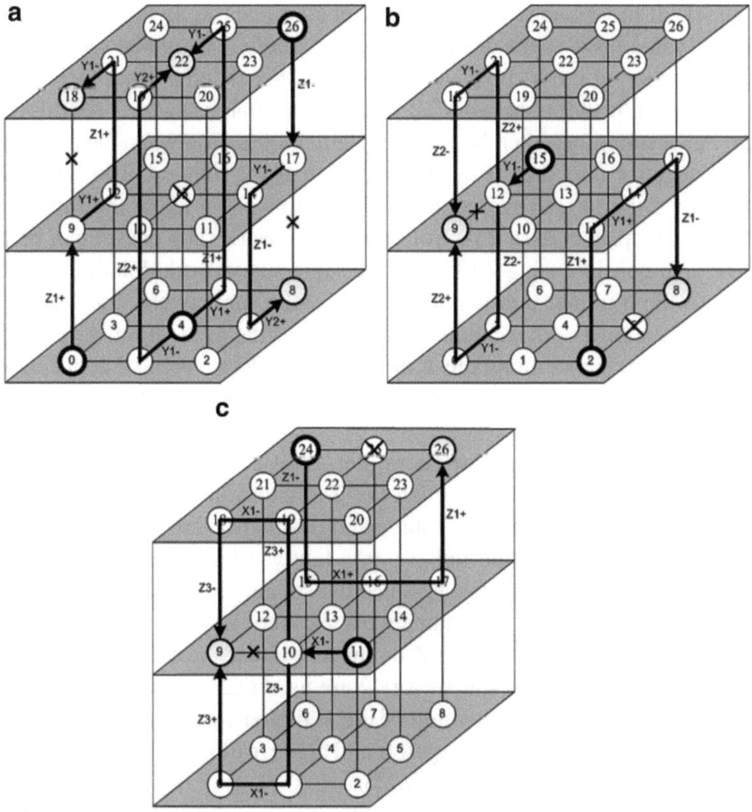

Fig. 9.17 RR-3D when there is only a single path between the source and destination routers

the lowest possible subnetwork and then they can switch to a higher subnetwork in the ascending order if needed. The rules of the RR-3D algorithm are as follows:

- Rule1: If a fault occurs on the Z dimension, packets are rerouted through the Y dimension.
- Rule2: If a fault occurs on the X or Y dimension, packets are rerouted through the Z dimension.

For performing these rules, packets may require to change their subnetworks. Let us consider the example of Fig. 9.17a where the source and destination are located at the router 4 and 22, respectively, and the router 15 is faulty. Since the fault has occurred along the Z dimension, according to Rule1, packets are rerouted through the Y dimension which can be in its positive or negative direction. Let us assume that rerouting takes place through the positive direction of the Y dimension. Since the subnetwork 1 covers Y1+ (shown in Fig. 9.13), the packet uses this channel. Then the packet should be routed along the positive direction of the Z dimension. The subnetwork 1 also covers Z1+, so the packet is still routed using the channels of the subnetwork 1. The packet continues along this direction until it reaches the same layer as the destination router where it should be sent through the negative direction of the Y dimension. However, this channel is not included in the subnetwork 1 while the subnetwork 2 covers Y1– so that the packet uses this channel to reach the destination router. Now, we explain the situation when the packet is sent along the negative direction of the Y dimension at the router 4. The subnetwork 1 does not cover the negative direction of the Y dimension while the subnetwork 2 covers it, so the packet uses Y1– from this subnetwork. The packet can be routed along the Z dimension using Z2+ from the same subnetwork. Finally, the positive direction of the Y dimension is not covered by the subnetwork 2 and the Y2+ is taken from the subnetwork 3 and the packet reaches the destination. As every router has at least one neighbor along the Y dimension, faults on vertical connections can be tolerated by rerouting packets through the Y dimension (see two more examples in Fig. 9.17a).

Figure 9.17b shows the cases where the fault occurs on the Y links and it is tolerated by rerouting packets through the Z dimension according to Rule2. We investigate the example where the source and destination are located at the routers 15 and 9, respectively, and the link 12–9 is faulty. At first the negative direction of the Y dimension should be taken. Since the subnetwork 1 does not cover it, the next subnetwork is checked. The subnetwork 2 covers Y1– and thus the packet uses this channel. For routing along the Z dimension either in the positive or negative direction, the channels of the subnetwork 2 are used (Z2*). Then the packet should be routed along the negative direction of the Y dimension which is covered by the subnetwork 2 (Y1−). Finally, the negative and positive directions of the Z dimension are covered by the same subnetwork (Z2*).

In Fig. 9.17c, according to Rule2, faults in the X dimension are tolerated by rerouting packets through the Z dimension. Let us assume the case where the source and destination are located at the router 11 and 9, respectively, and the link 10–9 is faulty. The negative direction of the X dimension is not covered by the subnetworks 1 and 2, and thus the X1− from the subnetwork 3 is used. The packet is rerouted along the Z dimension using Z3+ or Z3− from the same subnetwork. The packet needs to take the X1− and then Z3+ or Z3− to reach the destination router. All of these channels are covered by the subnetwork 3. As every router has at least one neighbor in the Z dimension, faults on the X or Y dimension can be tolerated by rerouting packets along the Z dimension (see additional examples in Fig. 9.17b, c).

9.5 Results and Discussion

To evaluate the efficiency of the proposed routing schemes, a 2D and 3D NoC simulator is developed with VHDL to model all major components of the on-chip network. For all the routers, the data width is set to 32 bits. Each input buffer can accommodate six flits in each channel. Moreover, the packet length is uniformly distributed between 1 and 8 flits. The request rate is defined as the ratio of the successful packet injection into the network over the total number of injection attempts. As a performance metric, we use latency defined as the number of cycles between the initiation of a packet issued by a processing element and the time when the packet is completely delivered to the destination.

For evaluating the performance in a 2D mesh network, RR-2D is compared with a reconfigurable routing presented in [19] (in simulation we call this method ReRS). As discussed in the related works, ReRS does not require any virtual channel and it is able to tolerate all locations of a single faulty router. However, using ReRS, unnecessary longer paths are taken to tolerate faults which results in creating congestion around the faulty regions. In addition, ReRS is based on deterministic routing and thus packets cannot be well distributed over the network. RR-2D is our proposed method which is based on a fully adaptive routing and utilizes one and two virtual channels along the X and Y dimensions, respectively. RR-2D is able to tolerate all locations of a single faulty link or router. To have a fair comparison we use the same number of virtual channels in both methods and for this purpose, an extra virtual channel is added to the ReRS approach. This virtual channel is used for the performance purposes.

For measuring the performance in a 3D mesh network, the proposed method (RR-3D) is compared with HamFA [27]. HamFA is a fault-tolerant method tolerating almost all one-faulty unidirectional links. It is able to tolerate faults either on vertical or horizontal link without using virtual channels. HamFA is a partially adaptive routing algorithm. On the other hand, RR-3D can tolerate both faulty links and routers while guaranteeing to tolerate all single faults wherever in the network. RR-3D is built upon a fully adaptive routing algorithm requiring two, two, and four virtual channels. To have a fair comparison, the same number of virtual channels is used in both methods.

We perform two sets of simulations: 1- to measure the performance of the proposed methods against the baseline methods in both 2D and 3D mesh networks. 2- to measure the reliability of the proposed method compared with baseline methods. In both sets of simulations, we perform the experiments on an 8×8 and a $4 \times 4 \times 4$ mesh network. For performance analysis, the simulator is warmed up for 20,000 cycles and then the average performance is measured over another 200,000 cycles.

9.5.1 Performance Analysis Under Uniform Traffic Profile

In the uniform traffic profile, each processing element generates data packets and sends them to another processing element using a uniform distribution [29]. In Fig. 9.18a, the average communication latencies of RR-2D and ReRS are measured for fault-free and a single faulty router cases. In addition, the performance of RR-2D is measured under six faults in the network, each randomly chosen to be a faulty link or router. As observed from the results, in fault-free cases, the ReRS method is performing the best as it is based on deterministic routing (i.e. similar to XY

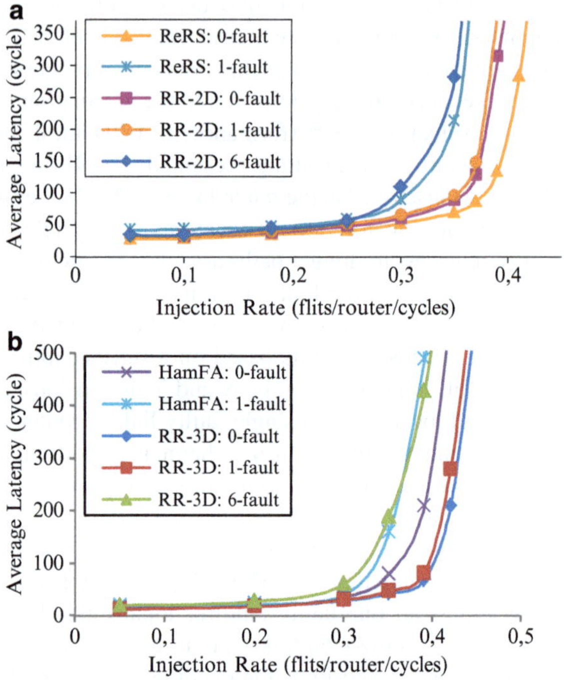

Fig. 9.18 Performance under uniform traffic profile in (**a**) 2D network, (**b**) 3D network

routing) which is well suited to uniform traffic. When a single fault occurs in the network, the performance of the ReRS method is considerably decreased while the RR-2D method maintains the performance in the presence of faults even if there are six faults in the network. The reason is that not only RR-2D avoids taking unnecessary non-minimal routes but also it is based on a fully adaptive routing and thus performing well in distributing packets over different routes.

In Fig. 9.18b, the average communication delay of RR-3D and HamFA schemes is plotted. RR-3D outperforms HamFA in the uniform traffic. Due to similar reasons as RR-2D, the network performance is maintained under the presence of faults in the network.

9.5.2 *Performance Analysis Under Hotspot Traffic Profile*

Under the hotspot traffic pattern, one or more routers are chosen as hotspots, receiving an extra portion of the traffic in addition to the regular uniform traffic. Given a hotspot percentage of H, a newly generated packet is directed to each hotspot router with an additional H percent probability. We simulate the hotspot traffic $H = 10\%$ with a single hotspot router at (4,4) in an 8×8 network and two hotspot routers at positions (2,1,2) and (3,2,2) in a $4 \times 4 \times 4$ network, respectively. In Fig. 9.19a, the performance of RR-2D and ReRS is measured for fault-free and

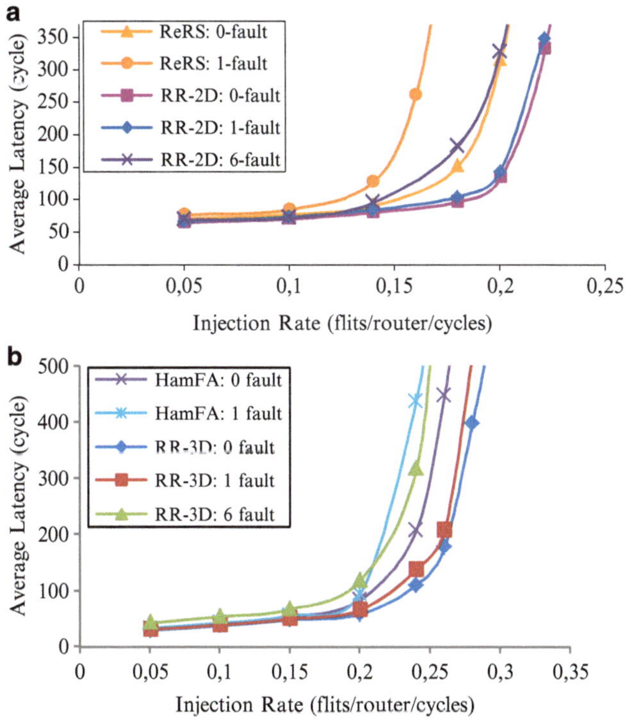

Fig. 9.19 Performance under hotspot traffic profile in (**a**) 2D network, (**b**) 3D network

one-faulty router cases while the performance of RR-2D is shown under six faults as well (mixture of faulty routers and links). RR-2D performs considerably better than ReRS even under the presence of six faults. This is due to the fact that RR-2D is an adaptive method and can balance the traffic over the network. On the other hand, RR-2D avoids using non-minimal routes as possible. The performance of the RR-3D and HamFA methods is illustrated in Fig. 9.19b. As observed from this figure, RR-3D leads to the best performance in fault-free cases. When there are six faults in the network, the performance is still high and it means RR-3D performs well in tolerating faults while maintaining the performance.

9.5.3 Reliability Evaluation Under Uniform Traffic Profile

As ReRS is able to tolerate faulty routers in its basic form, we measure the reliability of it by disabling only routers. RR-2D can tolerate both faulty routers and links and the reliability is measured under the mixture of them. We increase the number of faults from 1 to 6. All faults are selected using a random function. A network is reliable if all the injected packets reach their destinations. In other words, the

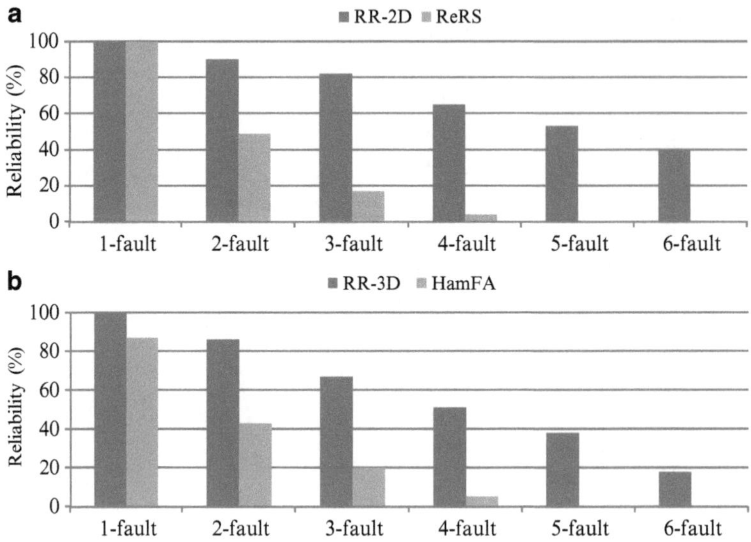

Fig. 9.20 Reliability measurement in (**a**) 2D network, (**b**) 3D network

network is counted as unreliable even if all packets reach the destinations except one packet. As shown in Fig. 9.20a, RR-2D can tolerate up to six faults by more than 40% reliability.

For a 3D network, we compare the reliability of RR-3D with HamFA. As HamFA is designed for tolerating unidirectional faulty links, its reliability is also measured based on these kinds of faults. On the other hand, RR-3D tolerates both faulty links and routers and the reliability value is obtained under the presence of both kinds of faults. We inject 1 to 6 faults into the network (mixture of faulty routers and links) to measure the reliability of RR-3D while 1 to 6 unidirectional faulty links are injected to measure the reliability of HamFA. As shown in Fig. 9.20b, the reliability of HamFA decreases significantly in comparison with RR-3D. Using RR-3D by the probability of 18%, the network performs normally without any packet loss when there are six faults in the network.

9.5.4 Hardware Analysis

To assess the area overhead and power consumption of on-chip implementation, the whole platform of each method is synthesized by Synopsys Design Compiler. We measured the area overhead and power consumption of the RR-2D and ReRS, methods. The power consumption is measured under a single faulty router for both methods. Each scheme includes network interfaces, routers, and communication channels. Both methods are synthesized using the TSMC 65 nm technology at the operating frequency of 500 MHz and supply voltage of 1V. We perform place-and-route, using Cadence Encounter, to have precise power and area estimations. The power dissipation is calculated using Synopsys PrimePower in an 8×8 mesh network. According to our analysis the area overheads of RR-2D and ReRS are almost the same as both of them use a simple routing unit and a similar number of wires to distribute the fault information. The power consumption of ReRS is slightly larger than RR-2D since RR-2D takes advantages of the fully adaptive routing and also utilizes the shortest paths as possible.

The whole platforms of the HamFA and RR-3D methods are also synthesized by Synopsys Design Compiler. The same numbers of channels are used in both methods and two faulty links are injected into the network. Depending on the technology and manufacturing process, the pitch of TSVs can range from 1 to 10 μm. In this work, the pad size for TSVs is assumed to be 5 μm square with pitch of around 8 μm. The layout areas of the HamFA and RR-3D schemes are almost similar while the area overhead of RR-3D is slightly larger than HamFA. This small difference is because of employing the monitoring and management technique which does not exist in the HamFA method. The average power consumption of the RR-3D scheme is 8% less than that of the HamFA scheme as it is able to balance the traffic over the network and delivering packets to destinations through the shortest paths as possible.

9.6 Summary and Conclusion

In this chapter, we proposed high-performance fault-tolerant methods for 2D and 3D NoCs. These methods are built upon fully adaptive algorithms, and thus avoiding congestion in the network. The proposed fault-tolerant algorithms are discussed for tolerating faulty routers and links. They utilize only the shortest paths to tolerate faults as long as such path exists. Non-minimal paths are necessitated when the source and destination are located in the same dimension and there is a fault between them. These kinds of faults are tolerated without creating cycles in the network. In a 3D network, we presented a fully adaptive routing algorithm by dividing the network into four disjoint subnetworks. The fault capability is improved by allowing packets to switch between subnetworks in the strictly ascending order.

References

1. M. Moadeli, A. Shahrabi, W. Vanderbauwhede, M. Ould-Khaoua, An analytical performance model for the Spidergon NoC, in *Proceedings of 21st Annual Conference on Advanced Networking and Applications* (2007), pp. 1014–1021
2. Y. Xie, G.H. Loh, B. Black, K. Bernstein, Design space exploration for 3D architectures. ACM J. Emerg. Technol. Comput. Syst. (JETC) **2**(2), 65–103 (2006)

3. M. Daneshtalab, M. Ebrahimi, P. Liljeberg, J. Plosila, H. Tenhunen, High-performance TSV architecture for 3-D ICs, in *Proceedings of IEEE Computer Society Annual Symposium on VLSI (ISVLSI)* 2010, pp. 467–468
4. A.Y. Weldezion, M. Grange, D. Pamunuwa, Z. Lu, A. Jantsch, R. Weerasekera, H. Tenhunen, Scalability of network-on-chip communication architecture for 3-D meshes, in *Proceedings of 3rd ACM/IEEE International Symposium on Networks-on-Chip (NOCS)* (2009), pp. 114–123
5. J. Hu, L. Wang, L. Jin, H.Z. JiangNan, Electrical modeling and characterization of through silicon vias (TSV), in *Proceedings of International Conference on Microwave and Millimeter Wave Technology (ICMMT)*, vol. 2 (2012), pp. 1–4
6. B.S. Feero, P.P. Pande, Networks-on-chip in a three-dimensional environment: A performance evaluation. IEEE Trans. Comput. **58**(1), 32–45 (2009)
7. C. Seiculescu, S. Murali, L. Benini, G. De Micheli, SunFloor 3D: A tool for networks on chip topology synthesis for 3D systems on chips, in *Proceedings of Design, Automation Test in Europe Conference Exhibition (DATE)* (2009), pp. 9–14
8. H. Matsutani, M. Koibuchi, H. Amano, Tightly-coupled multi-layer topologies for 3-D NoCs, in *Proceedings of 41st International Conference on Parallel Processing* (2007), p. 75
9. V.F. Pavlidis, E.G. Friedman, 3-D topologies for networks-on-chip. IEEE Trans. VLSI Syst. **15**(10), 1081–1090 (2007)
10. F. Li, C. Nicopoulos, T. Richardson, Y. Xie, V. Narayanan, M.K, Design and management of 3D chip multiprocessors using network-in-memory, in *Proceedings of ISCA-33* (2006), pp. 130–141
11. W. Dally, B. Towles, *Principles and Practices of Interconnection Networks* (Morgan Kaufmann, San Francisco, 2003)
12. L.M. Ni, P.K. McKinley, A survey of wormhole routing techniques in direct networks. Computer **26**(2), 62–76 (1993)
13. M. Li, Q.-A. Zeng, W.-B. Jone, DyXY – A proximity congestion-aware deadlock-free dynamic routing method for network on chip, in *Proceedings of 43rd ACM/IEEE Design Automation Conference* (2006), pp. 849–852
14. M. Ebrahimi, M. Daneshtalab, P. Liljeberg, J. Plosila, J. Flich, H. Tenhunen, Path-based partitioning methods for 3D networks-on-chip with minimal adaptive routing, *IEEE Transactions on Computers* (2012)
15. M. Ebrahimi, X. Chang, M. Daneshtalab, J. Plosila, P. Liljeberg, H. Tenhunen, DyXYZ: Fully adaptive routing algorithm for 3D NoCs, in *Proceedings of 21th IEEE Euromicro Conference on Parallel, Distributed and Network-Based Computing (PDP)* (2013), pp. 499–503
16. D. Fick, A. DeOrio, G. Chen, V. Bertacco, D. Sylvester, D. Blaauw, A highly resilient routing algorithm for fault-tolerant NoCs, in *Proceedings of Design, Automation Test in Europe Conference Exhibition (DATE)* (2009), pp. 21–26
17. D.A. Fabien Chaix, Fault-tolerant deadlock-free adaptive routing for any set of link and node failures in multi-cores systems (2010), pp. 52–59
18. W.-C. Tsai, D.-Y. Zheng, S.-J. Chen, Y.-H. Hu, A fault-tolerant NoC scheme using bidirectional channel, in *Proceedings of 48th ACM/EDAC/IEEE Design Automation Conference (DAC)* (2011), pp. 918–923
19. Z. Zhang, A. Greiner, S. Taktak, A reconfigurable routing algorithm for a fault-tolerant 2D-mesh network-on-chip, in *Proceedings of 45th ACM/IEEE Design Automation Conference (DAC)* (2008), pp. 441–446
20. M. Koibuchi, H. Matsutani, H. Amano, T. Mark Pinkston, A lightweight fault-tolerant mechanism for network-on-chip, in *Second ACM/IEEE International Symposium on Networks-on-Chip (NoCS)* (2008), pp. 13–22
21. M. Ebrahimi, M. Daneshtalab, J. Plosila, F. Mehdipour, MD: Minimal path-based fault-tolerant routing in on-chip networks, in *Proceedings of IEEE/ACM 18th Asia and South Pacific Design Automation Conference (ASP-DAC)* (2013), pp. 35–40
22. M. Ebrahimi, M. Daneshtalab, J. Plosila, H. Tenhunen, MAFA: Adaptive fault-tolerant routing algorithm for networks-on-chip, in *Proceedings of 15th Euromicro Conference on Digital System Design (DSD)* (2012), pp. 201–207

23. M. Ebrahimi, M. Daneshtalab, J. Plosila, High performance fault-tolerant routing algorithm for NoC-based many-core systems, in *Proceedings of 21th IEEE Euromicro Conference on Parallel, Distributed and Network-Based Computing (PDP)* (2013), pp. 463–469
24. M. Ebrahimi, M. Daneshtalab, J. Plosila, H. Tenhunen, Minimal-path fault-tolerant approach using connection-retaining structure in networks-on-chip, in *Proceedings of 7th International Symposium on Networks-on-Chip (NOCS)* (2013)
25. S. Pasricha, Y. Zou, A low overhead fault tolerant routing scheme for 3D networks-on-chip, in *2011 12th International Symposium on Quality Electronic Design (ISQED)* (2011), pp. 1–8
26. A.A. Chien, J.H. Kim, Planar-adaptive routing: Low-cost adaptive networks for multiprocessors, in *The 19th Annual International Symposium on Computer Architecture, 1992. Proceedings* (1992), pp. 268–277
27. M. Ebrahimi, M. Daneshtalab, J. Plosila, Fault-tolerant routing algorithm for 3D NoC using Hamiltonian path strategy, in *Proceedings of 16th ACM/IEEE Design, Automation, and Test in Europe (DATE)* (2013), pp. 1601–1605
28. C.J. Glass, C.J. Glass, L.M. Ni, L.M. Ni, Maximally fully adaptive routing in 2D meshes, in *Proceedings of International Conference on Parallel Processing* (1992), pp. 101–104
29. C.J. Glass, L.M. Ni, The turn model for adaptive routing, in *Proceedings of the 19th Annual International Symposium on Computer Architecture* (1992), pp. 278–287
30. M. Ebrahimi, M. Daneshtalab, P. Liljeberg, H. Tenhunen, "Fault-tolerant Method with Distributed Monitoring and Management Technique for 3D stacked mesh", in *Proceedings of the 17th International Symposium on Computer Architecture and Digital Systems*, 2013.

Part IV
Power/Energy and Thermal Issues

Chapter 10
Bufferless and Minimally-Buffered Deflection Routing

Chris Fallin, Greg Nazario, Xiangyao Yu, Kevin Chang, Rachata Ausavarungnirun, and Onur Mutlu

Abstract A conventional Network-on-Chip (NoC) router uses input buffers to store in-flight packets. These buffers improve performance, but consume significant power. It is possible to bypass these buffers when they are empty, reducing dynamic power, but static buffer power remains, and when buffers are utilized, dynamic buffer power remains as well. To improve energy efficiency, *bufferless deflection routing* removes input buffers, and instead uses deflection (misrouting) to resolve contention. Bufferless deflection routing is able to provide similar network performance to conventional buffered routing when the network carries light to moderate traffic, because deflections are relatively rare. However, at high network load, deflections cause unnecessary network hops, wasting power and reducing performance. In order to avoid some deflections and recover some performance, recent work has proposed to add a small buffer which holds only flits that contend with others and would have been deflected. This minimally-buffered deflection (MinBD) router improves performance relative to bufferless deflection routing without incurring the cost of a large buffer, because it can make more efficient use of a small buffer. The result is a router design which is more energy-efficient than prior buffered, bufferless, and hybrid router designs.

C. Fallin • G. Nazario • K. Chang • R. Ausavarungnirun • O. Mutlu (✉)
Carnegie Mellon University, Pittsburgh, PA, USA
e-mail: cfallin@cmu.edu; gnazario@cmu.edu; kevincha@cmu.edu; rausavar@cmu.edu; onur@cmu.edu

X. Yu
Massachusetts Institute of Technology, Cambridge, MA, USA
e-mail: yxy@mit.edu

M. Palesi and M. Daneshtalab (eds.), *Routing Algorithms in Networks-on-Chip*,
DOI 10.1007/978-1-4614-8274-1_10,

10.1 Introduction

A network-on-chip is a first-order component of current and future multicore and manycore CMPs (Chip Multiprocessors) [11], and its design can be critical for system performance. As core counts continue to rise, NoCs with designs such as 2D-mesh (e.g., Tilera [50] and Intel Terascale [28]) are expected to become more common to provide adequate performance scaling. Unfortunately, packet-switched NoCs are projected to consume significant power. In the Intel Terascale 80-core chip, 28 % of chip power is consumed by the NoC [28]; for MIT RAW, 36 % [45]; for the Intel 48-core SCC, 10 % [6]. NoC energy efficiency is thus an important design goal [4, 5].

Mechanisms have been proposed to make conventional input-buffered NoC routers more energy-efficient (i.e., use less energy per unit of performance). For example, bypassing empty input buffers [37, 49] reduces some dynamic buffer power, but static power remains.[1] Such bypassing is also less effective when buffers are not frequently empty. *Bufferless deflection routers* [17, 38] remove router input buffers completely (hence eliminating their static and dynamic power) to reduce router power. When two flits[2] contend for a single router output, one must be deflected to another output. Thus, a flit never requires a buffer in a router. By controlling which flits are deflected, a bufferless deflection router can ensure that all traffic is eventually delivered. Removing buffers yields simpler and more energy-efficient NoC designs: e.g., CHIPPER [17] reduces average network power by 54.9 % in a 64-node system compared to a conventional buffered router.

Unfortunately, at high network load, deflection routing reduces performance and efficiency. This is because deflections occur more frequently when many flits contend in the network. Each deflection sends a flit further from its destination, causing unnecessary link and router traversals. Relative to a buffered network, a bufferless network with a high deflection rate wastes energy, and suffers worse congestion, because of these unproductive network hops. In contrast, a buffered router is able to hold flits (or packets) in its input buffers until the required output port is available, incurring no unnecessary hops. Thus, a buffered network can sustain higher performance at peak load, but at the cost of large buffers, which can consume significant power and die area.

The best interconnect design would obtain the energy efficiency of the bufferless approach with the high performance of the buffered approach. Neither purely bufferless deflection routing nor conventional input-buffered routing satisfy this goal. Ideally, a router would contain only a small amount of buffering, and would use this buffer space only for those flits that actually require it, rather than all flits that arrive.

[1] One recent estimate indicates that static power (of buffers and links) could constitute 80–90 % of interconnect power in future systems [7].

[2] In a conventional bufferless deflection network, *flits* (several of which make up one packet) are independently routed, unlike most buffered networks, where a packet is the smallest independently-routed unit of traffic.

In this chapter, we discuss *minimally-buffered deflection* routing (MinBD) [20], a router design which combines both bufferless and buffered paradigms in a fine-grained and efficient way. MinBD uses deflection routing, but also incorporates a small buffer. The router does not switch between modes, but instead, always operates in a minimally-buffered deflection mode, and can buffer or deflect any given flit. When a flit first arrives, it does not enter a buffer, but travels straight to the routing logic. If two flits contend for the same output, the routing logic chooses one to deflect, as in a bufferless router. However, the router can choose to buffer up to one deflected flit per cycle rather than deflecting it. This fine-grained buffering-deflection hybrid approach significantly reduces deflection rate (by 54 % [20]), and improves performance, as we show. It also incurs only a fraction of the energy cost of a conventional buffered router, because only a relatively small fraction of flits are buffered (20 % of all flits). MinBD provides higher energy efficiency while also providing high performance, compared to a comprehensive set of baseline router designs. In this chapter, we will discuss:

- Bufferless deflection routing [17, 38], which uses deflection in place of buffering to resolve contention between flits. We will introduce the basic design of the BLESS [38] and CHIPPER [17] routers, and discuss the deflection arbitration, livelock freedom, and packet reassembly problems associated with bufferless routing.
- A new NoC router, *MinBD* (minimally-buffered deflection routing) [20], that combines deflection routing with minimal buffering. The router performs deflection routing, but can choose to buffer up to one flit per cycle in a small side buffer, which significantly reduces deflection rate and enhances performance compared to a pure bufferless design while requiring smaller buffer space than a conventional input-buffered design.
- An evaluation of MinBD against aggressive NoC router baselines: a two-cycle virtual channel buffered router [12] with empty buffer bypassing [37, 49] at three buffer capacities (with a sensitivity analysis over many more configurations), CHIPPER [17], and a hybrid bufferless-buffered design, AFC [29], with SPEC CPU2006 [44] multiprogrammed workloads on 16- and 64-node CMP systems. From our results, we conclude that MinBD has the best energy efficiency over all of these prior design points, while achieving competitive system throughput and logic critical path delay with the input-buffered router (the best-performing baseline) and competitive area and power consumption with the pure-bufferless router (the smallest and most power-efficient baseline).

10.2 Background

This section provides background on NoC-based cache-coherent CMPs, and on bufferless deflection routing. We assume the reader is familiar with the basic operation of conventional input-buffered routers. The key idea of such routers is to

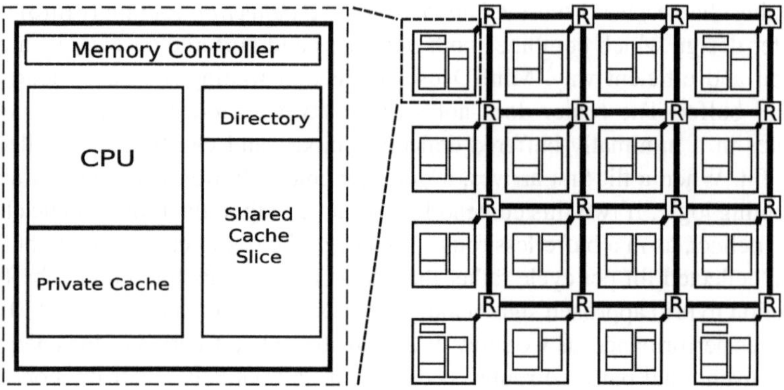

Fig. 10.1 An example Network-on-Chip (NoC)-based system: an on-chip packet-switched network connects nodes which often consist of cores, cache slices, and memory controllers

buffer every flit that enters the router from an input port before the flits can arbitrate for output ports. Dally and Towles [12] provide a good reference on these routers.

NoCs in cache-coherent CMPs: On-chip networks form the backbone of memory systems in many recently-proposed and prototyped large-scale CMPs (chip multiprocessors) [28, 45, 50]. Most such systems are cache-coherent shared memory multiprocessors. Packet-switched interconnect has served as the substrate for large cache-coherent systems for some time (e.g., for large multiprocessor systems such as SGI Origin [34]), and the principles are the same in a chip multiprocessor: each core, slice of a shared cache, or memory controller is part of one "node" in the network, and network nodes exchange packets that request and respond with data in order to fulfill memory accesses. A diagram of a typical system is shown in Fig. 10.1. For example, on a miss, a core's private cache might send a request packet to a shared L2 cache slice, and the shared cache might respond with a larger packet containing the requested cache block on an L2 hit, or might send another packet to a memory controller on an L2 miss. CMP NoCs are typically used to implement such a protocol between the cores, caches and memory controllers.

10.2.1 Bufferless Deflection Routing in NoCs: BLESS

Bufferless Deflection Routers: Bufferless deflection routing was first proposed by Baran [3]. Bufferless deflection routing operates without in-network buffering. Instead, a unit of traffic continuously moves between network nodes until it reaches its destination. When contention occurs for a network link, a bufferless deflection router sends some traffic to another output link instead, *deflecting* it. Hence, the use of buffers is replaced by occasional extra link traversals.

Bufferless deflection routing has found renewed interest in NoC design because on-chip wires (hence, network links) are relatively cheap, in contrast to buffers, which consume significant die area and leakage power [4, 5, 7, 29, 38]. Several evaluations of bufferless NoC design [17, 26, 29, 38] have demonstrated that removing the buffers in NoC routers, and implementing a routing strategy which operates without the need for buffers (such as the one we describe below), yield energy-efficiency improvements because occasional extra link traversals due to deflections consume relatively less energy than the dynamic energy used to buffer traffic at every network hop and the static energy consumed whenever a buffer is turned on. (Our motivational experiments in Sect. 10.3 demonstrate the performance and energy impact of such a network design in more detail.) Although other solutions exist to reduce the energy consumption of buffers, such as dynamic buffer bypassing [37, 49] (which we also incorporate into our baseline buffered-router design in this chapter), bufferless deflection routing achieves additional savings in energy and area by completely eliminating the buffers.

One recent work proposed BLESS [38], a router design that implements bufferless deflection routing, which we describe here. The fundamental unit of routing in a BLESS network is the *flit*, a packet fragment transferred by one link in one cycle. Flits are routed independently in BLESS.[3] Because flits are routed independently, they must be reassembled after they are received. BLESS assumes the existence of sufficiently-sized reassembly buffers at each node in order to reconstruct arriving flits into packets. (Later work, CHIPPER [17], addresses the reassembly problem explicitly, as we discuss below.)

Deflection Routing Arbitration: The basic operation of a BLESS bufferless deflection router is simple. In each cycle, flits arriving from neighbor routers enter the router pipeline. Because the router contains no buffers, flits are stored only in pipeline registers, and must leave the router at the end of the pipeline. Thus, the router must assign every input flit to some output port. When two flits request the same output port according to their ordinary routing function, the router deflects one of them to another port (this is always possible, as long as the router has as many outputs as inputs). BLESS performs this router output port assignment in two stages: flit ranking and port selection [38]. In each cycle, the flits that arrive at the router are first *ranked* in a priority order (chosen in order to ensure livelock-free operation, as we describe below). At the same time, the router computes a list of productive output ports (i.e., ports which would send the flit closer to its destination) for each flit. Once the flit ranking and each flits' productive output ports are available, the router assigns a port to each flit, starting from the highest-ranked flit and assigning ports to flits one at a time. Each flit obtains a productive output port if one is still available, and is "deflected" to any available output port otherwise. Because there

[3]BLESS' independent flit routing stands in contrast to conventional wormhole or VC (virtual-channel) routing, in which a packet's body flits always follow its head flits: because a deflection can occur in any cycle, any flit in a BLESS network could be separated from the rest of its packet and must carry its own routing information. This is described more fully in Moscibroda and Mutlu [38].

are as many output ports as input ports, and only the flits arriving on the input ports in a given cycle are considered, this process never runs out of output ports and can always assign each flit to some output. Hence, no buffering is needed, because every flit is able to leave the router at the end of the router pipeline.

Livelock freedom in BLESS: Although a BLESS router ensures that a flit is always able to take a network hop to *some* other router, a deflection takes a flit further from its destination, and such a flit will have to work its way eventually to its destination. In such a network design, explicit care must be taken to ensure that all flits eventually arrive at their destinations (i.e., that no flit circles, or *gets stuck*, in the network forever). This property is called *livelock freedom*. Note that conventional virtual channel-buffered routers, which buffer flits at every network hop, are livelock-free simply because they never deflect flits: rather, whenever a flit leaves a router and traverses a link, it always moves closer toward its destination (this is known as *minimal routing* [12]).

BLESS ensures livelock freedom by employing a priority scheme called *Oldest-First* [38]. Oldest-First prioritization is a total order over all flits based on each flit's age (time it has spent in the network). If two flits have the same age (entered the network in the same cycle), then the tie is broken with other header fields (such as sender ID) which uniquely identify the flit. This total priority order leads to livelock-free operation in a simple way: there must be one flit which is the oldest, and thus has the highest priority. This flit is always prioritized during flit-ranking at every router it visits. Thus, it obtains its first choice of output port and is never deflected. Because it is never deflected, the flit always moves closer toward its destination, and will eventually arrive. Once it arrives, it is no longer contending with other flits in the network, and some other flit is the oldest flit. The new oldest flit is guaranteed to arrive likewise. Inductively, all flits eventually arrive.

Flit injection and ejection: A BLESS router must inject new flits into the network when a node generates a packet, and it must remove a flit from the network when the flit arrives at its destination. A BLESS router makes a *local decision* to inject a flit whenever, in a given cycle, there is an empty slot on any of its input ports [38]. The router has an injection queue where flits wait until this injection condition is met. When a node is not able to inject, it is *starved*; injection starvation is a useful proxy for network congestion which has been used to control congestion-control mechanisms in bufferless deflection networks [8, 39, 40].

When a flit arrives at its destination router, that router removes the flit from the network and places it in a *reassembly buffer*, where it waits for the other flits from its packet to arrive. Flits in a packet may arrive in any order because each flit is routed independently, and might take a different path than the others due to deflections. Once all flits in a packet have arrived in that packet's reassembly buffer, the packet is delivered to the local node (e.g., core, cache, or memory controller). A BLESS router can eject up to one flit per cycle from its inputs to its reassembly buffer. Figure 10.2 depicts the reassembly buffers as well as the injection queue of a node in a BLESS NoC.

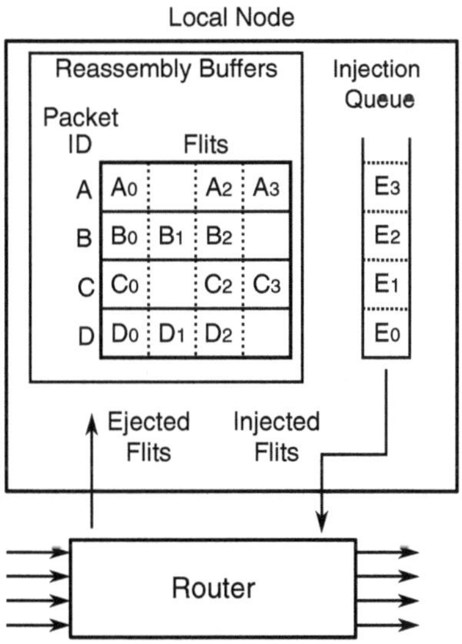

Fig. 10.2 Reassembly buffers and injection queue in a BLESS NoC

10.2.2 Low-Complexity Bufferless Deflection Routing: CHIPPER

CHIPPER [17], another bufferless deflection router design, was proposed to address implementation complexities in prior bufferless deflection routers (e.g., BLESS). The CHIPPER router has smaller and simpler deflection-routing logic than BLESS, which leads to a shorter critical path, smaller die area and lower power.

10.2.2.1 Problems in BLESS

The Oldest-First age-based arbitration in BLESS leads to slow routers with large hardware footprint [17, 26, 37] for several reasons, which we describe here.

Deflection arbitration: First, implementing deflection arbitration in the way that BLESS specifies leads to complex hardware. Routers that use Oldest-First arbitration must sort input flits by priority (i.e., age) in every cycle. This requires a three-stage sorting network for four inputs. Then, the router must perform port assignment in priority order, giving higher-priority flits first choice. Because a lower-priority flit might be deflected if a higher priority-flit takes an output port first, flits must be assigned output ports sequentially. This sequential port allocation leads to a *long critical path*, hindering practical implementation. This critical path (through priority sort and sequential port allocation) is illustrated in Fig. 10.3.

Packet reassembly: Second, as noted above, BLESS makes use of reassembly buffers to reassemble flits into packets. Reassembly buffers are necessary because each flit is routed independently and may take a different path than the others in a

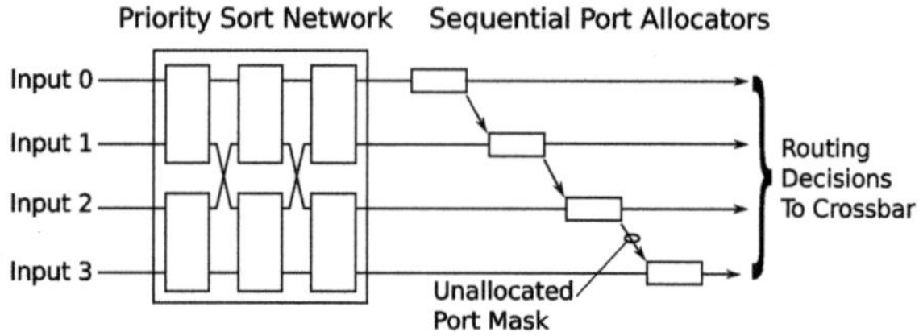

Fig. 10.3 In BLESS, a priority sort network and sequential port allocation are necessary to implement livelock-free routing, yielding a long critical path

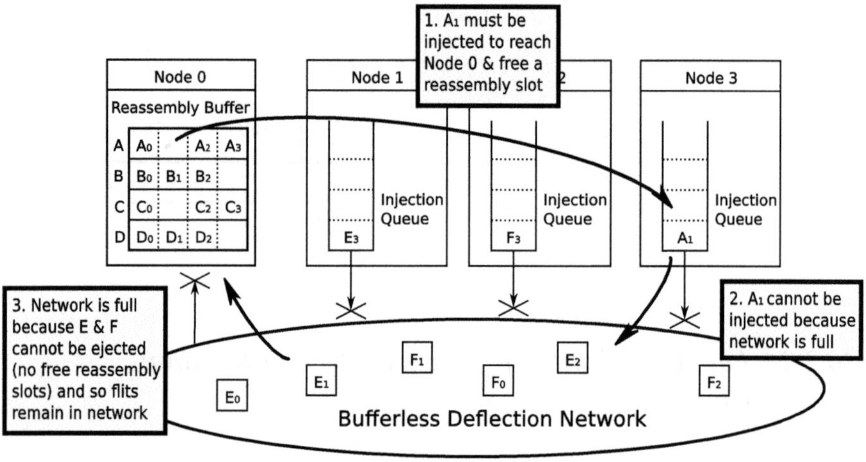

Fig. 10.4 Deadlock due to reassembly-buffer overflow in bufferless routing

packet, arriving at a different time. Moscibroda and Mutlu [38] evaluate bufferless deflection routing assuming a large enough reassembly buffer, and report maximum buffer occupancy.

However, with a limited reassembly buffer that is smaller than a certain size, deadlock will occur in the worst case (when all nodes send a packet simultaneously to a single node). To see why this is the case, observe the example in Fig. 10.4 (figure taken from Fallin et al. [17]). When a flit arrives at the reassembly buffers in Node 0, the packet reassembly logic checks whether a reassembly slot has already been allocated to the packet to which this flit belongs. If not, a new slot is allocated, if available. If the packet already has a slot, the flit is placed into its proper location within the packet-sized buffer. When no slots are available and a flit from a new packet arrives, the reassembly logic must prevent the flit from being ejected out of the network. In the worst case, portions of many separate packets arrive at Node 0, allocating all its slots. Then, flits from other packets arrive, but cannot be ejected, because no reassembly slots are free. These flits remain in the network, deflecting and retrying ejection. Eventually, the network will fill with these flits. The flits which are required to complete the partially-reassembled packets may have not yet been injected at their respective sources, and they cannot be injected, because the network is completely full. Thus, deadlock occurs. Without a different buffer management scheme, the only way to avoid this deadlock is to size the reassembly buffer at each node for the worst case when all other nodes in the system send a packet to that node simultaneously. A bufferless deflection router implementation with this amount of buffering would have significant overhead, unnecessarily wasting area and power. Hence, an explicit solution is needed to ensure deadlock-free packet reassembly in practical designs.

10.2.2.2 CHIPPER: A Low-Complexity Bufferless Deflection Router

We now outline the operation of CHIPPER [17], a bufferless deflection router design which makes bufferless deflection routing practical by providing for low-cost deflection arbitration and packet reassembly.

Golden Packet-based deflection arbitration: A bufferless deflection router must ensure that the network has *livelock freedom* by providing a strong guarantee that any flit eventually arrives at its destination. BLESS ensured that any flit arrives by enforcing a *total priority order* among all flits, such that the highest-priority (oldest) flit is delivered, then another flit attains the highest priority (becomes the oldest). However, enforcing a total priority order creates significant complexity in a BLESS router (as we described above).

The CHIPPER router design starts from the observation that *minimal* livelock-free routing requires only that one flit is prioritized until it arrives, and that any flit is eventually chosen to be this specially-prioritized flit if it remains in the network long enough. This priority scheme is called *Golden Packet*, and it allows the CHIPPER router to use a simpler design than the BLESS router.

The Golden Packet priority scheme globally prioritizes one *packet* in the network at a time. Flits in this packet (which we call *golden flits*) are prioritized over other flits in the network. (Within the packet, ties are broken by the flit sequence number within the packet.) The prioritization rules are shown in Ruleset 1. When a packet becomes the Golden Packet, it remains so for a *golden epoch*, which is set to a length L that is long enough so that the packet can reach any destination in the network from any source.

The Golden Packet is chosen *implicitly* (i.e., without the need for all routers to explicitly coordinate their choice). The CHIPPER network can uniquely name any packet with a *packet ID* (e.g., source ID and cache-miss MSHR number, or some other transaction identifier). One *packet ID* is designed as golden based on a predefined function of the current time (in clock cycles).[4] In particular, the packet ID which is currently golden is incremented once every L cycles (the golden epoch), and wraps around when all possible packet IDs have each been designated as golden for the length of a golden epoch. In this way, any packet eventually becomes golden if it remains in the network long enough.

[4]CHIPPER assumes that all routers are in a single clock domain, hence can maintain synchronized golden packet IDs simply by counting clock ticks.

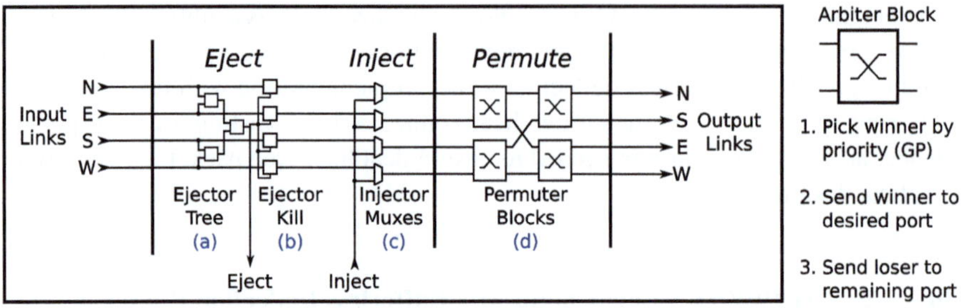

Fig. 10.5 CHIPPER router microarchitecture: router pipeline (*left*) and detail of a single arbiter block (*right*)

Ruleset 1 Golden Packet prioritization rules

Golden Tie: If two flits are golden, the lower-numbered flit (first in golden packet) wins.
Golden Dominance: If one flit is golden, it wins over any non-golden flit.
Common Case: Contests between two non-golden flits are decided pseudo-randomly.

The most important consequence of Golden Packet is that each router *only needs to correctly route the highest-priority flit.* This is sufficient to ensure that the first outstanding flit of the Golden Packet is delivered within L cycles. Because the packet will periodically become Golden until delivered, all of its flits are guaranteed delivery.

Because Golden Packet prioritization provides livelock freedom as long as the highest-priority flit is correctly routed, the deflection routing (arbitration) logic does not need to sequentially assign each flit to the best possible port, as the BLESS router's deflection routing logic does (Fig. 10.3). Rather, it only needs to recognize a golden flit, if one is present at the router inputs, and route that flit correctly if present. All other deflection arbitration is best-effort. Arbitration can thus be performed more quickly with simpler logic.

We now describe the CHIPPER router's arbitration logic here; the router's pipeline is depicted in Fig. 10.5 (see Fallin et al. [17] for more details, including the ejection/injection logic which is not described here). The CHIPPER router's arbitration logic is built with a basic unit, the *two-input arbiter block*, shown on the right side of Fig. 10.5. Each two-input arbiter block receives up to two flits every cycle and routes these two flits to its outputs. In order to route its input flits, the two-input arbiter block chooses one *winning* flit. If a golden flit is present, the golden flit is the winning flit (if two golden flits are present, the tie is broken as described by the prioritization rules). If no golden flit is present, one of the input flits is chosen randomly to be the winning flit. The two-input arbiter block then examines the winning flit's destination, and sends this flit toward the arbiter block's output which leads that flit closer to its destination. The other flit, if present, must then take the remaining arbiter block output.

The CHIPPER router performs deflection arbitration among four input flits (from the four inputs in a 2D mesh router) using a *permutation network* of four arbiter blocks, connected in two stages of two blocks each, as shown in the *permute* pipeline stage of Fig. 10.5. The permutation network allows a flit from any router input to reach any router output. When flits arrive, they arbitrate in the first stage, and winning flits are sent toward the second-stage arbiter block which is connected to that flit's requested router output. Then, in the second stage, flits arbitrate again. As flits leave the second stage, they proceed directly to the router outputs via a pipeline register (no crossbar is necessary, unlike in conventional router designs). This two-stage arbitration has a shorter critical path than the sequential scheme used by a BLESS router because the arbiter blocks in each stage work in parallel, and because (unlike in a BLESS arbiter) the flits need *not* be sorted by priority first. The arbiter-block permutation network cannot perform *all* possible flit permutations (unlike the BLESS router's routing logic), but because a golden flit (if present) is always prioritized, and hence always sent to a router output which carries the flit closer to its destination, the network is still livelock-free. Because the permutation network (i) eliminates priority sorting, and (ii) partially parallelizes port assignment, the router critical path is improved (reduced) by 29.1 %, performing within 1.1 % of a conventional buffered router design [17].

Addressing packet reassembly deadlock with Retransmit-Once: Fallin et al. [17] observe that the reassembly deadlock problem is fundamentally due to a lack of global flow control. Unlike buffered networks, which can pass tokens upstream to senders to indicate whether downstream buffer space is available, a bufferless deflection network has no such backpressure. Allowing receivers to exert backpressure on senders solves the problem. Thus, CHIPPER introduces a new low-overhead flow control protocol, *Retransmit-Once*, as its second major contribution.

Retransmit-Once *opportunistically* assumes that buffer space will be available, imposing no network overhead in the common case. When no space is available, any subsequent arriving packet is dropped at the receiver. However, the receiver makes note of this dropped packet. Once reassembly buffer space becomes available, the reassembly logic in the receiver *reserves* buffer space for the previously dropped packet, and the receiver then requests a retransmission from the sender. Thus, at most one retransmission is necessary for any packet. In addition, by dropping only short request packets (which can be regenerated from a sender's request state), and using reservations to ensure that longer data packets are never dropped, Retransmit-Once ensures that senders do not have to buffer data for retransmission. In our evaluations of realistic workloads, retransmission rate is 0.021 % maximum with 16-packet reassembly buffers, hence the performance impact is negligible. Fallin et al. [17] describe the Retransmit-Once mechanism in more detail and report that it can be implemented with very little overhead by integrating with cache-miss buffers (MSHRs) in each node.

10.3 Motivation: Performance at High Load

Previous NoC designs based on bufferless deflection routing, such as BLESS [38] and CHIPPER [17] which we just introduced, were motivated largely by the observation that many NoCs in CMPs are over-provisioned for the common-case network load. In this case, a bufferless network can attain nearly the same application performance while consuming less power, which yields higher energy efficiency. We now examine the buffered-bufferless comparison in more detail. Figure 10.6 shows (i) relative application performance (weighted speedup: see Sect. 10.5), and (ii) relative energy efficiency (performance per watt), when using a bufferless network, compared to a conventional buffered network. Both plots show these effects as a function of network load (average injection rate). Here we show a virtual channel buffered network (4 VCs, 4 flits/VC) (with buffer bypassing) and the CHIPPER bufferless deflection network [17] in a 4×4-mesh CMP (details on methodology are in Sect. 10.5).

For low-to-medium network load, a bufferless network has performance close to a conventional buffered network, because the deflection rate is low: thus, most flits take productive network hops on every cycle, just as in the buffered network. In addition, the bufferless router has significantly reduced power (hence improved energy efficiency), because the buffers in a conventional router consume significant power. However, as network load increases, the deflection rate in a bufferless deflection network also rises, because flits contend with each other more frequently. With a higher deflection rate, the dynamic power of a bufferless deflection network rises more quickly with load than dynamic power in an equivalent buffered network,

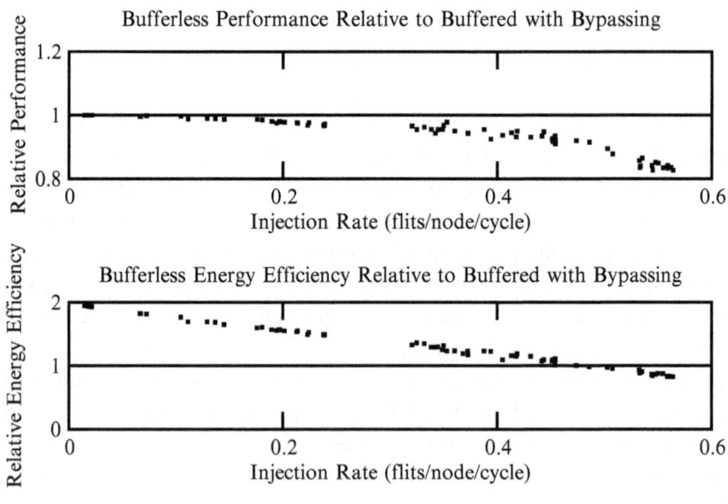

Fig. 10.6 System performance and energy efficiency (performance per watt) of bufferless deflection routing, relative to conventional input-buffered routing (4 VCs, 4 flits/VC) that employs buffer bypassing, in a 4×4 2D mesh. Injection rate (X axis) for each workload is measured in the baseline buffered network

because each deflection incurs some extra work. Hence, bufferless deflection networks lose their energy-efficiency advantage at high load. Just as important, the high deflection rate causes each flit to take a longer path to its destination, and this increased latency reduces the network throughput and system performance.

Overall, neither design obtains both good performance and good energy efficiency at all loads. If the system usually experiences low-to-medium network load, then the bufferless design provides adequate performance with low power (hence high energy efficiency). But, if we use a conventional buffered design to obtain high performance, then energy efficiency is poor in the low-load case, and even buffer bypassing does not remove this overhead because buffers consume static power regardless of use. Finally, simply switching between these two extremes at a per-router granularity, as previously proposed [29], does not address the fundamental inefficiencies in the bufferless routing mode, but rather, uses input buffers for all incoming flits at a router when load is too high for the bufferless mode (hence retains the relative energy-inefficiency of buffered operation at high load). We now introduce MinBD, the *minimally-buffered deflection router*, which combines bufferless and buffered routing to reduce this overhead.

10.4 MinBD: Minimally-Buffered Deflection Router

The *MinBD* (minimally-buffered deflection) router is a new router design that combines bufferless deflection routing with a small buffer, called the "side buffer." We start by outlining the **key principles** which the design follows to reduce deflection-caused inefficiency by using buffering:

1. When a flit would be deflected by a router, it is often better to buffer the flit and arbitrate again in a later cycle. Some buffering can avoid many deflections.
2. However, buffering every flit leads to unnecessary power overhead and buffer requirements, because many flits will be routed productively on the first try. The router should buffer a flit only if necessary.
3. Finally, when a flit arrives at its destination, it should be removed from the network (ejected) quickly, so that it does not continue to contend with other flits.

Basic High-Level Operation: The MinBD router does not use input buffers, unlike conventional buffered routers. Instead, a flit that arrives at the router proceeds directly to the routing and arbitration logic. This logic performs deflection routing, so that when two flits contend for an output port, one of the flits is sent to another output instead. However, unlike a bufferless deflection router, the MinBD router can also *buffer* up to one flit per cycle in a single FIFO-queue side buffer. The router examines all flits at the output of the deflection routing logic, and if any are deflected, one of the deflected flits is removed from the router pipeline and buffered (as long as the buffer is not full). From the side buffer, flits are re-injected into the network by the router, in the same way that new traffic is injected. Thus, some flits that would have been deflected in a bufferless deflection router are removed from the

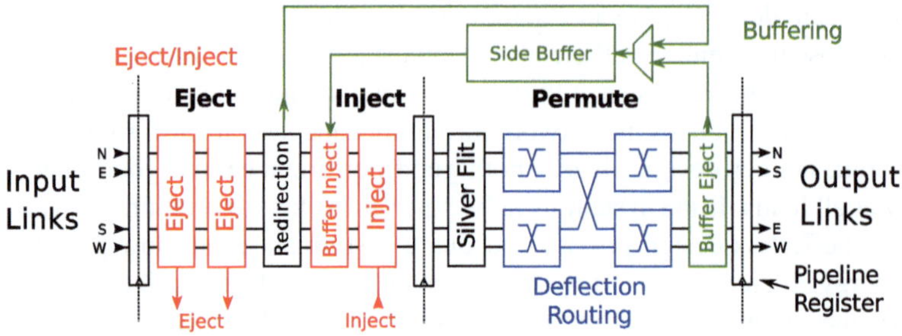

Fig. 10.7 MinBD router pipeline

network temporarily into this side buffer, and given a second chance to arbitrate for a productive router output when re-injected. This reduces the network's deflection rate (hence improves performance and energy efficiency) while buffering only a fraction of traffic.

We will describe the operation of the MinBD router in stages. First, Sect. 10.4.1 describes the deflection routing logic that computes an initial routing decision for the flits that arrive in every cycle. Then, Sect. 10.4.2 describes how the router chooses to buffer some (but not all) flits in the side buffer. Section 10.4.3 describes how buffered flits and newly-generated flits are injected into the network, and how a flit that arrives at its destination is ejected. Finally, Sect. 10.4.4 discusses correctness issues, and describes how MinBD ensures that all flits are eventually delivered.

10.4.1 Deflection Routing

The MinBD router pipeline is shown in Fig. 10.7. Flits travel through the pipeline from the inputs (on the left) to outputs (on the right). We first discuss the deflection routing logic, located in the Permute stage on the right. This logic implements deflection routing: it sends each input flit to its preferred output when possible, deflecting to another output otherwise.

MinBD uses the deflection logic organization first proposed in CHIPPER [17]. The *permutation network* in the Permute stage consists of two-input blocks arranged into two stages of two blocks each. This arrangement can send a flit on any input to any output. (Note that it cannot perform all possible permutations of inputs to outputs, but as we will see, it is sufficient for correct operation that at least one flit obtains its preferred output.) In each two-input block, arbitration logic determines which flit has a higher priority, and sends that flit in the direction of its preferred output. The other flit at the two-input block, if any, must take the block's other output. By combining two stages of this arbitration and routing, deflection arises as a distributed decision: a flit might be deflected in the first stage, or the second stage. Restricting the arbitration and routing to two-flit subproblems reduces complexity and allows for a shorter critical path, as demonstrated in [17].

Ruleset 2 MinBD prioritization rules (based on Golden Packet [17] with new rule 3)

Given: two flits, each *Golden*, *Silver*, or *Ordinary*. (Only one can be Silver.)

1. **Golden Tie**: Ties between two Golden flits are resolved by sequence number (first in Golden Packet wins).
2. **Golden Dominance**: If one flit is Golden, it wins over any Silver or Ordinary flits.
3. **Silver Dominance**: Silver flits win over Ordinary flits.
4. **Common Case**: Ties between Ordinary flits are resolved randomly.

In order to ensure correct operation, the router must arbitrate between flits so that every flit is eventually delivered, despite deflections. MinBD adapts a modified version of the Golden Packet priority scheme [17], which solves this *livelock-freedom problem*. This priority scheme is summarized in Ruleset 2. The basic idea of the Golden Packet priority scheme is that at any given time, at most one packet in the system is *golden*. The flits of this golden packet, called "golden flits," are prioritized above all other flits in the system (and contention between golden flits is resolved by the flit sequence number). While prioritized, golden flits are never deflected by non-golden flits. The packet is prioritized for a period long enough to guarantee its delivery. Finally, this "golden" status is assigned to one globally-unique packet ID (e.g., source node address concatenated with a request ID), and this assignment rotates through all possible packet IDs such that any packet that is "stuck" will eventually become golden. In this way, all packets will eventually be delivered, and the network is livelock-free. (See [17] for the precise way in which the Golden Packet is determined; MinBD uses the same rotation schedule.)

However, although Golden Packet arbitration provides correctness, a *performance* issue occurs with this priority scheme. Consider that most flits are not golden: the elevated priority status provides worst-case correctness, but does not impact common-case performance (prior work reported over 99 % of flits are delivered without becoming golden [17]). However, when no flits are golden and ties are broken randomly, the arbitration decisions in the two permutation network stages are not coordinated. Hence, a flit might win arbitration in the first stage, and cause another flit to be deflected, but then lose arbitration in the second stage, and also be deflected. Thus, unnecessary deflections occur when the two permutation network stages are uncoordinated.

In order to resolve this performance issue, we observe that it is enough to ensure that in every router, at least one flit is prioritized above the others in every cycle. In this way, at least one flit will certainly not be deflected. To ensure this when no golden flits are present, MinBD adds a "silver" priority level, which wins arbitration over common-case flits but loses to the golden flits. One silver flit is designated randomly among the set of flits that enter a router at every cycle (this designation is local to the router, and not propagated to other routers). This modification helps to reduce deflection rate. Prioritizing a silver flit at every router does not impact correctness, because it does not deflect a golden flit if one is present, but it ensures that at least one flit will consistently win arbitration at both stages. Hence, deflection rate is reduced, improving performance.

10.4.2 Using a Small Buffer to Reduce Deflections

The key problem addressed by MinBD is *deflection inefficiency at high load*: in other words, when the network is highly utilized, contention between flits occurs often, and many flits will be deflected. We observe that adding a small buffer to a deflection router can reduce deflection rate, because the router can choose to buffer rather than deflect a flit when its output port is taken by another flit. Then, at a later time when output ports may be available, the buffered flit can re-try arbitration.

Thus, to reduce deflection rate, MinBD adds a "side buffer" that buffers only some flits that otherwise would be deflected. This buffer is shown in Fig. 10.7 above the permutation network. In order to make use of this buffer, a "buffer ejection" block is placed in the pipeline after the permutation network. At this point, the arbitration and routing logic has determined which flits to deflect. The buffer ejection block recognizes flits that have been deflected, and picks up to one such deflected flit per cycle. It removes a deflected flit from the router pipeline, and places this flit in the side buffer, as long as the side buffer is not full. (If the side buffer is full, no flits are removed from the pipeline into the buffer until space is created.) This flit is chosen randomly among deflected flits (except that a golden flit is never chosen: see Sect. 10.4.4). In this way, some deflections are avoided. The flits placed in the buffer will later be re-injected into the pipeline, and will re-try arbitration at that time. This re-injection occurs in the same way that new traffic is injected into the network, which we discuss below.

10.4.3 Injection and Ejection

So far, we have considered the flow of flits from router input ports (i.e., arriving from neighbor routers) to router output ports (i.e., to other neighbor routers). A flit must enter and leave the network at some point. To allow traffic to enter (be injected) and leave (be ejected), the MinBD router contains injection and ejection blocks in its first pipeline stage (see Fig. 10.7). When a set of flits arrives on router inputs, these flits first pass through the ejection logic. This logic examines the destination of each flit, and if a flit is addressed to the local router, it is removed from the router pipeline and sent to the local network node.[5] If more than one locally-addressed flit is present, the ejection block picks one, according to the same priority scheme used by routing arbitration.

However, ejecting a single flit per cycle can produce a bottleneck and cause unnecessary deflections for flits that could not be ejected. In the workloads we evaluate, at least one flit is eligible to be ejected 42.8 % of the time. Of those cycles,

[5]Note that flits are reassembled into packets after ejection. To implement this reassembly, we use the Retransmit-Once scheme, as used by CHIPPER and described in Sect. 10.2.2.2.

20.4 % of the time, at least two flits are eligible to be ejected. Hence, in ∼8.5 % of all cycles, a locally-addressed flit would be deflected rather than ejected if only one flit could be ejected per cycle. To avoid this significant deflection-rate penalty, MinBD doubles the ejection bandwidth. To implement this, a MinBD router contains two ejection blocks. Each of these blocks is identical, and can eject up to one flit per cycle. Duplicating the ejection logic allows two flits to leave the network per cycle at every node.[6]

After locally-addressed flits are removed from the pipeline, new flits are allowed to enter. There are two injection blocks in the router pipeline shown in Fig. 10.7: (i) re-injection of flits from the side buffer, and (ii) injection of new flits from the local node. (The "Redirection" block prior to the injection blocks will be discussed in the next section.) Each block operates in the same way. A flit can be injected into the router pipeline whenever one of the four inputs does not have a flit present in a given cycle, i.e., whenever there is an "empty slot" in the network. Each injection block pulls up to one flit per cycle from an injection queue (the side buffer, or the local node's injection queue), and places a new flit in the pipeline when a slot is available. Flits from the side buffer are re-injected before new traffic is injected into the network. However, note that there is no guarantee that a free slot will be available for an injection in any given cycle. We now address this starvation problem for side buffer re-injection.

10.4.4 Ensuring Side Buffered Flits Make Progress

When a flit enters the side buffer, it leaves the router pipeline, and must later be re-injected. As we described above, flit re-injection must wait for an empty slot on an input link. It is possible that such a slot will not appear for a long time. In this case, the flits in the side buffer are delayed unfairly while other flits make forward progress.

To avoid this situation, MinBD implements *buffer redirection*. The key idea of buffer redirection is that when this side buffer starvation problem is detected, one flit from a randomly-chosen router input is *forced* to enter the side buffer. Simultaneously, the flit at the head of the side buffer is injected into the slot created by the forced flit buffering. In other words, one router input is "redirected" into the FIFO buffer for one cycle, in order to allow the buffer to make forward progress. This redirection is enabled for one cycle whenever the side buffer injection is starved (i.e., has a flit to inject, but no free slot allows the injection) for more than some

[6]For fairness, because dual ejection widens the datapath from the router to the local node (core or cache), we also add dual ejection to the baseline bufferless deflection network and input-buffered network when we evaluate performance, but not when we evaluate the power, area, or critical path of these baselines.

threshold $C_{threshold}$ cycles (in our evaluations, $C_{threshold} = 2$). Finally, note that if a golden flit is present, it is never redirected to the buffer, because this would break the delivery guarantee.

10.4.5 Livelock and Deadlock-Free Operation

MinBD provides livelock-free delivery of flits using Golden Packet and buffer redirection. If no flit is ever buffered, then Golden Packet [17] ensures livelock freedom (the "silver flit" priority never deflects any golden flit, hence does not break the guarantee). Now, we argue that adding side buffers does not cause livelock. First, the buffering logic never places a golden flit in the side buffer. However, a flit could enter a buffer and then become golden while waiting. Redirection ensures correctness in this case: it provides an upper bound on residence time in a buffer (because the flit at the head of the buffer will leave after a certain threshold time in the worst case). If a flit in a buffer becomes golden, it only needs to remain golden long enough to leave the buffer in the worst case, then progress to its destination. MinBD chooses the threshold parameter ($C_{threshold}$) and golden epoch length so that this is always possible. More details can be found in our extended technical report [18].

MinBD achieves deadlock-free operation by using Retransmit-Once [17], which ensures that every node always consumes flits delivered to it by dropping flits when no reassembly/request buffer is available. This avoids packet-reassembly deadlock (as described in [17]), as well as protocol level deadlock, because message-class dependencies [25] no longer exist.

10.5 Evaluation Methodology

To obtain application-level performance as well as network performance results, we use an in-house CMP simulator. This simulator consumes instruction traces of ×86 applications, and faithfully models the CPU cores and the cache hierarchy, with a directory-based cache coherence protocol (based on the SGI Origin protocol [9]) running on the modeled NoC. The CPU cores model stalls, and interact with the caches and network in a closed-loop way. The modeled network routers are cycle-accurate, and are assumed to run in a common clock domain. The instruction traces are recorded with a Pin-tool [36], sampling representative portions of each benchmark as determined by PinPoints [41]. We find that simulating 25M cycles gives stable results with these traces. Detailed system parameters are shown in Table 10.1.

Note that we make use of a *perfect shared cache* to stress the network, as was done in the CHIPPER [17] and BLESS [38] bufferless router evaluations. In this model, every request generated by an L1 cache miss goes to a shared cache slice,

Table 10.1 Simulated baseline system parameters

CPU cores	Out-of-order, 3-wide issue and retire (1 memory op/cycle), 16 MSHRs [33]
L1 caches	64 KB, 4-way associative, 32-byte blocks
L2 (shared) cache	Distributed across nodes; perfect (always hits) to penalize our design conservatively & isolate network performance from memory effects
Shared cache mapping	Consecutive cache blocks striped across L2 cache slices
Cache coherence scheme	Directory-based, perfect directory (SGI Origin protocol [34])
Data packet sizes	1-flit request packets, 4-flit data packets
Network Links	1-cycle latency (separate pipeline stage), 2.5 mm, 128 bits wide
Baseline bufferless router	CHIPPER [17], 2-cycle router latency; 64-cycle Golden Epoch; Retransmit-Once [17]
Baseline buffered router	(m VCs, n flits/VC): $(8,8)$, $(4,4)$, $(4,1)$; 2-cycle latency, buffer bypassing [37, 49]; Additional configurations evaluated in Fig. 10.9
AFC (Adaptive Flow Control)	As described in [29]: 4 VCs/channel, 4 flits/VC. 2-cycle latency (buffered & bufferless modes). Implements buffer bypassing as well
MinBD	2-cycle latency (Sect. 10.4); 4-flit side buffer (single FIFO); $C_{threshold} = 2$; 64-cycle Golden Epoch; Retransmit-Once [17]

and the request is always assumed to hit and return data. This potentially increases network load relative to a real system, where off-chip memory bandwidth can also be a bottleneck. However, note that this methodology is *conservative*: because MinBD degrades performance relative to the buffered baseline, the performance degradation that we report is an upper bound on what would occur when other bottlenecks are considered. We choose to perform our evaluations this way in order to study the true capacity of the evaluated networks (if network load is always low because system bottlenecks such as memory latency are modeled, then the results do not give many insights about router design). Note that the cache hierarchy details (L1 and L2 access latencies, and MSHRs) are still realistic. We remove only the off-chip memory latency/bandwidth bottleneck.

Baseline Routers: We compare MinBD to a conventional input-buffered virtual channel router [12] with buffer bypassing [37, 49], a bufferless deflection router (CHIPPER [17]), and a hybrid bufferless-buffered router (AFC [29]). In particular, we sweep buffer size for input-buffered routers. We describe a router with m virtual channels (VCs) per input and n flits of buffer capacity per VC as an (m,n)-buffered router. We compare to a $(8,8)$, $(4,4)$, and $(4,1)$-buffered routers in our main results. The $(8,8)$ point represents a very large (overprovisioned) baseline, while $(4,4)$ is a more reasonable general-purpose configuration. The $(4,1)$ point represents the *minimum* buffer size for deadlock-free operation (two message classes [25], times two to avoid routing deadlock [10]). Furthermore, though 1-flit VC buffers would reduce throughput because they do not cover the credit round-trip latency, we optimistically assume zero-latency credit traversal in our simulations (which benefits the baseline design, hence is conservative for our claims). Finally, we simulate smaller (non-deadlock-free) designs with 1 and 2 VCs per input for our power-performance tradeoff evaluation in Sect. 10.6.2 (Fig. 10.9), solely for completeness (we were able to avoid deadlock at moderate loads and with finite simulation length for these runs).

Application Workloads: We use SPEC CPU2006 [44] benchmarks (26 applications[7]) to construct 75 multiprogrammed workloads (which consist of many single-threaded benchmark instances that run independently). Each workload consists of 16 or 64 randomly-selected applications which are randomly mapped onto the mesh. Workloads are categorized by average network injection rate in the baseline (4,4)-buffered system (measured in flits/cycle/node). For 4×4 workloads, these injection rate categories are $(0,0.15)$, $(0.15,0.3)$, $(0.3,0.4)$, $(0.4,0.5)$, and >0.5; for 8×8 workloads, $(0,0.05)$, $(0.05,0.15)$, $(0.15,0.25)$, and >0.25 (an 8×8 mesh saturates at a lower load than a 4×4 mesh, due to limited bisection bandwidth). Each category contains 15 workloads.

Synthetic-Traffic Workloads: To show robustness under various traffic patterns, we evaluate 4×4 and 8×8 networks with uniform-random, bit-complement, and transpose synthetic traffic [12] in addition to application workloads. For each pattern, network injection rate is swept from zero to network saturation.

Performance Metrics: To measure system performance, we use the well-known *Weighted Speedup* metric [43]: $WS = \sum_{i=1}^{N} \frac{IPC_i^{shared}}{IPC_i^{alone}}$. All IPC_i^{alone} values are measured on the baseline bufferless network. Weighted speedup correlates to system throughput [16] and is thus a good general metric for multiprogrammed workloads.

Power Modeling: We use a modified and improved version of ORION 2.0 [48] (configured for 65 nm), as developed by Grot et al. [24], as well as Verilog models synthesized with a commercial 65 nm design library. We use Verilog models of router control logic, and add datapath area and power using ORION models for crossbars, buffers, links, and pipeline registers. CHIPPER and MinBD do not use a conventional crossbar, instead decoupling the crossbar into a permutation network, which we model using muxes. We rely on ORION's modeling for the baseline input-buffered router's control logic (e.g., arbitration). This hybrid synthesis/ORION approach models each portion of the router in a way that captures its key limitations. The control logic is logic- rather than wiring-dominated, and its arbitration paths determine the critical path; hence, standard-cell synthesis will model area, power and timing of MinBD and CHIPPER control logic with reasonable accuracy. The router datapath is wiring-dominated and relies on heavily-optimized components with custom layouts such as large crossbars and wide muxes, pipeline registers and memories. ORION explicitly models such router components.

[7] 410.bwaves, 416.gamess and 434.zeusmp were excluded because we were not able to collect representative traces from these applications.

We report energy efficiency as performance-per-watt, computed as weighted speedup divided by average network power.

10.6 Evaluation

In this section, we evaluate MinBD against a bufferless deflection router [17] and an input-buffered router with buffer bypassing [37,49], as well as a hybrid of these two, AFC [29], and demonstrate that by using a combination of deflection routing and buffering, MinBD achieves performance competitive with the conventional input-buffered router (and higher than the bufferless deflection router), with a smaller buffering requirement, and better energy efficiency than all prior designs.

10.6.1 Application Performance

Figure 10.8 (top pane) shows application performance as weighted speedup for 4×4 (16-node) and 8×8 (64-node) CMP systems. The plots show average results for each workload category, as described in Sect. 10.5, as well as overall average results. Each bar group shows the performance of three input-buffered routers: 8 VCs with 8 flits/VC, 4 VCs with 4 flits/VC, and 4 VCs with 1 flit/VC. Next is CHIPPER, the bufferless deflection router, followed by AFC [29], a coarse-grained hybrid router that switches between a bufferless and a buffered mode. MinBD is shown last in each bar group. We make several observations:

1. MinBD improves performance relative to the bufferless deflection router by 2.7 % (4.9 %) in the 4×4 (8×8) network over all workloads, and 8.1 % (15.2 %) in the highest-intensity category. Its performance is within 2.7 % (3.9 %) of the $(4, 4)$ input-buffered router, which is a reasonably-provisioned baseline, and within 3.1 % (4.2 %) of the $(8, 8)$ input-buffered router, which has large, power-hungry buffers. Hence, adding a side buffer allows a deflection router to obtain significant performance improvement, and the router becomes more competitive with a conventional buffered design.
2. Relative to the 4-VC, 1-flit/VC input-buffered router (third bar), which is the smallest deadlock-free (i.e., correct) design, MinBD performs nearly the same despite having less buffer space (4 flits in MinBD vs. 16 flits in $(4, 1)$-buffered). Hence, buffering only a portion of traffic (i.e., flits that would have been deflected) makes more efficient use of buffer space.
3. AFC, the hybrid bufferless/buffered router which switches modes at the router granularity, performs essentially the same as the 4-VC, 4-flit/VC input-buffered router, because it is able to use its input buffers when load increases. However, as we will see, this performance comes at an efficiency cost relative to our hybrid design.

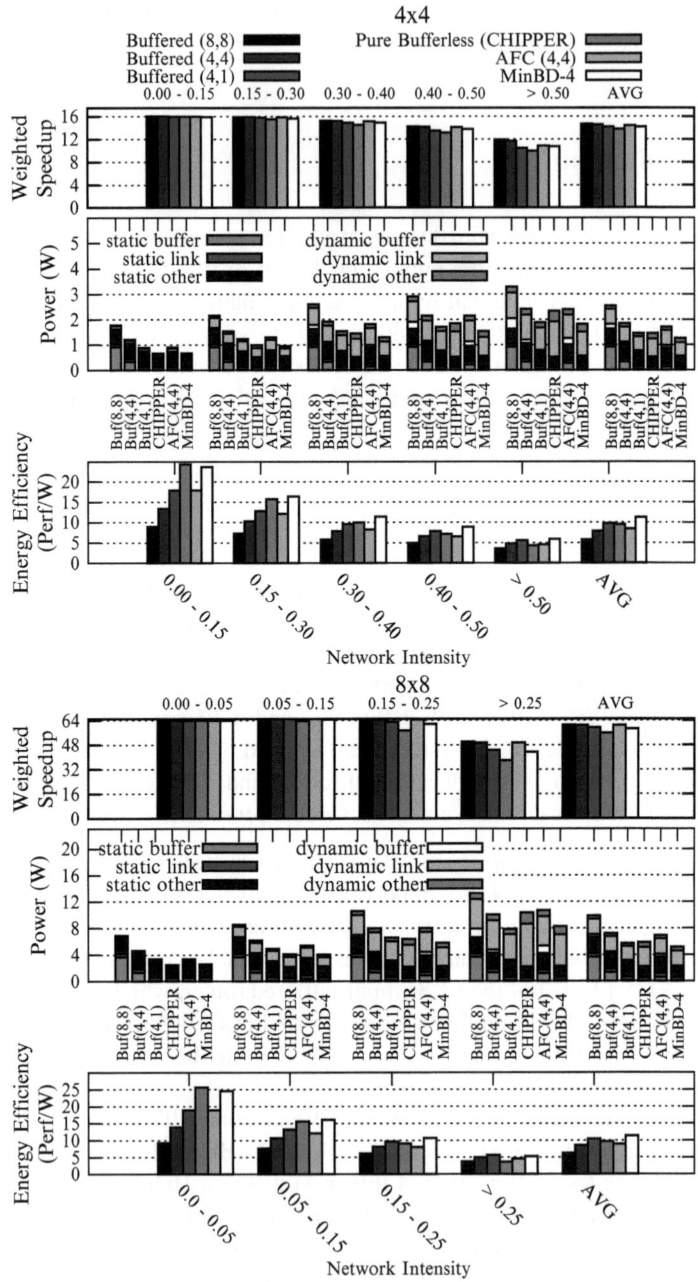

Fig. 10.8 Performance (weighted speedup), network power, and energy efficiency (performance per watt) in 16 (4 × 4) and 64 (8 × 8)-node CMP systems

10.6.2 Network Power and Energy Efficiency

Network Power: Figure 10.8 (middle pane) shows average total network power, split by component and type (static/dynamic), for 4×4 and 8×8 networks across the same workloads. Note that static power is shown in the bottom portion of each bar, and dynamic power in the top portion. Each is split into buffer power, link power, and other power (which is dominated by datapath components, e.g., the crossbar and pipeline registers). We make several observations:

1. Buffer power is a large part of total network power in the input-buffered routers that have reasonable buffer sizes, i.e., $(4,4)$ and $(8,8)$ (VCs, flits/VC), even with empty-buffer bypassing, largely because static buffer power (bottom bar segment) is significant. Removing large input buffers reduces static power in MinBD as well as the purely-bufferless baseline, CHIPPER.[8] Because of this reduction, MinBD's total network power never exceeds that of the input-buffered baselines, except in the highest-load category in an 8×8-mesh (by 4.7 %).
2. Dynamic power is larger in the baseline deflection-based router, CHIPPER, than in input-buffered routers: CHIPPER has 31.8 % (41.1 %) higher dynamic power than the $(4,4)$-buffered router in the 4×4 (8×8) networks in the highest-load category. This is because bufferless deflection-based routing requires more network hops, especially at high load. However, in a 4×4 network, MinBD consumes less dynamic power (by 8.0 %) than the $(4,4)$-buffered baseline in the highest-load category because reduced deflection rate (by 58 %) makes this problem relatively less significant, and allows savings in buffer dynamic energy and a simplified datapath to come out. In an 8×8 network, MinBD's dynamic power is only 3.2 % higher.
3. MinBD and CHIPPER, which use a permutation network-based datapath rather than a full 5×5 crossbar, reduce datapath static power (which dominates the "static other" category) by 31.0 %: the decoupled permutation network arrangement has less area, in exchange for partial permutability (which causes some deflections). Input-buffered routers and AFC require a full crossbar because they cannot deflect flits when performing buffered routing (partial permutability in a non-deflecting router would significantly complicate switch arbitration, because each output arbiter's choice would be limited by which other flits are traversing the router).
4. AFC, the coarse-grained hybrid, has nearly the same network power as the $(4,4)$ buffered router at high load: 0.6 % (5.7 %) less in 4×4 (8×8). This is because its buffers are enabled most of the time. At low load, when it can power-gate its buffers frequently, its network power reduces. However, AFC's network power

[8]Note that network power in the buffered designs takes buffer bypassing into account, which reduces these baselines' dynamic buffer power. The $(4,4)$-buffered router bypasses 73.7 % (83.4 %) of flits in 4×4 (8×8) networks. Without buffer bypassing, this router has 7.1 % (6.8 %) higher network power, and 6.6 % (6.4 %) worse energy-efficiency.

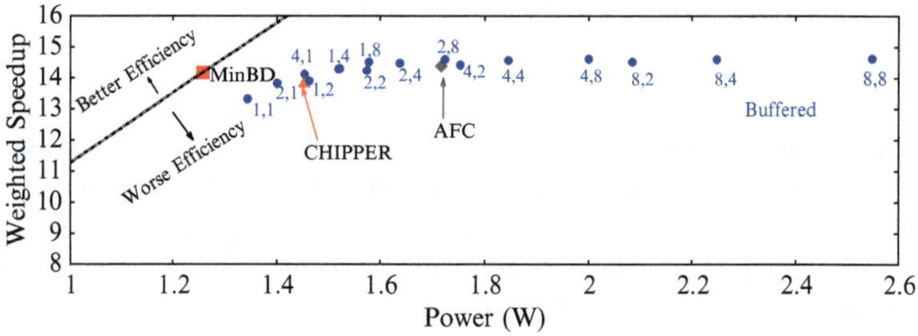

Fig. 10.9 Power (X) vs. application performance (Y) in 4×4 networks. The *line* represents all points with equivalent performance-per-watt to MinBD

is still higher than the pure bufferless router (CHIPPER) or MinBD because (i) it still spends some time in its buffered mode, and (ii) its datapath power is higher, as described above. On average, AFC still consumes 36.8 % (18.1 %) more network power than CHIPPER, and 33.5 % (33.0 %) more than MinBD, in the lowest-load category.

Energy efficiency: Figure 10.8 (bottom pane) shows energy efficiency. We make two key observations:

1. MinBD has the highest energy efficiency of any evaluated design: on average in 4×4 (8×8) networks, 42.6 % (33.8 %) better than the reasonably-provisioned (4, 4) input-buffered design. MinBD has 15.9 % (8.7 %) better energy-efficiency than the most energy-efficient prior design, the (4, 1)-buffered router.
2. At the highest network load, MinBD becomes less energy-efficient compared to at lower load, and its efficiency degrades at a higher rate than the input-buffered routers with large buffers (because of deflections). However, its per-category energy-efficiency is still better than all baseline designs, with two exceptions. In the highest-load category (near saturation) in an 8×8-mesh, MinBD has nearly the same efficiency as the (4, 1)-buffered router (but, note that MinBD is much more efficient than this baseline router at lower loads). In the lowest-load category in a 4×4 mesh, the purely-bufferless router CHIPPER is slightly more energy-efficient (but, note that CHIPPER's performance and efficiency degrade quickly at high loads).

We conclude that, by achieving competitive performance with the buffered baseline, and making more efficient use of a much smaller buffer capacity (hence reducing buffer power and total network power), MinBD provides better energy efficiency than prior designs. To summarize this result, we show a 2D plot of power and application performance in Fig. 10.9 for 4×4 networks, and a wider range of buffered router designs, as well as MinBD and CHIPPER. (Recall from Sect. 10.5 that several of the baseline input-buffered designs are not deadlock free (too few VCs) or have a buffer depth that does not cover credit round-trip latency, but we evaluate them anyway for completeness.) In this plot, with power on the X axis and application performance on the Y axis, a line through the origin represents a

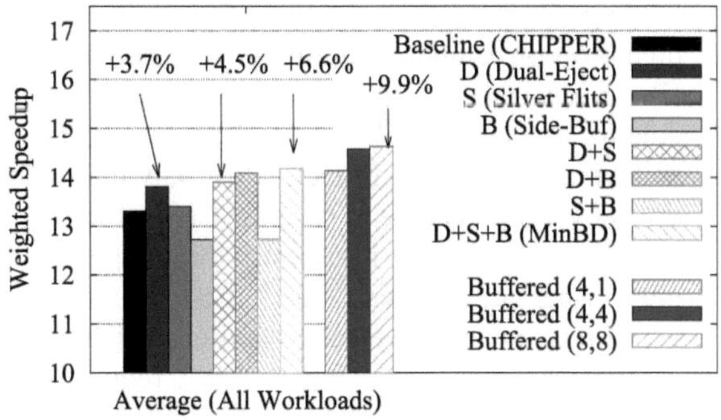

Fig. 10.10 Breakdown of performance gains for each mechanism in MinBD

Table 10.2 Average deflection rates for deflection-based routers

Baseline (CHIPPER)	D	S	B	D + S	D + B	S + B	D + S + B (MinBD)
0.28	0.22	0.27	0.17	0.22	0.11	0.16	0.10

fixed performance-per-watt ratio (the slope of the line). This equal-efficiency line is shown for MinBD. Points above the line have better efficiency than MinBD, and points below have worse efficiency. As shown, MinBD presents the best energy efficiency among all evaluated routers. The trend in an 8×8 network (not shown for space) is similar (see technical report [18]).

10.6.3 Performance Breakdown

To understand the observed performance gain in more detail, we now break down performance by each component of MinBD. Figure 10.10 shows performance (for 4×4 networks) averaged across all workloads for eight *deflection systems*, which constitute all possible combinations of MinBD's mechanisms added to the baseline (CHIPPER) router: dual-width ejection (**D**), silver-flit prioritization (**S**), and the side buffer (**B**), shown with the same three input-buffered configurations as before. The eighth bar (**D + S + B**), which represents all three mechanisms added to the baseline deflection network, represents MinBD. Table 10.2 shows deflection rate for the same set of systems.

We draw three main conclusions:

1. All mechanisms individually contribute to performance and reduce deflection rate. Dual ejection (**D**) increases performance by 3.7 % over baseline CHIPPER.[9]

[9]The main results presented in Fig. 10.8 use this data point (with dual ejection) in order to make a fair (same external router interface) comparison.

Adding silver-flit prioritization (**D** + **S**) increases performance by 0.7 % over dual-ejection alone. Adding a side buffer to the other two mechanisms (**D** + **S** + **B**) increases performance by 2.0 % on average.

2. Adding a side buffer by itself (**B**) is not sufficient to attain the performance that MinBD achieves. In fact, when only the side buffer is added, performance drops by 4.3 % relative to baseline CHIPPER. The degradation occurs primarily because the increased in-network throughput (enabled by a reduced deflection rate) in the network with side buffers places more ejection pressure on nodes, which exacerbates the ejection bottleneck that we observed in Sect. 10.4.1. This performance reduction comes despite a reduced deflection rate: even though flits are not deflected as frequently, they must still be removed from the network efficiently for performance to increase.
3. Adding dual ejection to the side buffered system (**D** + **B**) to address the ejection bottleneck increases performance to 5.8 % above baseline. Silver-flit prioritization in addition to this configuration point yields the MinBD router (eighth bar), which attains 6.6 % performance above baseline (2.7 % above dual-ejection alone) on average for all workload intensities. Overall, deflection rate reduces by 64 % from baseline CHIPPER to MinBD (and 54 % from CHIPPER with dual-ejection (**D**) to MinBD, as shown in our primary results).

10.6.4 Synthetic Traffic Performance

We study the network-level performance of MinBD and baseline designs by applying synthetic traffic patterns: uniform random, bit-complement, and transpose [12]. Figure 10.11 shows latency curves with injection rate swept from zero to saturation for the bufferless deflection router, MinBD, and the $(4,1)$ and $(8,8)$ input-buffered router (other input-buffered routers are omitted for clarity; these are the smallest and largest evaluated input-buffered routers, to show lower and upper bounds). Latency curves are shown for uniform random, bit complement, transpose, and hotspot patterns. In the "hotspot" pattern, nodes send requests to a single node at the center of the mesh with 20 % probability and to random locations otherwise. (Under a 100 % hotspot pattern, the performance of all designs converges because the receiving node's ejection bandwidth is the bottleneck. A 20 % skewed hotspot pattern is more realistic because it tests the network's capacity to handle unbalanced load.)

Under uniform random traffic (which most closely resembles our multiprogrammed application workloads with striped data mapping), MinBD performs better than the bufferless baseline, with a higher saturation point. MinBD has a lower network saturation point than the input-buffered network with large input buffers, but very similar saturation point to the small $(4,1)$ input-buffered router, as our earlier results indicated. We conclude that with only 4 flits of buffering per router, MinBD closes nearly half of the gap in network saturation throughput between the bufferless router (CHIPPER) and the largest input-buffered router (with 256 flits of total buffer space), and performs similarly to a smaller input-buffered router with 16 flits of total buffer space.

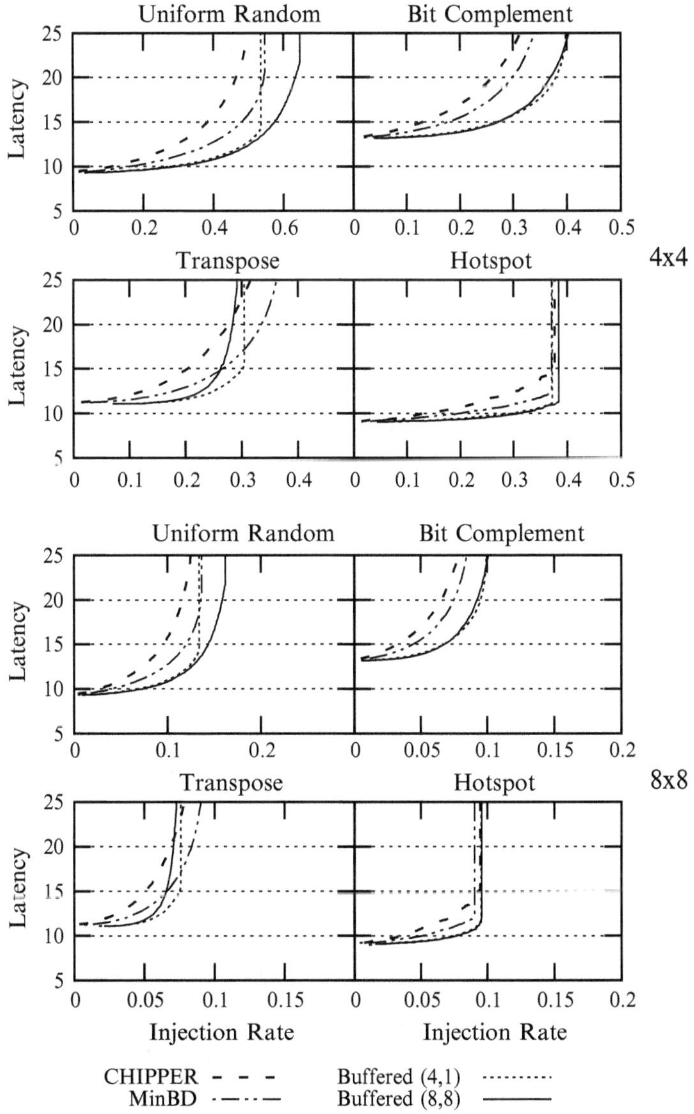

Fig. 10.11 Synthetic traffic evaluations for MinBD, CHIPPER and input-buffered routers (with small and large input buffers), in 4×4 and 8×8 meshes

In addition, non-uniform traffic patterns demonstrate the robustness and adaptivity of deflection routing: in particular, the transpose traffic pattern demonstrates a lower saturation point in the input-buffered router than either deflection-based router (MinBD or CHIPPER). This advantage occurs because deflection routing is *adaptive* (a flit's path can change based on network conditions). Deflections spread traffic away from hotspots and balance load in unbalanced traffic patterns.

While adaptive routing is also possible in input-buffered routers, it is more complex because it requires the router to track network load or congestion (locally or globally) and make decisions accordingly. In contrast, deflection routing provides adaptivity by virtue of its ordinary operation.

10.6.5 Sensitivity to Parameters

Side Buffer Size: As side buffer size is varied from 1 to 64 flits, mean weighted speedup (application performance) changes only 0.2 % on average across all workloads (0.9 % in the highest-intensity category) in 4×4 networks. We conclude that the *presence* of the buffer (to buffer at least one deflected flit) is more important than its size, because the average utilization of the buffer is low. In a 4×4 MinBD network with 64-flit side buffers at saturation (61 % injection rate, uniform random), the side buffer is empty 48 % of the time on average; 73 % of the time, it contains 4 or fewer flits; 93 % of the time, 16 or fewer. These measurements suggest that a very small side buffer captures most of the benefit. Furthermore, total network power increases by 19 % (average across all 4×4 workloads) when a 1-flit buffer per router is increased to a 64-flit buffer per router. Hence, a larger buffer wastes power without significant performance benefit.

We avoid a 1-flit side buffer because of the way the router is pipelined: such a single-flit buffer would either require for a flit to be able to enter, then leave, the buffer in the same cycle (thus eliminating the independence of the two router pipeline stages), or else could sustain a throughput of one flit only every two cycles. (For this sensitivity study, we optimistically assumed the former option for the 1-flit case.) The 4-flit buffer we use avoids this pipelining issue, while increasing network power by only 4 % on average over the 1-flit buffer.

Note that although the size we choose for the side buffer happens to be the same as the 4-flit packet size which we use in our evaluations, this need not be the case. In fact, because the side buffer holds deflected *flits* (not packets) and deflection decisions occur at a per-flit granularity, it is unlikely that the side buffer will hold more than one or two flits of a given packet at a particular time. Hence, unlike conventional input-buffered routers which typically size a buffer to hold a whole packet, MinBD's side buffer can remain small even if packet size increases.

Packet Size: Although we perform our evaluations using a 4-flit packet size, our conclusions are robust to packet size. In order to demonstrate this, we also evaluate MinBD, CHIPPER, and the (4,4)- and (8,8)-input-buffered routers in 4×4 and 8×8 networks using a data packet size of 8 flits. In a 4×4 (8×8) network, MinBD improves performance over CHIPPER by 17.1 % (22.3 %), achieving performance within 1.2 % (8.1 %) of the (4,4)-input-buffered router and within 5.5 % (12.8 %) of the (8,8)-input-buffered router, while reducing average network power by 25.0 % (18.1 %) relative to CHIPPER, 16.0 % (9.4 %) relative to the (4,4)-input-buffered router, and 40.3 % (34.5 %) relative to the (8,8)-input-buffered router, respectively. MinBD remains the most energy-efficient design as packet size increases.

Table 10.3 Normalized router area and critical path for bufferless and buffered baselines, compared to MinBD

Router design	CHIPPER	MinBD	Buffered (8,8)	Buffered (4,4)	Buffered (4,1)
Normalized area	1.00	1.03	2.06	1.69	1.60
Normalized critical path length	1.00	1.07	0.99	0.99	0.99

10.6.6 Hardware Cost: Router Area and Critical Path

We present normalized router area and critical path length in Table 10.3. Both metrics are normalized to the bufferless deflection router, CHIPPER, because it has the smallest area of all routers. MinBD adds only 3 % area overhead with its small buffer. In both CHIPPER and MinBD, the datapath dominates the area. In contrast, the large-buffered baseline has 2.06× area, and the reasonably-provisioned buffered baseline has 1.69× area. Even the smallest deadlock-free input-buffered baseline has 60 % greater area than the bufferless design (55 % greater than MinBD). In addition to reduced buffering, the reduction seen in CHIPPER and MinBD is partly due to the simplified datapath in place of the 5 × 5 crossbar (as also discussed in Sect. 10.6.2). Overall, MinBD reduces area relative to a conventional input-buffered router both by significantly reducing the required buffer size, and by using a more area-efficient datapath.

Table 10.3 also shows the normalized critical path length of each router design, which could potentially determine the network operating frequency. MinBD increases critical path by 7 % over the bufferless deflection router, which in turn has a critical path 1 % longer than an input-buffered router. In all cases, the critical path is through the flit arbitration logic (the permutation network in MinBD and CHIPPER, or the arbiters in the input-buffered router). MinBD increases critical path relative to CHIPPER by adding logic in the deflection-routing stage to pick a flit to buffer, if any. The buffer re-injection and redirection logic in the first pipeline stage (ejection/injection) does not impact the critical path because the permutation network pipeline stage has a longer critical path.

10.7 Related Work

To our knowledge, MinBD is the first NoC router design that combines deflection routing with a small side buffer that reduces deflection rate. Other routers combine deflection routing with buffers, but do not achieve the efficiency of MinBD because they either continue to use input buffers for all flits (Chaos router) or switch all buffers on and off at a coarse granularity with a per-router mode switch (AFC), in contrast to MinBD's fine-grained decision to buffer or deflect each flit.

Buffered NoCs that also use deflection: Several routers that primarily operate using buffers and flow control also use deflection routing as a secondary mechanism under high load. The Chaos Router [32] deflects packets when a packet queue becomes full to probabilistically avoid livelock. However, all packets that pass through the router are buffered; in contrast, MinBD performs deflection routing first, and only buffers some flits that would have been deflected. This key aspect of our design reduces buffering requirements and buffer power. The Rotary Router [1] allows flits to leave the router's inner ring on a non-productive output port after circulating the ring enough times, in order to ensure forward progress. In this case, again, deflection is used as an escape mechanism to ensure probabilistic correctness, rather than as the primary routing algorithm, and all packets must pass through the router's buffers.

Other bufferless designs: Several prior works propose bufferless router designs [17, 21, 26, 38, 47]. We have already extensively compared to CHIPPER [17], from which we borrow the deflection routing logic design. BLESS [38], another bufferless deflection network, uses a more complex deflection routing algorithm. Later works showed BLESS to be difficult to implement in hardware [17, 26, 37], thus we do not compare to it in this work. Other bufferless networks drop rather than deflect flits upon contention [21, 26]. Some earlier large multiprocessor interconnects, such as those in HEP [42] and Connection Machine [27], also used deflection routing. The HEP router combined some buffer space with deflection routing (Smith, 2008, Personal communication). However, these routers' details are not well-known, and their operating conditions (large off-chip networks) are significantly different than those of modern NoCs.

More recently, Fallin et al. [19] applied deflection routing to a hierarchical ring topology, allowing most routers (those that lie within a ring) to be designed without any buffering or flow control, and using only small buffers to transfer between rings. The resulting design, HiRD, was shown to be more energy-efficient than the baseline hierarchical ring with more buffering. HiRD uses many of the same general ideas as MinBD to ensure forward progress, e.g., enforcing explicit forward-progress guarantees in the worst case without impacting common-case complexity.

Improving high-load performance in bufferless networks: Some work has proposed *congestion control* to improve performance at high network load in bufferless deflection networks [8, 39, 40]. Both works used source throttling: when network-intensive applications cause high network load which degrades performance for other applications, these intensive applications are prevented from injecting network traffic some of the time. By reducing network load, source throttling reduces deflection rate and improves overall performance and fairness. These congestion control techniques and others (e.g., [46]) are orthogonal to MinBD, and could improve MinBD's performance further.

Hybrid buffered-bufferless NoCs: AFC [29] combines a bufferless deflection router based on BLESS [38] with input buffers, and switches between bufferless deflection routing and conventional input-buffered routing based on network load at each router. While AFC has the performance of buffered routing in the highest-load case, with better energy efficiency in the low-load case (by power-gating buffers when not needed), it misses opportunity to improve efficiency because it switches buffers on at a coarse granularity. When an AFC router experiences high load, it performs a mode switch which takes several cycles in order to turn on its buffers. Then, it pays the buffering energy penalty for every flit, whether or not it would have been deflected. It also requires buffers as large as the baseline input-buffered router design in order to achieve equivalent high-load performance. As a result, its network power is nearly as high as a conventional input-buffered router at high load, and it requires fine-grained power gating to achieve lower power at reduced network load. In addition, an AFC router has a larger area than a conventional buffered router, because it must include both buffers and buffered-routing control logic as well as deflection-routing control logic. In contrast, MinBD does not need to include large buffers and the associated buffered-mode control logic, instead using only a smaller side buffer. MinBD also removes the dependence on efficient buffer power-gating that AFC requires for energy-efficient operation at low loads. We quantitatively compared MinBD to AFC in Sect. 10.4 and demonstrated better energy efficiency at all network loads.

Reducing cost of buffered routers: Empty buffer bypassing [37, 49] reduces buffered router power by allowing flits to bypass input buffers when empty. However, as our evaluations (which faithfully model the power reductions due to buffer bypassing) show, this scheme reduces power less than our new router design: bypassing is only effective when buffers are empty, which happens more rarely as load increases. Furthermore, buffers continue to consume static power, even when unused. Though both MinBD and empty-buffer-bypassed buffered routers avoid buffering significant traffic, MinBD further reduces router power by using much smaller buffers.

Kim [30] proposed a low-cost buffered router design in which a packet uses a buffer only when turning, not when traveling straight along one dimension. Unlike our design, this prior work does not make use of deflection, but uses deterministic X-Y routing. Hence, it is not adaptive to different traffic patterns. Furthermore, its performance depends significantly on the size of the turn-buffers. By using deflection, MinBD is less dependent on buffer size to attain high performance, as we argued in Sect. 10.6.5. In addition, [30] implements a token-based injection starvation avoidance scheme which requires additional communication between routers, whereas MinBD requires only per-router control to ensure side buffer injection.

10.8 Conclusion

In this chapter, we introduced deflection routing and discussed two bufferless deflection routers, BLESS [38] and CHIPPER [17]. We then described MinBD [20], a minimally-buffered deflection router design. MinBD combines deflection routing with a small buffer, such that some network traffic that would have been deflected is placed in the buffer instead. Previous router designs which use buffers typically place these buffers at the router inputs. In such a design, energy is expended to read and write the buffer for every flit, and buffers must be large enough to efficiently handle all arriving traffic. In contrast to prior work, a MinBD router uses its buffer for only a fraction of network traffic, and hence makes more efficient use of a given buffer size than a conventional input-buffered router. Its average network power is also greatly reduced: relative to an input-buffered router, buffer power is much lower, because buffers are smaller. Relative to a bufferless deflection router, dynamic power is lower, because deflection rate is reduced with the small buffer.

We evaluate MinBD against a comprehensive set of baseline router designs: three configurations of an input-buffered virtual-channel router [12], a bufferless deflection router, CHIPPER [17], and a hybrid buffered-bufferless router, AFC [29]. Our evaluations show that MinBD performs competitively and reduces network power: on average in a 4×4 network, MinBD performs within 2.7 % of the input-buffered design (a high-performance baseline) while consuming 31.8 % less total network power on average relative to this input-buffered router (and 13.4 % less than the bufferless router, which performs worse than MinBD). Finally, MinBD has the best energy efficiency among all routers which we evaluated. We conclude that a router design which augments bufferless deflection routing with a small buffer to reduce deflection rate is a compelling design point for energy-efficient, high-performance on-chip interconnect.

10.9 Future Work

We believe promising future research directions exist in bufferless and minimally-buffered deflection routers. One promising direction is to develop more effective packet prioritization mechanisms that are applicable to such routers. For example, to improve system performance and fairness, mechanisms have been proposed to perform packet prioritization in a manner that is aware of application characteristics [13] or packet latency slack [14]. Similarly, mechanisms have been proposed to provide quality of service to different applications in buffered routers (e.g., [23,24,35]). Development of similar mechanisms for bufferless and minimally buffered routers is an open research problem. Another promising direction is to develop techniques to increase the applicability of bufferless and minimally buffered deflection routing to different topologies, such as express cube topologies [22, 31] and hybrid and hierarchical topologies [2, 15, 24]. Adaptation of bufferless and minimally-buffered deflection routing in a manner that guarantees correctness in different topologies, and the evaluation of resulting techniques is an important area for future work. As a first step, our recent research [19] has developed new techniques to use deflection routing in hierarchical ring topologies, with promising results that show that a minimally-buffered hierarchical ring design significantly improves system energy efficiency. We intend to explore these research directions in the future and hope that this chapter serves as an inspiration to other researchers as well.

Acknowledgements We thank the anonymous reviewers of our conference papers CHIPPER [17] and MinBD [20] for their feedback. We gratefully acknowledge members of the SAFARI group and Michael Papamichael at CMU for feedback. Chris Fallin is currently supported by an NSF Graduate Research Fellowship (Grant No. 0946825). Rachata Ausavarungnirun is currently partially supported by the Royal Thai Government Scholarship. Onur Mutlu is partially supported by the Intel Early Career Faculty Honor Program Award. Greg Nazario and Xiangyao Yu were undergraduate research interns while this work was done. We acknowledge the generous support of our industrial partners, including AMD, HP Labs, IBM, Intel, NVIDIA, Oracle, Qualcomm, and Samsung. This research was partially supported by grants from NSF (CAREER Award CCF-0953246, CCF-1147397 and CCF-1212962). This article is a significantly extended and revised version of our previous conference papers that introduced CHIPPER [17] and MinBD [20].

References

1. P. Abad et al., Rotary router: an efficient architecture for CMP interconnection networks, in *ISCA-34*, San Diego, 2007
2. J. Balfour, W.J. Dally, Design tradeoffs for tiled CMP on-chip networks, in *ICS*, Cairns, 2006
3. P. Baran, On distributed communications networks. IEEE Trans. Commun. Syst. **CS-12**, 1–9 (1964)
4. S. Borkar, Thousand core chips: a technology perspective, in *DAC-44*, San Diego, 2007
5. S. Borkar, Future of interconnect fabric: a contrarian view, in *SLIP'10*, Anaheim, 2010
6. S. Borkar, NoCs: What's the point? in *NSF Workshop on Emerging Technologies for Interconnects (WETI)*, Washington, DC, Feb 2012
7. P. Bose, The power of communication: trends, challenges (and accounting issues), in *NSF WETI*, Washington, DC, Feb 2012
8. K. Chang et al., HAT: heterogeneous adaptive throttling for on-chip networks, in *SBAC-PAD*, New York, 2012
9. D.E. Culler et al., *Parallel Computer Architecture: A Hardware/Software Approach* (Morgan Kaufmann, San Francisco, 1999)
10. W. Dally, C. Seitz, Deadlock-free message routing in multiprocessor interconnection networks. IEEE Trans. Comput. **36**, 547–553 (1987)
11. W.J. Dally, B. Towles, Route packets, not wires: on-chip interconnection networks, in *DAC-38*, Las Vegas, 2001
12. W. Dally, B. Towles, *Principles and Practices of Interconnection Networks* (Morgan Kaufmann, Amsterdam/San Francisco, 2004)
13. R. Das, O. Mutlu, T. Moscibroda, C. Das, Application-aware prioritization mechanisms for on-chip networks, in *MICRO-42*, New York, 2009
14. R. Das et al., Aérgia: exploiting packet latency slack in on-chip networks, in *ISCA-37*, Saint-Malo, 2010
15. R. Das et al., Design and evaluation of hierarchical on-chip network topologies for next generation CMPs, in *HPCA-15*, Raleigh, 2009
16. S. Eyerman, L. Eeckhout, System-level performance metrics for multiprogram workloads. IEEE Micro **28**, 42–53 (2008)
17. C. Fallin et al., CHIPPER: a low-complexity bufferless deflection router, in *HPCA-17*, San Antonio, 2011
18. C. Fallin et al., MinBD: minimally-buffered deflection routing for energy-efficient interconnect. SAFARI technical report TR-2011-008: http://safari.ece.cmu.edu/tr.html (2011)
19. C. Fallin et al., HiRD: a low-complexity, energy-efficient hierarchical ring interconnect. SAFARI technical report TR-2012-004: http://safari.ece.cmu.edu/tr.html (2012)
20. C. Fallin et al., MinBD: minimally-buffered deflection routing for energy-efficient interconnect, in *NOCS-4*, Copenhagen, 2012

21. C. Gómez et al., Reducing packet dropping in a bufferless NoC, in *Euro-Par-14*, Las Palmas de Gran Canaria, 2008
22. B. Grot, J. Hestness, S. Keckler, O. Mutlu, Express cube topologies for on-chip interconnects, in *HPCA-15*, Raleigh, 2009
23. B. Grot, S. Keckler, O. Mutlu, Preemptive virtual clock: a flexible, efficient, and cost-effective QOS scheme for networks-on-chip, in *MICRO-42*, New York, 2009
24. B. Grot et al., Kilo-NOC: a heterogeneous network-on-chip architecture for scalability and service guarantees, in *ISCA-38*, San Jose, 2011
25. A. Hansson et al., Avoiding message-dependent deadlock in network-based systems-on-chip. VLSI Des. **2007**, 1–10 (2007)
26. M. Hayenga et al., SCARAB: a single cycle adaptive routing and bufferless network, in *MICRO-42*, New York, 2009
27. W. Hillis, *The Connection Machine* (MIT, Cambridge, 1989)
28. Y. Hoskote et al., A 5-GHz mesh interconnect for a teraflops processor. IEEE Micro **27**, 51–61 (2007)
29. S.A.R. Jafri et al., Adaptive flow control for robust performance and energy, in *MICRO-43*, Atlanta, 2010
30. J. Kim, Low-cost router microarchitecture for on-chip networks, in *MICRO-42*, New York, 2009
31. J. Kim, J. Balfour, W. Dally, Flattened butterfly topology for on-chip networks, in *MICRO-40*, Chicago, 2007
32. S. Konstantinidou, L. Snyder, Chaos router: architecture and performance, in *ISCA-18*, Toronto, 1991
33. D. Kroft, Lockup-free instruction fetch/prefetch cache organization, in *ISCA-8*, Minneapolis, 1981
34. J. Laudon, D. Lenoski, The SGI Origin: a ccNUMA highly scalable server, in *ISCA-24*, Boulder, 1997
35. J. Lee, M. Ng, K. Asanovic, Globally-synchronized frames for guaranteed quality-of-service in on-chip networks, in *ISCA-35*, Beijing, 2008
36. C.K. Luk et al., Pin: building customized program analysis tools with dynamic instrumentation, in *PLDI*, Chicago, 2005
37. G. Michelogiannakis et al., Evaluating bufferless flow-control for on-chip networks, in *NOCS*, Grenoble, 2010
38. T. Moscibroda, O. Mutlu, A case for bufferless routing in on-chip networks, in *ISCA-36*, Austin, 2009
39. G. Nychis et al., Next generation on-chip networks: what kind of congestion control do we need? in *HotNets-IX*, Monterey, 2010
40. G. Nychis et al., On-chip networks from a networking perspective: congestion and scalability in many-core interconnects, in *SIGCOMM*, Helsinki, 2012
41. H. Patil et al., Pinpointing representative portions of large Intel Itanium programs with dynamic instrumentation, in *MICRO-37*, Portland, 2004
42. B. Smith, Architecture and applications of the HEP multiprocessor computer system, in *SPIE*, San Diego, 1981
43. A. Snavely, D.M. Tullsen, Symbiotic jobscheduling for a simultaneous multithreaded processor, in *ASPLOS-9*, Cambridge, 2000
44. Standard Performance Evaluation Corporation: SPEC CPU2006 (2006), http://www.spec.org/cpu2006
45. M. Taylor et al., The Raw microprocessor: a computational fabric for software circuits and general-purpose programs. IEEE Micro **22**, 25–35 (2002)
46. M. Thottethodi, A. Lebeck, S. Mukherjee, Self-tuned congestion control for multiprocessor networks, in *HPCA-7*, Nuevo Leone, 2001
47. S. Tota et al., Implementation analysis of NoC: a MPSoC trace-driven approach, in *GLSVLSI-16*, Philadelphia, 2006

48. H. Wang et al., Orion: a power-performance simulator for interconnection networks, in *MICRO-35*, Istanbul, 2002
49. H. Wang et al., Power-driven design of router microarchitectures in on-chip networks, in *MICRO-36*, San Diego, 2003
50. D. Wentzlaff et al., On-chip interconnection architecture of the tile processor. IEEE Micro **27**(5), 15–31 (2007)

Chapter 11
A Thermal Aware Routing Algorithm for Application-Specific Network-on-Chip

Zhiliang Qian and Chi-Ying Tsui

Abstract In this chapter, we propose a routing algorithm to reduce the hotspot temperature for application-specific Network-on-chip (NoC). Using the traffic information of the applications as well as the mapping of the tasks onto the NoC, we design a routing scheme which can achieve a higher adaptivity than the generic ones and at the same time distribute the traffic more uniformly across the chip. A set of admissible paths which is deadlock free for all the communications is first obtained. To reduce the hotspot temperature, we propose to route packets using the optimal distribution ratio of the communication traffic among the set of candidate paths. The problem of finding this optimal distribution ratio is formulated as a linear programming (LP) problem and is solved in the offline phase. A router microarchitecture which supports the ratio-based selection policy is also proposed. From the simulation results, the peak energy reduction considering the energy consumption of both the processor elements and routers can be as high as 20 % for synthetic traffic and real benchmarks.

11.1 Introduction

11.1.1 Background

Embedded systems have led to the advent of multiprocessor System-on-Chip (MPSoC) design. With the decrease in the size of on-chip cores, more and more homogeneous and heterogeneous cores can be integrated on a single chip. A scalable and modular solution to the interconnecting problem becomes critically important

Z. Qian (✉) • C.-Y. Tsui
VLSI Research Laboratory, Department of Electronic and Computer Engineering, The Hong Kong University of Science and Technology, Hong Kong, China
e-mail: qianzl@ust.hk; eetsui@ust.hk

M. Palesi and M. Daneshtalab (eds.), *Routing Algorithms in Networks-on-Chip*,
DOI 10.1007/978-1-4614-8274-1_11, © Springer Science+Business Media New York 2014

[16]. Network-on-chip (NoC) has been proposed to tackle this distinctive challenge [2]. NoCs utilize the interconnected routers instead of buses or point-to-point wires to send and receive packets between processor elements (PE), which overcome the scalability limitations of the buses and bring significant performance improvement in terms of communication latency, power consumption and reliability etc.

11.1.2 Thermal Issues in NoC

For the design of NoC based high performance MPSoC, power and temperature have become the dominant constraints [27]. Higher power consumption will lead to higher temperature and at the same time, the uneven power consumption distribution will create thermal hotspots. These thermal hotspots have adverse effect on the carrier mobility, the meantime between failure (MTBF), and also the leakage current of the chip. As a result, they will degrade the performance and reliability dramatically. Consequently, it is highly desirable to have an even power and thermal profile across the chip [26]. This imposes as a design constraint for NoC to avoid uneven power consumption profile so as to reduce the hotspot temperature.

In a typical NoC-based MPSoC design flow, we need to allocate and schedule the tasks on the available processor cores and map these processors to the NoC platform first. After the task and processor mapping, routing algorithm is developed to decide the physical paths (i.e., the routers to be traversed) for sending the packets from the sources to the destinations. Each phase of the NoC design affects the total power consumption as well as the power profile across the chip [25]. Previously, several task mapping and processor core floorplan algorithms [15] have been proposed to achieve a thermal balanced NoC design. However, the routing algorithm is rarely exploited for this purpose. Since the communication network (including the routers and the physical links) consumes a significant portion of the chip's total power budget (such as 39 % of total tile power in [29]), the decision on the routing path of the packets will greatly affect the power consumption distribution and hence the overall hotspot temperature across the chip [27]. Therefore, it is important to consider the thermal constraint in the routing phase of the NoC design.

In this chapter, we exploit the adaptive routing algorithm to achieve an even temperature distribution for application-specific MPSoCs. Given an application described by a task flow graph and a target NoC topology, we assume that the tasks are already scheduled and allocated to the processors and the processor mapping is also done. We then utilize the traffic information of the application specified in the task flow graph, which can be obtained through profiling [4, 14], to decide how to split the traffic among different physical routes for a balanced power distribution. Figure 11.1 shows an example of the task flow graph and the corresponding mesh topology mapping for a video object plane decoder (VOPD) application [4]. As shown in Fig. 11.1, each tile is made up of a processor element (PE) which executes certain operations and a router which is connected to its neighbors as well as a local PE. The routing algorithm is designed to determine the routers to be traversed for

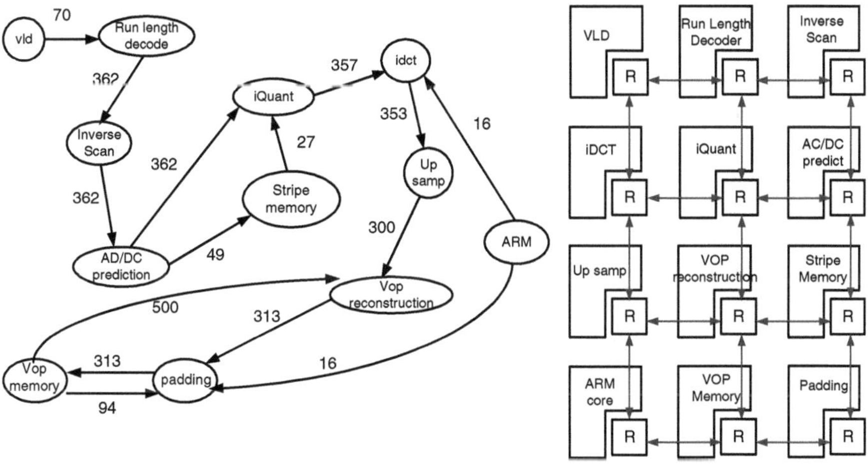

Fig. 11.1 Task graph and tile mapping for the Video Object Plane Decoder (VOPD) [4] application on 4 × 3 NoC

each communication. Of note, when we decide the routing for each traffic pair, we have to make sure all the routing paths are deadlock free [16].

Similar to [9], we use the peak power (i.e., the peak energy under a given time window) metrics to evaluate the effectiveness in reducing the hotspot temperature. Through simulation-based evaluation, we demonstrate the proposed algorithm can reduce the peak energy of the tiles by 10–20 % while improving or maintaining the NoC performance in terms of throughput and latency.

11.2 Related Work

In the area of temperature-aware NoC design, many previous works focus on the power consumption distribution of the processor cores. In [18], a dynamic task migration algorithm was proposed to reduce the hotspot temperature due to the processor core (i.e., PE). In [1], a thermal management hardware infrastructure was implemented to adjust the frequency and voltage of the processing elements according to the temperature requirements at run time.

Since the power consumption of the routers is as significant as the processor core, the thermal constraint should also be addressed in the routing algorithm design. There have been a lot of works on NoC routing algorithms for various purposes including low power routing [19], fault-tolerant routing [11] and congestion avoidance routing to improve latency [17]. However, there are only a few works [7,9,27] taking temperature issue into account.

In [9], an ant-colony-based dynamic routing algorithm was proposed to reduce the peak power. Heavy packet traffics are distributed in the routing based on this dynamic algorithm to minimize the occurrence of hot spots. However this dynamic routing algorithm is generic in nature and does not take into account the application-specific traffic information. Therefore it may not be capable to achieve an optimal path distribution. Special control packets are sent among the routers to implement the algorithm which increases the power overhead. Also, two additional forward and backward ant units are needed in the router which results in a large area overhead. Moreover, this work only minimizes the peak power of the routers but does not consider the effect of the processor core power on the temperature. In [27], a run-time thermal-correlation based routing algorithm is developed. When the peak temperature of the chip exceeds a predefined threshold, the NoC is under thermal emergency and the dynamic algorithm will throttle the load or re-route the packets using the paths of the least thermal correlation with the run time hottest regions. The algorithm also does not consider the specific traffic information of the applications. It may be inefficient if multiple hotspots occur at the same time. Moreover, it is not very clear how to do the re-routing while still guaranteeing the deadlock free property. In [7], a new routing-based traffic migration algorithm VDLAPR and the buffer allocation scheme are proposed to trade-off between the load balanced and the temperature balanced routing for 3D NoCs. In particular, the VDLAPR algorithm is designed for 3D NoCs by distributing the traffic among various layers. For routing within each layer, a thermal-aware routing algorithm such as the one introduced in this chapter is still needed.

In this chapter, we propose a thermal-aware routing for application specific NoC. To guarantee deadlock-free property, generic routing schemes use turn model based algorithms such as X-Y routing, odd-even routing or forbidden turn routing [16]. However, this will limit the flexibility of re-distributing the traffic to achieve an even power consumption profile. Here, we further utilize the characterized application traffic information to achieve a larger path set for routing and at the same time provide deadlock avoidance. Higher adaptivity and hence better performance can be achieved since more paths can be used for the re-distribution of the traffic. Given the set of possible routing paths, we formulate the problem of allocating the optimal traffic among all paths as a mathematical programming problem. At run time, the routing decisions will be made distributively according to the calculated traffic splitting ratios. We demonstrate the effectiveness of the proposed routing strategy on peak energy reduction using both synthetic and real application traffics.

11.3 Methodology Overview

11.3.1 Assumptions and Preliminaries

We aim to achieve an even power consumption distribution and reduce the hotspot temperature for application specific MPSoC using NoC connection. We assume that the given application is specified by a task flow graph which characterizes the communication dependencies and the bandwidth requirements among the tasks of the application (such as Fig. 11.1). In this chapter, we use a tile-based mesh topology for the NoC due to its popularity and simplicity. However, our methodology can be easily extended to other irregular topologies. Based on the task flow graph and the target topology, task allocation and processor mapping are assumed to have been determined before the routing phase. The energy consumption of each processor core is then estimated according to the tasks mapped to it. Taking all the above information as input, we address the issue of reducing the peak temperature via exploring the routing algorithm design. We adopt an adaptive routing strategy instead of a static routing scheme because it provides path diversity to distribute traffic among the sources and the destinations [16] and achieves better performance in latency, throughput and congestion avoidance. We also assume minimum path routing is used throughout this chapter.

To facilitate the discussion of the proposed methodology and algorithms, we borrow some of the definitions presented and discussed in [22].

Definition 1. A Task graph $TG = T(T,C)$ is a directed graph where the vertex $t_i \in T$ represents a task in the application and the edge $c_{ij} = (t_i, t_j) \in C$ represents the communication from t_i to t_j.

Definition 2. A core communication graph $CCG = (P,E)$ is a directed graph where the vertex $p_i \in P$ represents the ith processor in the mesh and the edge $e_{ij} = (p_i, p_j) \in E$ represents the communication from p_i and p_j. Given an application-specific task graph and its corresponding mapping function, CCG can be generated accordingly.

Definition 3. A channel dependency graph $CDG = (L,R)$ is a directed graph where the vertex $L_{mn} \in L$ represents a physical link (m,n) in the mesh topology pointing from tile m to tile n. An edge $R_{\mu\nu} = (L_\mu, L_\nu)$ $(\mu = mn,\ \nu = np) \in R$ exists if there is a dependency from L_μ to L_ν which indicates a path passing through the two links (m,n) and (n,p) consecutively.

From Duato's theorem [10], a routing algorithm is deadlock free for an application if there are no cycles in its channel dependency graph (CDG).

Definition 4. Strongly connected components (SCC): A directed graph is strongly connected if there is a path from each vertex to every other. The strongly connected components of a graph are its maximally connected sub-graphs.

We illustrate the definitions of CCG, CDG and SCC by an example shown in Fig. 11.2. Here we use a 3×3 mesh NoC (Fig. 11.2b) as the communication backbone for an application characterized by the core communication graph shown in Fig. 11.2a. For each communication pair $e_{ij} = (p_i, p_j)$ in the CCG, assume the Manhattan distance (i.e. the minimum number of hops) between the source and destination tile pair (i,j) is $d(i,j) = |i_x - j_x| + |i_y - j_y|$ and the distance between the two tiles in the x-direction is $d_x(i,j) = |i_x - j_x|$, the total number of

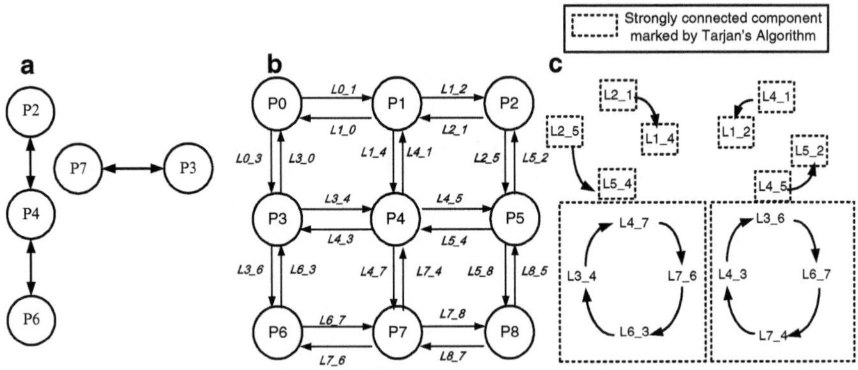

Fig. 11.2 An example of (**a**) Core communication graph (CCG) (**b**) NoC architecture (**c**) Channel dependency graph (CDG) and its strongly connected components (SCC) after mapping (**a**) onto (**b**)

minimum length paths connecting the two tiles is given by $N_{ij} = C_{d(i,j)}^{d_x(i,j)}$[1] since we need to traverse along the x direction $d_x(i,j)$ times among total $d(i,j)$ number of hops. Let $l(i,j,k)\,(k \leq N_{ij})$ denote the kth path connecting the two tiles. If $l(i,j,k)$ traverses two network links (m,n) and (n,p) consecutively, then an edge is added to connect the two nodes L_{mn} and L_{np} in the CDG. By inspecting all the feasible paths, we can construct the whole channel dependency graph. In Fig. 11.2, taking the communication pair $(2,4)$ as an example, two minimum length paths exist: $(2 \rightarrow 5 \rightarrow 4)$ and $(2 \rightarrow 1 \rightarrow 4)$. In the CDG, we add two edges connecting (L_{2-5}, L_{5-4}) and (L_{2-1}, L_{1-4}), respectively. As shown in Fig. 11.2c, there are in total 10 strongly connected components in the resulting CDG which can be found by the Tarjan's algorithm [28]. We can find two circles in Fig. 11.2c, i.e., $L_{3-4} \rightarrow L_{4-7} \rightarrow L_{7-6} \rightarrow L_{6-3} \rightarrow L_{3-4}$ and $L_{4-3} \rightarrow L_{3-6} \rightarrow L_{6-7} \rightarrow L_{7-4} \rightarrow L_{4-3}$, where some edges are required to be removed by the path set finding algorithm in Sect. 11.4.1 to avoid deadlocks.

Definition 5. Routing adaptivity: Routing adaptivity is defined as the ratio of the total number of available paths provided by a routing algorithm to the number of all possible minimum length paths between the source and destination pairs. Higher adaptivity will improve the capability of avoiding congestion and redistributing the traffic to reduce thermal hotspot. However, we need to meet the deadlock-free constraints while improving the routing adaptivity.

For a given application, we first use the task graph TG (Definition 1) to characterize its traffic patterns as well as the communication volume between the tasks. Then based on the mapping algorithm on the target NoC platform, the task graph TG is transformed to the core communication graph CCG (Definition 2). The channel

[1] C_n^r is the combination notation, where $C_n^r = \frac{n!}{r!(n-r)!}$

a

b

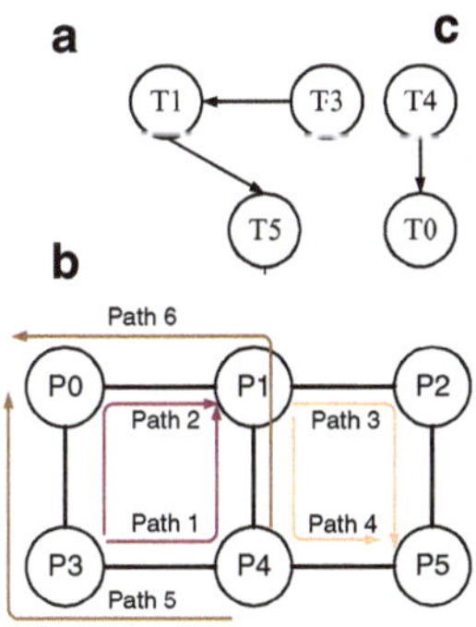

c

Communication pair	Path used in routing	Number of packets sent over the path	
		Strategy 1 (uniform traffic distribution)	Strategy 2 (allocating traffic non-uniformly)
Tile P3 to P1	Path 1: P3->P4->P1	500	400
	Path 2: P3->P0->P1	500	600
Tile P1 to P5	Path 3: P1->P2->P5	500	850
	Path 4: P1->P4->P5	500	150
Tile P4 to P0	Path 5: P4->P3->P0	500	900
	Path 6: P4->P1->P0	500	100
Router energy distribution		Strategy 1	Strategy 2
Tile 0		1500 E_1	1600 E_1
Tile 1		2500 E_1	2100 E_1
Tile 2		500 E_1	850 E_1
Tile 3		1500 E_1	1900 E_1
Tile 4		2000 E_1	1550 E_1
Tile 5		1000 E_1	1000 E_1
Peak Energy		2500 E_1	2100 E_1
Maximum energy difference		2000 E_1	1250 E_1

d

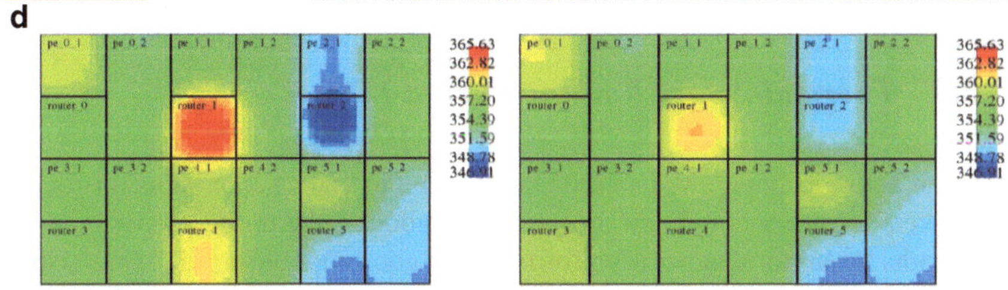

Fig. 11.3 A motivation example of allocating traffic among paths (**a**) the core communication graph (CCG) (**b**) routing paths allocation on 2×3 NoC (**c**) two strategies of using the routing paths (**d**) thermal profile comparison (*left*: strategy 1, *right*: strategy 2)

dependency graph *CDG* (Definition 3) is built based on the routing paths assigned to each edge in *CCG*. In this chapter, we propose to find the strongly connected components *SCC* (Definition 4) of the underlying *CDG* to remove the circular dependency among the channel resources to avoid potential deadlock. Finally, routing adaptivity (Definition 5) is the metric that reflects the capability of redistributing traffic for different cycle breaking algorithms in *SCC*.

11.3.2 Motivation for Thermal-Aware Routing

For adaptive routing, usually there will be more than one path available for every communication pair. If traffic is distributed equally on all the paths, some of the routers may have more paths passing through them and hence more packets to receive and send. We show the need to allocate traffic properly among the paths by an example illustrated in Fig. 11.3.

In Fig. 11.3, three communication pairs are assumed to occur concurrently: from tile P3 to P1, tile P1 to P5 and tile P4 to P0. We consider the case of 1,000 packets that are generated for each communication pair and sent over the network in this example. The energy consumption of processing a single packet in a router is denoted as E_1. We compare two routing strategies in the figure. Strategy 1 uses uniform traffic distribution among all paths between the source and the destination nodes. Strategy 2 allocates traffic non-uniformly among the candidate paths. The total number of packets handled by each router is different in these two schemes. The router energy distribution is shown in Fig. 11.3c. By properly allocating the traffic, we can reduce the peak energy of the tile by 16 % and the energy difference among the tiles by 37.5 %. Moreover, we can further assume the average energy consumption of the processors and the routers are about the same and then use Hotspot [13] to simulate and evaluate the thermal profile across the chip. As shown in Fig. 11.3d, strategy 2 indeed makes the thermal profile more uniform and reduces the hotspot temperature.

11.3.3 Thermal-Aware Routing Design Flow

From the above example, we can see that we need to find a set of paths for routing the packets for every communication pair and allocate the traffic properly among these paths so as to achieve a uniform power consumption profile. One critical issue of determining the path set is to provide deadlock avoidance. In generic routing algorithms, deadlock can be prevented by disallowing various turns [16]. For application specific NoCs, it will be too conservative and unnecessarily prohibit some legitimate paths to be used [22] as some disallowed turns can actually be used because the application does not have traffics interacting with these turns to form circular dependencies in the CDG. By using the information specified by the task flow graph and its derived communication graph, we can find more paths available for routing which increases the flexibility of distributing traffic among the sources and the destinations. In this chapter, we use a similar approach as [22] to find the set of admissible paths for each communication pair while still satisfying the deadlock free requirement.

After obtaining the admissible path set for routing, we then find the optimal traffic allocation to each path based on the bandwidth requirement for the purpose of achieving a uniform power profile. The problem is formulated as a mathematical optimization problem and solved by a LP solver. These phases of design are done offline at the design time. After that, the optimal distribution ratio of each path is obtained. For each particular source-destination communication pair, the ratios of the paths are converted into the probabilities of using each port to route to the destination for each router. At run time, these probability values are stored into the routing tables in the routers. For each incoming packet, the router will inquiry its routing table and return the candidate output ports for this packet according to the input direction and the destination. The final output port will be chosen according to the probability value of each candidate.

Figure 11.4 summarizes the whole design flow of our proposed methodology. In the next section, the details of the algorithm will be presented.

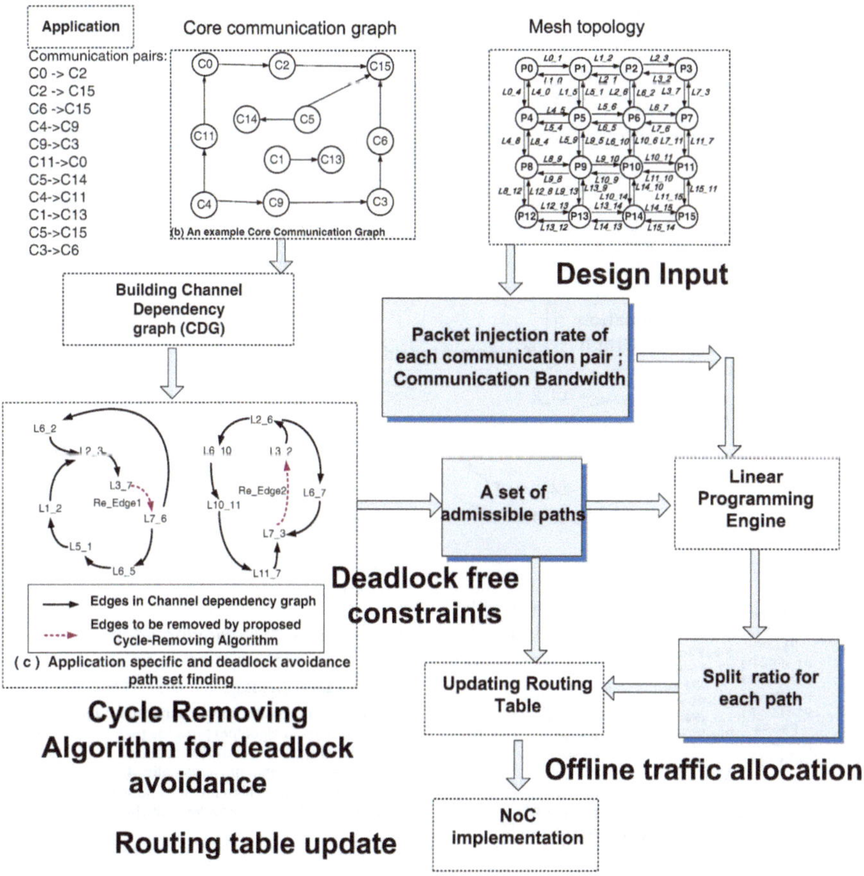

Fig. 11.4 Proposed design flow for the thermal-aware routing

11.4 Main Algorithm

In this section, we present the details of the offline routing algorithms. We first discuss the algorithm of finding the set of admissible paths for each source-destination communication pair. The admissible paths avoid the circular dependency among any paths and hence provide the deadlock free property. Then we present the optimal traffic allocation problem formulation.

11.4.1 Application-Specific Path Set Finding Algorithm

In [22], a dynamic routing algorithm which increases the average routing adaptivity while maintaining the deadlock-free property at the same time is proposed. The

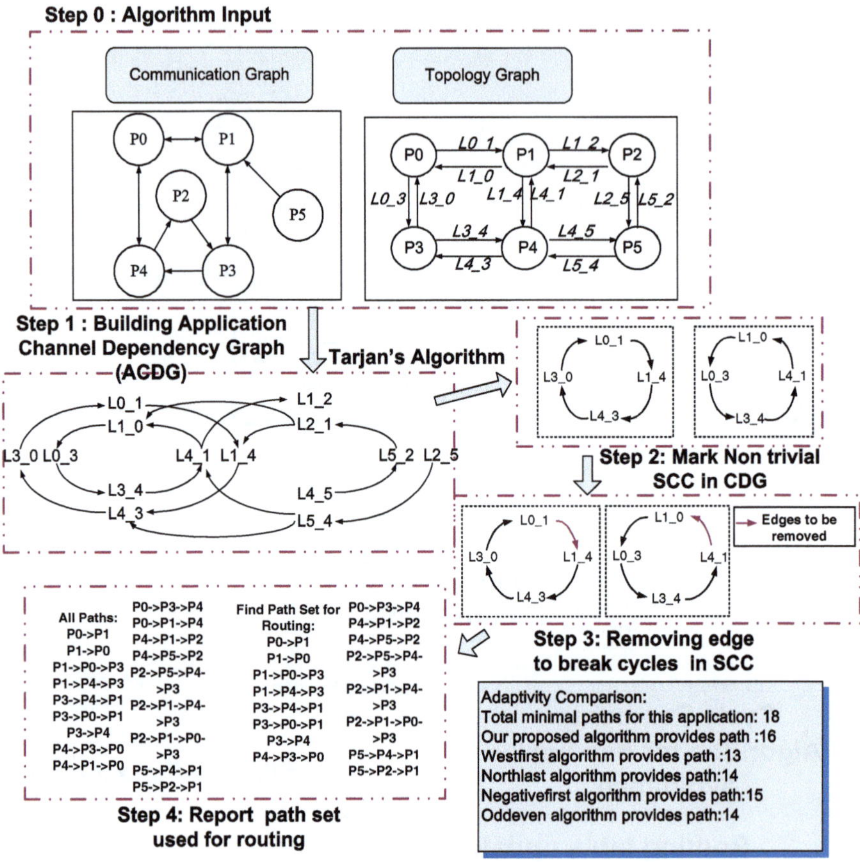

Fig. 11.5 Application specific and deadlock free path finding algorithm

average routing adaptivity is often used to represent the degree of adaptiveness and flexibility of a routing algorithm. Here we use a similar approach. However, different from [22] which focuses on maximizing the average routing adaptivity to improve the latency, we aim to maximize the flexibility to re-divert the traffic so as to even out the power consumption distribution. Therefore we need to consider the bandwidth requirement of each communication also since the amount of packet processed in each flow contributes differently to the overall energy distribution.

Figure 11.5 shows the main flow of our path finding algorithm. Similar to [22], based on the application's task flow graph, we examine all the paths between the source and destination pairs to build the application channel dependency graph (ACDG). Most likely, there will be cycles in the ACDG so that some edges have to be removed to break these cycles to guarantee deadlock free. In [22], a branch and bound algorithm is used to select the set of edges to remove all the cycles while

Table 11.1 Notations of application specific path set finding algorithm

Notations	Description
p	A path from tile $src(p)$ to tile $dst(p)$
c	A communication edge in core communication graph
$W(c)$	Bandwidth of communication c
C	Whole set of communication c in application
$\Phi(c)$	Set of minimum paths for communication c
S_{edge}	Set of edges to be removed for breaking cycles
$R_{\mu\nu}$	An edge joining two link vertex μ, ν in CDG
$P_{\mu\nu}$	Set of paths which introduce $R_{\mu\nu}$in CDG
$\Phi_{S_{edge}}(c)$	Set of paths for c after edges in S_{edge} removed

maintaining all the connectivity and maximizing the average adaptivity. The average adaptivity α is defined as:

$$max\,\alpha = max\,\frac{1}{|C|}\sum_{c\in C}\alpha_c\,;\ \alpha_c = \frac{|\Phi_{S_{edge}}(c)|}{|\Phi(c)|} \tag{11.1}$$

where

$$\Phi_{S_{edge}}(c) = \Phi(c)\setminus\{p|p\in\Phi(c)\bigwedge p\in P_{\mu\nu}\,\forall R_{\mu\nu}\in S_{edge}\} \tag{11.2}$$

The notations used in the above equations are summarized in Table 11.1. Connectivity is guaranteed for every communication c by making sure that at least one path exists between the source and destination nodes. So we have $|\Phi_{S_{edge}}(c)| \geqslant 1; \forall c \in C$. In this chapter, instead of finding all cycles in the ACDG and breaking each cycle separately, as done in [22], we apply Tarjan's algorithm [28] to find all the strongly connected components (SCC) and try to eliminate cycles within each nontrivial components (i.e., components with more than one vertex). One important property of SCC is that cycles of a directed graph are contained in the same component. The cycles can then be eliminated within each nontrivial components to achieve the deadlock free. Tarjan's algorithm is used due to its lower complexity ($O(|V|+|E|)$). In addition, in many cases, several edges are shared among different cycles (as illustrated in Fig. 11.4, the two edges (L_{3-7},L_{7-6}) and (L_{7-3},L_{3-2}) are shared among several cycles). If we inspect each cycle separately, we may consider these edges more than once. On the other hand, when we employ Tarjan's algorithm, cycles with common edges are in the same component and hence decision can be made more efficiently if we remove some shared edges to break these cycles simultaneously. When we choose edges to break the cycle, instead of optimizing the average routing adaptivity, we maximize the following objective function:

$$max\,\frac{1}{|C|}\sum_{c\in C}\alpha_c W_c = max\,\frac{1}{|C|}\sum_{c\in C}W(c)\times\frac{|\Phi_{S_{edge}}(c)|}{|\Phi(c)|} \tag{11.3}$$

Table 11.2 Parameters and notations in LP formulation

Notations	Network on chip parameters
N	Total number of tiles in NoC
C_{ij}	Link capacity between tiles i and j
L_{ij}	Number of minimum paths between tiles i and j
$l(i,j,k)$	The kth path joining tiles i and j; $1 \leqslant k \leqslant L_{ij}$
T_i	Set of paths pass tiles i $T_i = \{l(m,n,t) \| i \in l(m,n,t)\}$
$p(i,j)$	Packet injection rate between tiles i and j
S_{packet}	Number of flits in a packet
S_{bit}	Number of bits in a packet
Notations	Tile energy parameters
E_{p-i}	Average processor energy of the ith tile
$E_{forward}$	Energy of forwarding a flit in the router
$E_{buffer-rw}$	Buffer read and write energy for a flit in the router
E_{sw}	Switch allocation energy for a flit in the router
E_{rc}/E_{sel}	Routing computation/output port selection energy
E_{vc}	Virtual channel allocation energy

Here in Eq. 11.3, we weight the routing adaptivity of each communication with its corresponding bandwidth requirement ($W(c)$ is the bandwidth of communication c). The rationale is that for the communications with higher bandwidth requirement, usually more packets need to be processed and routed. Therefore, their impact on the power consumption distribution is higher. We should desire higher flexibility for these communications to re-divert the traffic and hence higher adaptivity.

11.4.2 Optimal Traffic Allocation

In the following, we use the notations summarized in Table 11.2 and the energy model described in Sect. 11.4.2.1 to obtain the linear programming (LP) formulation of the optimal traffic allocation problem. In the LP engine, the inputs are the application task flow graphs and the admissible path set. The outputs are the path ratios to be used in the routing at run time.

11.4.2.1 Energy Consumption Model

We assume the energy consumption of each processor i (E_{p-i}) is available after task mapping. Wormhole routing is used in our routing scheme. In wormhole routing, each packet is divided into several flits which are the minimum units for data transmission and flow control. For every data packet, the head flit sets up the path directions for the body and the tail flits. Thus, E_{rc}, E_{sel} and E_{vc} in Table 11.2 only incur when the head flit is being processed by the router. Total energy consumption for processing a single packet in router i can be represented as:

$$\triangle E_{r-i} = (E_{buffer-rw} + E_{forward} + E_{sw}) \times S_{packet} + E_{rc} + E_{sel} + E_{vc} \quad (11.4)$$

Let n_i denote the number of packets that received by router i, then the total router energy consumption is equal to $E_{r-i} = \triangle E_{r-i} \times n_i$. The total energy of each tile i (E_i) is equal to $E_{p-i} + E_{r-i}$.

11.4.2.2 LP Problem Formulation

Given the set of admission paths for every communication pair which is deadlock free, we want to obtain the ratio of traffic allocated to each path so as to minimize the maximum energy consumption among all the tiles. We formulate the following linear programming problem to obtain the optimal path ratios:

1. **Variables** $r(i,j,k)$**:** portion of traffic allocated to the kth path $l(i,j,k)$ between tiles i and j among all the L_{ij} paths, where $1 \leq i \leq N_{tile}$, $1 \leq j \leq N_{tile}$, $1 \leq k \leq L_{ij}$.
2. **Objective functions:** The energy consumption of the ith tile, E_i, is given by

$$E_i = E_{p-i} + \triangle E_{r-i} \times n_i \quad (11.5)$$

where n_i is the total number of packets received by router i per unit time and is equal to the summation of the packets received from all paths passing through tile i, i.e., T_i . More specifically, n_i is given by:

$$n_i = \sum_{\forall l(a,b,k) \in T_i} r(a,b,k) \times p(a,b) \quad (11.6)$$

In order to balance the tile energy E_i, our objective function is written in a min-max form as follows:

$$obj:\ min(max(E_i))\ for\ 1 \leq i \leq N_{tile} \quad (11.7)$$

It is equivalent to:

$$min(E)\ s.t.\ E \geq E_i\ for\ 1 \leq i \leq N_{tile} \quad (11.8)$$

3. **Problem constraints:** The objective function should be optimized with regard to the following constraints:

3-1. Traffic split constraints: summation of all the traffic allocation ratios between a given pair (i,j) should equal to 1.

$$\sum_{k=1}^{L_{i,j}} r(i,j,k) = 1\ \forall (i,j) \in C \quad (11.9)$$

$$r(i,j,k) \geq 0 \forall i,j \in [1, N_{tile}],\ k \leq L_{ij} \quad (11.10)$$

3-2. Bandwidth constraints: the aggregate bandwidth used for a specific link should not exceed the link capacity imposed by the physical NoC platform.

The communication bandwidth = packet injection rate × packet size × clock frequency. Assume T is the cycle time and (i, j) is a physical link in the mesh NoC, a path $l(a,b,k)$ will traverse this link if and only if $l(a,b,k) \in T_i \cap T_j$. So we have

$$\sum_{\forall l(a,b,k) \in T_i \cap T_j} \frac{r(a,b,k) \times p(a,b) \times S_{bit}}{T} \leq C_{ij} \tag{11.11}$$

Taking the application's task flow graph and the task mapping as input, the packet injection rate $p(a,b)$ can be calculated by summing the bandwidth volume from all the communication pairs where the source tasks are mapped onto tile a and destination tasks are mapped onto tile b.

The above formulation is a typical linear programming problem and can be solved efficiently using MATLAB CVX optimization toolbox [8].

11.4.2.3 Using the Path Ratios in the Routers

After solving the LP problem, we obtain a set of admissible paths and their corresponding traffic allocation ratios. To use these information in the implementation of the NoC routing, we can use two schemes. The simplest way is source routing. In the source processor, the header flit of each packet contains the entire path information. The source processor i decides which path to use to send the packet to destination j according to the traffic allocation ratios $r(i, j,k)$ of the set of admissible paths. Intel Teraflops chip [16] uses this source-based routing scheme as the router will be simple. However, this will create a large overhead on the effective bandwidth as the packet needs to contain additional payload to record the entire routing path.

A more efficient way of implementing the thermal aware routing is to use routing tables stored in each router. One of the major advantages of table-based routing is that it can be dynamically reconfigured or reloaded [22] to allow modifications in the communication requirements. However since the routing decision is made within each router without knowing the entire path information, the path traffic-allocation ratios obtained previously cannot be directly used. In the following, we will show how to convert these ratios into local probabilities for the router to select which output port to send out the packet.

To support the thermal-aware routing, the routing table is organized as follows: for each router t in the mesh topology and each of its input direction $d \in D_{in}(t)$, there is a routing table $RT(t,d)$ [22]. For each output port o, there is a set of corresponding entries in the routing table. Each entry is made up of the tile id numbers of the source (s), and destination (b) pair, as well as the probability values $p(o)$ of using o to route to b. Formally, $RT(t,d) = \{(s,b,o,p(o)) | o \in D_{out}(t), 0 \leq p(o) \leq 1\}$.

When using routing tables within the routers, the final routing path for a packet of a specific traffic flow is composed by the output port selected in each router along the

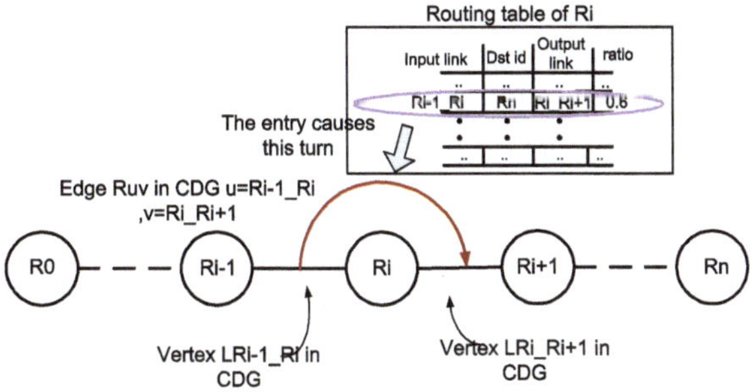

Fig. 11.6 Proof of deadlock free inheritance

path distributively. One problem may hereby arise: will the selected output ports in the routers form a new path that is not included in the admissible path set and hence introduce the possibility of deadlock? In the following, we prove that path generated at run-time using the table-based routing is indeed deadlock free provided the path set used for generating the routing table $RT(t,d)$ contains no circles.

Lemma 1 (deadlock-free property in distributive routing). *For the table-based routing, after generating the distributed routing table for each router according to the deadlock free path set P, the actual route generated at run time will not contain a disallowed path not included in P. Deadlock-free property is hence inherited from the path set P.*

Proof (by contradiction). We prove that a disallowed path R cannot exist if all the entries of the routing table are created according to P. Assume $R =< r_0, \cdots, r_{i-1}, r_i, r_{i+1} \cdots r_n >$ is an actual path used to route a packet from node r_0 to r_n at run time (Fig. 11.6). R is obtained by sequentially looking up the routing tables from routers r_0 to r_{n-1}. Assume this path $R \notin P$ is a disallowed path which may cause circular dependencies with the other routing paths in P, then at least one edge $R_{\mu\nu}(\mu = r_{i-1}r_i,\ \nu = r_i r_{i+1})$ is an element of S_{edge} in Table 11.1 in order to generate the circular dependencies. However, since R is built from the routing tables in router r_i, which means there is an entry in the routing table of r_i using link $r_{i-1}r_i$ as input and link $r_i r_{i+1}$ as output. It is already known the entries of the routing tables are created according to the paths in the admissible path set P. Therefore, there is another path $p \in P$ while $p \neq R$ traversing through r_i from the input link $r_{i-1}r_i$ to the output link $r_i r_{i+1}$, hence p also contains the edge $R_{\mu\nu}(\mu = r_{i-1}r_i,\ \nu = r_i r_{i+1})$ in its CDG. On the other hand, since $R_{\mu\nu} \in S_{edge}$, it will create cycles in the CDG and P is actually not a deadlock-free path set. This contradicts with the assumption. □

Now the issue is how to obtain the probability values $p(o)$ for each output o from the path traffic allocation ratios. For each router t and each of its input port d, we obtain a subset of admissible paths (i.e., $T_{t,d}$) that pass through t using input port d. We denote the set $T_{t,d}(o,s,b) \subseteq T_{t,b}$ which contains the paths that depart from router t from output port o with the packet source s and destination b. For a given destination tile b, in minimum path routing, only two candidate output ports (o_1 and o_2) are feasible. The probabilities of selecting ports o_1 and o_2 as the output port are represented as $p_{t,d}(o_1,s,b)$ and $p_{t,d}(o_2,s,b)$. They are calculated by comparing the aggregate traffic of the paths locating in $T_{t,d}$ that use ports o_1 and o_2, respectively, to route to the tile b. Following the notations in Table 11.1, let paths $l(s,b,k) \in T_{t,d}(o_1,s,b)$ and $l(s,b,l) \in T_{t,d}(o_2,s,b)$, we have:

$$p_{t,d}(o_1,s,b) + p_{t,d}(o_2,s,b) = 1 \tag{11.12}$$

$$\frac{p_{t,d}(o_1,s,b)}{p_{t,d}(o_2,s,b)} = \frac{\sum_{\forall l(s,b,k)\in T_{t,d}(o_1,s,b)} p(s,b) \times r(s,b,k)}{\sum_{\forall l(s,b,l)\in T_{t,d}(o_2,s,b)} p(s,b) \times r(s,b,l)} \tag{11.13}$$

After enumerating all the routers and the communication pairs, the port probability values are obtained and stored into the routing tables offline.

In Eqs. 11.12 and 11.13, in order to maintain the global path traffic allocation ratios, each entry in the routing table $RT(t,d)$ needs to distinguish the source tile location s of the packet. Therefore the number of entries in the routing table is increased. We can optimize the routing table size by grouping the entries of different sources but the same destinations together. The new format of the routing table becomes $RT(t,d) = \{(b,o,p(o))|o \in D_{out}(t), 0 \leq p(o) \leq 1\}$. In this case, the port probability value $p(o,b)$ in $RT(t,d))$ is calculated as:

$$p_{t,d}(o_1,b) + p_{t,d}(o_2,b) = 1 \tag{11.14}$$

$$\frac{p_{t,d}(o_1,b)}{p_{t,d}(o_2,b)} = \frac{\sum_s \sum_{\forall l(s,b,k)\in T_{t,d}(o_1,s,b)} p(s,b) \times r(s,b,k)}{\sum_s \sum_{\forall l(s,b,l)\in T_{t,d}(o_2,s,b)} p(s,b) \times r(s,b,l)} \tag{11.15}$$

In Eqs. 11.14 and 11.15, although the exact path traffic allocation ratios cannot be kept, the traffic distribution on each router and each link can still be maintained which is more important to delivery the energy distribution more uniformly. From the simulation results which will be presented in Sect. 11.6, this routing table implementation achieves similar improvement in peak energy reduction and latency performance compared to the routing table using the source-destination pair.

11.5 Router Microarchitecture

The block diagram of NoC router design to support thermal aware routing is illustrated in Fig. 11.7a. For minimum path routing, if the input direction and the destination are fixed, there are at most two candidate output ports, o_1 and o_2, and $p(o_1) + p(o_2) = 1$. The routing selection unit in the router selects the output port O_{dir} by comparing the probability value $p(o_1)$ with a random number τ in $[0,1]$. It will select port O_1 $if\ \tau \leq p(o_1)$, otherwise port O_2 will be chosen.

A pseudo random number generator using linear feedback shift register (LFSR) is employed to generate the random number at run time.

The header flits at each input ports are first decoded by a parser (the HPU module shown in Fig. 11.7b) to extract their destinations. Then by inquiring $RT(t,d)$, two candidate output ports are returned with the corresponding probability values. The

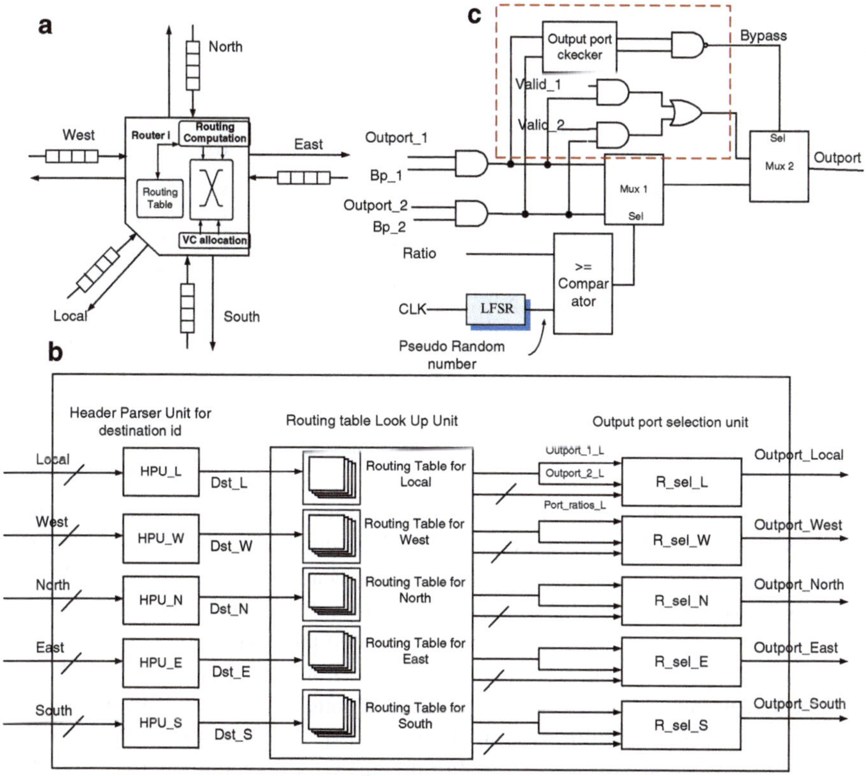

Fig. 11.7 Block diagram of the router supporting ratio-based routing (**a**) Router microarchitecture (**b**) Routing computation unit (**c**) Ratio-based output port selection

output port selection unit then make a decision on the output port. In addition to the probability selection, backpressure information (Bp_1 and Bp_2 in Fig. 11.7c) from downstream routers are also taken into consideration. If one candidate output port is not available due to limited buffer space, the backpressure signal will disable this output port in the selection.

11.6 Experimental Results

11.6.1 Simulation Environment Setup

We implemented and evaluated the proposed thermal-aware routing scheme for mesh-based NoC architecture. A C++ program is developed to analyze the traffic parameters automatically and generate the corresponding LP formulation for a given application. The LP problem is then solved by CVX [8] which is

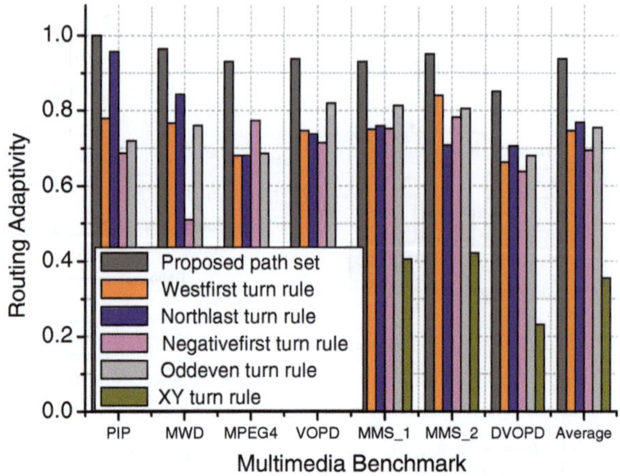

Fig. 11.8 Adaptivity comparisons for different benchmarks

a toolbox embedded in MATLAB for convex optimization. In order to evaluate the NoC performance, a systemC based cycle-accurate simulator was developed combining the features of several widely adopted academic tools (including Noxim [21], Nirgam [20] and Booksim [5]). We adopted both synthetic traffic and real benchmarks to evaluate the performance and compare with other deadlock-free turn model based routing algorithms (westfirst, northlast, negativefirst, oddeven and X-Y routing [16]). The real application patterns include PIP (Picture-In-picture) [4], MWD (Multi-Window Display) [4], MPEG4 [3], VOPD (Video Object Plane Decoder) [4], MMS_1(Multimedia system mapping to 16 cores) [14], MMS_2(Multimedia system mapping to 25 cores) [23] and DVOPD (Dual Video Object Plane Decoder) [24]. For synthetic traffic, we use eight different traffic scenarios, namely Uniform random, Transpose-1 [12], Transpose-2 [12], Hotspot in center (Hotspot-C) [6], Hotspot in top-right corner (hotspot-Tr) [22], Hotspot in right-side (Hotspot-Rs) [22], Butterfly [21] and Bursty traffic [21]. 3×3, 4×4 and 5×5 NoC meshes are used for different benchmarks and 4×4 meshes are used for all synthetic traffics.

11.6.2 Adaptivity Improvement

First we compare the routing adaptivity of the admissible path sets generated by the proposed cycle removal algorithm with other turn model based routing algorithm under several real benchmarks. The results are shown in Fig. 11.8. It can be seen that 20–30 % more paths are available if we take the application traffic into consideration, which indicates a better opportunity of using more paths to even out the NoC power profile.

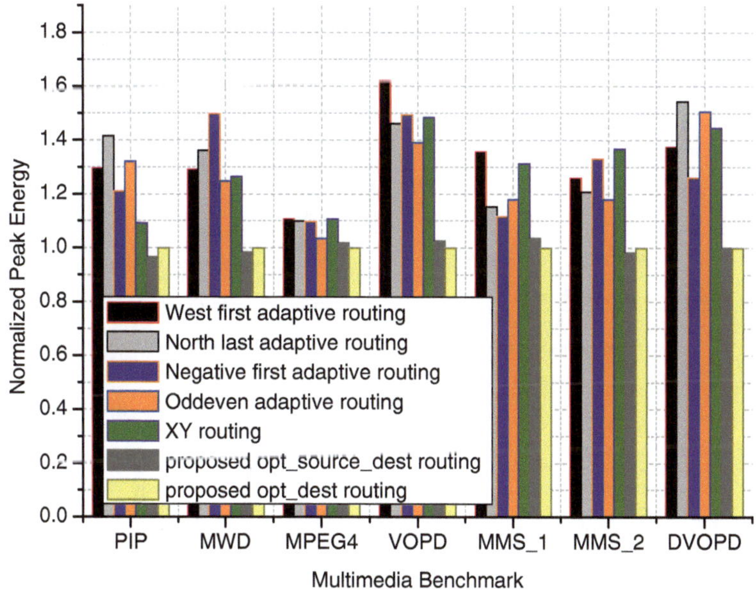

Fig. 11.9 Peak energy simulation for real benchmarks

11.6.3 Peak Energy Simulation Results

Next we evaluate the effectiveness of the proposed algorithm in peak energy reduction. In the simulation, energy parameters ($E_{buffer-rw}$, E_{rc} etc.) are adopted from Noxim and Booksim simulators. After 10,000 warm-up cycles, we carry out energy simulation using a fixed time window for all cases. Ten different simulations were carried out for each packet injection rate to obtain the average peak energy consumption. The peak energy comparisons for real benchmarks and synthetic benchmarks are shown in Figs. 11.9–11.12, respectively.

In the figures, the proposed source-destination routing table and destination-only routing table schemes are denoted as opt_source_dest and opt_dest, respectively. The theoretical optimal energy consumption which is obtained by solving the LP formulation using CVX is also included as a reference and is denoted as opt_peak.

From Figs. 11.9 and 11.10, we can see that comparing with other adaptive routing algorithms, the proposed routing algorithm can achieve more than 20 % peak energy reduction in most traffic patterns. At the same time, the latency performance can also be improved or maintained for all the cases.

In Fig. 11.13, we illustrate a scenario of the tile energy distribution under Hotspot-C traffic (i.e. four center cores are traffic hotspots). Figure 11.13a is the energy distribution using the proposed thermal-aware routing algorithm, and Fig. 11.13b–d show the energy distribution using odd-even, negative-first and XY routing, respectively. For all the four cases, the total energy consumption of the NoC

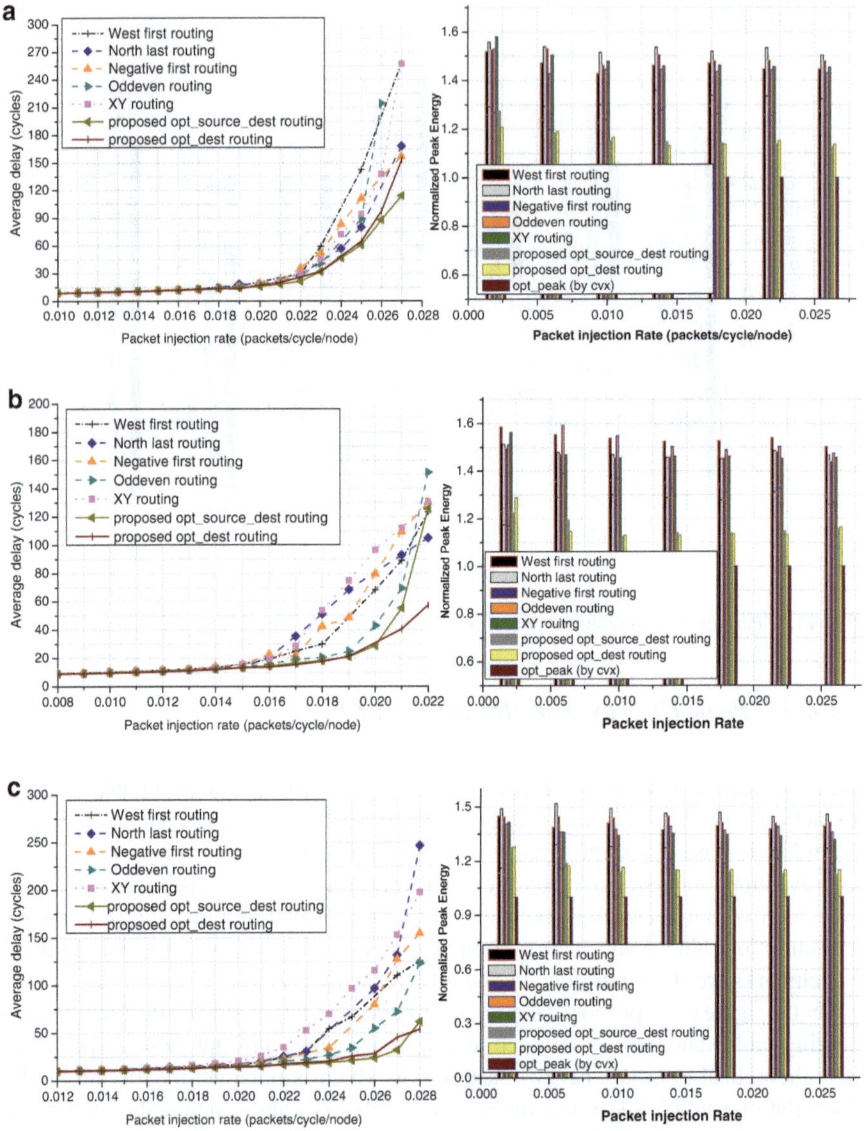

Fig. 11.10 Latency and peak energy simulation for (**a**) Random traffic, (**b**) Transpose-1 traffic and (**c**) Transpose-2 traffic

is same, i.e., 4.55×10^{-5} J. However, the peak tile energy of the proposed, odd-even, negative-first and XY routing are 5.95×10^{-6} J, 8.09×10^{-6} J, 7.53×10^{-6} J and 7.72×10^{-6} J respectively. From the figures, we can see that our proposed scheme indeed leads to a more even energy profile across the NoC chip.

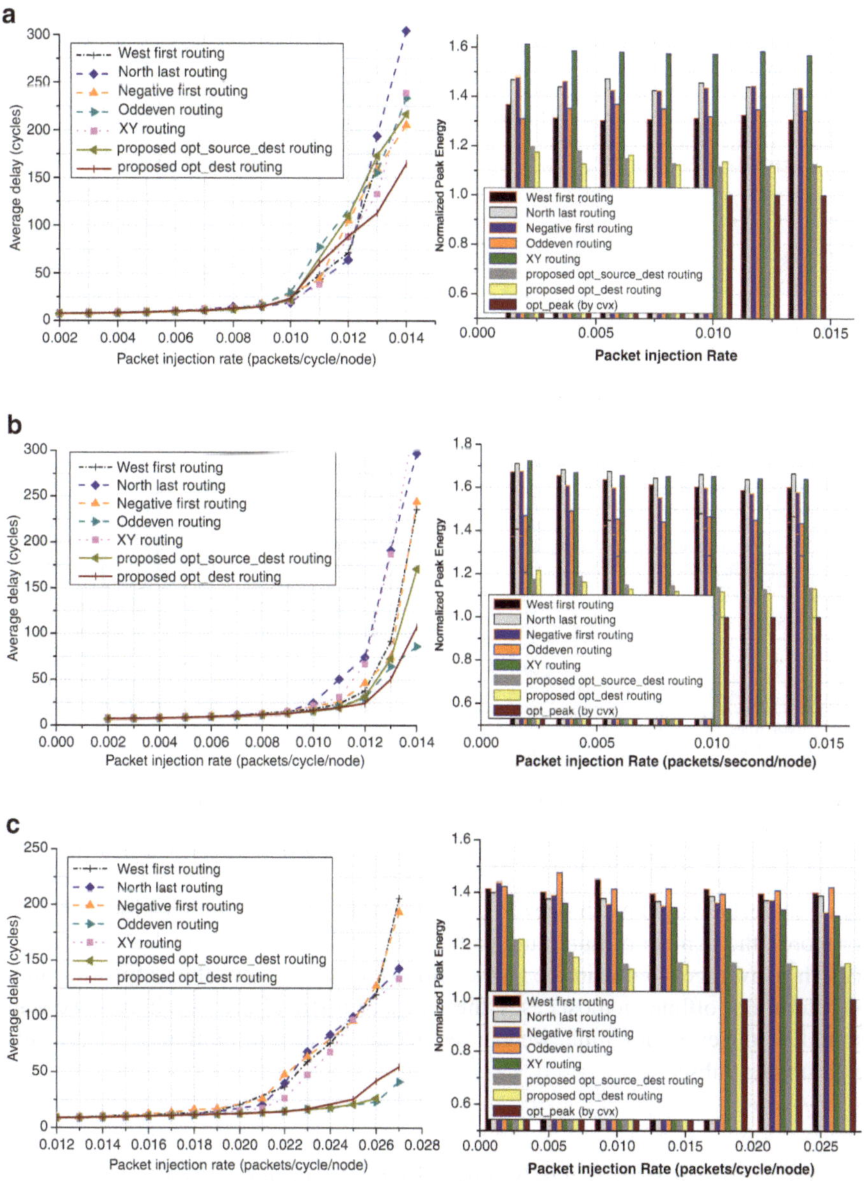

Fig. 11.11 Latency and peak energy simulation for (a) Hotspot in right-side (Hotspot-Rs) traffic, (b) Hotspot in top-right corner (Hotspot-Tr) traffic and (c) Bursty traffic

In Table 11.3, we summarize the execution time of our off-line routing algorithm (including path generation and LP solving) under various mesh size and communication density ρ (ρ is defined as the ratio of the total number of communications pairs to the number of processors in mesh). It can be seen that the execution time is

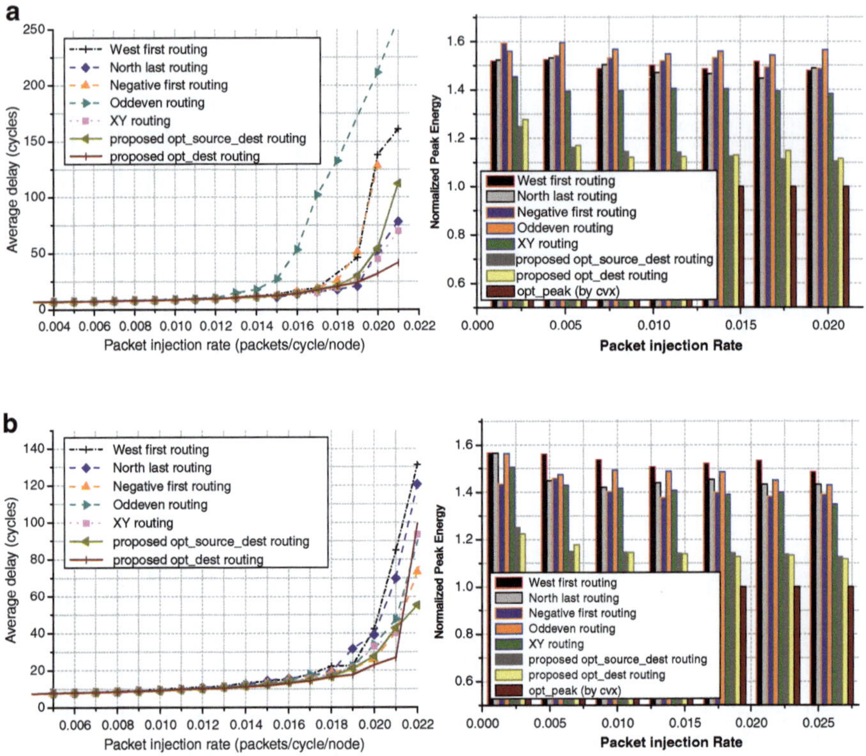

Fig. 11.12 Latency and peak energy simulation for (**a**) Hotspot in center (Hotspot-C) traffic and (**b**) Butterfly traffic

reasonable. For larger mesh size (7 × 7 or more) and more communication pairs (100 or more), the number of minimum paths increase dramatically. It takes longer time (2–3 h in average) to obtain the traffic allocation ratios. Since the traffic allocation is determined offline, in most cases, the time cost is still affordable. In case we want to reduce the execution time, we can restrict the number of minimum-length paths to a smaller subset.

11.6.4 Simulation Results with Different Processor/Router Energy Ratio

The energy consumption of a tile consists of that of the processor and the router. For different applications, the power contribution of the router and the processor differs greatly. Our thermal-aware routing algorithm can only re-distribute the router power consumption. Under this constraint, we next evaluate the effectiveness of our algorithm on the peak energy reduction when the ratio of the energy contribution

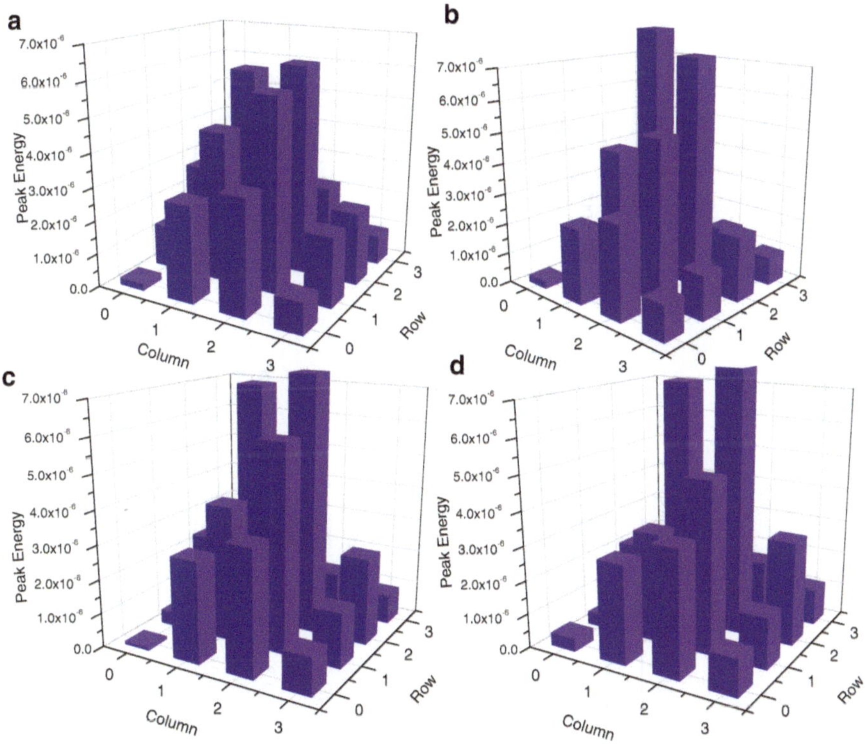

Fig. 11.13 An example of the NoC energy profile under hotspot-4c traffic (**a**) proposed routing (**b**) oddeven routing (**c**) negativefirst routing (**d**) XY routing

Table 11.3 Execution time for different mesh sizes

	Running time (s)					
Mesh size	3 × 3		4 × 4		5 × 5	
Com density (ρ)	2	4	2	4	2	4
Time (s)	5	6	18	20	227	243

from the router and the processor varies. We define $r_e = \frac{Average\,processor\,energy}{Average\,router\,energy}$ to reflect this ratio. The experimental results shown in the previous sections assume $r_e = 1$. In this sub-section, we simulate the peak energy reduction using our routing scheme for different r_e values. Tables 11.4–11.6 summarize the results for synthetic traffic and real benchmarks, respectively. From the results we can see that when r_e increases, the peak energy reduction is smaller because the relative contribution of the router energy reduces. Overall, we can achieve an average 6–17 % peak energy reduction over XY and odd-even routing schemes when r_e ranges from 0.67 to 4 across all the benchmarks.

Table 11.4 Peak energy reduction under various r_e (for Random, Transpose-1, Transpose-2 and Hotspot-C traffic)

	Peak energy reduction							
Benchmark	Random		Transpose-1		Transpose-2		Hotspot-C	
Average energy ratio (r_e)	vs. XY (%)	vs. OE (%)	vs. XY (%)	vs. OE (%)	vs. XY (%)	vs. OE (%)	vs. XY (%)	vs. OE (%)
0.67	17.4	12.8	15.6	17.9	9.5	12.1	15.7	17.6
1.00	15.7	15.3	13.3	14.2	10.3	10.3	14.4	16.2
1.33	13.3	9.5	12.5	14.5	7.2	9.4	13.2	14.9
1.67	10.6	8.6	11.2	13.8	6.4	9.4	12.3	14.0
2.00	11.9	9.3	11.1	13.8	6.5	9.1	11.6	15.0
2.33	10.5	8.2	8.5	10.7	5.5	7.0	10.8	13.9
2.67	9.4	7.7	8.9	11.1	4.3	6.9	10.2	13.0
3.00	8.8	7.0	8.8	10.6	3.7	5.6	9.6	10.9
3.33	7.9	5.9	7.1	9.9	4.2	6.6	9.1	10.5
3.67	8.1	6.5	7.8	9.2	4.1	5.8	8.7	9.9
4.00	7.9	6.1	7.0	9.0	4.6	6.4	8.3	9.4
Average	11.04	8.81	10.16	12.25	6.03	8.05	11.26	13.21

Table 11.5 Peak energy reduction under various r_e (for Butterfly, Hotspot-Tr, Hotspot-Rs and Bursty traffic)

	Peak energy reduction							
Benchmark	Butterfly		Hotspot-Tr		Hotspot-Rs		Bursty	
Average energy ratio (r_e)	vs. XY (%)	vs. OE (%)	vs. XY (%)	vs. OE (%)	vs. XY (%)	vs. OE (%)	vs. XY (%)	vs. OE (%)
0.67	14.7	20.2	16.6	9.9	21.4	11.2	12.6	16.1
1.00	13.1	19.1	15.4	9.2	19.9	10.5	10.4	13.0
1.33	10.8	15.2	13.6	7.9	18.5	9.7	8.1	11.3
1.67	9.8	14.4	12.5	7.3	17.2	8.9	7.5	9.9
2.00	9.9	13.7	11.6	6.7	16.3	9.4	5.6	9.1
2.33	8.9	13.0	11.0	6.5	15.5	8.2	7.1	8.7
2.67	8.8	12.8	10.1	5.8	14.6	7.8	5.3	7.9
3.00	8.7	11.2	9.5	5.4	13.8	7.2	5.5	7.6
3.33	7.4	9.8	8.9	5.1	13.1	6.9	4.9	7.2
3.67	5.8	9.4	8.4	4.8	12.5	6.6	4.3	6.0
4.00	6.0	8.8	8.0	4.6	12.0	6.3	4.5	6.4
Average	9.45	13.42	11.42	6.65	15.89	8.43	6.89	9.38

Table 11.6 Peak energy reduction under various r_e (for real benchmark traffic)

Benchmark	Peak energy reduction					
	MMS-1		MPEG4		VOPD	
Average energy ratio (r_e)	vs. XY (%)	vs. OE (%)	vs. XY (%)	vs. OE (%)	vs. XY (%)	vs. OE (%)
0.67	17.7	11.8	7.8	10.2	16.5	28.9
1.00	15.7	10.4	6.9	9.0	12.3	23.6
1.33	14.5	9.8	6.1	8.0	10.3	20.5
1.67	12.5	8.1	5.5	7.3	9.9	19.0
2.00	10.6	6.5	5.4	7.1	9.3	17.7
2.33	9.3	5.6	5.0	6.5	8.3	16.1
2.67	10.4	7.0	4.6	6.0	7.6	14.8
3.00	9.7	6.5	4.3	5.6	7.6	14.2
3.33	8.5	5.6	4.0	5.5	7.7	13.8
3.67	7.9	5.2	4.2	5.4	7.1	13.0
4.00	7.3	4.7	3.9	5.0	6.5	12.0
Average	11.28	7.38	5.25	6.87	9.37	17.6

11.7 Conclusions

NoC has been widely adopted to handle the complicate communications for future MPSoCs. As temperature becomes one key constraint in NoC, in this chapter, we propose an application-specific and thermal-aware routing algorithm to distribute the traffic more uniformly across the chip. A deadlock free path set finding algorithm is first utilized to maximize the routing adaptivity. A linear programming (LP) problem is formulated to allocate traffic properly among the paths. A table-based router is also designed to select output ports according to the traffic allocation ratios. From the simulation results, the peak energy reduction can be as high as 20 % for both synthetic traffic and real benchmarks.

References

1. D. Atienza, E. Martinez, Inducing thermal-awareness in multicore systems using networks-on-chip, in *IEEE Computer Society Annual Symposium on VLSI, ISVLSI'09*, Tampa, 13–15 May 2009, pp. 187–192
2. L. Benini, G. De Micheli, Networks on chips: a new SoC paradigm. Computer **35**(1), 70–78 (2002)
3. D. Bertozzi, L. Benini, Xpipes: a network-on-chip architecture for gigascale systems-on-chip. IEEE Circuits Syst. Mag. **4**, 18–31 (2004)
4. D. Bertozzi, A. Jalabert, S. Murali, R. Tamhankar, S. Stergiou, L. Benini, G. De Micheli, NoC synthesis flow for customized domain specific multiprocessor systems-on-chip. IEEE Trans. Parallel Distrib. Syst. **16**(2), 113–129 (2005)
5. Booksim simulator, (2010) http://nocs.stanford.edu/
6. R.V. Boppana, S. Chalasani, A comparison of adaptive wormhole routing algorithms, in *Proceedings of the 20th Annual International Symposium on Computer Architecture*, San Diego, 16–19 May 1993, pp. 351–360
7. C.-H Chao, K.-C. Chen, A.-Y. Wu, Routing-based traffic migration and buffer allocation schemes for three-dimensional network-on-chip systems with thermal limit. IEEE Trans. VLSI. no.99, pp.1,1, 0 (2013)
8. CVX: Matlab software for disciplined convex programming, (2013) http://cvxr.com/cvx
9. M. Daneshtalab, A. Sobhani, A. Afzali-Kusha, O. Fatemi, Z. Navabi, NoC hot spot minimization using AntNet dynamic routing algorithm, in *International Conference on Application-Specific Systems, Architectures and Processors, ASAP'06*, Steamboat Springs, Sept 2006, pp. 33–38
10. J. Duato, A necessary and sufficient condition for deadlock-free adaptive routing in wormhole networks. IEEE Trans. Parallel Distrib. Syst. **6**(10), 1055–1067 (1995)
11. D. Fick, A. DeOrio, G. Chen, V. Bertacco, D. Sylvester, D. Blaauw, A highly resilient routing algorithm for fault-tolerant NoCs, in *Design, Automation & Test in Europe Conference & Exhibition, DATE'09*, Nice, 20–24 Apr 2009, pp. 21–26
12. C.J. Glass, L.M. Ni, The turn model for adaptive routing, in *The 19th Annual International Symposium on Computer Architecture*, Queensland, Australia, 1992, pp. 278–287
13. Hotspot 5.0 temperature modeling tool, (2011) http://lava.cs.virginia.edu/HotSpot
14. J. Hu, R. Marculescu, Energy- and performance-aware mapping for regular NoC architectures. IEEE Trans. Comput. Aided Des. Integr. Circuits Syst. **24**(4), 551–562 (2005)

15. W. Hung, C. Addo-Quaye, T. Theocharides, Y. Xie, N. Vijakrishnan, M.J. Irwin, Thermal-aware IP virtualization and placement for networks-on-chip architecture, in *IEEE International Conference on Computer Design: VLSI in Computers and Processors*, San Jose, 11–13 Oct 2004, pp. 430–437
16. N.E. Jerger, L-S. Peh, *On-Chip Networks* (Morgan & Claypool, San Rafael, 2009)
17. Y.-C. Lan, M.C. Chen, A.P. Su, Y.-H. Hu, S.-J. Chen, Fluidity concept for NoC: a congestion avoidance and relief routing scheme, in *IEEE SOC Conference*, Newport Beach, 17–20 Sept 2008, pp. 65–70
18. G.M. Link, N. Vijaykrishnan, Hotspot prevention through runtime reconfiguration in network-on-chip, in *Proceedings of the Design, Automation and Test in Europe*, Munich, vol. 1, 7–11 Mar 2005, pp. 648–649
19. M.B. Marvasti, M. Daneshtalab, A. Afzali-Kusha, S. Mohammadi, PAMPR: power-aware and minimum path routing algorithm for NoCs, in *15th IEEE International Conference on Electronics, Circuits and Systems*, St. Julien's, 2008, pp. 418–421
20. Nigram simulator, (2010) http://nigram.ecs.soton.ac.uk
21. Noxim simulator User Guide, (2010) http://www.noxim.org
22. M. Palesi, R. Holsmark, S. Kumar, V. Catania, Application specific routing algorithms for networks on chip. IEEE Trans. Parallel Distrib. Syst. **20**(3), 316–330 (2009)
23. M. Palesi, S. Kumar, V. Catania, Bandwidth-aware routing algorithms for networks-on-chip platforms. Comput. Digit. Tech. IET **3**(5), 413–429 (2009)
24. A. Pullini, F. Angiolini, P. Meloni, D. Atienza, S. Murali, L. Raffo, G. De Micheli, L. Benini, NoC design and implementation in 65 nm technology, in *First International Symposium on Networks-on-Chip, NOCS 2007*, Princeton, 7–9 May 2007, pp. 273–282
25. Z. Qian, C.-Y. Tsui, A thermal-aware application specific routing algorithm for network-on-chip design, in *16th Asia and South Pacific IEEE/ACM Design Automation Conference (ASP-DAC)*, Yokohama, 25–28 Jan 2011, pp. 449,454
26. B.C. Schafer, T. Kim, Hotspots elimination and temperature flattening in VLSI circuits. IEEE Trans. Very Large Scale Integr. (VLSI) Syst. **16**(11), 1475–1487 (2008)
27. L. Shang, L.-S. Peh, A. Kumar, N.K. Jha, Temperature-aware on-chip networks. IEEE Micro **26**(1), 130–139 (2006)
28. R. Tarjan, Depth-first search and linear graph algorithms. SIAM J. Comput. **1**(2), 146–160 (1972)
29. S. Vangal, A. Singh, J. Howard, S. Dighe, N. Borkar, A. Alvandpour, A 5.1 GHz 0.34 mm^2 router for network-on-chip applications, in *IEEE Symposium on VLSI Circuits*, Kyoto, 14–16 June 2007, pp. 42–43

Part V
Emerging Technologies

Chapter 12
Traffic- and Thermal-Aware Routing Algorithms for 3D Network-on-Chip (3D NoC) Systems

Kun-Chih Chen, Chih-Hao Chao, Shu-Yen Lin, and An-Yeu (Andy) Wu

Abstract Three-dimensional Network-on-Chip (3D NoC) has been proposed to solve the complex on-chip communication issues in future 3D multicore systems. However, the thermal problems of 3D NoC are more serious than 2D NoC due to stacking dies. To keep the temperature below a certain thermal limit, many approaches of run-time thermal management were proposed. In this chapter, we will introduce some design concepts of traffic- and thermal-aware routing algorithms, which aim at minimize the performance impact caused by the run-time thermal managements. The investigative approaches can mitigate the design challenges of 3D NoC systems. Without the enhancement of cooling devices, the 3D NoC system can still be thermal-safe. Besides, the advantages of 3D integration are preserved, because the thermal-limited performance back-off is reduced.

12.1 Introduction

12.1.1 Background

As the advances of semiconductor technology, there are more and more components needs to be integrated within a chip. For high-performance chip multi-processors (CMPs) or multi-processor SoCs (MPSoCs), tens or even hundreds of processor

K.-C. Chen (✉) • C.-H. Chao • S.-Y. Lin • A.-Y. Wu
Graduate Institute of Electronics Engineering, National Taiwan University,
Room: E2-232, No.1, Sec. 4, Roosevelt Road, Taipei 106, Taiwan
e-mail: kunchih@access.ee.ntu.edu.tw; chihhao@access.ee.ntu.edu.tw;
linyan@access.ee.ntu.edu.tw; andywu@cc.ee.ntu.edu.tw

M. Palesi and M. Daneshtalab (eds.), *Routing Algorithms in Networks-on-Chip*,
DOI 10.1007/978-1-4614-8274-1_12, © Springer Science+Business Media New York 2014

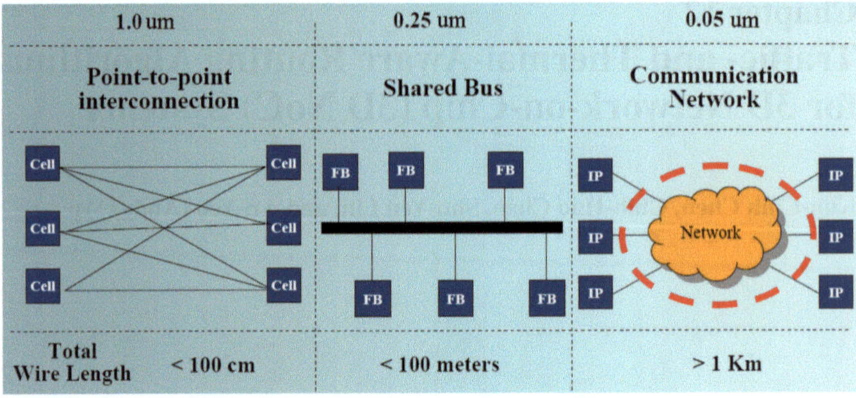

Fig. 12.1 The trend of on-chip interconnections

cores, memories, and other components are required to be connected. On-chip interconnection gradually becomes a major challenge for optimization of performance, cost, and power consumption in system level [1].

Figure 12.1 shows the trend of the on-chip interconnection in the view of architecture. The traditional point-to-point interconnection suffers from the high complexity of wire routing, which leads to large layout area and long transmission delay. The shared bus architecture suffers from its limited bandwidth. The low scalability of these two traditional paradigms make them insufficient to accommodate the communication requirements with predictable performance and design effort. By viewing the on-chip interconnection as a micro-network, Network-on-Chip (NoC) has been viewed as a novel and practical solution [2]. Many topologies, such as ring, two-dimensional mesh (2D mesh), two-dimensional torus (2D torus), star, octagon, were discussed in the literatures for NoC. Among these topologies, mesh-based topologies are very popular in research and commercial fields. Due to the regularity, 2D mesh has small layout overhead and good electrical properties. Therefore, mesh-based topologies are popular for homogeneous multi-/many-core systems [3–5].

The multicore systems will support the performance improvement until the interconnection becomes the bottleneck [6]. The limitation had been addressed in [1]. In order to remain fully wirable. Figure 12.2 provides a projection of number of wire levels in an SoC system [1]. Obviously, as following the Moore's Law, the improbable 90 metal levels will be required, which is an impractical solution. Recently, the emerging Through Silicon Via (TSV)-based die stacking three-dimensional (3D) IC process provides a novel connecting approach, which can reduce the wire delay between dies [7]. By combining with the 3D IC technology, three-dimensional Network-on-Chip (3D NoC) has the following three advantages over traditional 2D NoC, which can achieve higher performance with less power and smaller form factor for on-chip data transference [8]:

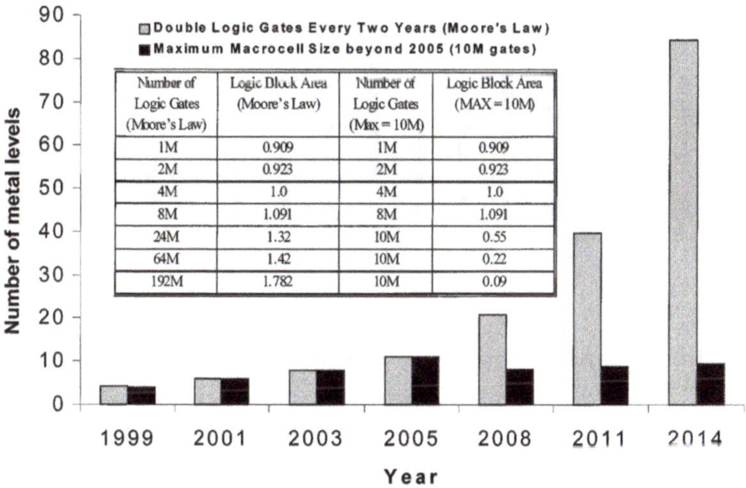

Number of Logic Gates (Moore's Law)	Logic Block Area (Moore's Law)	Number of Logic Gates (Max = 10M)	Logic Block Area (MAX = 10M)
1M	0.909	1M	0.909
2M	0.923	2M	0.923
4M	1.0	4M	1.0
8M	1.091	8M	1.091
24M	1.32	10M	0.55
64M	1.42	10M	0.22
192M	1.782	10M	0.09

Fig. 12.2 Projection of wire levels as the technology scaling down (©2001 IEEE. Reprinted, with permission, from [1])

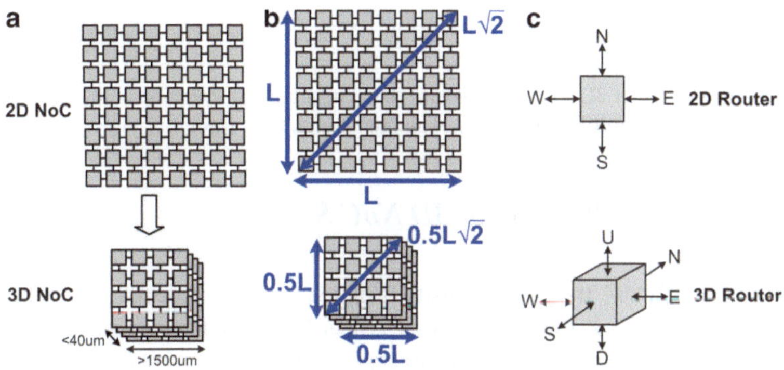

Fig. 12.3 Advantages of 3D NoC comparing to 2D NoC

- **Smaller layout footprint:** As the example shown in Fig. 12.3a, an 8×8 2D NoC can be stacked to a $4 \times 4 \times 4$ 3D NoC. The form factor of the chip becomes smaller, in which the IP mapping density is higher.
- **Shorter physical distance and hop count:** As the example shown in Fig. 12.3b, the original longest physical distance is reduced from $L\sqrt{2}$ to $0.5L\sqrt{2}$, and the L is the length/width of the 2D NoC die. The hop count between the farthest pair of routers is reduced from 15 to 10.
- **More directions per router:** As shown in Fig. 12.3c, two extra directions, U and D, are available for transmission, which leads to higher path diversity and higher bandwidth. Therefore, the network throughput can be improved.

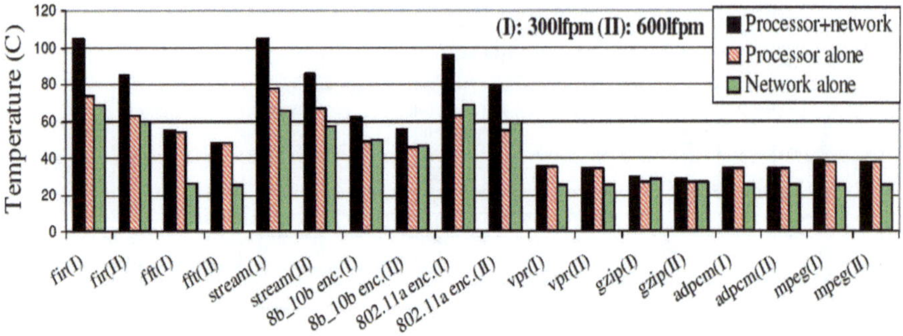

Fig. 12.4 Thermal impact of processor cores and on-chip network (©2004 IEEE/ACM. Reprinted, with permission, from [9])

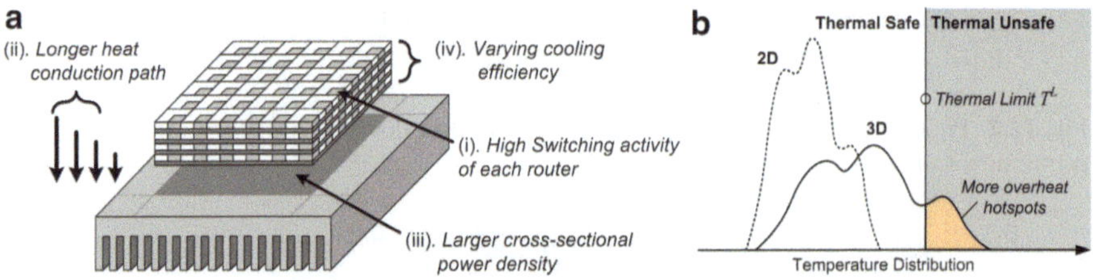

Fig. 12.5 (**a**) Factors of 3D thermal problem; (**b**) Temperature distribution is higher and wider, making more hotspots (©IEEE. Reprinted, with permission, from [16])

12.1.2 Thermal Problem of 3D NoC Systems

In conventional 2D IC, the power density is increased with respect to the technology scaling down, which makes the thermal issues have been major factors [6]. The increasing power density increases the heat generation rate of unit chip area, leading to high temperature. High temperature results in slower circuit switching, larger leakage power, and higher vulnerability of thermal run-away. In traditional 2D NoC systems, routers have been proven to have comparable thermal impact as processors, contributing significant thermal overhead to overall chip [9–11], as shown in Fig. 12.4. The reason is that the power density of the router is similar to or even higher than the average power density of the processor [11], owing to the high switching activity in the routers. Therefore, NoC router is one of the sources generating thermal hotspots [11, 12].

For a k-tier 3D IC platform, the problem of power density will be k times higher than the traditional 2D IC with the same footprint and process technology. Therefore, the thermal problem is severer and has been viewed as a major issue of 3D IC. Because of the stacking structure of 3D IC, the factors that worsen the thermal problem of 3D NoC include: (i) the high switching activity of each router, (ii) the longer heat conduction path, (iii) the larger cross-sectional power density, (iv) varying cooling efficiency, which are shown in Fig. 12.5a. The mean temperature of the NoC is pushed higher by the larger power density and the longer average heat conduction path. The temperature variance is increased due to the varying cooling efficiency in different layers. Figure 12.5b shows that the 3D NoC inevitably tends to have more routers' temperature exceed the thermal limit because of the higher mean and larger variance of the temperature distribution. The crucial high temperature produces unsafe working status of the chip.

For thermal safety, the system generally requires a better cooling solution. However, the cost of the heat sink grows exponentially as its cooling capability. Consequently, it is infeasible to eliminate hotspot solely by enhancing the cooling device. In this chapter, we will investigate some thermal-aware routing algorithms, which makes the 3D NoC system can still be thermal-safe without the enhancement of cooling devices.

12.2 Design Issues of Traffic- and Thermal-Aware Routing

The traditional design issues of traffic- and thermal-aware routing can be categorized into two different types: (i) off-line type and (ii) on-line type, which are for two different system types. The key difference of these two types is that whether the transient temperature profile of the system is obtainable during operation. We will address the two different types of design problems in the section.

12.2.1 Design Issues of Off-line Traffic- and Thermal-Aware Routing

For off-line thermal management, the configuration of the system is determined in design stage. The thermal behavior of the system is verified through simulation in off-line design phase. Figure 12.6 shows the design space of the off-line thermal-aware routings, which can be categorized into the Temperature Balancing Design (*TBD*) scheme and Load Balancing Design (*LBD*).

Traditionally, improving the throughput of a NoC relies on balancing the loading of channels, which can be viewed as the *LBD* scheme. As the thermal limit T^L is very high (or without T^L), *LBD* leads to the optimal performance because the stress of the heavily loaded channels can be relieved. However, the varying cooling efficiency of the different layers of 3D NoC results in unbalanced temperature profile for the *LBD* scheme. The upper layers are thermal dominant and have higher temperature. For the design case of high T^L, the channel bandwidth can support the performance until reaches the bandwidth limit (BW Limit) of the channel. By this way, the maximal temperature of the system can easily become higher than the thermal limit of the system. Therefore, traffic loading has to be small to prevent 3D NoC from overheating, resulting that the traffic loading is very limited by temperature.

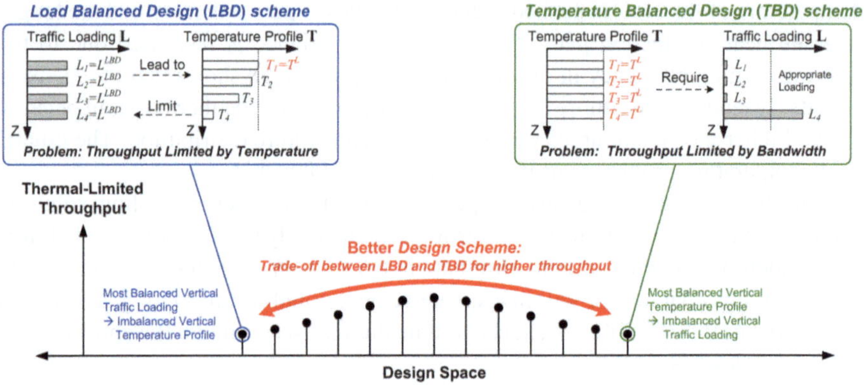

Fig. 12.6 Design space of off-line thermal-aware routing algorithm

In contrast, the Temperature Balancing Design (*TBD*) scheme tries to balance the network temperature. In 2D NoC, *TBD* can reduce the temperature variance for lower maximal temperature, allowing higher throughput without violating the temperature constraint [9,10]. However, the varying cooling efficiency in the vertical direction of 3D NoC reveals the unexpected shortage of the traditional *TBD* scheme, which is the originally design for 2D systems. Adopting *TBD* scheme in 3D NoC requires an extreme imbalance of vertical traffic loading. When thermal limit T^L is low, the traffic loading is only limited by temperature. However, in typical cases, the unbalanced traffic loading results in traffic congestion in the bottom layer, because the bandwidth of the channels in the bottom layer is insufficient for the heavy traffic loading. Moreover, the traffic imbalance also results in low channel utilization in the upper layers. Therefore, the performance of a temperature-balanced 3D NoC design is usually not the optimum. In Sect. 12.3, a routing-based traffic-migration method will be introduced to find the optimum design space of the system without embedded thermal sensor.

12.2.2 Design Issues of On-line Traffic- and Thermal-Aware Routing

To ensure thermal safety for high-performance 3D NoCs, the run-time thermal management (*RTM*) is required [9, 13–15]. As shown in Fig. 12.7, the node of 3D NoC starts from the ambient temperature T_A and heats up toward its steady state temperature T_{SS}, which is usually higher than the thermal limit T_L. The monitoring unit senses the temperature of the node and reports it to the temperature-aware controller. When the temperature exceeds the trigger level T_T, the controller changes the control policy to throttle the near-overheated node. When the temperature falls beneath the trigger level for coming back to normal working, the controller stops throttling. Then, the router can work normally.

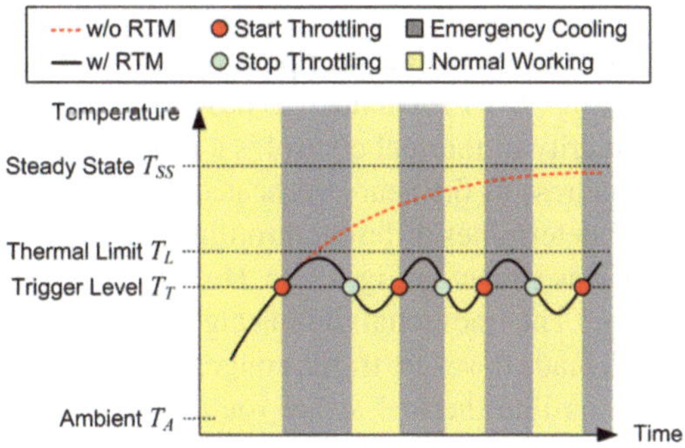

Fig. 12.7 Transient temperature trace with and without *RTM* (©ACM. Reprinted, with permission, from [15])

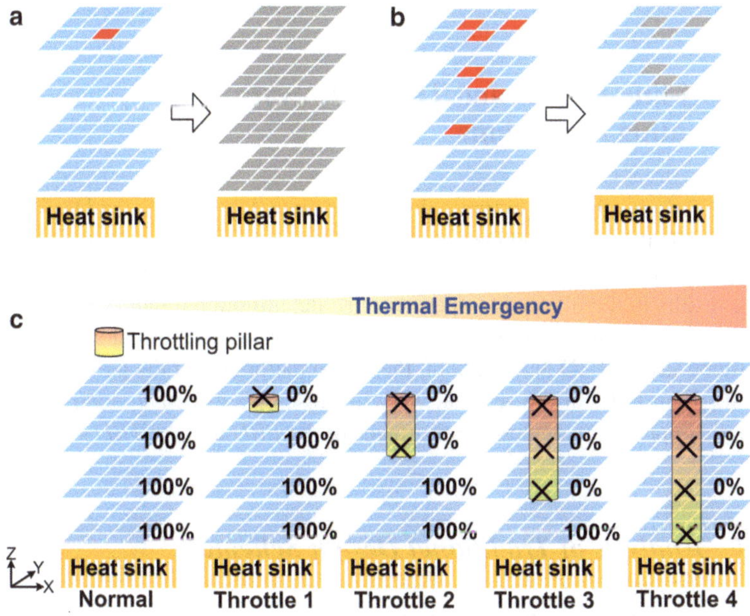

Fig. 12.8 (**a**) Global throttling (*GT*) scheme throttles entire network; (**b**) Distributive throttling (*DT*) scheme only throttles the overheated node; (**c**) Thermal-aware vertical throttling (*TAVT*) determines the throttling state based on the level of thermal-emergence

The simplest *RTM* uses global throttling (*GT*) scheme to cool down the network [13]. When any node is near overheat (i.e., the temperature of the node exceeds the T_T), the entire network will be slowed down, as shown in Fig. 12.8a. Although the *GT* has short throttling time, the impact of availability is huge. To reduce the performance impact as applying the *GT* scheme, a distributive and collaborative throttling scheme, *ThermalHert*, was proposed [9]. The distributed traffic throttling (*DT*) controls the quota of incoming traffic of the node, while the temperature exceeds the T_T, as shown in Fig. 12.8b. For the 3D NoC systems, the heterogeneous thermal conductance may result in long cooling time for the nodes, which are far away from the heat sink, while employing the *DT* scheme. To provide an effective heat conductance path within short response time, thermal-aware vertical throttling (*TAVT*) scheme was proposed in [15]. Different from the *DT* scheme, the granularity of thermal control is a pillar, which consists of the nodes with the same XY address as the near-overheated node. Based on the level of thermal emergency, *TAVT* determines different throttling states (i.e., the number of nodes in a throttling pillar), as shown in Fig. 12.8c. However, the *GT*, *DT*, and *TAVT* results in topology change. The traditional routing algorithms result in blocking packets in the network, which leads to severe traffic congestion and large performance impact. In Sect. 12.4, some on-line thermal-aware routing algorithms will be introduced to reduce the performance impact while the *RTM* is triggered.

12.3 Off-line Traffic- and Thermal-Aware Routing Algorithms

Figure 12.6 shows the two extreme design schemes, Load Balanced Design (*LBD*) scheme and Traffic Balanced Design (*TBD*). The problem of *TBD* is the consequent requirement of unbalanced vertical traffic loading. On the other hand, the *LBD* scheme results in unbalanced temperature profile, although it can leads to the most balanced vertical traffic loading. Due to the heterogeneous thermal conductance of each silicon layer of 3D NoC system, these two factors resist the performance gain of 3D stacking [14]. In the section, we will presents a effective, controllable, and systematic approach to amend the shortages of the *LBD* and the *TBD* schemes.

12.3.1 Analysis of Traditional Design Schemes

12.3.1.1 Notations and Scope Definition

Heat conduction in 3D NoC system can be modeled using Fourier's heat flow analysis. Fourier's analysis has been the academic and industrial standard method for circuit-level, architecture-level, and chip-package thermal analysis in the past decades. The method is analogous to Ohm's method of modeling electrical circuit. Heat flow is analogous to electrical current, and temperature is analogous to voltage. Each element of temperature profile is determined by power, thermal conductance, and thermal capacitance. Therefore, the thermal model of single tile, which contains one router, one processor, and one memory, can be constructed as Fig. 12.9a. The temperature of router, memory, and processor at node (x, y, z) are $T^R_{x,y,z}$, $T^M_{x,y,z}$, and $T^P_{x,y,z}$; the corresponding power are $P^R_{x,y,z}$, $P^M_{x,y,z}$, and $P^P_{x,y,z}$. As shown in Fig. 12.9b, the *X by Y by Z* 3D NoC is composed of identical single tiles. The involved notations in the analysis of this section are shown in Table 12.1.

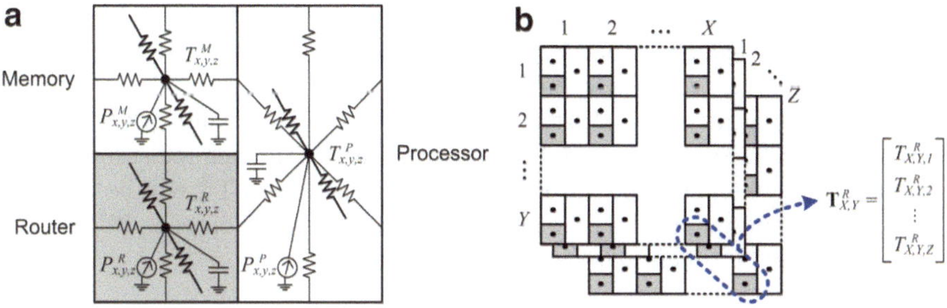

Fig. 12.9 (**a**) Thermal model of single tile, and (**b**) The abstracted model of an X by Y by Z 3D NoC systems (©IEEE. Reprinted, with permission, from [16])

Table 12.1 Parameter notations

Parameter	Description
Superscript	*R*: router; *M*: memory; *P*: processor; *BW*: bandwidth; *LBD*: load balanced design; *TBD*: traffic balanced design
Subscript	*x,y,z*: location variable; *X,Y,Z*: upper bound of *x,y,z*
T, *T*	Matrix/vector and scalar of temperature; $T^R_{x,y,z}$: temperature of the router at *(x,y,z)*
P, *P*	Matrix/vector and scalar of power; $P^R_{x,y,z}$: power consumption of the router at *(x,y,z)*
L, *L*	Matrix/vector and scalar of traffic loading (rate); $L_{x,y,z}$: traffic loading of the router at *(x,y,z)*
$g_{x,y,z}$	Thermal conductance between the routers at *(x,y,z)* and *(x,y,z+1)*; $g_{x,y,Z}$ is the effective thermal conductance between bottom layer and ambient
f	Function describes traffic and power, $routerpower = f(channelloading)$
T^L, T^A	Thermal limit and ambient temperature

For a 3D NoC system, the temperature profile $\mathbf{T}^R$ of routers and the power profile $\mathbf{P}^R$ of routers can be represented as follows:

$$\mathbf{T}^R = \begin{bmatrix} \mathbf{T}^R_{1,1} & \cdots & \mathbf{T}^R_{X,1} \\ \vdots & \ddots & \vdots \\ \mathbf{T}^R_{1,Y} & \cdots & \mathbf{T}^R_{X,Y} \end{bmatrix}, \text{ and } \mathbf{T}^R_{x,y} = \begin{bmatrix} T^R_{x,y,1} \\ \vdots \\ T^R_{x,y,Z} \end{bmatrix} \tag{12.1}$$

$$\mathbf{P}^R = \begin{bmatrix} \mathbf{P}^R_{1,1} & \cdots & \mathbf{P}^R_{X,1} \\ \vdots & \ddots & \vdots \\ \mathbf{P}^R_{1,Y} & \cdots & \mathbf{P}^R_{X,Y} \end{bmatrix}, \text{ and } \mathbf{P}^R_{x,y} = \begin{bmatrix} T^P_{x,y,1} \\ \vdots \\ T^P_{x,y,Z} \end{bmatrix} \tag{12.2}$$

$\mathbf{T^R_{x,y}}$ is the 1D vertical temperature profile of the routers with the same (x,y) address, and $\mathbf{T^R}$ is the entire 3D temperature profile of routers. $\mathbf{P^R_{x,y}}$ is the 1D vertical power profile of the routers with the same (x,y) address, and $\mathbf{P^R}$ is the entire 3D power profile of routers. Similarly, the temperature profile and power profile of memory part are represented by $\mathbf{T^M}$ and $\mathbf{P^M}$, and the temperature profile and power profile of processor part are represented by $\mathbf{T^P}$ and $\mathbf{P^P}$.

The design goal of the thermal-aware routing is to maximize the network throughput and keep temperature below thermal limit. Through redistributing the traffic load $\mathbf{L}$, the power profile $\mathbf{P^R}$ of the overheated case $\mathbf{T^R}$ will be reduced. Therefore, the power profiles of the memories and the processors (i.e., $\mathbf{P^M}$ and $\mathbf{P^P}$) are not changed. $\mathbf{L}$ is defined as:

$$\mathbf{L} = \begin{bmatrix} \mathbf{L_{1,1}} & \cdots & \mathbf{L_{X,1}} \\ \vdots & \ddots & \vdots \\ \mathbf{L_{1,Y}} & \cdots & \mathbf{L_{X,Y}} \end{bmatrix}, \text{ and } \mathbf{L}_{x,y} = \begin{bmatrix} L_{x,y,1} \\ \vdots \\ L_{x,y,Z} \end{bmatrix}. \tag{12.3}$$

12.3.1.2 Analysis of Load Balanced Design (LBD) Scheme

For a conventional 3D NoC system, the throughput maximization relies on balancing channel loading of the network. Therefore, the channel loading matrix $\mathbf{L}$ is balanced and identical for all routers as:

$$\mathbf{L}_{x,y}|\forall x,y = \begin{bmatrix} L_{x,y,1} \\ \vdots \\ L_{x,y,Z} \end{bmatrix} = \begin{bmatrix} L^{LBD} \\ \vdots \\ L^{LBD} \end{bmatrix}, \ (LBD \text{ scheme}). \tag{12.4}$$

The channel bandwidth L^{BW} is an upper bound that limits the transfer rate. The channel loading is bounded as:

$$L^{LBD} \leq L^{BW-LBD}, \tag{12.5}$$

where the "$BW - LBD$" represents the bandwidth-bound of LBD. Because the channel loading affects the switching activities of each router, we assume that the power of a router is an increasing function f of the channel loading (i.e., $P^{LBD} = f_{L\to P}(L^{LBD})$). As shown in (12.6), $\mathbf{P}^R$ is balanced while $\mathbf{L}$ is balanced.

$$\mathbf{P}^R_{x,y}|\forall x,y = \begin{bmatrix} P^R_{x,y,1} \\ \vdots \\ P^R_{x,y,Z} \end{bmatrix} = \begin{bmatrix} f_{L\to P}(L^{LBD}) \\ \vdots \\ f_{L\to P}(L^{LBD}) \end{bmatrix} = \begin{bmatrix} P^{LBD} \\ \vdots \\ P^{LBD} \end{bmatrix}. \tag{12.6}$$

According to Fourier's Heat Conduction Law, the most heat is dissipated through the vertical direction in 3D stacking chip [16]. Therefore, the vertical temperature profile $\mathbf{T}^R_{x,y}$ of each (x,y) is mainly determined by the vertical power profile $\mathbf{P}^R_{x,y}$. In LBD, the vertical temperature profile results in the thermal differences as:

$$\mathbf{T}^R_{x,y} = \begin{bmatrix} T^R_{x,y,1} \\ \vdots \\ T^R_{x,y,Z-1} \\ T^R_{x,y,Z} \end{bmatrix} = \begin{bmatrix} T^R_{x,y,2} + P^{LBD} \cdot g^{-1}_{x,y,1} \\ \vdots \\ T^R_{x,y,Z} + P^{LBD} \cdot g^{-1}_{x,y,Z-1} \\ T^A + P^{LBD} \cdot g^{-1}_{x,y,Z} \end{bmatrix}. \tag{12.7}$$

Because $P^{LBD} \cdot g^{-1}_{x,y,z}$ is never negative, the following thermal gradient is always true:

$$T^A \leq T^R_{x,y,Z} \leq T^R_{x,y,Z-1} \leq \cdots \leq T^R_{x,y,2} \leq T^R_{x,y,1} \leq T^L. \tag{12.8}$$

Since we must keep the temperatures of all routers not above the thermal limit, the channel loading L^{LBD} is bounded. By combining (12.7) and (12.8), we can derive the thermal-limited bound of channel loading as using *LBD*, which can be described as:

$$L^{LBD} = f^{-1}_{L\to P}(P^{LBD}) \leq f^{-1}_{L\to P}((T^L - T^A)/\sum_{z=1}^{Z} g^{-1}_{x,y,z}) = L^{TL-LBD}, \tag{12.9}$$

where the "$TL - LBD$" represents the thermal-limit bound of *LBD*. The optimal criterion of *LBD* scheme is that the channel loading of the bandwidth bound L^{BW-LBD} is a tighter bound than the one of the thermal-limited bound L^{TL-LBD}. Because T^L is not extremely high in the most cases, the L^{TL-LBD} is usually smaller than L^{BW-LBD}. Therefore, *LBD* scheme cannot guarantee the optimal network throughput.

12.3.1.3 Analysis of Thermal Balanced Design (TBD) Scheme

The goal of *TBD* scheme is to balance the temperature profile for eliminating hotspots. Assume there is only one heat sink at the bottom of the chip, as shown in Fig. 12.8. The optimum result of temperature balancing scheme is that the every $T^R_{x,y,z}$ in the temperature profile is equal to thermal limit T^L, which is shown as follows:

$$\mathbf{T}^R_{x,y}|_{\forall x,y} = \begin{bmatrix} T^R_{x,y,1} \\ \vdots \\ T^R_{x,y,Z} \end{bmatrix} = \begin{bmatrix} T^{TBD} \\ \vdots \\ T^{TBD} \end{bmatrix}, (TBD\ scheme). \tag{12.10}$$

In usual, the T^{TBD} is less than the T^L. By following (12.10), the heat transfer toward ambient is maximum, if the temperature difference between bottom layer and ambience is $T^L - T^A$, and the temperature difference between vertical neighbor routers is zero. Such result leads to the extreme unbalanced vertical power profile for each $\mathbf{P}^R_{x,y}$, as follows:

$$\mathbf{P}^R_{x,y} = \begin{bmatrix} P^R_{x,y,1} \\ \vdots \\ P^R_{x,y,Z-1} \\ \\ P^R_{x,y,Z} \end{bmatrix} = \begin{bmatrix} (T^R_{x,y,1} - T^R_{x,y,2}) \cdot g_{x,y,1} \\ \vdots \\ (T^R_{x,y,Z-1} - T^R_{x,y,Z}) \cdot g_{x,y,Z-1} \\ \\ (T^R_{x,y,Z} - T^A) \cdot g_{x,y,Z} \end{bmatrix}$$
$$= \begin{bmatrix} 0 \cdot g_{x,y,1} \\ \vdots \\ 0 \cdot g_{x,y,Z-1} \\ \\ (T^L - T^A) \cdot g_{x,y,Z} \end{bmatrix} = \begin{bmatrix} 0 \\ \vdots \\ 0 \\ \\ P^{TBD} \end{bmatrix} \tag{12.11}$$

To fit such power profile, all the traffic must concentrate on the bottom silicon layer and lead to huge traffic congestion. Therefore, all the channel bandwidth in the routers in the upper $Z-1$ silicon layers is not utilized, which is shown as follows:

$$\mathbf{L}_{x,y} = \begin{bmatrix} L_{x,y,1} \\ \vdots \\ L_{x,y,Z-1} \\ \\ L_{x,y,Z} \end{bmatrix} = \begin{bmatrix} f^{-1}_{L\to P}(P^R_{x,y,1}) \\ \vdots \\ f^{-1}_{L\to P}(P^R_{x,y,Z-1}) \\ \\ f^{-1}_{L\to P}(P^R_{x,y,Z}) \end{bmatrix} = \begin{bmatrix} f^{-1}_{L\to P}(0) \\ \vdots \\ f^{-1}_{L\to P}(0) \\ \\ f^{-1}_{L\to P}(P^{TBD}) \end{bmatrix} = \begin{bmatrix} 0 \\ \vdots \\ 0 \\ \\ L^{TBD} \end{bmatrix}. \tag{12.12}$$

The channel loading is subjected to the thermal-limited bound in *TBD* scheme, which can be derived as:

$$L^{TBD} \le f^{-1}_{L\to P}((T^L - T^A) \cdot g_{x,y,z}) = L^{TL-TBD} \tag{12.13}$$

Similarly, the channel loading in *TBD* scheme is subjected to channel bandwidth:

$$L^{TBD} \le L^{BW-TBD}. \tag{12.14}$$

The optimal criterion of *TBD* scheme is that the channel loading of thermal limited bound L^{TL-TBD} is a tighter bound than the one of the bandwidth bound L^{BW-TBD}. By observing (12.13), if T^L is not very low, L^{TL-TBD} will be relaxed, resulting in the possibility that L^{TL-TBD} becomes larger than L^{BW-TBD}. Therefore, the *TBD* scheme cannot guarantee to achieve the optimal throughput.

To discuss whether *LBD* or *TBD* can achieve maximal throughput for 3D NoC, we have to compare the bandwidth bound and the thermal-limited bound of *LBD* and *TBD*. Table 12.2 shows all the four cases. In Case 1, both criteria are satisfied, so *LBD* and *TBD* are both optimal. However, by comparing (12.9) and (12.13), Case 1 only happens when $Z = 1$, which is a 2D NoC case with very high thermal limit.

Table 12.2 Parameter notations

	$L^{BW-LBD} \leq L^{TL-LBD}$	$L^{BW-LBD} > L^{TL-LBD}$
$L^{TL-TBD} \leq L^{BW-TBD}$	Case 1 *LBD* is optimal, *TBD* is optimal	Case 2 *LBD* is non-optimal, *TBD* is optimal
$L^{TL-TBD} > L^{BW-TBD}$	Case 3 *LBD* is optimal, *TBD* is non-optimal	Case 4 *LBD* is non-optimal, *TBD* is non-optimal

Case 2 shows the situation that thermal limit is very low, leading to non-optimal *LBD* and optimal *TBD*. Case 3 shows the situation that thermal limit is very high, resulting in optimal *LBD* and non-optimal *TBD*. Case 4 shows the situation that a middle thermal limit is given. In this case, none of *LBD* and *TBD* can guarantee maximal throughput. Therefore,we need a new design scheme to amend the shortages of the *LBD* scheme and the *TBD* scheme, which will be introduced latter.

12.3.2 Routing-Based Power Migration

12.3.2.1 Desired Traffic Loading of Different Thermal Limits

In [16], the authors proposed a routing-based traffic migration method, which is a proactive routing that changes the traffic loading **L** for improving the thermal-limited throughput. Because the key difference between 2D NoC and 3D NoC is the varying cooling efficiency of routers in different silicon layers, 3D NoC has the lateral-homogeneous vertical-heterogeneous thermal characteristics. For each silicon layer (i.e., XY-plane), the traditional *LBD* scheme still works for eliminating hotspots. However, the vertical-heterogeneous thermal characteristics of 3D NoC causes the *LBD* scheme and the *TBD* scheme fail to reach maximal throughput. To solve this problem, the routing path is decomposed into the lateral paths on XY plane and the vertical paths along Z direction. Then, we partition the proactive routing into two parts: the lateral routing and the vertical routing. The design goal of the lateral routing is to balance the lateral traffic loading on each XY plane; the design goal of the vertical routing is to accommodate the various vertical traffic loadings for different thermal limits.

For different thermal limit, the desired traffic loading is different. If the thermal limit is high, the temperature bounded traffic load L^{LBD} can be high. The throughput maximization problem is similar to traditional NoC performance optimization problem. Therefore, the *LBD* scheme is preferred, and the target vertical traffic distribution is (12.4), as shown in Fig. 12.10a. In contrast, if the thermal limit is low, the throughput is more limited by temperature. Hence eliminating overheated hotspots is primary objective for proactive routing. Therefore, *TBD* scheme is

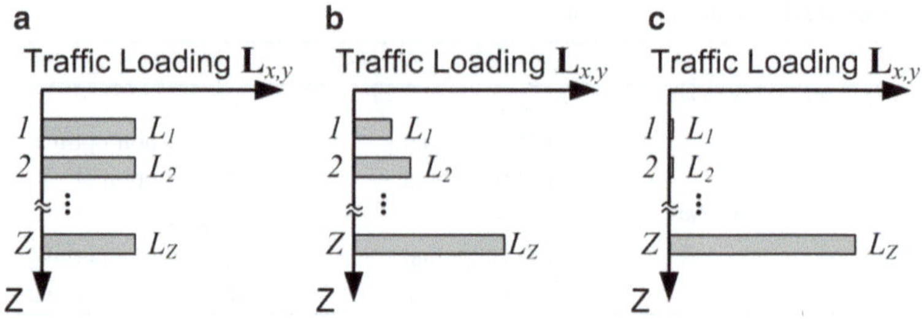

Fig. 12.10 Desired traffic loading for (**a**) high thermal limit cases, adopting *LBD* scheme; (**b**) middle thermal limit cases, adopting a mixture of LBD and TBD scheme; (**c**) low thermal limit cases, adopting TBD scheme (©IEEE. Reprinted, with permission, from [16])

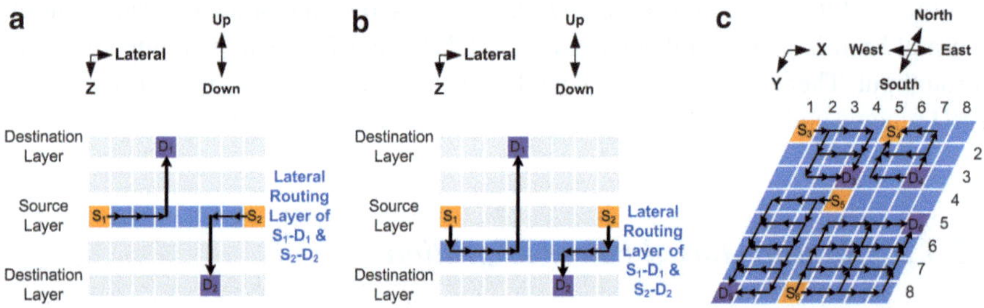

Fig. 12.11 (**a**) In traditional dimension-ordered routing, the lateral routing layer is identical to the source layer. (**b**) With traffic migration $D = 1$, the lateral routing layer is one layer below the source layer. (**c**) The lateral routing is an adaptive routing (©IEEE. Reprinted, with permission, from [16])

preferred, and the target vertical distribution is (12.12), as shown in Fig. 12.10c. If the thermal limit is in the middle, both temperature and channel loading may limit the throughput. Therefore, a vertical traffic distribution between (12.4) and (12.12) is preferred to achieve maximum throughput, as shown in Fig. 12.10b.

12.3.2.2 Vertical-Downward Lateral-Adaptive Proactive Routing (VDLAPR) Algorithm

To accommodate these three traffic loadings shown in Fig. 12.10, the *VDLAPR* algorithm was proposed in [16]. Figure 12.11 shows the routing scheme of traffic migration. Traditional dimension-ordered routing algorithms, such as XYZ routing in Fig. 12.11a, cannot accommodate all the vertical traffic requirements of different thermal limits shown in Fig. 12.10. We relieve the minimal routing constraint in vertical direction and define downward level *D*, a scalar control parameter representing the vertical misroute distance, for gradually redistributing the traffic. Assume the source address of a packet is (X^S, Y^S, Z^S), and the destination address is $(\mathrm{X^D,Y^D,Z^D})$. With given downward level, the address of target lateral routing layer of the packet, defined as Z^{TLRL}, can be found by the following equation:

$$Z^{TLRL} = Z^S + D. \tag{12.15}$$

Because the routing layer is within the 3D mesh, the actual lateral routing layer Z^{TLRL} is bounded by [1,Z], as:

$$Z^{LRL} = \begin{cases} 1 & \text{, if } Z^{TLRL} < 1 \\ Z & \text{, if } Z^{TLRL} > Z \\ Z^{TLRL} & \text{, otherwise} \end{cases} \tag{12.16}$$

To deliver a packet from source to destination, it is firstly injected to the source router, which is at (X^S,Y^S,Z^S). The source router routes the packet to the intermediate router at (X^S,Y^S,Z^{LRL}), and then the intermediate router routes the packet to (X^D,Y^D,Z^{LRL}). When the packets reach the router that has the same XY address as the destination, they are directly routed to the destination (X^D,Y^D,Z^D). Figure 12.11b illustrates an example. $D = 1$ represents the case that the lateral routing path is migrated down by one layer.

For the lateral routing, XY routing is adopted for simplicity. However, for non-uniform random traffic patterns, XY routing is not able to balance the loading of the network. Many 2D adaptive routing algorithms had been presented for balancing the traffic loading on XY plane. These adaptive routing algorithms adaptively select the routing paths from the set of output channels according to the channel status. To determine the set of output channels, Odd-Even turn model [17] is well known and widely employed. In comparison with the other turn models like West-First, North-Last, and Negative-First, Odd-Even turn model can provide more even adaptiveness to non-uniform traffic patterns. The selection strategy we adopted in this section is Neighbor-on-Path (*NOP*) [21], which outperforms than traditional Output-Buffer-Level (*OBL*) selection strategy. The examples of path diversity are shown in Fig. 12.11c. For the eastward source-destination pairs like $\mathrm{S_3 - D_3}$ and $\mathrm{S_6 - D_6}$, the east-north (EN) turn and east-south (ES) turn are only allowed at even columns (even X addresses). For the westward pairs like $\mathrm{S_4 - D_4}$ and $\mathrm{S_5 - D_5}$, the west-north (WN) turn and west-south (WS) turn are only allowed at odd columns.

If thermal limit is high, setting $D = 0$ can keep the lateral routing layer unchanged. If thermal limit is low, setting a larger D can form a traffic distribution more like Fig. 12.10c. From (12.15) and (12.16), setting $D \geq Z$ leads to the result that all the lateral routing paths lie on the bottom layer, which is the extreme case that makes the vertical traffic loading of the lateral routing closes to (12.12). The power profile of the 3D NoC becomes the *TBD* scheme, approaching (12.11). For medium thermal limits, setting suitable D is essential to improve the thermal-limited throughput. Because the vertical traffic loading is redistributed by the proposed routing-based traffic migration method, the channel loading will be unbalanced. In Sect. 12.3.3, we will introduce the vertical buffer allocation method to solve this problem.

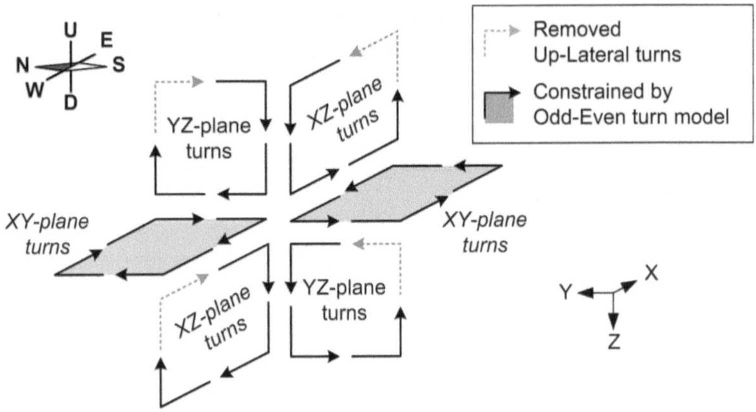

Fig. 12.12 Turn limitation in 3D space (©IEEE. Reprinted, with permission, from [16])

The *VDLAPR* resist the turn model based approach to prevent deadlock. As shown in Fig. 12.12, the deadlock-freedom of the turns on XY-plane is achievable through the Odd-Even turn model. The circular waiting condition is never true on XZ-plane and YZ-plane since we remove the turns of Up-then-East (UE), Up-then-West (UW), Up-then-North (UN), and Up-then-South (US). Therefore, we can guarantee the routing algorithm is deadlock-free.

12.3.3 Vertical Buffer Allocation Method

For supporting the method of routing-based traffic migration, we extend the queuing model proposed in [18], which adopts M/M/1/K queues. Several assumptions are made for simplification of the derivation: Poisson packet arrival time and exponential service time for each channel are required. Besides, each packet is viewed as an atomic data in this derivation, and the network system is near full-loading (i.e., the throughput of each router is near maximum). We use the north input channel of the router at (x,y,z) as an example. Assume the network is not overloaded in steady state, the main target of buffer allocation is to calculate the full probability for each channel. For channel $C_{x,y,z,N}$, the full probability $b_{x,y,z,N}$ is:

$$b_{x,y,z,N} = \frac{1-\rho_{x,y,z,N}}{1-\rho_{x,y,z,N}^{1+k_{x,y,z,N}}} \times \rho_{x,y,z,N}^{k_{x,y,z,N}}, \text{ where } \rho_{x,y,z,N} = \frac{\lambda_{x,y,z,N}}{\mu_{x,y,z,N}}. \qquad (12.17)$$

The full probability depends on two parameters: arrival rate $\lambda_{x,y,z,N}$ and service rate $\mu_{x,y,z,N}$, as shown by (12.17). However, the $\lambda_{x,y,z,N}$ becomes unpredictable, if the path selection depends on the channel status. Besides, the $\mu_{x,y,z,N}$ of one router depends on the full probability of all its downstream channels because of the connection of an NoC system. Consider some of the downstream packets are delivered toward east, which is to $C_{x+1,y,z,W}$. The full probability $b_{x+1,y,z,W}$ will affect the effective service rate of $C_{x,y,z,W}$. The effective service rate of $C_{x+1,y,z,W}$ can be approximated by $1/b_{x+1,y,z,W}$. When $C_{x+1,y,z,W}$ is full, the reciprocal of the decreasing rate of the occupation in queue equals to the average waiting time for entering the queue. Hence, the average waiting time for entering the queue of $C_{x+1,y,z,W}$ can be approximated by:

$$W_{x+1,y,z,W} = \left(\frac{1}{b_{x+1,y,z,W}} - \lambda_{x+1,y,z,W}\right)^{-1}. \tag{12.18}$$

At steady state, by applying Little′s formula, the average waiting time for entering the queue of $C_{x+1,y,z,W}$ can be written as:

$$W_{x+1,y,z,W} = \left(\bar{\mu}^{W}_{x,y,z,N} - p^{W}_{x,y,z,N} \times \lambda_{x,y,z,N}\right)^{-1}. \tag{12.19}$$

Substituting $W_{x+1,y,z,W}$ in (12.18) and (12.19), we get the effective service rate toward east of $C_{x,y,z,N}$ as follows:

$$\bar{\mu}^{E}_{x,y,z,N} = \frac{1}{b_{x+1,y,z,W}} - \lambda_{x+1,y,z,W} + p^{E}_{x,y,z,N} \times \lambda_{x,y,z,N}, \tag{12.20}$$

and

$$\bar{\mu}_{x,y,z,N} = \sum_{\forall dir} p^{dir}_{x,y,z,N} \times \bar{\mu}^{dir}_{x,y,z,N}. \tag{12.21}$$

Because of the interconnection between each node of NoC system, by merging the cascaded the M/M/1/K queues, we can get a simplified model for average queue length of $C_{x,y,z,N}$, which can be approximated by:

$$q_{x,y,z,N} = \frac{\lambda_{x,y,z,N}}{\mu_{x,y,z,N} - \lambda_{x,y,z,N}}. \tag{12.22}$$

By viewing the two queues as two independent queues, the average queuing length is equal to the sum of the length of the two queues, which can be described as:

$$q_{x,y,z,N} = \frac{\lambda_{x,y,z,N}}{S^{-1} - \lambda_{x,y,z,N}} + \frac{\lambda_{x,y,z,N}}{\bar{\mu}_{x,y,z,N} - \lambda_{x,y,z,N}}, \tag{12.23}$$

where S is the service time of current router. Substituting $q_{x,y,z,N}$ in (12.22) and (12.23), we can get the channel service rate $\mu_{x,y,z,N}$:

$$\mu_{x,y,z,N} = \lambda_{x,y,z,N} + \left(\frac{1}{S^{-1} - \lambda_{x,y,z,N}} + \frac{1}{\bar{\mu}_{x,y,z,N} - \lambda_{x,y,z,N}}\right)^{-1}. \tag{12.24}$$

Table 12.3 Channel buffer depth from results of vertical buffer allocation

D Level	*D*=1				*D*=2				*D*=3			
Layer	Z=0	Z=1	Z=2	Z=3	Z=0	Z=1	Z=2	Z=3	Z=0	Z=1	Z=2	Z=3
Lateral: E,S,W,N	1	4	4	7	1	1	5	9	1	1	1	13
Up	1	2	6	7	1	2	5	8	1	3	3	9
Down	5	5	5	1	5	5	5	1	5	5	5	1

Then, we can compute the full probability $b_{x,y,z,N}$ for $C_{x,y,z,N}$. Similar computation is applied to all the channels in the network. The marginalized service rate $\mu_{z,N}$ for the north input channel in layer z can be described as:

$$\mu_{z,N} = \lambda_{z,N} + \left(\frac{1}{S^{-1} - \lambda_{z,N}} + \frac{1}{\bar{\mu}_{z,N} - \lambda z,N}\right)^{-1}, \tag{12.25}$$

which is dependent on the marginalized arrival rate $\lambda_{z,N}$ and the marginalized effective service rate $\bar{\mu}_{z,N}$. Here we simplify the computation of $\lambda_{z,N}$ by directly taking average of $\lambda_{x,y,z,N}$ in each layer:

$$\lambda_{z,N} = \frac{1}{XY} \sum_{\forall x,y} \lambda_{x,y,z,N}. \tag{12.26}$$

Similarly, $\mu_{z,N}$ is calculated by taking average of $\mu_{x,y,z,N}$. Therefore, we can get the $\mu_{z,N}$ for calculating the full probability $b_{z,N}$ for layer z as:

$$b_{z,N} = \frac{1 - \rho_{z,N}}{1 - \rho_{z,N}^{1+k_{z,N}}} \times \rho_{z,N}^{k_{z,N}}, \text{ where } \rho_{z,N} = \frac{\lambda_{z,N}}{\mu_{z,N}}. \tag{12.27}$$

12.3.4 Design Examples of Systematic Design Scheme

In this section, we refer the experimental setup in [16] and demonstrate the design examples by applying the introduced design scheme. Because the latency in real application is not infinite, the maximum average latency is set to 500 cycles for estimating the saturation throughput. The thermal limit is set from 100 to 150 °C to cover Case 2 and Case 4 of Table 12.2, and a 200 °C case is used to demonstrate the Case 3 of Table 12.2. The buffer constraint N_B is set to 16 flits for vertical buffer allocation. Uniform random traffic is offered as an example here, and the results as applying other traffic patterns are shown in [16]. With known traffic pattern and given thermal limit, the method of vertical buffer allocation produces the optimal buffer depth for each downward level, as in Table 12.3. The detail algorithm of buffer allocation and proper downward level determination are described in [16]. Then, the achievable throughput is obtained by simulation with applying each configuration with increasing injection rate.

We use the Table 12.4 to show the achievable throughput comparison for uniform random traffic. First, we observe the achievable throughput of the traditional

Table 12.4 Random traffic design examples, achievable throughput (flit/node/cycle)

Design case	*TBD* is optimal		*TBD* is non-optimal, and *LBD* is non-optimal				*LBD* is optimal
Thermal limit	100 °C	110 °C	120 °C	130 °C	140 °C	150 °C	200 °C
LBD scheme	0.0554	0.0660	0.0767	0.0873	0.0978	0.1085	0.1695
TBD scheme	0.0642	0.0764	0.0850	0.0860	0.0860	0.0860	0.0860
Introduced scheme	0.0642	0.0764	0.0885	0.1018	0.1058	0.1173	0.1695
Downward level	$D=3$	$D=3$	$D=3$	$D=2$	$D=1$	$D=1$	$D=0$
Improv. over *LBD*	15.8%	15.6%	15.5%	16.7%	8.1%	8.1%	0.0%
Improv. over *TBD*	0.0%	0.0%	4.2%	18.4%	23.0%	36.3%	97.1%

LBD scheme. The achievable throughput increases as the thermal limit becomes higher, but *LBD* is not the scheme achieves maximal throughput. The reason is that the thermal-limited bound of *LBD* is tighter than the bandwidth bound of *LBD*, so the optimal criterion of *LBD* is not satisfied. Therefore, *LBD* is optimal only for the design case of 200 °C, in which the optimal criterion of *LBD* scheme is satisfied. Second, we observe the achievable throughput of the traditional *TBD* scheme. The achievable throughput increases as the thermal limit becomes higher and then saturates at 0.086 flit/node/cycle. For the 100 and 110 °C cases, *TBD* achieves maximal throughput, because the optimal criterion of *TBD* scheme is satisfied. However, as the thermal limit becomes higher, the thermal-limited bound of *TBD* is relaxed. Then, the optimal criterion of *TBD* scheme is not satisfied. The bandwidth bound of *TBD* scheme is tighter.

Comparing to *LBD* and *TBD* scheme, the introduced design scheme always achieves maximal throughput, because it covers both *LBD* and *TBD*. For the cases with 100 and 110 °C thermal limits, the proposed scheme is identical to *TBD*; for the case with 200 °C thermal limit, the proposed scheme is identical to *LBD*. Besides, for the cases in which the optimal criteria of *LBD* and *TBD* are not satisfied, the proposed scheme can reach the optimal configuration in the design space expanded between *LBD* and *TBD*. For these *TBD*-non-optimal and *LBD*-non-optimal cases, the introduced design scheme improves the achievable throughput from 4.2 to 36.3%. The design examples also shows that although *VDLAPR* increase the latency in vertical direction, the thermal-limited bound is a more serious factor that limits performance.

12.4 On-line Traffic- and Thermal-Aware Routing Algorithms

As mentioned in Sect. 12.2.2, to ensure thermal safety for high-performance 3D NoCs, run-time thermal management (*RTM*) is required. To consider the both cooling efficiency and the performance maintenance, based on the level of thermal emergency, the thermal-aware vertical throttling (*TAVT*) scheme was proposed in [15], as shown in Fig. 12.8c. Because the near-overheated node of NoC system are throttled in online operation, the network topology is changed during operation. Such time-varying topology can be defined as a Non-Stationary Irregular mesh (*NSI-mesh*), as shown in Fig. 12.13. The main problem of *NSI-mesh* is that traditional routing algorithms cannot sustain successful data delivery in such topology. In this section, we analyze the routing problem of fail packet delivery and introduce some advanced thermal-aware routings for the 3D NoC systems, which employ the *TAVT* as the run-time thermal management.

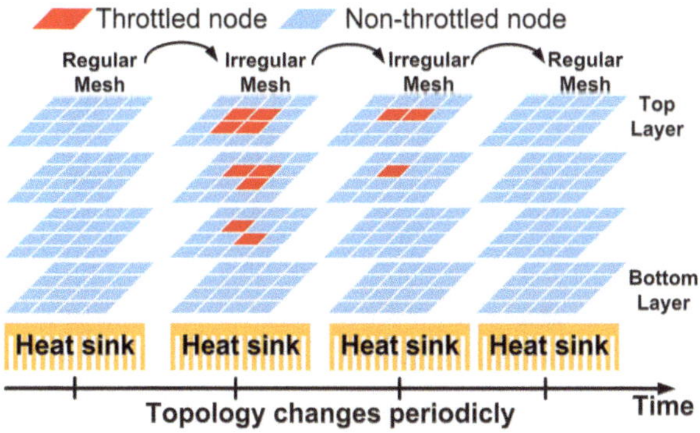

Fig. 12.13 The topology transforms in online operation because of the *RTM*. The topology transforms in online operation because of the *RTM* (©IEEE. Reprinted, with permission, from [19])

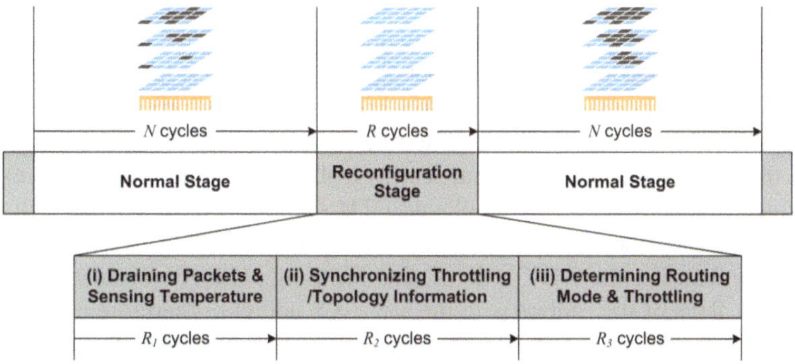

Fig. 12.14 Topology changing, the network operation stages, and the processing stages in reconfiguration (©ACM. Reprinted, with permission, from [15])

12.4.1 Review of Thermal-Aware Vertical Throttling (TAVT) Scheme

In [15], the authors proposed a framework of run-time thermal management, as shown in Fig. 12.14. We assume a distributed thermal sensing mechanism is embedded in the network for each router to obtain its own temperature, and each tile has a timer for synchronizing their operation stages. In online operation, the 3D NoC is iteratively switching between the normal stage and the reconfiguration stage. Most data transfer of application layer is done in normal stage. In the normal stage, the topology may be irregular mesh or regular one, depending on the temperature of the routers. The reconfiguration stage is used for setting the necessary control registers for the temperature control and the data delivery of the next normal stage. There are three processing stages in reconfiguration stage, which are described below.

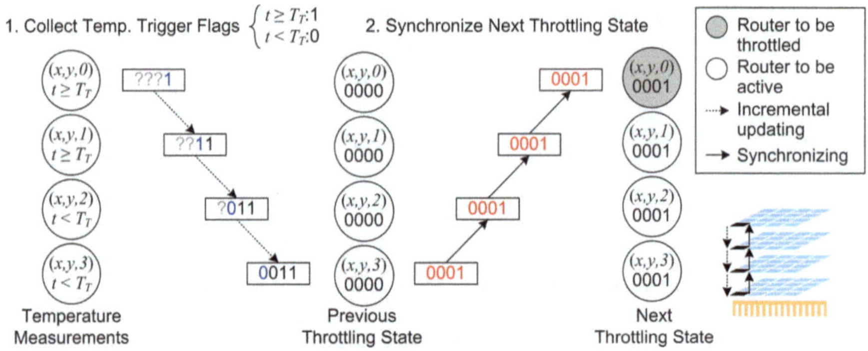

Fig. 12.15 Z-direction synchronization of topology information in a pillar of a $4 \times 4 \times 4$ mesh (©ACM. Reprinted, with permission, from [15])

In reconfiguration stage (i), the network is drained out of packets to prevent data getting blocked in the next normal stage. The transport layer stops accepting more payloads from the application layer and waits for the currently queued packet injection process to finish. Meanwhile, the thermal-aware controller of each tile reads the temperature from thermal sensor. If the temperature of a router exceeds the trigger temperature of the employed *RTM* scheme, the trigger flag will be set to one; otherwise the trigger flag is set to zero. The throttling policy is related to the trigger flag. For the *TAVT* scheme, the throttling policy of a router depends on the trigger flags of the routers with the same XY address. Therefore, the throttling policy of each router has to be determined in the synchronization stage.

In reconfiguration stage (ii), the topology information is synchronized through the proposed dimensional decomposed parallel synchronizations. Because the determination of the throttling policy relies on the trigger flags of the routers with the same XY address, the trigger flags are collected in Z-direction by packets. Figure 12.15 illustrates a $4 \times 4 \times 4$ 3D mesh example. The top node initiates the collecting operation by injecting a packet with its own trigger flag, and the other nodes update the packet, which contains trigger flags. When the bottom node receives the packet of the trigger flags, it determines the throttling policy of the next normal stage according to the previous throttling state and the trigger flags. The next throttling state is passed from the bottom to the top. Then, the 4-bit throttling state in the Z-direction is synchronized for each pillar. After Z-direction synchronization, the routers do X-direction synchronization and Y-direction synchronization. As shown in Fig. 12.16, the 4-bit throttling states are collected and synchronized in X-direction, and then the 16-bit XZ-plane topologies are collected and synchronized in Y-direction. Finally, all the nodes have the entire 64-bit topology (or throttling) information. The router will be shut down and become inactive one as the trigger flag is equal to one. Otherwise, the router is still active in the next normal operation.

In reconfiguration stage (iii), throttling is applied according to the throttling state determined in reconfiguration stage (ii). With the topology information, the routing

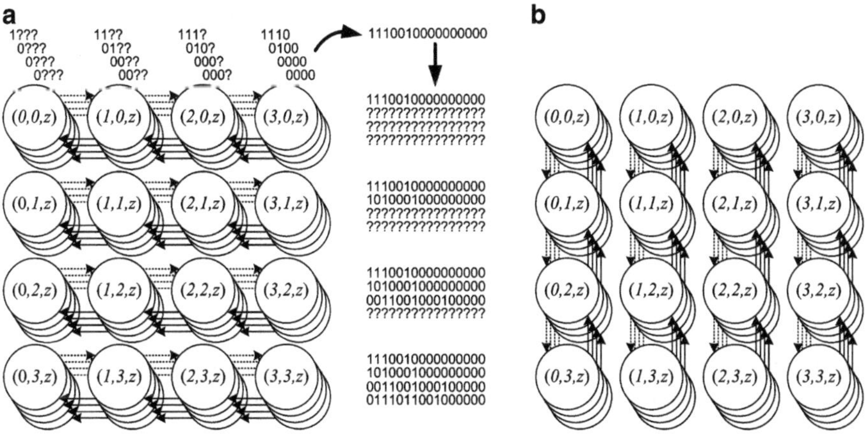

Fig. 12.16 (**a**) In X-direction, the 16-bit topologies of each X-Z plane are collected and synchronized. (**b**) In Y-direction, the entire 64-bit topology is collected and synchronized (©ACM. Reprinted, with permission, from [15])

mode for data delivery toward each destination can be determined. The details of the routing mode determination will be introduced in Sect. 12.4.3. After reconfiguration stage (iii), the network goes to the normal stage, and the data transfer for application layer restarts.

12.4.2 Problem Description and Analysis

We define the time-variant irregular-mesh topology as a non-stationary irregular mesh (*NSI-mesh*), which is shown in Fig. 12.13. A mesh-based thermal-aware 3D NoC adopting *TAVT* can be categorized in *NSI-mesh* topology. *NSI-mesh* is different to traditional irregular mesh, which determines topology in the offline stage. For the *NSI-mesh* of *TAVT*-based 3D NoC systems, there are three characteristics:

- If a router is throttled, all the routers above it are throttled.
- If a router is not throttled, all the routers below it are not throttled.
- The routers in the bottom layer are never throttled, because the bottom layer is applied to form the guaranteed bypassing layer.

The small interval of topology transformation and the large range of the number of throttled routers make conventional algorithms infeasible in *NSI-mesh*. To ensure the success of packet delivery in an *NSI-mesh* network, we should prevent the occurrence of all the following four cases. In the first one, as shown in Fig. 12.17a, the source router is fully throttled. In the second one, as shown in Fig. 12.17b, the destination router is fully throttled. In the third case, as shown in Fig. 12.17c, at least one router on the routing path is fully throttled. The last one is shown in Fig. 12.17d,

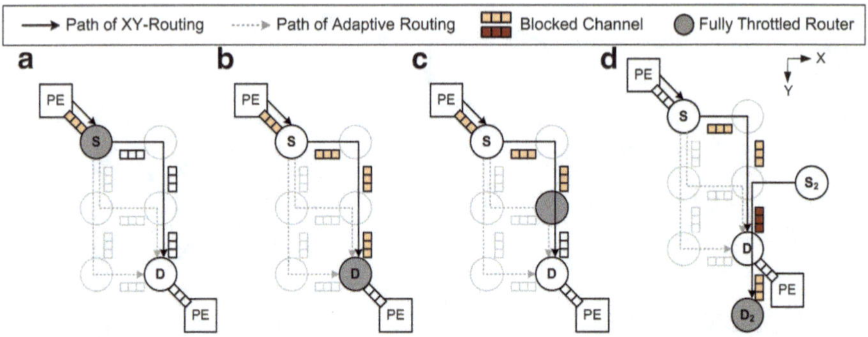

Fig. 12.17 Fail delivery cases that block packets. (**a**) Source-throttled case. (**b**) Destination-throttled case. (**c**) Path-throttled case. (**d**) Long-term Head of Line (HoL) blocking caused by the previous three cases (©ACM. Reprinted, with permission, from [15])

where the channels on the routing path are occupied by other blocked packets. If the source router is fully throttled, the packetized message will be blocked in the network interface. If any case in Fig. 12.17b or c occurs, the injected packets will be blocked somewhere on the routing path and form a congestion-tree. The other packets will be blocked as in Fig. 12.17d.

To eliminate the source-throttled case in Fig. 12.17a and the destination-throttled case in Fig. 12.17b, the throttling information of all routers are required for each node. The Head-of-Line (HoL) problem in Fig. 12.17b traditionally results from the congestion in the switch, and the probability of occurrence can be reduced by applying Virtual Channel (*VC*) flow control or output buffering router architectures. Due to the source-throttled case, the destination-throttled case, and the path-throttled case, a new type long-term HoL blocking may occurs. The long-term HoL blocking has to be eliminated by preventing the occurrence of the aforementioned three cases. However, the path-throttled case in Fig. 12.17c is dependent on the routing path. We must guarantee that there is at least one non-fully throttled path toward destination router before we inject the packet, which is routed on the guaranteed path.

12.4.3 Transport-Layer Assisted Routing (TLAR) Scheme

To deliver data successfully in *NSI-mesh*, Transport Layer Assisted Routing (*TLAR*) scheme was proposed in [15]. The key idea of *TLAR* scheme is that the topology information in transport layer is used to assist the determination of routing. Figure 12.18 shows the block diagram of the transport layer in *TLAR* framework. There are five major components: (i) transmitter queue and packetizer (*Tx*), (ii) receiver queue and depacketizer (*Rx*), (iii) transport layer controller (*TLC*), (iv) topology table (*TT*), and (v) routing mode memory (*RMM*). *Tx* and *Rx* are used to transmit and receive the packets, respectively. *TLC* handles the requests from

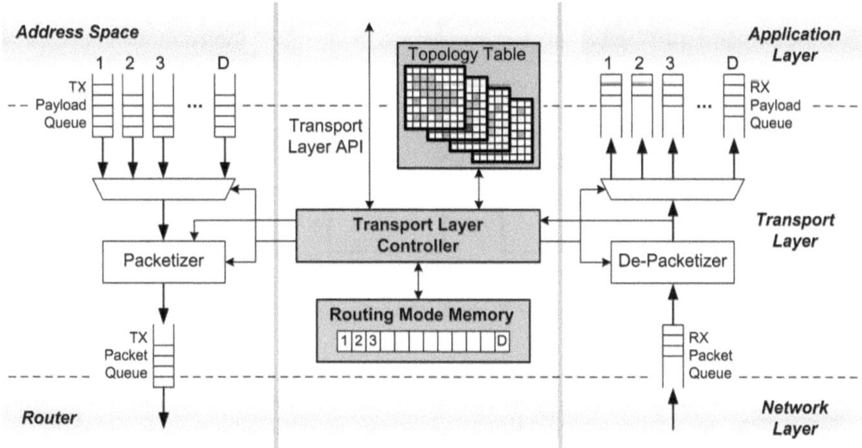

Fig. 12.18 Block diagram of transport layer in the tile of thermal-aware 3D NoC (©ACM. Reprinted, with permission, from [15])

Fig. 12.19 Flow chart of the transport layer assisted routing scheme (©ACM. Reprinted, with permission, from [15])

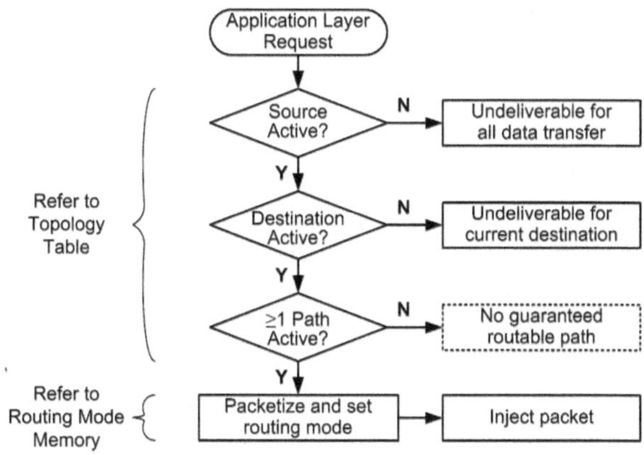

the application layer and the request of the network layer. *TT* stores the throttling information of the entire network, which represents the topology of the network. The results of path selection, what we defined as the routing mode, are saved in the *RMM*.

Figure 12.19 shows the flow chart for handling the application-layer requests. When the application layer requests to deliver data from the current node to the destination node, *TLC* first checks whether the source router is active. If the source router is throttled, it is undeliverable for all data transfer in the period. Then, *TLC* checks whether the destination router is active. If the destination router is throttled, it is undeliverable for current data transfer. Because the downward path as the guaranteed routable path is left, it is unconditionally routable if both the source router and the destination router are active. However, if a general *RTM* is adopted, the path-throttled case may happen. There could be no guaranteed routable path,

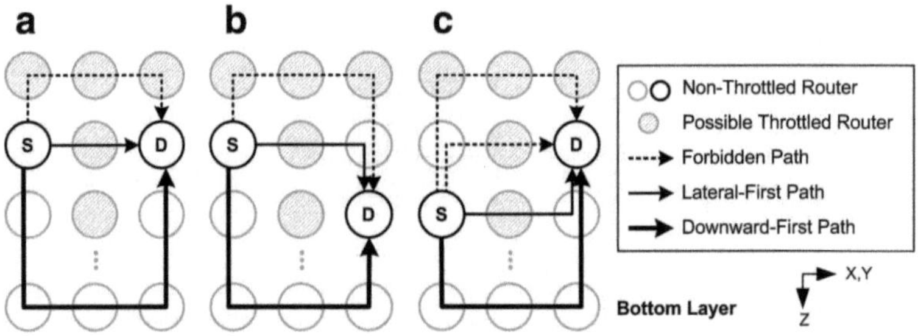

Fig. 12.20 Path selection examples of *TLAR* (©ACM. Reprinted, with permission, from [15])

and the checking cannot be ignored. *TLC* refers to *RMM* for setting the routing mode flag in the packet header. After packetization, the packet is injected to the *Tx* packet queue. The routers in the network layer follow the routing mode to prevent the path-throttled case.

Figure 12.20 shows an example of the routing mode selection of the *TLAR* scheme. The routing contains the downward routing [16], which is a combination of vertical routing and lateral routing. As shown in Fig. 12.20a–c, only the lateral-first path and the downward-first path are allowed at source router. The up-then-lateral turns (i.e., Up-North, Up-East, Up-South, and Up-West) are prohibited to prevent deadlocks. The downward-first path is guaranteed routable because of the characteristics of the *NSI-mesh* topology. However, the lateral-first path is guaranteed routable if and only if all the routers on the lateral-first path are active. Therefore, in the reconfiguration stage of Fig. 12.14, the lateral-first path has to be checked. In this sections, we investigate three downward-lateral routing algorithms and the corresponding checking methods for determination of the routability of the lateral-first path.

12.4.3.1 Algorithm 1: Downward-Lateral Deterministic Routing (DLDR)

The baseline algorithm in *TLAR* is the combination of downward routing and deterministic routing. The downward routing is used for moving packets up and down in the vertical direction. The lateral deterministic routing (*LDR*) is used for routing packets in the source layer. The path diversity (i.e., number of routable path) is one, no matter source layer or bottom layer is chosen. The selection of deterministic routing algorithm affects the computation complexity for checking whether the source layer is laterally routable, as shown in Fig. 12.19. For simplicity, XY routing can be adopted, a dimension-ordered routing (*DOR*), for the deterministic routing.

Checking is done in incremental style for each destination during the reconfiguration stage of *RTM*. The *TLC* checks if there is any fully-throttled router on the path toward every destination based on the table that stores the throttling information. The checking of all XY locations in the source layer can be done in $O(N^2)$ by using the incremental checking flow as the sequence number shown in Fig. 12.21b.

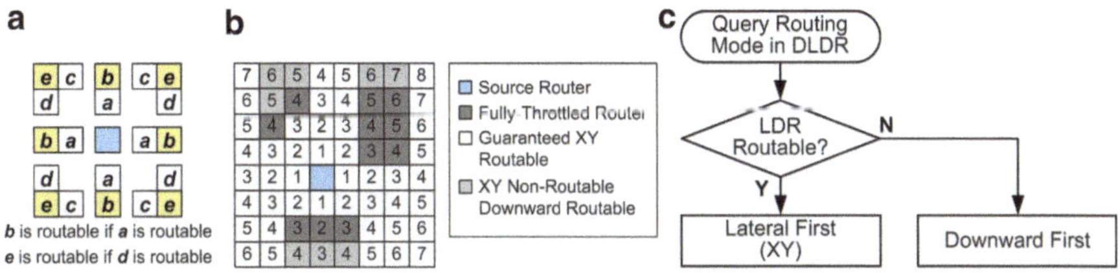

Fig. 12.21 (**a**) Dependency for incremental routability checking of *DLDR*. (**b**) An example of checking in *DLDR*. (**c**) Operation flow for setting the routing mode in *DLDR* (©ACM. Reprinted, with permission, from [15])

The dependency of routability is shown in Fig. 12.21a. Because the lateral routing algorithm is XY routing, node *e* will be routable if *d* is routable. Node *b* is routable if node *a* is routable. The operation flow of *DLDR* is shown in Fig. 12.21c. If a packet is *LDR* routable, it first traverses through the lateral path in the source layer and then going up or down to its destination router. Otherwise, the downward path is chosen, and the packet traverses the lateral routing in the bottom layer.

12.4.3.2 Algorithm 2: Downward-Lateral Adaptive Routing (DLAR)

Downward-lateral adaptive routing (*DLAR*) is an improved version of *DLDR*. The problem of the *DLDR* is that the path diversity for lateral routing is too small. For a source-destination pair, once a router is fully throttled on the routing path, there is no other lateral path available. All the packets are forced to traverse the network in the bottom layer, which easily becomes the bottleneck. Therefore, the throughput is severely limited and the latency increases rapidly. To increase the path diversity of the lateral routing, we replace the lateral deterministic routing by a lateral adaptive routing evolved from [20], which is a throttling- and traffic-aware adaptive routing algorithm. The Odd-Even turn model is adopted for deadlock prevention in the lateral routing. The selection function is consisted of the Neighbor-On-Path (*NOP*) method [21] and the Regional Congestion Awareness (*RCA*) method [22].

Checking if a path is routable for an adaptive routing is much more complex than for deterministic routing. The first problem is that the number of possible path is very large. For a minimal adaptive routing with V and H diversity in y- and x-direction respectively, the number of possible path is $(V+H)!/V!H!$, which brings high computation complexity. The second problem is that the real routing path depends on the results of channel selections in the routers, which is distributed. Considering the computation complexity and feasibility of real implementation, in [15], the authors proposed to only check whether there is any fully-throttled router in the minimal-path region. If there are one or more fully-throttled routers in the minimal-path region, the probability of packets being blocked is not zero. We view this case as lateral non-guaranteed and use downward routing for packet delivery. Checking if there is a fully-throttled router in the minimal-path region for fully adaptive routing can be completed in an incremental style, as shown in Fig. 12.22b.

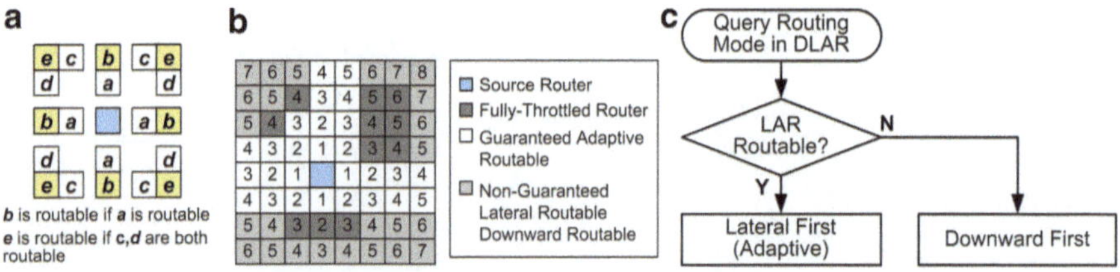

Fig. 12.22 (**a**) Dependency for incremental routability checking of *DLAR*. (**b**) An example of checking in *DLAR*. (**c**) Operation flow for setting the routing mode of a packet in *DLAR* (©ACM. Reprinted, with permission, from [15])

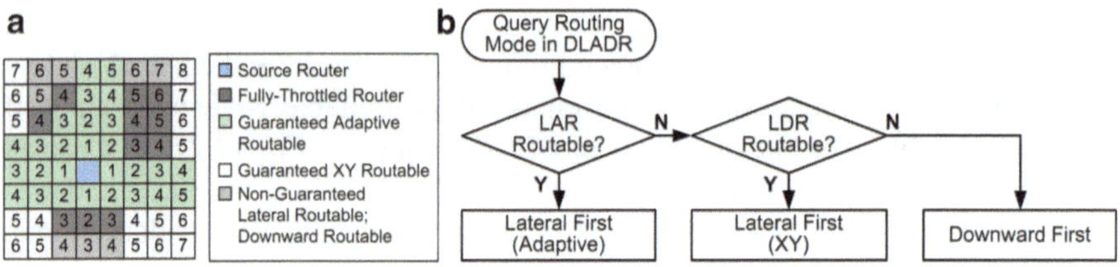

Fig. 12.23 (**a**) An example of checking in *DLADR*. (**b**) Operation flow for setting the routing mode of a packet in *DLADR* (©ACM. Reprinted, with permission, from [15])

The checking dependency is shown in Fig. 12.22a, where node *e* will be routable if *c* and *d* are both routable. Similarly, node *b* is routable if node *a* is routable. As shown in Fig. 12.22c, if the lateral path is guaranteed routable, the packet is first laterally routed on the source layer. Otherwise the lateral routing is completed in the bottom layer.

12.4.3.3 Algorithm 3: Downward-Lateral Adaptive-Deterministic Routing (DLADR)

The advantages of *DLAR* are the increased path diversity and the capability of load balancing. However, if there is a throttled router on one of the lateral-first paths, *DLAR* chooses downward-first path to prevent the occurrence of the path-throttled case. Therefore, the *LAR* routable destinations are fewer than *LDR* routable destinations. To increase the number of lateral-first path routable destinations, we introduce the downward-lateral adaptive-deterministic routing (*DLADR*). The idea of *DLADR* is to combine the *DLDR* and *DLAR*, as shown in Fig. 12.23b. The destinations are categorized into three types: (i) the guaranteed adaptive routable (*LAR* routable), (ii) the guaranteed XY routable (*LDR* routable), and (iii) the non-guaranteed lateral routable, which is downward routable. If a destination is guaranteed adaptive routable, it is guaranteed XY routable. If a destination is guaranteed XY routable, it is downward routable. Therefore, the downward routable destination set is a super set of the *LDR* routable set, and the *LDR* routable set is a super set of the *LAR* routable set.

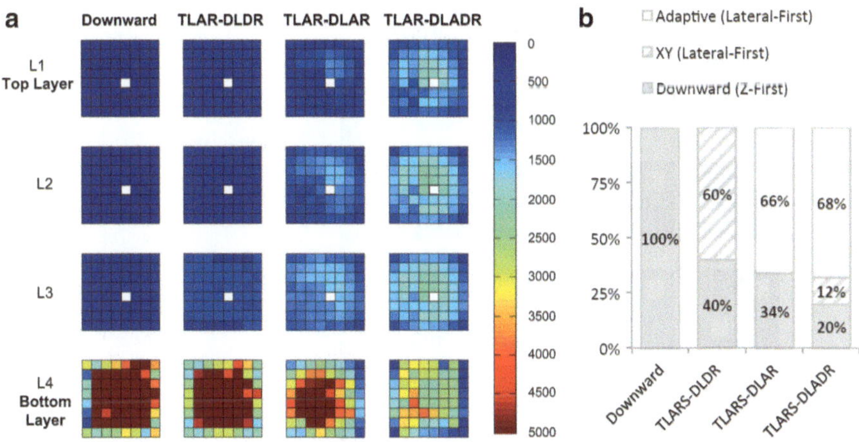

Fig. 12.24 (**a**) Statistical traffic load distribution (*STLD*) and (**b**) distribution of the routing mode decision with one $1 \times 1 \times 3$ pillar fully throttled in the center under uniform traffic pattern (©ACM. Reprinted, with permission, from [15])

Figure 12.23b shows the operation flow for setting the routing mode of the packets in *DLADR*. Because the lateral-first adaptive routing is able to balance the traffic loading, we should use the lateral-first adaptive routing as much as possible. The lateral-first deterministic routing results in less traffic congestion in the bottom layer than downward-first routing, so the priority of lateral-first deterministic routing is higher than downward-first routing. In *DLADR*, we use simpler adaptive routing instead of the throttling- and traffic-aware routing. West-first turn model is adopted for the routing function, and the selection of output channel depends on which channel has more unallocated flit buffers. By combining essences of *DLDR* and *DLAR*, the vertical traffic loading is more balanced. Therefore, *DLADR* is able to achieve higher throughput and lower latency with a simpler architecture. Figure 12.23a shows an example of checking order in *DLADR*.

12.4.3.4 Performance Evaluation of TLAR Algorithms

To demonstrate the system performance as applying the three *TLAR* algorithms, the cycle-accurate NoC platform, *Noxim*, is adopted and modified for using vertical links to connect each router in different silicon layer [12]. An $8 \times 8 \times 4$ mesh-based 3D NoC is set as a design example in this section. For each router, the queueing depth of each input channel is 8 flits without virtual channel, and the wormhole flow control is employed. Besides, the packet length is randomly from 2 to 10 flits.

For the cases of irregular mesh, we use two vertical throttling cases: (i) one $1 \times 1 \times 3$ throttled pillar and (ii) two $2 \times 2 \times 3$ throttled pillars. The location of throttled pillars can be found in Figs. 12.24 and 12.25. First we use statistical traffic load distribution (*STLD*) and decision distribution to show the network loading. All the

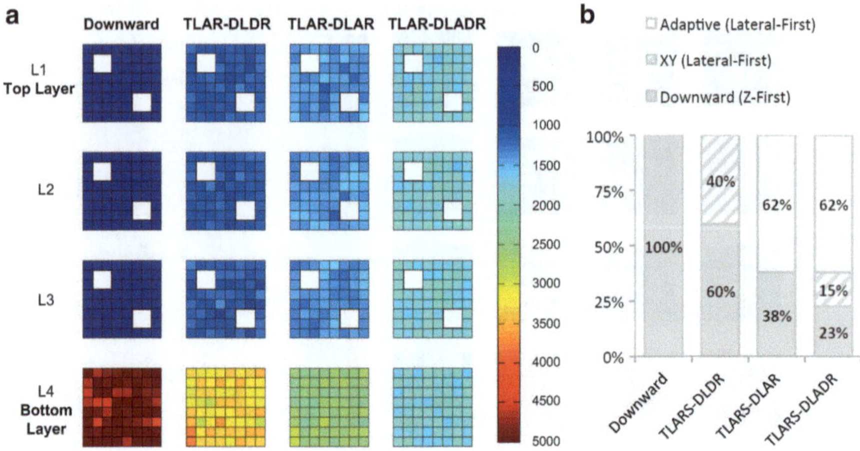

Fig. 12.25 (**a**) Statistical traffic load distribution (*STLD*) and (**b**) distribution of the routing mode decision with one $2 \times 2 \times 3$ pillar fully throttled on the diagonal under uniform traffic pattern (©ACM. Reprinted, with permission, from [15])

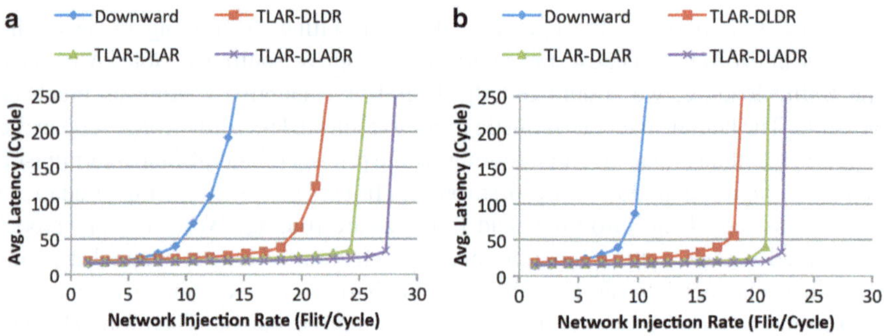

Fig. 12.26 Average latency vs injection rate with (**a**) one $1 \times 1 \times 3$ pillar throttled and (**b**) two $2 \times 2 \times 3$ pillars throttled (©ACM. Reprinted, with permission, from [15])

experiments in Figs. 12.24 and 12.25 use the same injection rate that makes average latency of *TLAR-DLDR* twice the zero load latency.

Figure 12.24a shows the *STLD* of the baseline downward routing and the other three *TLAR* algorithms. Though there is only one throttled pillar in the network, many packets have to be routed downward through the bottom layer as applying the downward routing. The congestion degree of *DLDR* in the bottom layer is slightly reduced but is still the second largest. The *DLAR* can balance network traffic, because the congestion in the bottom layer is relaxed by applying traffic- and throttling-aware adaptive routing algorithm. For the *DLADR*, more packets are routed laterally in the source layer because of larger routing path diversity. Figure 12.24b shows the distribution of the routing mode decision. Because of the larger lateral path diversity provided by the *TLAR*, the total ratio of downward routing is significantly reduced. Figure 12.25 shows similar results when we increase the number of disconnected throttled region and the size of the region. Because more lateral path is blocked, the ratio of Z-first downward decision increases. The performance evaluation of the *TLAR* is shown in Fig. 12.26. Because of larger path diversity, the *TLAR* can help to reduce the average latency by around 34.8% $\sim$ 70.4%.

12.5 Conclusions

In this chapter, the traffic- and thermal-aware routing problems on 3D NoC systems are categorized into: off-line type and on-line type. For the off-line traffic- and thermal-aware routing, we introduced an novel design scheme to find an optimal design space of the system without *RTM*. For the thermal-aware 3D NoC system with *RTM*, the topology may become the *NSI-mesh*, which makes the traditional routing algorithms cannot sustain successful data delivery. To amend the shortages, some on-line traffic- and thermal-aware routing algorithms were investigated in this chapter. By these introduced routing algorithms, the benefits of 3D NoC system can be preserved without the enhancement of cooling devices.

References

1. J. Davis, R. Venkatesan, A. Kaloyeros et al., Interconnect limits on Gigascale Integration (GSI) in the 21st century. Proc. IEEE **89**(3), 305–324 (2001)
2. S. Kumar, A. Jantsch, J.P. Soininen et al., A network on chip architecture and design methodology, in *Proceedings of the IEEE International Symposium on VLSI*, Vancouver, 2002, pp. 105–112
3. Y. Hoskote, S. Vangal, A. Singh et al., A 5-GHz mesh interconnect for a teraflops processor. IEEE Micro **27**(5), 51–61 (2007)
4. J. Howard, S. Dighe, Y. Hoskote et al., A 48-Core IA-32 Message-passing processor with DVFS in 45nm CMOS, in *IEEE International Solid-State Circuit Conference Digest of Technical Papers (ISSCC)*, San Francisco, 2010, pp. 70–11
5. S. Bell, B. Edwards, J. Amann et al., TILE64-Processer: A 64-Core SoC Mesh Interconnect, in *IEEE International Solid-State Circuit Conference Digest of Technical Papers (ISSCC)*, San Francisco, 2008, pp. 3–7
6. Y. Xie, J. Cong, S. Sapatnekar, *Three-Dimensional Integrated Circuit Design* (Springer, Heidelberg, 2009)
7. A.W. Topol, D. Tulipe, L. Shi et al., Three-dimensional integrated circuits. IBM J. Res. Dev. **50**(4.5), 491–506 (2006)
8. B.S. Feero, P.P. Pande, Networks-on-chip in a three dimensional environment: a performance evaluation. IEEE Trans. Comput. **58**(1), 32–45 (2009)
9. L. Shang, L. Peh, A. Kumar, N.K. Jha, Thermal modeling, characterization and management of on-chip networks, in *Proceedings of the IEEE/ACM International of Symposium on Microarchitecture (Micro)*, Portland, 2004, pp. 67–78

10. W. Hung, C. Addo-Quaye, T. Theocharides et al., Thermal-aware IP virtualization and placement for networks-on-chip architecture, in *Proceedings of the IEEE International Conference on Computer Design (ICCD)*, San Jose, 2004, pp. 430–437
11. G.M. Link, N. Vijaykrishnan, Hotspot prevention through runtime reconfiguration in network-on-chip, in *Proceedings of the IEEE Conference on Design, Automation, and Test in Europe (DATE)*, Munich, 2005, pp. 648–649
12. K.Y. Jheng, C.H. Chao, H.Y. Wang, A.Y. Wu, Traffic-thermal mutual-coupling co-simulation platform for three-dimensional network-on-chip, in *Proceedings of the IEEE International Symposium on VLSI Design, Automation, and Test (VLSI-DAT)*, Hsinchu, 2010, pp. 135–138
13. G. Hinton, D. Sager, M. Upton et al., The micro-architecture of the Pentium 4 processor, *Intel Technology Journal*, 2001
14. C.H. Chao, K.Y. Jheng, H.Y. Wang et al., Traffic- and thermal-aware run-time thermal management scheme for 3D NoC system, in *IEEE International Symposium on Network-on-Chip (NOCS)*, Grenoble, 2010, pp. 223–230
15. C.H. Chao, K.C. Chen, T.C. Yin et al., Transport layer assisted routing for run-time thermal management of 3D NoC systems, *ACM Trans. Embedd. Comput. Syst.*, 13, 1, Article 11 (August 2013), 22 pages
16. K.C. Chen, S.-Y. Lin, H.-S. Hung, A.-Y. Wu, Topology-Aware Adaptive Routing for Non-Stationary Irregular Mesh in Throttled 3D NoC Systems, *IEEE Transactions on Parallel and Distributed Systems*, **24**(10), 2109–2120 (2013)
17. G.M. Ghiu, The odd-even turn model for adaptive routing. IEEE Trans. Parallel Distrib. Syst. **11**(7), 729–738 (2000)
18. U.Y. Ogras, J. Hu, R. Marculescu, Key research problems in NoC design: a holistic perspective, in *Proceedings of the IEEE/ACM/IFIP International Conference on Hardware/Software Codesign and Systems Synthesis (CODES+ISSS)*, Jersey, 2005, pp. 69–74
19. K.C. Chen, S.Y. Lin, H.S. Hung, A.Y. Wu, Topology-aware adaptive routing for non-stationary irregular mesh in throttled 3D NoC systems (early access on *IEEE Transactions on Parallel and Distributed Systems*, 2013)
20. S.Y. Lin, T.C. Yin, H.Y. Wang, A.Y. Wu, Traffic-and thermal-aware routing for throttled three-dimensional network-on-chip system, in *Proceedings of the IEEE International Symposium VLSI Design, Automation, and Test (VLSI-DAT)*, Hsinchu, 2011, pp. 320–323
21. G. Ascia, V. Catania, M. Palesi, D. Patti, Neighbors-on-path: a new selection strategy for on-chip networks, in *Proceedings of the IEEE/ACM/IFIP Workshop Embedded Systems for Real Time Multimedia*, Seoul, 2006, pp. 79–84
22. P. Gratz, B. Grot, S.W. Keckler, Regional congestion awareness for load balance in networks-on-chip, in *Proceedings of the IEEE Symposium High Performance Computer Architecture (HPCA)*, Salt Lake City, 2008, pp. 203–214

Chapter 13
Scalable Architecture for All-Optical Wavelength-Routed Networks-on-Chip

Somayyeh Koohi and Shaahin Hessabi

Abstract Scaling the transistor feature size along with the increase of chip operation frequency leads to the growth of circuit complexity, which makes the design of electrical interconnects increasingly difficult in large chips. Optics provides low power dissipation which remains independent of capacity and distance, as well as wavelength parallelism, ultra-high throughput, and minimal access latencies. Additionally, wavelength routing, bit rate transparency, high-capacity, low propagation loss, and low power dissipation of silicon photonics are attractive for realizing *optical Networks-on-Chip* (ONoC) in Chip Multi-Processors (CMPs). In this chapter, we propose a new architecture for nanophotonic NoC, named 2D-HERT, which consists of optical data and control planes. The proposed data plane is built upon a new topology and all-optical switches that passively route optical data streams based on their wavelengths. Utilizing wavelength routing method, the proposed deterministic routing algorithm, and Wavelength Division Multiplexing (WDM) technique, the proposed data plane eliminates the need for optical resource reservation at the intermediate nodes. For resolving end-point contention, we propose an all-optical request-grant arbitration architecture which reduces optical losses compared to the alternative arbitration schemes. By performing a series of simulations, we study the efficiency of the proposed architecture, its power and energy consumption, and the data transmission delay.

13.1 Silicon Photonics

One of the most serious challenges for future high-performance Multi-Processor Systems-on-Chip (MPSoC) is to realize the enormous bandwidth capacities and rigorous latency requirements when interconnecting a large number of processing

S. Koohi (✉) • S. Hessabi
Department of Computer Engineering, Sharif University of Technology, Tehran, Iran
e-mail: koohi@sharif.edu; hessabi@sharif.edu

M. Palesi and M. Daneshtalab (eds.), *Routing Algorithms in Networks-on-Chip*, 339
DOI 10.1007/978-1-4614-8274-1_13, © Springer Science+Business Media New York 2014

cores. To deal with this challenge, significant research activity has recently focused on intrachip global communication using packet-switched micro networks, referred to as NoC [1]. Since performance-per-watt is expected to remain the fundamental design metric for high-performance multi-processor systems [2], on-chip interconnection networks will have to satisfy communication bandwidth and latency requirements in a power efficient fashion.

Scaling the transistor feature size along with the increase of chip operation frequency leads to growth of the circuit complexity, which makes the design of electrical interconnects (EI) increasingly difficult in large chips. This problem that has been predicted for about three decades [3] stems from many limitations associated with metallic interconnects. In addition to quantified limitations of metallic interconnect, such as latency, throughput, bandwidth and power consumption, there are a variety of other problems that cannot be quantified easily, such as interline crosstalk, wave reflection phenomena, Electromagnetic Interference (EMI) and the difficulty of voltage isolation and timing accuracy [4]. Although using low resistance metals such as copper and low dielectric constant materials decreases the interconnect delay, bandwidths will be insufficient for long interconnects for future operating frequencies, and power budget cannot be confined to the package power constraints. While NoC, as a new architectural trend, can improve bandwidth of electrical interconnections, it is unclear how electrical NoCs will continue to satisfy future bandwidth and latency requirements within the package power budget [5]. Other physical approaches that can improve the metallic interconnections, such as cooling the chips and/or circuits (to decrease interconnect resistance), or using superconducting lines [6] cannot address qualitative problems like voltage isolation, timing accuracy and EMI.

Optics is a very different physical approach that can address most of the problems associated with electrical interconnects such as bandwidth, latency, crosstalk, voltage isolation, and wave reflection [4]. Additionally, bit rate transparency [7] of optical switching elements and low propagation loss of optical waveguides lead to low power dissipation of silicon photonics. Importance of power dissipation in NoCs along with power reduction capability of on-chip optical interconnects, offers *optical network-on-chip* (ONoC) as a novel technology solution which can introduce on-chip interconnection architecture with high transmission capacity, low power consumption, and low latency. While traditional NoCs enforce unaffordable power dissipation in high performance MPSoCs, the unique advantages of ONoC offer considerable power efficiency and also performance-per-watt scaling as the most critical design metric.

Several on-chip interconnect architectures have been proposed that leverage CMOS-compatible photonics for future multicore microprocessors. Most of the proposed optical interconnect architectures are bus-based. For example, the Cornell hybrid electrical/optical interconnect architecture [8] comprises an optical ring that assigns unique wavelengths per node in order to implement a multi-bus. Firefly [9], as a hybrid electrical/optical network, proposes the implementation of reservation-assisted single-write-multi-read buses. Moreover, HP Corona crossbar architecture [10] is in fact several multiple-writer, single-reader buses routed in a snake pattern among the nodes.

The Columbia optical network [5] is one of the few that proposes on-chip optical switches. The proposed architecture combines a high-speed photonic circuit-switched network with an electronic packet-switched control network. The electrical sub-network sets up the switches in advance of data transmission and tears down the network thereafter. Hence, the network must transmit a large amount of data to amortize the relatively high latency of the electrical setup/teardown network. Optical NoC proposed by Koohi et al. [11], referred to as CONoC, improves the hybrid architecture proposed by Shacham et al. [5]. CONoC utilizes wavelength routing method, instead of electrical methods, for optical packet ejection while retaining the scalability of the network. Despite its simpler electrical routers and reduced setup latency compared to previously proposed ONoCs, CONoC cannot eliminate the role of electrical transactions. Cianchetti et al. [12] have presented Phastlane, a hybrid electrical/optical routing network, for future multicore microprocessors. Phastlane network is built upon a low latency optical crossbar for data transmission under contention-less conditions. When contention exists, the router makes use of electrical buffers and, if necessary, a high speed drop signaling network.

Almost all of the previously proposed hybrid architectures in [5, 8, 9, 11, 12] and [13] suffer from high latency and power overheads for electrically resolving optical contentions. Therefore, an all-optical NoC can overcome the limitations of electrically-assisted ONoCs. Briere et al. [14] have developed an all-optical contention-free NoC which routes optical signals according to their wavelengths. However, the proposed contention-free structure is obtained at the cost of large arrays of fixed-wavelength light sources and fast switches for wavelength selection, which limit the scalability, and also severely increase power consumption and area issues.

The selected topology of the photonic on-chip interconnect plays prime role in the performance of ONoC architecture as well as routing and switching techniques that can be used. Although recent studies on the design of optical on-chip networks have addressed various network topologies, almost all of these optical architectures [5, 9, 11, 15–17] are built upon traditional topologies initially introduced for electrical NoCs, such as Mesh, Torus, Spidergon, Fat tree, Honeycomb, and crossbar. Moreover, previously developed optical architectures adopt routing algorithms of the corresponding electrical topologies for optical data routing through the network. Due to the inherently different properties of light transmission through optical waveguides and photonic switching elements, novel topologies specifically developed for optical infrastructure are inevitable to fully utilize advantages of the silicon photonic technology.

This chapter proposes a novel topology for ONoC architectures, which concerns physical properties of light transmission, and examines the advantages and limitations of routing data streams through photonic switching elements. Some of the most interesting characteristics of the proposed topology are: (i) regularity, (ii) vertex symmetry, (iii) scalability to large scale networks, (iv) constant node degree, and (v) simplicity. Moreover, this chapter proposes a general all-optical routing algorithm which can be adapted to efficiently route optical data streams through various topologies. The key advantages of the proposed routing algorithm are minimalism and simplicity, which lead to small and simple optical routers.

Built upon our novel network topology, we propose a scalable all-optical NoC as a global communication medium, which offers all-optical on-chip routing of data streams. Passive routing is adopted in the proposed optical architecture, which is performed by routing optical data streams based on their wavelengths. Utilizing wavelength routing method, our proposed optical NoC eliminates the need for electrical resource reservation and the corresponding latency and area overheads. Taking advantage of Wavelength Division Multiplexing (WDM) technique, the proposed architecture avoids packet congestion scenarios (discussed in [5, 11, 12]), and guarantees contention-free operation of the network. While the number of switching elements in the contention-free ONoC proposed in [14] is quadratically proportional to the number of IP blocks, the optical on-chip architecture introduced in this chapter reduces the required number of routing elements to the number of processing cores. The later property along with the reduced number of wavelength channels in the proposed ONoC leads to a scalable architecture for on-chip routing of optical packets. Moreover, combining wavelength routing method with WDM technique improves the functionality and total performance of the proposed optical NoC, and enables the network to transmit multicast and unicast packets efficiently.

13.2 Data Plane: Wavelength Routing

This section introduces the data plane of the proposed architecture, built upon an all-optical switch architecture. Before exploring the architecture of the proposed ONoC, we discuss prominent criteria of an efficient optical topology.

13.2.1 Topology Exploration for ONoCs

Main issues in the design of optical network on-chip include the topology of the network, its flexibility for extension with minimal links and with no, or slight, modification of the edge nodes, short diameter to achieve low latency, simplicity of the photonic router design, etc. Prior to presenting our novel topology, we discuss these influential metrics in more detail.

13.2.1.1 Node Degree

Node degree in an optical on-chip network plays a critical role in the simplicity of the photonic router design and its implementation cost. Although implementation of high-degree electronic crossbar is simple, a crossbar is not a scalable photonic topology [18] because the number of microring resonators required for photonic crossbars increases quadratically with the node degree. Specifically, it is quite difficult to construct optical crossbars larger than 4×4 using the existing 2×2

photonic switching elements. Therefore, maintaining a small node degree in an optical NoC, especially for large scale networks, facilitates photonic router design and reduces fabrication complexity. On the other hand, due to the trade-off between the degree of connectivity and the network diameter, the proposed topology should address small node degree while retaining small network diameter.

13.2.1.2 Planar Layout

Planar layout of the proposed topology enables single layer construction of optical transmission network above the metal stack [5], thus reducing fabrication complexity, the chip dimensions, and the total cost. Moreover, two-dimensional (2-D) optical network reduces number of waveguide crossings, thus improving total waveguide intersection crosstalk.

13.2.1.3 Scalability

Ever-increasing number of processing cores enforces us to build a network interconnecting hundreds of cores in future MPSoCs. Therefore, scalable topology for optical NoCs is inevitable to preserve affordable communication cost. According to unique features of silicon photonic technology, a scalable topology for ONoCs should address the following properties:

- Constant node degree despite increases in network size;
- Extensible (i.e. capable of inserting arbitrary number of photonic routers to the ONoC) with no, or slight, modification to the existing nodes;
- Simple 2-D layout for large numbers of processing cores.

13.2.1.4 Symmetry

Symmetrical and regular optical on-chip architecture facilitates the design of photonic routers and optical routing algorithm, and also reduces fabrication complexity of the optical network.

13.2.2 Two Dimensional Hierarchical Expansion of Ring Topology

Most recent optical NoC architectures have been implemented on top of Mesh or Torus topologies. Despite their efficiency for electrical NoCs, node degree of five (including connection to the local IP) complicates photonic switch design in these architectures. Implementing Torus topology, Shacham et al. [5] solved the problem by introducing extra injection and ejection switches. These extra switches increase design complexity, and add to power and area overheads. ONoC architecture proposed by Koohi et al. [11] is built upon Spidergon topology. Despite constant node degree of four, network diameter of the Spidergon topology limits efficiency of the large scale ONoCs.

To overcome the drawbacks of the traditional topologies in optical NoCs, this chapter addresses a new topology for photonic on-chip networks. The proposed topology, which is referred to as 2D-HERT, benefits from high degree of connectivity along with small node degree. 2D-HERT is built upon clusters of processing cores locally connected by optical rings. Different clusters in the proposed topology are interconnected by global optical 2D rings. A sample 2D-HERT interconnecting 64 optical switches is shown in Fig. 13.1a, where all connections are made through optical links. As depicted in this figure, although 2D-HERT can be *viewed* as a two dimensional hierarchical expansion of Ring topology, various optical routers are interconnected at the same level of hierarchy, and each super node only presents a group of four optical routers.

Schematically, the proposed topology consists of k diagonals each interconnecting m (even) local clusters of four processing cores. For brevity, we will refer to the local clusters as super-nodes (*SN*). Hence, for a 2D-HERT architecture interconnecting N processing cores, we have $N = 4 \times k \times m$. Based on this notation, each IP (and its corresponding optical switch) is uniquely determined with triplet *(d,s,p)* $0 \leq d < k$, $0 \leq s < m$, $0 \leq p < 4$, where d and s refer to the index of the corresponding diagonal and super-node, respectively, and p represents the index of the processing core within the super-node. Figure 13.1a shows pair *(s,d)* for each super-node and triplet *(s,d,p)* for a sample IP. The main advantages of this topology are:

- Constant node degree of four, similar to Spidergon, for arbitrary numbers of processing cores;
- High degree of connectivity, similar to Mesh (Torus) topology, which leads to small network diameter;
- Regularity;
- Scalability for large scale optical networks;
- Local optical data transmission between neighbor nodes to reduce power and delay metrics for local traffic distribution.

2-D layout of the 64-node 2D-HERT is depicted in Fig. 13.1b.

13.2.3 Routing Algorithm

This chapter proposes Circular-First routing (CF) as a deterministic routing algorithm for 2D-HERT architecture. According to this routing scheme, for routing optical data streams from the source node (d_s, s_s, p_s) to the destination

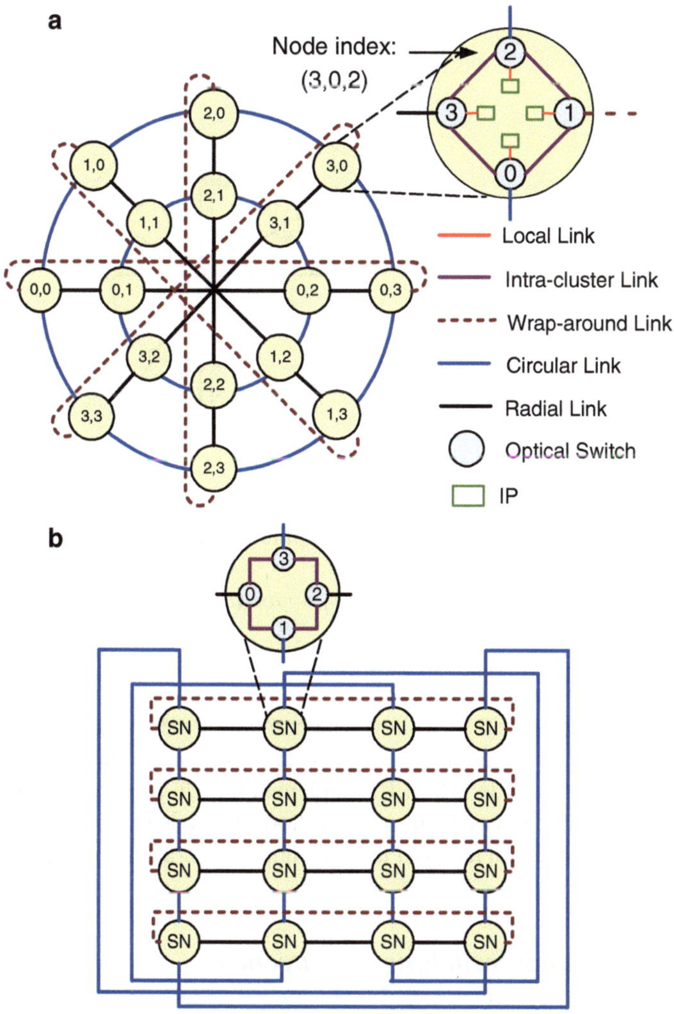

Fig. 13.1 (**a**) 2D-HERT architecture, (**b**) 2D-HERT layout (*SN* super node)

node (d_d, s_d, p_d), circular links are traversed first to reach the destination diagonal. Then, radial links are taken through this diagonal to reach the target super-node. For minimal data routing, the proposed routing scheme may take advantage of the wrap-around links, depending on the position of the target super-node. Finally, within the destination cluster, intra-cluster links are utilized to reach the target node. Figure 13.2 illustrates an example of the path taken by circular-first routing in a 144-node 2D-HERT.

It is straightforward to show that CF routing scheme takes the minimal path through 2D-HERT architecture. Considering the proposed algorithm, distance from

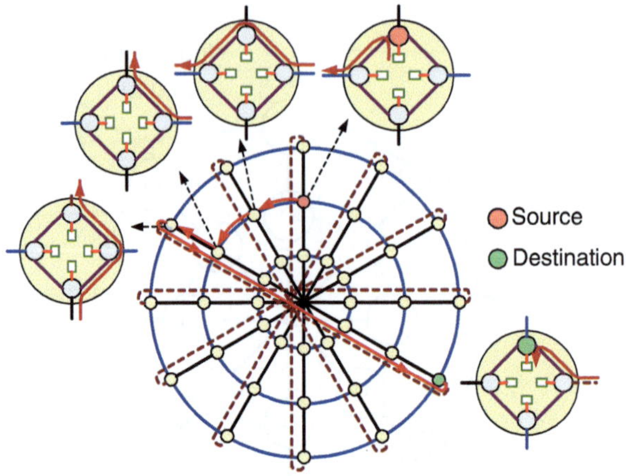

Fig. 13.2 Sample routing path

the source node (d_s, s_s, p_s) to the destination node (d_d, s_d, p_d) in terms of long hops, i.e. interconnecting super-nodes, can be computed as follows:

$$D = min\left(abs\left(d_s - d_d\right), k - abs\left(d_s - d_d\right)\right) + min\left(abs\left(s_s - s_d\right), m - abs\left(s_s - s_d\right)\right) \tag{13.1}$$

where, the first term represents the number of circular links traversed by the optical data stream, and the second term shows the number of radial links taken by the CF routing through the destination diagonal to reach the target super-node. Based on this equation, in an N-node 2D-HERT implementing CF routing scheme, network diameter, which is defined as the maximum shortest path length between any pair of nodes, equals $k/2 + m/2$ in terms of the long hops interconnecting super-nodes.

13.2.4 All-Optical Switching Architecture

According to the 2D-HERT topology, two-dimensional 4×4 switches are needed to interconnect local IPs and neighbor nodes with each other. Silicon microring resonator-based electro-optic switch array coupled with waveguide crossings can constitute a potential architecture for on-chip photonic crossbar interconnection. However, one key challenge in realizing such design in silicon is the considerable insertion loss and crosstalk imposed at the high-index contrast waveguide intersections with sub-micrometer size [19]. In this regard, various studies have focused on the design of efficient optical switch architectures customized for existing topologies. However, almost all of the previously proposed optical switches benefit from electrically controlled resonators [5, 11, 13, 19, 20].

As a means of wavelength selective filtering, SOI-based microring resonator structures have been explored as passive Optical Add/Drop (OAD) filters [21]. Figure 13.3a depicts an OAD element which adds (drops) optical wavelength λ_1

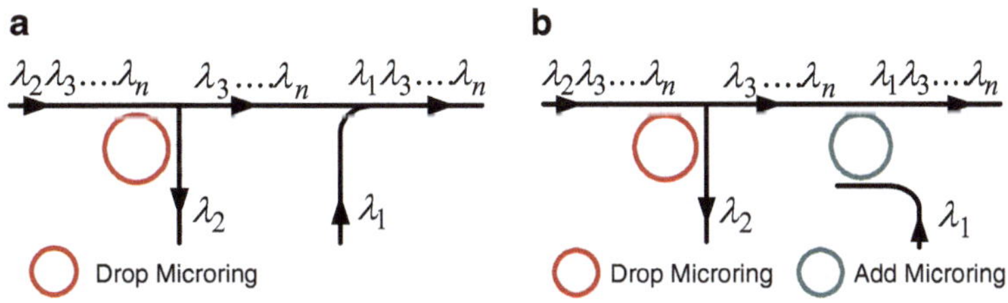

Fig. 13.3 (**a**) Optical Add/Drop Element (OAD), (**b**) modified OAD element

(λ_2) to (from) the optical stream without any electronic processing. To reduce the coupling loss, ring resonators can be used to inject a single wavelength from one waveguide into another. If the ring is placed between two waveguides and the coupling between the ring and the waveguides is properly chosen, the wavelength resonating within the ring will be transferred from one waveguide to the other [10]. Figure 13.3b depicts the modified OAD element realized in this way.

Utilizing OAD filters, we propose a passive optical switch, named as WaROS (Wavelength-Routed Optical Switch). The proposed switch eliminates optical resource reservation at the intermediate node, which is required in most of the previously introduced ONoCs [5, 11, 15, 16].

13.2.4.1 All-Optical Data Routing

2D-HERT associates one wavelength to each router in the network, while the optical streams targeted to a specific node are modulated on its dedicated wavelength. As a key idea, the modulation wavelength of the optical packet is utilized as the destination address to uniquely determine the routing path from the source to the destination node. In other words, the address of the target is not contained in the packet; rather, it is embedded in the wavelength of the optical signal.

Although some of the optical bus-based architectures, such as Corona [10], associate one wavelength to each optical router, they do not route optical data streams based on the modulated wavelength. In a bus-based architecture, various nodes are simply connected through multiple optical buses and modulation wavelength of the optical stream is solely used to eject the data at the destination node. Since on-chip data routing is not performed in the bus-based architectures, these architectures suffer from high network diameter, which increases data transmission power and delay. However, 2D-HERT architecture is a network of optical switches, in which a deterministic routing algorithm determines an efficient path for optical packets from the source to the destination node. This path may go through various intermediate nodes. For this purpose, 2D-HERT utilizes the modulation wavelength of the packet, as the destination address, to make proper turns at the intermediate nodes.

The minimum number of wavelength channels that guarantees contention-free routing of optical packets considerably impacts the performance of the optical NoC. Hence, WaROS, as a basic building block of the proposed ONoC, should be developed to minimize the required number of wavelength channels in the network. Regarding WDM technique, the number of required channels depends on the maximum number of multiplexed optical flows on a waveguide, referred to as *maximum degree of multiplexing* (MDM). Assuming that only one data stream can be targeted to each node at a time, the MDM metric for the radial and circular links can be derived as follows.

According to the CF routing scheme, optical streams passing through the radial link connecting SN_i to SN_{i+1} on a diagonal, consist of optical packets targeted to any super-node on the same diagonal at the distance d_1 of SN_i, where $d_1 \le m/2$. Hence, the MDM for the radial links equals:

$$MDM_{Radial} = 4 * m/2 = 2m \tag{13.2}$$

On the other hand, optical data streams passing through the circular link connecting ith diagonal (D_i) to i + 1th diagonal (D_{i+1}) consist of optical packets targeted to any super-node located on the diagonals at the distance d_2 of D_i, where $d_2 \le k/2$. Hence, the MDM for circular links can be calculated as follows:

$$MDM_{Circular} = 4 * m * k/2 = N/2 \tag{13.3}$$

From the above discussion, maximum degree of multiplexing in 2D-HERT architecture is computed as the maximum of $MDM_{Circular}$ and MDM_{Radial} metrics, which equals $N/2$. Consequently, at least $N/2$ distinct wavelength channels are required in an N-node 2D-HERT architecture to guarantee contention-free operation of the network under CF routing scheme. For this purpose, each wavelength is associated with two different routers represented by (d_1,s_1,p_1) and (d_2,s_2,p_2) in the case of $d_1 \equiv d_2 \ (mod\ k/2)$, $s_1 = s_2$, and $p_1 = p_2$. In other words, k diagonals of an N-node 2D-HERT architecture are partitioned into two groups, referred to as *wavelength groups*, each containing $k/2$ consequent diagonals and $N/2$ distinct wavelength channels. Hence, it is easy to show that λ_k is assigned to node (d,s,p), where $k = 4 \times m \times d \ (mod\ k/2) + 4 \times s + p$. As an example, Fig. 13.4 depicts two wavelength groups (with different node colors) in a 144-node 2D-HERT ONoC with $k = 6$ and $m = 6$.

Although the proposed wavelength assignment approach devotes each wavelength channel to two different nodes in the network, the proposed routing algorithm avoids interference of different optical data streams, targeted to different routers with the same associated wavelengths, on the inter super-node links. However, due to multiple existing choices for intra super-node data routing, the proposed routing algorithm may lead to contention scenarios inside the local clusters. As an example, Fig. 13.5a depicts optical routing of two different data streams shown with red and green paths in the figure. Based on the proposed wavelength assignment approach, the same wavelength channel is associated to both destination nodes. Hence, two different paths are modulated on the same wavelength and may interfere on a local waveguide, shown with the dotted line, inside a cluster.

Generally, an optical flow initiated by a router inside a super-node may interfere with another optical flow passing through the same super-node in the case that

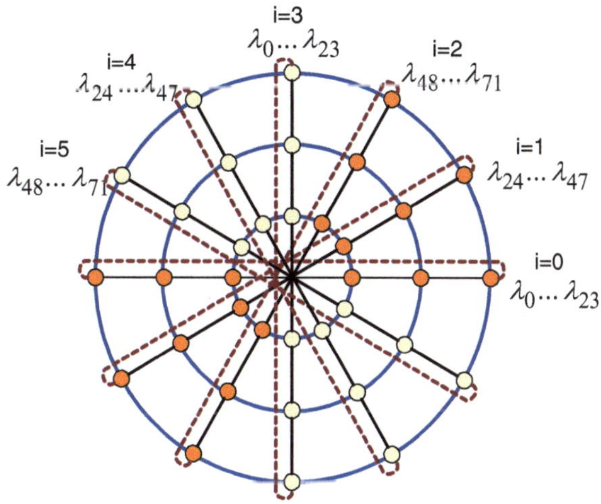

Fig. 13.4 Wavelength groups in 2D-HERT

the two data streams are targeted to different destination nodes, e.g. N_1 and N_2, associated with the same wavelength. Assuming only one data stream can be targeted to a specific node at each time, it is easy to show that optical interference is possible if N_1 and N_2 are located at different wavelength groups. Based on this fact, to prevent interference scenarios, the proposed ONoC architecture utilizes the path direction to distinguish optical data streams targeted to the different wavelength groups. Specifically, clockwise direction is taken within a super node for routing data streams to the destination nodes at the same wavelength group, while counter-clockwise direction is chosen for routing data packets targeted to the other wavelength group. Therefore, in the case of different wavelength groups for source and destination nodes, path direction changes from counter-clockwise to clockwise at the boundary of wavelength groups, where data streams enter the target group. As shown later, the proposed optical switch is developed to passively choose proper path direction for each optical data passing based on its modulated wavelength. While this property may lead to non-minimal paths inside the clusters, the proposed routing algorithm guarantees minimal routing through the long hops, interconnecting super nodes. Figure 13.5b illustrates the non-interfering routing paths inside source, intermediate, and target super-nodes for two pairs of source and destination nodes depicted in Fig. 13.5a.

13.2.4.2 WaROS Micro-architecture

According to the proposed topology, a 4 × 4 optical switch, represented by *(d,s,p)*, interconnects the corresponding local IP to the two local neighbor nodes and the adjacent super node. As shown in Fig. 13.1a, for odd values of p (either one or three), optical switch is connected to the neighbor cluster on the same diagonal through radial links, while for even values of p (either zero or two), it is connected

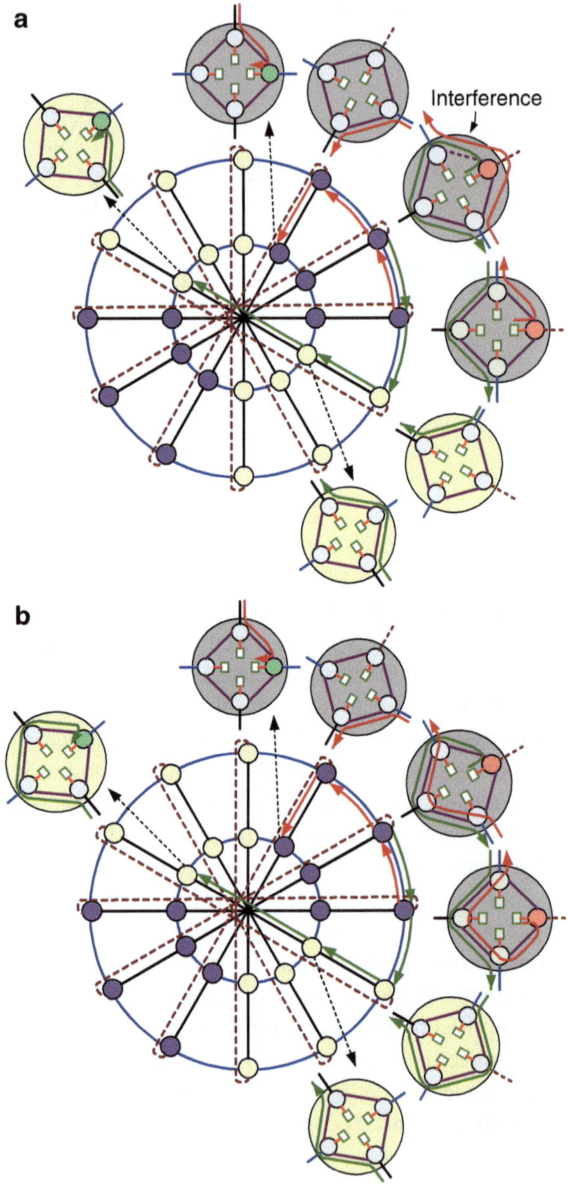

Fig. 13.5 (**a**) Intra-cluster interference, (**b**) resolving interference

to the neighbor cluster on the same circle through circular links. In this regard, we partition optical switches in 2D-HERT optical NoC into two groups; radial and circular switches. According to the CF routing scheme, routing role differs for different types of optical switches, which leads to different architectures for radial and circular switches.

13.2.4.2.1 Injection

One of the key advantages of the 2D-HERT ONoC is its capability of optical data multicasting. Due to the low-loss and ultra-high bandwidth of the optical waveguides, implementation of multicast communication with multiple unicast transmissions turns into an efficient scheme in optical networks. In this regard, each router can inject simultaneous packets to n destination nodes, where $1 \leq n \leq N-1$. However, since each wavelength channel is devoted to two different nodes in the network, to enable simultaneous data transmission to N nodes at the worst case, WaROS uses two injection ports, i.e., I_1 and I_2. In this manner, each injection port in an optical switch is responsible for transmitting optical messages to $N/2$ destination nodes. Assuming N_i as the source node, the first (I_1) and second (I_2) injection ports of the optical switch are utilized for optical data transmission to the same and the other wavelength groups, respectively. It is worth noting that the appropriate injection port for each optical packet is chosen by the processing core at the source node.

13.2.4.2.2 Ejection

We augment WaROS with Ejection Microring Resonators (EMRs) to extract optical data streams targeted to this IP from those passing through the switch. Hence, EMRs are activated when the modulated wavelength of the optical stream matches the dedicated wavelength of the router.

13.2.4.2.3 Routing

Utilizing OAD elements, WaROS architecture allows for multiplexing different optical data streams, targeted to different destinations, on the same optical waveguide and then demultiplexing them according to their wavelengths. To enable data demultiplexing, all possible destination nodes (d_s, s_s, p_s) for each source node (d_s, s_s, p_s) are partitioned into seven different groups, represented by $D_{1,\alpha}$, $D_{1,\beta}$, D_l, D_{d1}, D_{d2}, $D_{2,\alpha}$, and $D_{2,\beta}$ in Fig. 13.6. As follows, we discuss various destinations groups in more detail.

$D_{1,\alpha}$, $D_{1,\beta}$, $D_{2,\alpha}$, and $D_{2,\beta}$ represent destination nodes with $d_d \neq d_s$. As shown in Fig. 13.6, when source and destination nodes are located on the different diagonals, four different destination groups are defined according to the destination diagonal index. For example, $D_{1,\alpha}$ represents destination nodes with diagonal index smaller than d_s (i.e., $0 \leq d_d < d_s$). On the other hand, D_l, D_{d1}, and D_{d2} partition destination nodes when source and destination processing cores are located on the same diagonal (i.e., $d_d = d_s$). In this case, three different destination groups are defined according to the destination super-node index (s_d). For example, D_l represents destination nodes with the same super-node index as the source node (i.e., $s_s = s_d$).

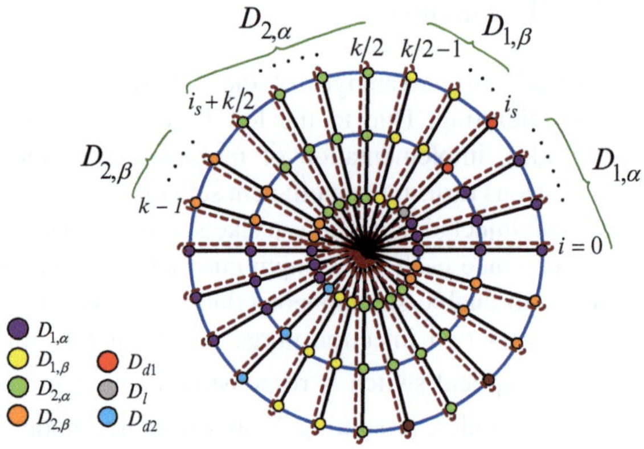

Fig. 13.6 Different groups of destination nodes

Mathematical definitions of different destination group with respect to the source node (d_s,s_s,p_s) follow for $0 \leq d_s < k/2$. Definitions in the case of $k/2 \leq d_s < k$ can be simply derived with a shift of diagonal indices by $k/2$.

$$D_{1,\alpha} = \left\{ R(d_d, s_d, p_d) \middle| 0 \leq d_d < d_s,\ 0 \leq s_d < m,\ 0 \leq p_d < 4 \right\} \tag{13.4a}$$

$$D_{1,\beta} = \left\{ R(d_d, s_d, p_d) \middle| d_s < d_d < {}^k/_2,\ 0 \leq s_d < m,\ 0 \leq p_d < 4 \right\} \tag{13.4b}$$

$$D_l = \left\{ R(d_d, s_d, p_d) \ \middle| \ d_d = d_s,\ s_s = s_d,\ p_d \neq p_s \right\} \tag{13.4c}$$

$$D_{d1} = \left\{ R(d_d, s_d, p_d) \ \middle| \ d_d = d_s,\ \left\{ \left(s_s - {}^m/_2 \right) \ mod\ m \right\} < s_d < s_s,\ 0 \leq p_d < 4 \right\} \tag{13.4d}$$

$$D_{d2} = \left\{ R(d_d, s_d, p_d) \ \middle| \ d_d = d_s,\ s_s < s_d \leq \left\{ \left(s_s + {}^m/_2 \right) \ mod\ m \right\},\ 0 \leq p_d < 4 \right\} \tag{13.4e}$$

$$D_{2,\alpha} = \left\{ R(d_d, s_d, p_d) \middle| {}^k/_2 \leq d_d \leq d_s + {}^k/_2,\ 0 \leq s_d < m,\ 0 \leq p_d < 4 \right\} \tag{13.4f}$$

$$D_{2,\beta} = \left\{ R(d_d, s_d, p_d) \middle| d_s + {}^k/_2 < d_d < k,\ 0 \leq s_d < m,\ 0 \leq p_d < 4 \right\} \tag{13.4g}$$

where *R(d,s,p)* represents the optical router connected to the *pth* processing core inside the *sth* super-node on the *dth* diagonal.

For appropriately demultiplexing optical data streams targeted to different destination groups, defined above, WaROS utilizes optically controlled microring

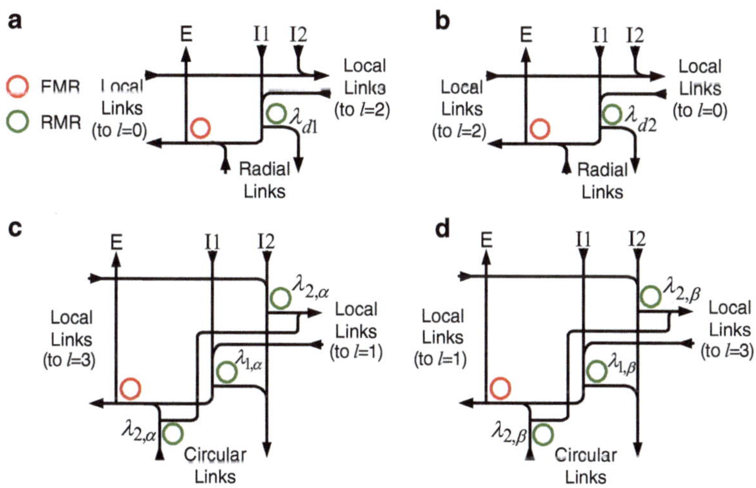

Fig. 13.7 WaROS architecture (**a**) p =1, (**b**) p =3, (**c**) p = 0, (**d**) p =2

resonators. These elements, referred to as Routing Microring Resonators (RMRs), also multiplex optical data streams transmitted from the source node with those passing through the optical switch.

The proposed architectures for WaROS are shown in Fig. 13.7 for radial and circular switches for different values of processor indices (*p*) depicted in Fig. 13.1. While EMR of a switch only drops its associated wavelength, RMRs are labeled with their respective resonant wavelength modes. Specifically, RMRs of switch *(d,s,p)* labeled with $\lambda_{1,\alpha}$, $\lambda_{1,\beta}$, λ_l, λ_{d1}, λ_{d2}, $\lambda_{2,\alpha}$, and $\lambda_{2,\beta}$ drop wavelength channels associated with the destination groups $D_{1,\alpha}$, $D_{1,\beta}$, D_l, D_{d1}, D_{d2}, $D_{2,\alpha}$, and $D_{2,\beta}$, respectively, defined by Eqs. 13.4a, 13.4b, 13.4c, 13.4d, 13.4e, 13.4f, 13.4g. As discussed by Small et al. [22], the diameter of microring resonators that switch multiple optical flows increases due to their small free spectral range (FSR). Hence, as an alternative implementation approach, we can replace each microring resonator with multiple small rings with larger FSR each tuned to a specific wavelength channel.

13.2.5 Advantages of the Proposed Data Plane

13.2.5.1 All-Optical Architecture

Similar to the architecture proposed by Briere et al. [14], 2D-HERT ONoC is built upon a passive optical architecture without requiring either electrical nor optical reservation of the optical resources at the intermediate switches (only reservation of the ejection channel at the destination node is required).

13.2.5.2 General Routing Algorithm

CF routing scheme organizes the processing cores into two groups, while optical data streams targeted to different groups are distinguished by the path direction which is either a clockwise or counter-clockwise. Hence, circular-first routing scheme can be adopted in any topology where clockwise and counter-clockwise directions can be defined, such as Torus, Spidergon, and Ring.

13.2.5.3 Data Multicasting

The proposed architecture can benefit from WDM technique for simultaneous transmission of unicast and multicast optical packets and assigns distinct wavelengths to each of unicast and multicast communication patterns in every node. For transmitting a multicast packet, multiple unicast optical packets are generated from the initial multicast electrical packet. In this regard, destination nodes of the electrical multicast packet determine the modulation wavelengths of the generated unicast optical packets.

13.2.5.4 Scalability and Compactness

The number of optical switching elements in our proposed architecture is proportional to the number of processing cores, while it is quadratically related to the number of IP blocks in [14].

13.3 Control Plane: All-Optical Request-Grant

Wavelength routing of optical packets in ONoCs [10, 11] can possibly avoid network contention. However, to resolve end-point contention, different control plane architectures [5, 10, 11, 16] have been proposed so far to prevent two or more source nodes from transmitting simultaneous data to the same destination. In the electrical control planes [5, 11, 16], electrical control packets are transmitted through an electrical sub-network to the destination node to reserve the required optical resources. While electrical control planes can be easily implemented [5, 11, 16], they undergo considerable power and delay overheads for electrical packet transmission.

As an optical solution, Koohi et al. [23] have proposed a request retransmission scheme to resolve contention at the destination node. For this purpose, in the case of busy destination, the source node waits for a time interval and then reattempts to transmit the optical data. Despite its simplicity, the proposed retransmission scenario increases the total number of control packets, which leads to considerable power overhead. Moreover, since the time interval between the consequent requests is determined at the source node, optical data transmission may not be initiated immediately after the previous transmitter releases the destination node. Therefore, data transmission latency increases and optical resource utilization decreases.

Corona architecture [10] proposes a distributed, all-optical, token-based arbitration scheme to resolve end-point contention. In this approach, a token for each node, which represents the right to modulate on each node's wavelength, is passed around all the nodes on a dedicated arbitration waveguide.

In this chapter, we propose an all-optical request-grant arbitration architecture, referred to as ORG, which reduces optical losses compared to the Corona's token-based architecture. As a key point, the proposed control network uses one wavelength per sender to manage end-point contention. Hence, wavelength routing of the data packet through Multiple-Write Single-Read (MWSR) waveguides in the data network along with the utilization of Single-Write Multiple-Read (SWMR) waveguides in the proposed control network offers an efficient all-optical approach for ONoCs.

13.3.1 Control Protocol

Before initiating optical data transmission at the source node, an optical *request* packet is routed to the target node to check the status of the corresponding ejection channels. In the case of free ejection channel, an optical *grant* packet is sent back to the source node to initiate its optical data transmission. While in the case of busy destination, a request flag is set at the destination node, and a *grant* packet is sent back to the source node only after the receiver channels are released. To guarantee fairness, multiple requests are granted in a round-robin fashion.

Once the existence of a free ejection channel at the destination node has been verified, an optical message can be transmitted through the optical waveguides and switches without buffering. The modulated wavelengths of the data streams optically control the switching state of the resonators on the path.

13.3.2 Topology

Processing cores in each super node are electrically connected to a control unit monitoring status of the ejection channels in the cluster. Moreover, optical control packets, i.e. *request* and *grant* packets, are transmitted and received, respectively, by the control unit in each super node. Considering 2D-HERT optical NoC, each control unit (CU) is uniquely determined with a pair of indices *(d,s)* $0 \leq d < k,\ 0 \leq s < m$ and is connected to the four optical switches indexed with *(d,s,p)* $0 \leq p < 4$. Hence, in an N-node 2D-HERT, the number of control units is limited to $N/4$.

As a simple version of the proposed control architecture, various control units are interconnected through an optical SWMR bus [9]. As a key point, the proposed control architecture associates one wavelength channel to each control unit. Therefore, for injecting control packets to the network, optical stream initiated by a specific control unit is modulated on its dedicated wavelength and is transmitted through the control bus. Hence, utilizing WDM technique, concurrent control packets can be targeted to the same control unit.

Considering SWMR architecture for the optical bus, data transmitted through the control bus is detected and received by all control units, where it is converted to electrical signals and examined to determine whether the control packet is targeted to the corresponding control unit. In this regard, taking advantage of a single waveguide for control packet transmission leads to a simple and low cost implementation. However, this simplicity comes at the price of increased transmission power for control packets, since packets should be properly received by all control units. As another alternative, we may dedicate an optical waveguide to each control unit, which means that control packets targeted to a specific control unit are transmitted through its associated waveguide. In this case, the number of photonic detectors and receiver circuits located on each control waveguide reduces to one, which leads to minimum optical power at the price of high design complexity and implementation cost.

From the above discussion, there is a trade-off between the number of control waveguides and the optical power dissipated for control packet transmission. As a power-efficient low-cost implementation approach, ORG architecture partitions $N/4$ control units to multiple disjoint sets, while control packets targeted to each set are transmitted through a dedicated optical control waveguide. In this regard, optical transmission power reduces compared to the single-waveguide approach, at the price of higher, yet reasonable, implementation cost. As a case study, we consider one optical control waveguide for all control units located on the same row in a planar layout. Each control unit has a splitter that transfers a fraction of the light from the corresponding control waveguide to a short dead-end waveguide that is populated with detectors. Therefore, in an N/4 -node control architecture arranged in a K × L planar layout, control units are interconnected by K different control waveguides. To reduce network diameter, defined as the maximum shortest path length between any pairs of nodes, each control waveguide is replaced with two unidirectional waveguides to route the optical control packets through the minimal path. As an example, Fig. 13.8 illustrates a 16-node control network interconnecting different CUs in a 64-node data network. In each CU, the passive microring resonators, shown in blue, are tuned to resonate at the associated wavelength channel of the corresponding CU. On the other hand, the active microring resonators, shown in green, modulate electrical control packets on the optical light stream, and only one of them is active at the time. As depicted in this figure, the proposed control architecture is built upon eight SWMR buses routed in a snake pattern among the nodes.

Finally, since the data transmission network does not share optical waveguides with the control buses, the wavelength channels devoted for data packet modulation

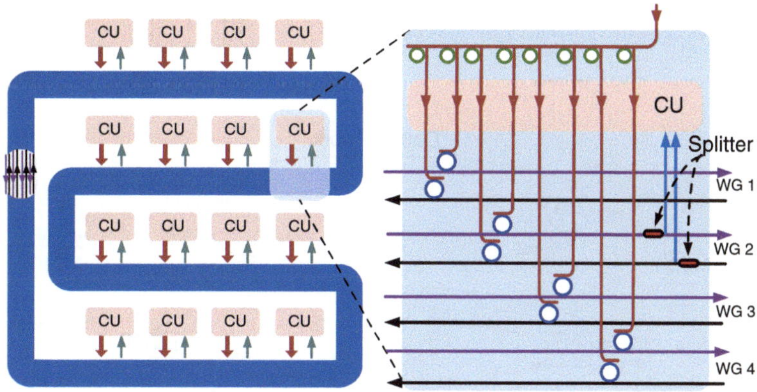

Fig. 13.8 Control plane architecture for a 64-node network

can be reused in the proposed control architecture. Hence, ORG architecture does not increase the minimum number of wavelength channels required for all-optical operation of the network.

13.3.3 Control Packet Format

Unlike the data transmission approach in the data plane, the modulation wavelength of an optical control packet in the control plane is set to the associated wavelength channel of the transmitter. Hence, the address of the target should be embedded in the control packet to enable control packet processing at the respective destination node. In the case of multicast communication, only one *request* packet is generated for all target CUs located on the same row, and the packet is transmitted through the corresponding optical waveguide. However, before the initiation of multicast data transmission, individual *grant* packets should be received from the destination control units. In other words, to ensure in-order data delivery, multicast data transmission is delayed until the existence of a free ejection channel is acknowledged by all respective multicast destinations.

Control packet is formatted as follows; a one-bit identifier, represented by T, specifies the type of the control packet (i.e. *request* or *grant*). The n field ($1 \leq n \leq N$ in an N-node network) contains the number of destinations, which equals one for unicast data transmission. The *Add* field is an array of destination addresses; its *ith* entry contains the address of the *ith* destination control unit (Add_{CU_i}) and the two-bit address of the target IP inside the corresponding local cluster (ADD_{IP_i}). For the *grant* packet ($T = 1$), a one-bit flag, represented by S, specifies the status of the respective ejection channel at the target node. Figure 13.9 depicts the control packet in the case of unicast data transmission.

Fig. 13.9 Control packet format

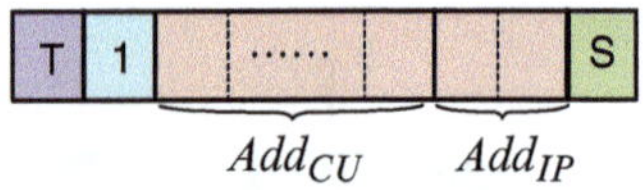

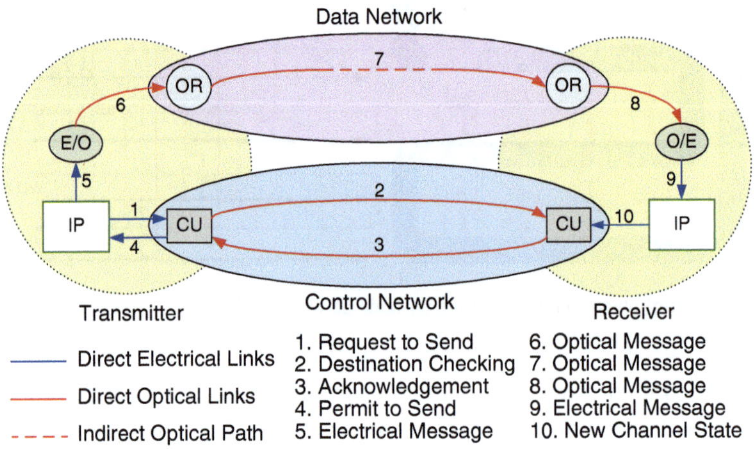

Fig. 13.10 Data transmission scenario

13.3.4 Advantages of the Proposed Control Plane

Figure 13.10 shows data transmission scenario in 2D-HERT ONoC. The key advantages of the ORG are summarized as follows.

13.3.4.1 All-Optical Architecture

All-optical control and data transmission phases, proposed in this chapter, considerably reduce power and delay metrics compared to previously proposed electrically-assisted ONoCs [5, 8, 9, 11–13] which suffer from high latency and power overheads for electrical reservation of optical resources.

Optical NoC proposed by Gu et al. [15] takes advantage of optical control packets for path reservation. However, control units in FONoC architecture [15] use electrical signals to configure the switching fabric according to the routing requirement of each packet. Therefore, the control interfaces utilize optical-electrical converter (OE) and electrical-optical converter (EO) to convert optical control packets into electrical signals and vice versa, respectively. However, these opto-electrical conversions take place at each intermediate router on the path of control packets, which leads to considerable power increase.

In addition to all-optical architecture for control packet transmission through TOC, the proposed control structure eliminates optical resource reservation at intermediate routers. These key advantages improve power and delay metrics of the 2D-HERT ONoC architecture compared to the previously proposed ONoCs.

13.3.4.2 Elimination of Path Tear-Down Phase

Reservation-assisted ONoCs [5, 8, 9, 11, 12, 15] utilize path tear-down packets to free up the path resources to be used by other optical messages. However, since our proposed architecture eliminates resource reservation at the intermediate switches, the path tear-down phase is eliminated from the control phase.

13.3.4.3 Scalability

The request-grant control architecture reduces the number of control units to $N/4$ in an N-node 2D-HERT, while it does not impact the minimum number of wavelengths required in the proposed architecture. Moreover, the width of the optical control bus is flexible and can be adjusted in the range of $[1, N/4]$. These architectural advantages improve scalability of the ONoC.

13.4 Simulation Environment

Behavioral simulation makes it possible to simultaneously take into account the electrical and optical phenomena in an optical on-chip interconnect. In this regard, to evaluate the advantages of the proposed ONoC architecture, we have developed a behavioral simulator and investigated the efficiency of the network along with its power consumption, data transmission delay, and energy dissipation for various traffic patterns. In this section, we describe general specifications of the implemented simulator.

13.4.1 Network Simulator

We have implemented a parameterized 2D-HERT optical NoC based on OMNeT++ simulation framework [24], in which routers are organized in a two-dimensional layout, as shown in Fig. 13.1b. As a case study, we analyze a 64-node topology which consists of four diagonals, each interconnecting four local clusters of four processing cores. Assuming a chip size of 1.8 cm $\times$ 1.8 cm [25], the proposed architecture is arranged in an 8×8 planarlayout and would be built in a future 22 nm CMOS process technology. According to the chip dimensions and 2D layout for the 64-node 2D-HERT from Fig. 13.1b, optical data links between adjacent super nodes have an approximate length of 1.5 mm. Assuming shorter intra-cluster links, between adjacent switches within a super node, compared to those interconnecting adjacent local networks, intra-super node optical links have an approximate length of 0.5 mm. We assume a propagation velocity of 15.4 ps/mm in a silicon waveguide for the optical signals [11]. Consequently, the optical delays for the inter-super node and intra-super node links are calculated as 23.1 ps and 7.7 ps, respectively. Moreover, since for each switch, the local link to the IP block is shorter than the three other links, we assume length of 0.2 mm for the local links, which leads to optical delay of 3 ps between each switch and its associated processing core.

According to the ORG architecture from Fig. 13.7, optical links between adjacent control units on a control waveguide have an approximate length of 2 mm, which leads to optical delay value of 30.8 ps between adjacent CUs. Finally, 5GHz clock speed [5] is assumed for data modulation, transmission, and demodulation in the network interface.

Finally, in this case study, we assumed an off-chip optical light source which feeds an on-chip filter bank of $N/2$ band-pass optical filters. In this case, outputs of the filter bank ($N/2$ wavelength channels) are routed to all optical transmitters across the chip to provide required wavelength channels.

13.4.2 Traffic Generation

Traffic pattern in an on-chip network is defined by the packet size distribution, packet injection process, and distribution of the packet destination. Packet sizes are uniformly distributed in the range of [128B, 1 KB]. We model inter-message gap as a Poisson random variable with the parameter μ. Based on this parameter, the load offered to the network (α) is defined as [5]:

$$\alpha = \frac{Message\ Duration}{Message\ Duration + \mu} \tag{13.5}$$

Finally, the traffic pattern in the network highly depends on the distribution of the packet destination. For investigating efficiency of the proposed ONoC, we perform various simulations under different synthetic workloads: uniform, hotspot, local, and first matrix transpose (FMT) [26]. In the case of local traffic, neighbor nodes for each processing core are assumed to be located within the same local cluster (super-node). Although 2D-HERT is capable of data multicasting, in this case study, we only discuss unicast data transmission.

13.4.3 Wavelength Allocation Scheme

Assuming per-port injection bandwidth of 640 Gbps, each optical data stream should be modulated on 64 wavelengths at the rate of 10 Gbps [27]. In a 64-node 2D-HERT architecture, maximum degree of multiplexing is computed as $N/2 = 32$. Hence, assuming 128 available wavelength channels, each optical data packet

Table 13.1 2D-HERT configuration parameters

	Parameter	Value
	Clock speed	5 GHz
	Modulation speed	10 Gb/s
Data plane	#Waveguides	16
	#Wavelengths	128
	#Per-packet wavelength/waveguide	4
Control plane	Packet length	8 bits
	Cycle/packet	1
	#Waveguides	4
	#Wavelengths	128
	#Per-packet wavelength/waveguide	8

should be modulated on 16 different waveguides, each transmitting four wavelength channels of the packet. Implementing WDM approach for data transmission through a waveguide considerably reduces data transmission latency, compared to the single-wavelength data transmission approach, at the cost of larger number of microring resonators.

On the other hand, regarding the proposed control architecture, the control packets targeted to $N/4 = 16$ different CUs can be multiplexed on the same control waveguide. Hence, assuming 128 distinct wavelength channels, each control packet should be modulated on four waveguides, each transmitting eight wavelength channels of the packet. Table 13.1 summarizes configuration parameters of the proposed ONoC architecture.

13.5 Simulation Results and Analysis

Since the impact of traffic pattern is analyzed in Sect. 13.6, in this section, we assume a uniform distribution for destinations.

13.5.1 Delay Analysis

Table 13.2 lists the main contributors to the optical latency for data and control packet transmission at the future 22 nm CMOS process technology and 5 GHz clock frequency [11]. Based on these parameters, Fig. 13.11a depicts the average optical latencies for control and data phases for varying values of α. Simulation results of average total latency for data transmission through the proposed optical architecture, computed as the sum of control and data transmission latencies are also shown in Fig. 13.11a. As depicted in this figure, for all traffic conditions, total latency for the

Table 13.2 Delay contributors

Parameter	Value	Unit
Modulator driver	9.5	ps
Modulator	14.3	ps
Photo-detector	0.2	ps
Amplifier	4	ps
Waveguide delay	15.4	ps/mm

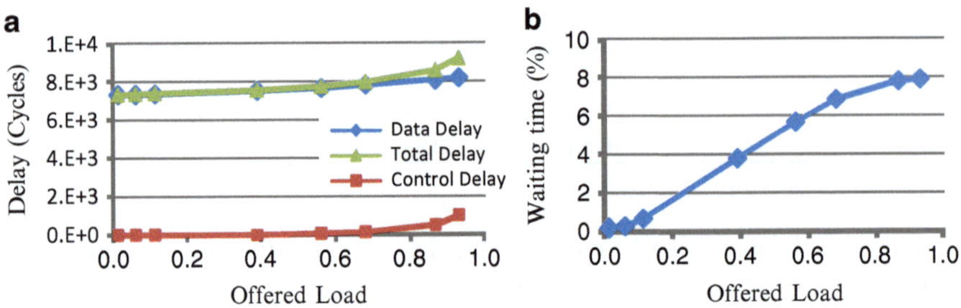

Fig. 13.11 (**a**) Delay values of 2D-HERT, (**b**) waiting interval as the percentage of total delay

data transmission is dominated by the optical data routing through the waveguides. Specifically, in the case of light traffic conditions, the latency of the control phase is negligible compared to the total latency and this ratio retains small values even for high traffic conditions.

Regarding the ORG architecture, in the case of busy destination, optical data transmission is postponed and the data packet is buffered at the source node. Data queuing latency at the transmitter side, referred to as *transmitter waiting interval*, impacts average latency of the optical control phase and hence, total performance of the 2D-HERT ONoC. Figure 13.11b shows average value of the transmitter waiting interval as the percentage of total latency for varying values of α. As shown in this figures, for low traffics, the waiting interval is inconsiderable, and this impact remains tolerable for high traffic loads. The latter property stems from the fact that the ORG architecture takes advantages of small optical control packets transmitted through the optical waveguides without imposing resource reservation at the intermediate nodes. Moreover, for receiving optical packets at the destination node, ejection channel of the destination optical switch is only occupied for a short period of time as a result of ultra-high bandwidth of optical waveguides.

13.5.2 Power Analysis

Total power consumption for transmitting an optical message through the 2D-HERT architecture is computed as the sum of optical and electrical losses for both the data and control packet transmissions. While optical power is dissipated in the waveguides and microring switches, electrical power is consumed in electro-optical converters:

$$P_{Total} = P_{Laser,Data} + P_{Laser,\ Cnt} + P_{E/O/E,\ Data} + P_{E/O/E,\ Cnt} \tag{13.6}$$

With an off-chip light source, the input laser power is constant and determined by the worst case optical loss in the network. The optical power loss for a single-wavelength data transmission through a source-destination path is computed as follows:

$$\begin{aligned} P_{Loss,Data} = {} & P_{MR,DP} \times N_{on} + P_{MR,TP} \times N_{off} + (P_B \times N_B) + P_W \\ & \times (L_l \times HopCount_l + L_s \times HopCount_s) + P_{IL,WC} \times N_{WC} + P_{CR} \end{aligned} \tag{13.7}$$

where N_{on} and N_{off} represent the number of resonators, passed by the optical message, in the ON and OFF states, respectively. $P_{MR,DP}$ and $P_{MR,TP}$ stand for the drop-port and through-port insertion loss for a passive microring, respectively. L_l and L_s are inter and intra super-node optical link lengths which are approximately 1.5 and 0.5 mm, respectively, $HopCount_l$ and $HopCount_s$ are the number of long hops and short hops (intra super-node hops), respectively, passed by the optical message, P_B is the waveguide bending loss, P_W is the waveguide propagation loss per unit distance, $P_{IL,WC}$ is the waveguide crossing insertion loss, N_B and N_{WC} stand for the number of waveguide bendings and crossings, and P_{CR} is the coupling loss from the waveguide to the optical receiver. For a bit error rate of 10^{-15}, the minimum power required by the receiver is −22.3dBm [28]. Hence, the power required for a multi-wavelength data transmission equals:

$$P_{Laser,Data} = WDM \times (mWFn(-22.3 + P_{Loss,Data})) / (P_{LE} \times P_{CW}) \tag{13.8}$$

where *mWFn()* represents a mathematical function performing unit conversion from dBm to mW, *WDM* is the number of wavelengths on which the corresponding optical data is modulated, P_{CW} is the coupling coefficient between the photonic source and the optical waveguide, and P_{LE} represents the laser efficiency for the conversion of electrons to photons.

In addition to optical losses in the waveguides and microring switches, electrical power is consumed in electro-optical converters, which is computed as follows:

$$P_{E/O/E,\ Data} = \mathrm{WDM} \times (\mathrm{P_T} + \mathrm{P_R}) \tag{13.9}$$

where P_T and P_R represent the power consumed by the transmitter and receiver circuits, respectively, at the 22 nm technology. Table 13.3 shows the values of these parameters in the network, which are extracted from the literature [11, 12, 29–31].

In addition to power dissipation in the data plane, control packet transmission through the optical control plane imposes additional optical losses. Based on the ORG architecture, a control packet transmitted through an optical control

Table 13.3 Power parameters

	Parameter	Value	Reference
Laser	P_{LE}	30%	[31]
	P_{CW}	90%	[31]
Distribution	P_{CR}	0.6 dB	[8]
	P_{W}	0.5 dB/cm	[30]
	$P_{MR,DP}$	0.5 dB	[9]
	$P_{MR,TP}$	0.01 dB	[9]
	P_{B}	0.005 dB	[9]
	$P_{IL,WC}$	0.12 dB	[9]
	$P_{Splitter}$	0.1 dB	[31]
E/O/E converters	P_{T}	5 mW	[29]
	P_{R}	0.3 mW	[29]

waveguide is detected and received by all control units located on the corresponding row. Therefore, the optical power loss for a single-wavelength control packet transmission through the control plane is computed as follows:

$$\begin{aligned} P_{Loss,Cnt} = {} & P_{MR,TP} \times N/4 + P_W \times L_{CW} + P_{CR} + (P_B \times N_B) \\ & + P_{IL,WC} \times N_{WC} + \sqrt{N/4} \times P_{Splitter} \end{aligned} \tag{13.10}$$

where L_{CW} represents the length of the optical control waveguide, $P_{Splitter}$ is the power loss per split, $\sqrt{N/4}$ presents the number of control units located on a control waveguide, and the remaining parameters are defined in the same way as those in Eq. 13.7. After consideration of receiver sensitivity, the minimum amount of power required for multi-wavelength control packet transmission is computed as:

$$P_{Laser,Cnt} = WDM_{CU} \times (mWFn(-22.3 + P_{Loss,Cnt})) / (P_{LE} \times P_{CW}) \tag{13.11}$$

where WDM_{CU} is the number of wavelengths on which the corresponding optical control packet is modulated. The electrical power consumed in electro-optical converters for transmitting a control packet is computed as:

$$P_{E/O/E,\ Cnt} = WDM_{CU} \times (P_T + P_R) \tag{13.12}$$

When the optical and electrical losses for both the data and control packet transmissions are taken into account, total power consumption for transmitting an optical message through the all-optical 2D-HERT architecture is computed form Eq. 13.6.

Finally, ring filters and modulators have to be thermally tuned to maintain their resonance wavelength modes under on-die temperature variations. Monolithic integration gives the most optimistic value for ring heating efficiency of all approaches (due to in-plane heaters and air-undercut), estimated at 1 μW per ring per K [18]. To calculate the required power for thermal tuning, we assume that, under typical conditions, the rings in the system would experience a temperature range of 20 K which leads to a thermal tuning power of 20 μW per ring.

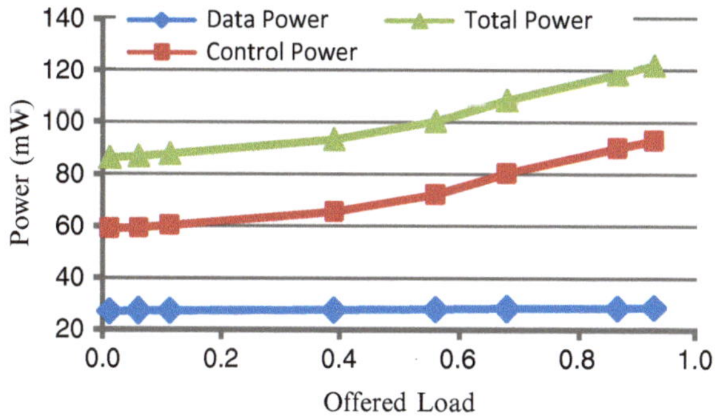

Fig. 13.12 Power components of 2D-HERT ONoC

13.5.2.1 Control Plane Power Analysis

In this section, we compare the worst-case on-chip power loss in the ORG and Corona's token-based arbitration architectures. For this purpose, the maximum insertion loss for the source-destination path is calculated in each of these architectures utilizing the power parameters from Table 13.3. This value, which is traffic-independent, equals 8 and 9.96 dB in the request-grant and token-based architectures, respectively. In other words, the proposed control architecture reduces on-chip optical loss by 37% over the corona's control plane. In the ORG architecture, the power reduction mainly stems from the reduced number of off-resonance microring resonators passed by each control packet. Specifically, in the token-based architecture, an optical token conveys the right to send data to a specific destination node. Hence, each token should be continuously routed through all optical nodes to enable probable data transmission form a source node to the corresponding destination node. However, in the request-grant control scheme, optical control packets are generated and transmitted only prior to the data transmission, and they are not required to pass through all optical nodes. For example, in the ORG architecture shown in Fig. 13.8, each control packet passes through four CUs located on the corresponding control waveguide.

13.5.2.2 Input Laser Power

Utilizing the extracted power models, the 2D-HERT simulator computes the amount of power consumption for each optical packet received at the destination node. The maximum value for different source-destination pairs is considered to be the required input laser power. Figure 13.12 shows the power consumption for data and control packet transmissions through the all-optical 2D-HERT for varying values of α. Simulation results of the total power dissipated for data transmission through the proposed optical architecture is also shown in Fig. 13.12. As shown in this figure, routing control packets optically through the network consumes approximately 68% and 76% of the total power dissipated for optical data transmission in the case of low

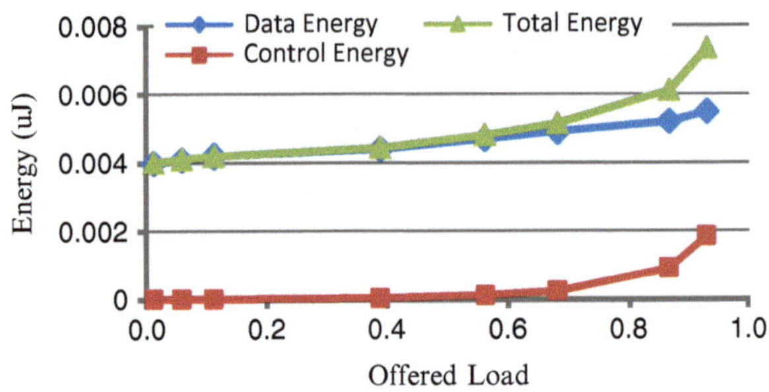

Fig. 13.13 Energy components of 2D-HERT ONoC

and high offered loads, respectively. Therefore, we conclude that power consumed in optical and electro-optical devices for routing control packets dominates total power consumption of the network, which emphasizes the importance of control design optimization in all-optical NoC architectures.

13.5.3 Energy Analysis

Figure 13.13 depicts average amounts of energy dissipated for control and data packet transmission through the 2D-HERT ONoC for varying values of α. As depicted in this figure, although power consumption of control architecture dominates that of the data network, total energy of the proposed optical network is dominated by the energy dissipation for data packet, rather than the control packet, transmission. The latter property stems from the small latency for control packet transmission which reduces the total energy dissipated in the proposed control architecture.

13.6 Synthetic Workload Evaluations

13.6.1 System-Level Analysis of 2D-HERT

This section analyzes system-level metrics of the 2D-HERT optical NoC compared to those of the ENoC, built upon the same topology, under various workloads. Based on the predictions made by Shacham et al. [5], we assumed 168-bit flits for data transmission between adjacent routers in the ENoC under 5-GHz clock frequency. Router processing delay is assumed to be 600 ps [5]. Moreover, we suppose propagation velocity of 131 ps/mm in an optimally repeated wire at 22 nm technology [11]. The estimated power consumption per unit length for electrical wires is about 1 mW/mm [32]. Shacham et al. [5] have reported energy

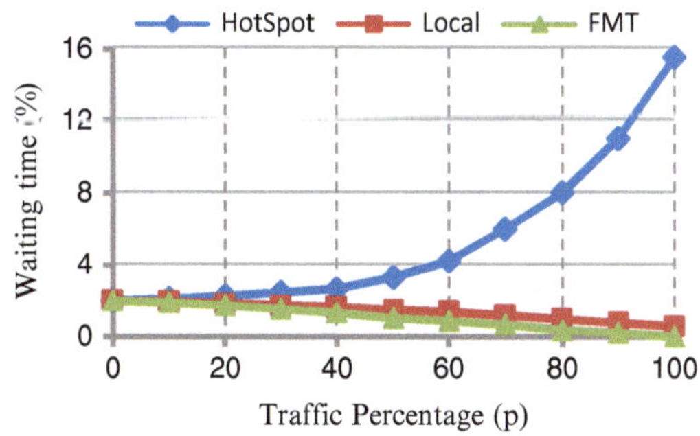

Fig. 13.14 Waiting time percentage (%)

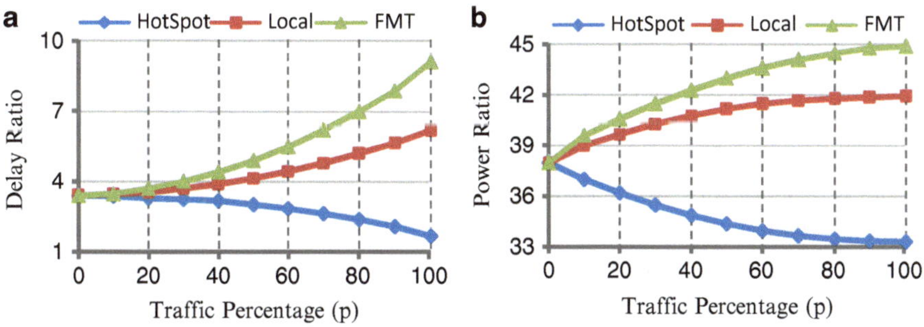

Fig. 13.15 Traffic analysis (**a**) $Delay_{ENoC}/Delay_{2D-HERT}$, (**b**) $Power_{ENoC}/Power_{2D-HERT}$

values in header processing operations. Assuming 5-GHz clock frequency, power consumption of buffer and crossbar units and amount of static power consumption are equal to 0.6 mW/bit, 1.8 mW/bit, and 1.75 mW/bit, respectively.

Figure 13.14 shows average duration of the transmitter waiting interval as the percentage of the total data transmission latency in 2D-HERT optical NoC for varying values of traffic percentage (*p*) of local, hotspot, and FMT traffic models assuming $\alpha = 0.4$. As shown in this figure, in the case of hotspot traffic pattern, waiting time interval significantly depends on the traffic percentage. This observation arises from the limited number of ejection wavelength channels at the hot node which considerably increases the probability of data transmission postponement at the transmitter under high traffic percentages. In the case of FMT traffic pattern, optical packets originated from a router are targeted to a specific destination node. Therefore, under high percentages of FMT traffic, most of the packets received by a router are sent from a single core. This latter property decreases the contention probability at the destination nodes, and therefore the duration of transmitter waiting interval. Similarly, by increasing local traffic percentage, optical packets received by each router are mostly sent by its neighbor nodes in the same local cluster. This locality reduces the probability of simultaneous transmission to a router.

Figure 13.15a, b depict ratios of average data transmission delay and power consumption of the ENoC, respectively, to the corresponding values of 2D-HERT optical NoC for varying values of traffic percentage (*p*) of local, hotspot, and FMT traffic patterns assuming $\alpha = 0.4$. As depicted in these figures, ONoC prominence over the traditional NoC is reduced for the large values of hotspot traffic percentage due to the busy ejection channel at the hot node. However, owing to the proposed all-optical data and control packet transmission approaches, 2D-HERT architecture, compared to the ENoC, still leads to about 1.7 and 33 times less delay and power, respectively, in the case of $p = 100\%$. As depicted in these figures, although the performance degradation is considerable in terms of total data transmission latency, total power consumption is slightly impacted by the traffic pattern. The latter property is a result of the proposed control mechanism in which the number of control packets transmitted prior to optical data transmission is constant and only data buffering at the source node, in the case of busy destination, increases power consumption.

Unlike hotspot traffic workload, increasing local and FMT traffic percentages improves power and delay values of the 2D-HERT optical NoC compared to those of the ENoC. This property stems from the limited number of source nodes capable of data transmission to each destination node under FMT and local traffic patterns. Specifically, in the case of FMT and local traffics, one and three nodes, respectively, are allowed to transmit optical data packet to each destination node. Hence, reduction of data buffering probability reduces arbitration delay and total power consumption. Moreover, under local workload, the power and delay metrics are further reduced due to the short intra-cluster links between neighbor nodes.

13.6.2 Design Comparison

In this section, we evaluate the performance of the proposed architecture against alternative optical NoCs using synthetic workloads. For this purpose, we compare 2D-HERT with Corona crossbar architecture [10], λ-router [14] (referred to as LR), Firefly [9], Phastlane [12], and a traditional electrical Torus (ETorus). We evaluate a CMP with 64 cores, which leads to the 64-node network for 2D-HERT, λ-router, Phastlane, and electrical torus. However, Corona and Firefly architectures implement a concentration factor of 4; i.e. four processor nodes share a single switch such that the topologies result in a 16-node network.

All topologies are evaluated under uniform, local, hotspot, and FMT traffic patterns with $p = 0.5$, $\alpha = 0.4$, and 128-bit data packets. We assumed separate optical control and data planes to reduce waveguide crossings insertion loss. Assuming per-port injection bandwidth of 640 Gbps in the ONoC architectures, we use design parameters for 2D-HERT, Corona, Firefly, and Phastlane from Table 13.1, [9, 10, 12], respectively.

Corona [10] is implemented as an all-optical 16×16 crossbar topology based on multiple-write single-read optical buses. It benefits from an all-optical arbitration ring-based network. A token for each node, which represents the right to modulate on each node's wavelength, is passed around all the nodes on a dedicated arbitration

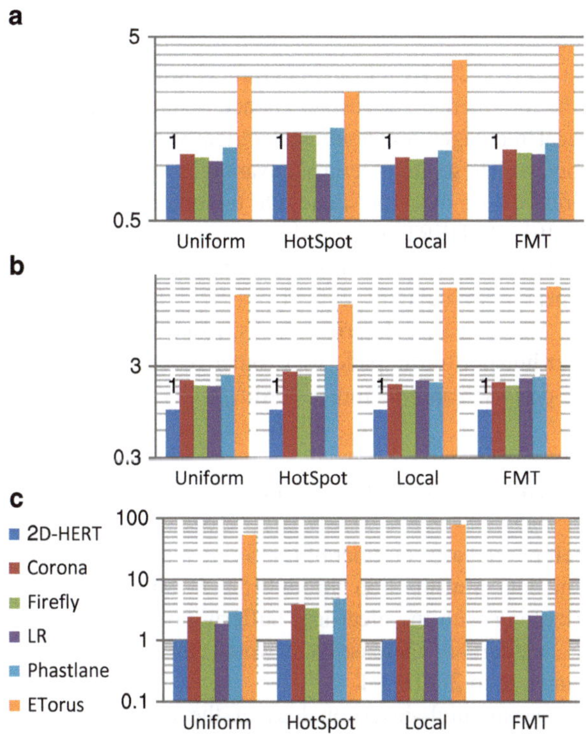

Fig. 13.16 Normalized system-level metrics (with respect to 2D-HERT) for various synthetic workloads (**a**) delay, (**b**) power, (**c**) energy

waveguide. If a node can grab a token, it absorbs the token, transmits the packet, and then releases the token to allow other nodes to obtain the token.

Firefly architecture [9] is implemented as multiple, smaller crossbars, and prevents global arbitration by using localized, electrical arbitration performed among smaller number of ports. Instead of using multi-write optical buses, the Firefly topology uses multi-read optical buses assisted with broadcast communication for path reservation and channel arbitration. In the Phastlane architecture [12], as a hybrid optical/electrical network, routers utilize electrical buffers for resolving optical contention at the intermediate routers; in the case of fully occupied buffers, optical packets are dropped and a high speed drop signal is sent back to the transmitter.

Figure 13.16a–c depict normalized data transmission delay, power consumption, and energy dissipation of the various architectures to those of 2D-HERT optical NoC, respectively. As depicted in these figures, in general, 2D-HERT is the most efficient optical architecture under various traffic patterns, while ETorus leads to the worse power and delay metrics compared to the optical architectures. As follows, we discuss the simulation results in details.

13.6.2.1 Reduced Data Transmission Delay

Unlike Phastlane [12], 2D-HERT architecture eliminates the need for resource reservation at the intermediate nodes, and hence, reduces the arbitration overhead. Moreover, optical messages are neither electrically buffered nor dropped at the intermediate routers. These architectural advantages improve total performance of the proposed architecture over Phastlane. On the other hand, compared to Corona [10] and Firefly [9], 2D-HERT reduces data transmission delay due to the smaller network diameter. In all, averaged across different traffic patterns, 2D-HERT ONoC reduces data transmission delay by 24%, 15%, 18%, 4%, and 70% over Phastlane, Firefly, Corona, λ-router, and electrical Torus, respectively. However, it is worth noting that under hotspot traffic pattern, λ-router reduces average packet delay by 11% against the proposed architecture according to its arbitration-free architecture.

13.6.2.2 Reduced Power and Energy Consumption

As discussed before, the large number of microring resonators results in high power consumption for data transmission through the λ-router. In Phastlane, electrical control network along with the electrical buffering of optical packets, in the case of contention, considerably increases the worst case power consumption in this architecture compared to 2D-HERT.

In Corona crossbar, high insertion loss of the optical crossbars, along with the global arbitration phase and the large number of off-resonance microrings passed by an optical packet, increases the worst case power consumption per-packet transmission in the network. With the assumption of a single-cycle router, Firefly reduces per-packet power consumption by approximately 5% over Corona crossbar topology due to its localized arbitration scheme. However, Firefly uses local electrical meshes, which increase the total power consumption due to the associated router and electrical link traversal power. This impact, along with the reservation broadcasting in Firefly architecture, results in higher worst case power consumption, compared to the all-optical 2D-HERT architecture.

In all, the 2D-HERT architecture achieves average per-packet power reduction of 58%, 47%, 52%, 45%, and 95% over Phastlane, Firefly, Corona, λ-router, and electrical Torus, respectively, for various traffic patterns. Finally, simulation results from Fig. 13.1c show that 2D-HERT results in 68%, 55%, 61%, 46%, and 98% lower per-packet energy consumption compared to Phastlane, Firefly, Corona, λ-router, and electrical Torus, respectively.

Evaluating the impact of traffic distribution on the efficiency of 2D-HERT architecture, Fig. 13.16a through Fig. 13.16c confirm that when locality is introduced in the traffic, the proposed architecture leads to much lower per-packet delay, power, and energy consumption which is due to the smaller waiting interval in the case of local or permutation traffic patterns. While in Phastlane, optical contention at any node on the source-destination path prevents optical data transmission, in 2D-HERT, only end-point contention postpones optical data transmission. Moreover, the proposed architecture considerably outperforms Corona for hotspot workload. The latter property stems from the global arbitration required for data transmission through the optical crossbars, such as Corona [10]. Token-based global arbitration enforces considerable power and delay overheads in the case of highly saturated network under hotspot traffic. Specifically, the proposed architecture achieves power reduction of 52%, 61%, 47%, and 50% over Corona crossbar under uniform, hotspot, local, and FMT workloads, respectively.

13.7 Silicon Photonics: Challenges and Future

In today's highly parallel computing systems integrated on a chip, performance is tightly dependent on the efficiency of on-chip data communications, while current bus-based on-chip communication architecture no longer satisfies communication bandwidth and latency requirements. To overcome these limitations, various researchers have proposed to implement on-chip global communication with packet switched NoCs. In high performance multiprocessor architectures, on-chip communications infrastructure emerges as a major power consumer. On the other hand, there are a broad range of limitations associated with metallic interconnects in future high performance computing systems, such as latency, throughput, bandwidth, power consumption, inter-line crosstalk, wave reflection phenomena, and EMI. On-chip optical interconnect has been proposed to overcome most of these problems. Silicon photonics [33] is an emerging area, and involves integrating optical and electronic circuits on silicon. It presents tremendous opportunities for both optical communication and data transfer in computers in the future. In this regard, during the past few years a lot of attention has been paid to develop silicon-based photonic architectures that can be integrated with multicore processors.

In this chapter, we proposed a new architecture for nanophotonic networks-on-chip, named 2D-HERT, which consists of optical data and control planes. The proposed architecture is built upon a novel topology, wavelength-routed optical switches (WaROS) which perform passive routing of optical data streams based on their wavelengths, and an optical request-grant arbitration architecture (ORG). 2D-HERT ONoC utilizes per-receiver wavelengths in the data network to prevent network contention and per-sender wavelengths in the control network to resolve end-point contention. Hence, it eliminates the need for electrical resource reservation and the corresponding latency and area overheads.

Although the process of introducing optics into computing has already started, there are many problems to be solved before developing a fully integrated optical architecture for core-to-core and core-to-memory interconnects. Probably the most important ones are manufacturing CMOS compatible light source, decreasing of heat dissipation, decreasing of overall power consumption and price [34]. Some of these challenges are briefly discussed as follows.

13.7.1 Laser

The main challenge for silicon photonics is growing the laser on a silicon chip, because silicon is a poor laser material [35]. In February 2011, however, researchers at the University of California announced that they had overcome this problem by taking advantage of the properties of nanostructures and by carefully controlling the growth process [36]. This is the first time that researchers have grown lasers from high-performance materials directly on silicon. The breakthrough can facilitate the process of growing nano-lasers directly on silicon, thus paving the way for integrating optical components on silicon chips.

13.7.2 Layout

Prior work has advocated both same-die and separate-die integration of optical components [17, 37, 38]. Monolithic integration has less interfacing overhead and higher yield than 3D stacking, but requires the optical components, which are relatively large, to consume active die area. 3D stacking, on the other hand, follows trends of future interconnects occupying separate layers and allows the CMOS and photonic processes to be independently optimized. The optical layer need not have any transistors, consisting only of patterning the waveguides and rings, diffusion to create the junctions for the modulators, germanium for the detectors, and a metalization layer to provide contacts between layers.

13.7.3 Area

Unlike electrical devices, optical devices are not readily scalable with technology node due to the light wavelength constraint. For instance, while transistor sizes are largely determined by the technology scale, ring resonator dimensions are largely determined by the coupling wavelength. Therefore, compact photonic switching elements are inevitable to build an optical on-chip network in future MPSoC designs. Ring sizes may shrink with improvements in wavelength, but improvements are limited; estimates indicate that rings start losing effectiveness at radii $<1.5\ \mu m$ [39]. Hence, it is unclear how hundreds of resonator-based photonic switches will be integrated on a single chip without considerable area overhead.

13.7.4 Thermal Sensitivity

A ring resonator's dimensions determine its coupling wavelength. To resonate with a very high extinction ratio, the ring must be an integral number of half-wavelengths in circumference. However, the circumference of the ring can change with temperature by ≈0.11 nm/K, causing a mismatch in resonance [40]. Moreover, the refractive index (n) of silicon changes due to changes in ambient temperature (ΔT), which can be modeled as $\Delta n \approx 1.84 \times 10^{-6} \times \Delta T$. As a result, microring resonators are very sensitive to temperature and drift spectrally approximately 0.09 nm/°C.

The resonance frequency of a microring can be changed by heating it, which causes a shift towards the red end of the spectrum, or by electrically injecting current (to shift the resonance towards the blue) [41]. This dynamic modification of resonance frequency is referred to as trimming. Trimming gives the architect the ability to precisely and adaptively control the wavelength at which a ring resonates [42]. However, employing either of heating or current injection techniques will increase the overall power consumption of the network.

Temperature fluctuations in the environment external to the chip will occur in the real world, and it is vital to understand how temperature fluctuations affect the amount of power necessary to support trimming. In current literature, researchers usually model the required microring trimming power as a fixed additional cost and typically estimate it by multiplying the estimated average trimming power per microring by the number of microrings [10, 18, 41, 43]. However, the practical issues underlying the system level implementation of trimming, especially in the context of networks that contain hundreds of thousands of rings, have not been studied.

References

1. L. Benini, G. De Micheli, Networks on chips: A new SoC paradigm. IEEE Comp. **35**(1), 70–80 (2002)
2. A. Shacham, K. Bergman, L.P. Carloni, Maximizing GFLOPS-per-Watt: High-bandwidth, low power photonic on-chip networks, in *$P = ac^2$ Conference* (New York, 2006), pp. 12–21
3. K.C. Saraswat, F. Mohammadi, Effect of scaling of interconnections on the time delay of VLSI circuits. IEEE Trans. Electron. Dev. **ED-29**, 645–650 (1982)
4. D. Miller, Rationale and challenges for optical interconnects to electronic chips. Proc. IEEE. **88**(6), 728–749 (2000)
5. A. Shacham, K. Bergman, L.P. Carloni, Photonic networks-on-chip for future generations of chip multi-processors. IEEE Trans. Comput. **57**, 1–15 (2008)
6. F. Adam, R. Gutiérrez-Castrejón, I. Tomkos, B. Hallock, R. Vodhanel, A. Coombe, W. Yuen, R. Moreland, B. Garrett, C. Duvall, C. Chang-Hasnain, Transmission performance of a 1.5-μm 2.5-Gb/s directly modulated tunable VCSEL. Photonics Technol. Lett. **15**(4), 599–601 (2003)
7. C. Guillemot, M. Renaud, P. Gambini, C. Janz, I. Andonovic, R. Bauknecht, B. Bostica, M. Burzio, F. Callegati, M. Casoni, D. Chiaroni, F. Clerot, S.L. Danielsen, F. Dorgeuille, A. Dupas, A. Franzen, P.B. Hansen, D.K. Hunter, A. Kloch, R. Krahenbuhl, B. Lavigne, A. Le Corre, C. Raffaelli, M. Schilling, J.-C. Simon, L. Zucchelli, Transparent optical packet switching: The European ACTS KEOPS project approach. J. Lightwave Technol. **16**(12), 2117–2134 (1998)
8. N. Kirman, M. Kirman, R.K. Dokania, J.F. Martinez, A.B. Apsel, M.A. Watkins, D.H. Albonesi, Leveraging optical technology in future bus-based chip multiprocessors, in *IEEE/ACM Annual International Symposium on Microarchitecture* (Florida, USA, 2006), pp. 492–503

9. Y. Pan, P. Kumar, J. Kim, G. Memik, Y. Zhang, A. Choudhary, Firefly: Illuminating future network-on-chip with nanophotonics, in *International Symposium on Computer Architecture (ISCA)* (Austin, Texas, USA, 2009), pp. 429–440
10. D. Vantrease, R. Schreiber, M. Monchiero, M. McLaren, N.P. Jouppi, M. Fiorentino, A. Davis, N.L. Binkert, R.G. Beausoleil, J.H. Ahn, Corona: System implications of emerging nanophotonic technology, in *International Symposium on Computer Architecture (ISCA)* (Beijing, China, 2008), pp. 153–164
11. S. Koohi, S. Hessabi, Contention-free on-chip routing of optical packets, in *International Symposium on Networks-on-Chip (NOCS)* (San Diego, CA, USA, 2009), pp. 134–143
12. M.J. Cianchetti, J.C. Kerekes, D.H. Albonesi, Phastlane: A rapid transit optical routing network, in *International Symposium on Computer Architecture (ISCA)* (Austin, Texas, USA, 2009), pp. 441–450
13. H. Gu, K.H. Mo, J. Xu, W. Zhang, A Low-power Low-cost optical router for optical networks-on-chip in multiprocessor systems-on-chip, in *IEEE Symposium on VLSI* (Kyoto, Japan, 2009), pp. 19–24
14. M. Briere, B. Girodias, Y. Bouchebaba, G. Nicolescu, F. Mieyeville, F. Gaffiot, I. O'Connor, System level assessment of an optical NoC in an MPSoC platform, in *Design, Automation and Test in Europe (DATE)* (Nice, France, 2007), pp. 1084–1089
15. H. Gu, J. Xu, W. Zhang, A low-power fat tree-based optical network-on-chip for multiprocessor system-on-chip, in *Design, Automation and Test in Europe (DATE)* (Nice, France, 2009), pp. 3–8
16. H. Gu, J. Xu, Z. Wang, A novel optical mesh network-on-chip for gigascale systems-on-chip, in *Asia Pacific Conference on Circuits and Systems (APCCAS)* (Macao, China, 2008), pp. 1728–1731
17. I. Stojmenovic, Honeycomb networks: Topological properties and communication algorithms. IEEE Trans. Parall. Distr. Syst. **8**(10), 1036–1042 (1997)
18. A. Joshi, C. Batten, Y.-J. Kwon, S. Beamer, I. Shamim, K. Asanovic, V. Stojanovi, Silicon-photonic clos networks for global on-chip communication, in *International Symposium on Networks-on-Chip (NOCS)* (San Diego, CA, USA, 2009), pp. 124–133
19. F. Xu, A.W. Poon, Multimode-interference waveguide crossing coupled microring-resonator-based switch nodes for photonic networks-on-chip, in *Lasers and Electro-Optics Conference and Quantum Electronics and Laser Science (CLEO/QELS) Conference* (San Jose, CA, 2008), pp. 1–2
20. N. Sherwood-Droz, H. Wang, L. Chen, B.G. Lee, A. Biberman, K. Bergman, M. Lipson, Optical 4×4 hitless silicon router for optical networks-on-chip (NoC). Opt. Express **16**(20), 15915–15922 (2008)
21. P. Koonath, T. Indukuri, B. Jalali, Add-drop filters utilizing vertically coupled microdisk resonators in silicon. J. Appl. Phys. Lett. **86**, 091102-1–091102-3 (2005)
22. B.A. Small, B.G. Lee, K. Bergman, Q. Xu, M. Lipson, Multiple-wavelength integrated photonic networks based on microring resonator devices. J. Opt. Netw. **6**(2), 112–120 (2007)
23. S. Koohi, S. Hessabi, All-optical wavelength-routed NoC based on a novel hierarchical topology, in *International Symposium on Networks-on-Chip (NOCS)* (Pittsburgh, Pennsylvania, USA, 2011), pp. 97–104
24. OMNeT++ discrete event simulation system, Available online at http://www.omnetpp.org/
25. ITRS, The international technology roadmap for semiconductors – 2009 edition, (2009), Available at http://public.itrs.net
26. S. Koohi, M.M. Aghatabar, S. Hessabi, Evaluation of traffic pattern effect on power consumption in mesh and torus-based network-on-chips, in *International Symposium on Integrated Circuits (ISIC)*, (Orchard Hot, Singapore, 2007)
27. S. Manipatruni, Q. Xu, M. Lipson, PINIP based high-speed high-extinction ratio micron-size silicon electrooptic modulator. Opt. Express **15**(20), 13035–13042 (2007)
28. I. O'Connor, F. Gaffiot, On-chip optical interconnect for low-power, in *Ultra-Low Power Electronics and Design*, ed. by E. Macii (Kluwer, Dordrecht, 2004)

29. G. Chen, H. Chen, M. Haurylau, N. Nelson, P.M. Fauchet, E.G. Friedman, D.H. Albonesi, Predictions of CMOS compatible on-chip optical interconnect. VLSI J. Integr. **40**(4), 434–446 (2007)
30. M. Lipson, Guiding, modulating, and emitting light on silicon-challenges and opportunities. J. Lightwave Technol. **23**(12), 4222 (2005)
31. D.M. Vantrease, Optical tokens in many-core processors, University of Wisconsin (2010)
32. M. Haurylau, C.Q. Chen, H. Chen, J.D. Zhang, N.A. Nelson, D.H. Albonesi, E.G. Friedman, P.M. Fauchet, On-chip optical interconnect roadmap: Challenges and critical directions. IEEE J. Sel. Top. Quant. Electron. **12**(6), 1699–1705 (2006)
33. K. Greene, A record-breaking optical chip, MIT Technol. Rev. Available online at http://www.technologyreview.com/Infotech/21005/?aâĂĽ=âĂĽf
34. L. Wosinski, W. Zhechao, Integrated silicon nanophotonics: A solution for computer interconnects, in *IEEE International Conference on Transparent Optical Networks (ICTON)* (Stockholm, Sweden, 2011), pp. 1–4
35. ITU-T Technology Watch Report, The optical world, June 2011, Available online at http://www.itu.int/ITU-T/techwatch
36. K. Bourzac, Laser-Quick Data Transfer, MIT Technol. Rev. Available online at http://www.technologyreview.com/computing/32324/
37. B. Black, M. Annavaram, N. Brekelbaum, J. DeVale, L. Jiang, G. Loh, D. McCaule, P. Morrow, D. Nelson, D. Pantuso, P. Reed, J. Rupley, S. Shankar, J. Shen, C. Webb, Die stacking (3D) microarchitecture, in *IEEE/ACM International Symposium on Micro-architecture* (Florida, USA, 2006), pp. 469–479
38. C. Batten, A. Joshi, J. Orcutt, A. Khilo, B. Moss, C. Holzwarth,M. Popovic, H. Li, H. Smith, J. Hoyt, F. Kartner, R. Ram, V. Stojanovic, K. Asanovic, Building manycore processor-to-dram networks with monolithic silicon photonics, in *IEEE/ACM International Symposium on Micro-architecture* (New York, USA, 2009), pp. 8–21
39. Q. Xu, D. Fattal, R. Beausoleil, Silicon microring resonators with 1.5 μm radius. Opt. Express.**16**(6), 4309–4315 (2008)
40. B. Guha, B. Kyotoku, M. Lipson, CMOS-compatible athermal silicon microring resonators. Opt. Express **18**, 3487–3493 (2010)
41. J. Ahn, M. Fiorentino, R.G. Beausoleil, N. Binkert, A. Davis, D. Fattal, N.P. Jouppi, M. McLaren, C.M. Santori, R.S. Schreiber, S.M. Spillane, D. Vantrease, Q. Xu, Devices and architectures for photonic chip-scale integration. Appl. Phys. Mater. Sci. Process. **95**(4), 989–997 (2009)
42. C. Nitta, M. Farrens, V. Akella, Addressing system-level trimming issues in on-chip nanophotonic networks, in *IEEE International Symposium on High Performance Computer Architecture (HPCA)* (San Antonio, Texas, USA, 2011), pp. 122–131
43. Y. Pan, J. Kim, G. Memik. Flexishare, Channel sharing for an energy-efficient nanophotonic crossbar, in *IEEE International Symposium on High Performance Computer Architecture (HPCA)* (Bangalore, India, 2010), pp. 1–12

Part VI
Industrial Case Study

Chapter 14
On Chip Network Routing for Tera-Scale Architectures

Aniruddha S. Vaidya, Mani Azimi, and Akhilesh Kumar

Abstract The emergence of Tera-scale architectures features the interconnection of tens to several hundred general purpose cores to each other and with other IP blocks. The high level requirements of the underlying interconnect infrastructure include low latency, high-throughput, scalable performance, flexible and adaptive routing, support for isolated partitions, fault-tolerance, and support for irregular or partially enabled configurations. This chapter presents the architecture and routing algorithms for supporting these requirements in the overall framework of mesh and torus-based point-to-point interconnect topologies. The requirements and desired attributes for tera-scale interconnects are outlined. This is followed by an overview of the interconnect architecture and micro-architecture framework. The descriptions of various routing algorithms supported are at the heart of the chapter, and include various minimal deterministic and adaptive routing algorithms for mesh and torus networks, a novel load-balanced routing algorithm called pole-routing, and performance-isolation routing in non-rectangular mesh partitions. The implementation aspects of these topics is covered through an overview of the environment for prototyping, debugging, performance evaluation and visualization in the context of specific interconnect configurations of interest. Overall, this chapter aims to illustrate a comprehensive approach in architecting (and micro-architecting) a scalable and flexible on-die interconnect and associated routing algorithms that are applicable to a wide range of applications in an industry setting.

This chapter includes material adapted from our earlier publications [1–3].

A.S. Vaidya (✉)
Nvidia Corporation, 2701 San Tomas Expy, Santa Clara, CA 95050, USA
e-mail: aniv@nvidia.com

M. Azimi • A. Kumar
Intel Corporation, 2200 Mission College Blvd, Santa Clara, CA 95052, USA
e-mail: mani.azimi@gmail.com; akhilesh.kumar@intel.com

M. Palesi and M. Daneshtalab (eds.), *Routing Algorithms in Networks-on-Chip*, 379
DOI 10.1007/978-1-4614-8274-1_14, © Springer Science+Business Media New York 2014

14.1 Introduction

Designing processors with many cores has been widely accepted in the industry as the primary approach for delivering ever increasing performance under hard power and area constraints. General purpose processors already have several tens of cores and can be expected to increases to a few hundred in this decade. For example, the Intel® Xeon Phi™ coprocessor code-named Knights Corner [6] which became available in late 2012 has 60 cores and can compute up to 1 teraFLOPS of double precision peak performance [18]. Such tera-scale processors provide a platform for use across a wide array of application domains, taking advantage of increasing device densities offered by Moore's law.

A typical implementation of such a processor includes tens of general-purpose cores today (and possibly a hundred plus cores in the near future), multiple levels of cache memory hierarchy to mitigate memory latency and bandwidth bottlenecks, and interfaces to off-chip memory and I/O devices. Most many-core chips use various building blocks connected through an on-chip interconnect to realize specific products. Scaling this architecture to future process generations requires a flexible, capable and optimized on-chip interconnect. In this chapter, we detail the technology development, research and prototyping environments for high-performance on-chip interconnects with application to scalable server and high-performance compute processor architectures.

On-chip interconnects can take advantage of abundant wires, smaller clock synchronization overhead, lower error rate, and lower power dissipation in links compared to off-chip networks, where chip pin counts and power dissipation in the transceivers and links dominate design considerations. However, efficient mapping to a planar substrate places restrictions on suitable topology choices [10]. Furthermore, a need to support performance isolation, aggressive power-performance management using dynamic voltage-frequency scaling (DVFS) techniques, and handling within-die process variation effectively places additional requirements on topology selection at design time [11].

In addition to physical design considerations that affect topology choices as above, there are workload considerations. A general-purpose many-core processor must efficiently run diverse workloads spanning legacy and emerging applications from domains as varied as scientific computing, transaction processing, visual computing, and cloud computing. Such workloads may exhibit communication characteristics with transitory hot-spots, jitters, and congestion among various functional blocks. Therefore, it is an imperative design requirement for on-chip interconnects to respond to these conditions gracefully.

There are several plausible approaches for designing the on-chip interconnect for a many-core tera-scale processor chip. Our work has focused on a flexible interconnect architecture based on two dimensional mesh and torus topologies. This architecture is further augmented by a rich set of routing algorithms for supporting various features. These architecture details and our earlier motivations have been documented in [1–3]. In this chapter, we primarily focus on the routing algorithms.

We present algorithms for deadlock-free routing in regular and irregular topologies including both deterministic and adaptive routing techniques; the novel pole-routing family of algorithms for load-balancing support in both fault-free and faulty networks; and finally algorithms for routing in partitions with performance-isolation. In an industry setting, implementability, validation and workload-driven performance evaluations of these routing algorithms and interconnect choices are key requirements; and this requires detailed prototyping and simulation environments and ability to visualize various scenarios. We have developed various innovative techniques and tools that we also describe and provide insight into.

This chapter is organized as follows. In Sect. 14.2, we set the context for our work including reviewing the requirements for tera-scale interconnects with expected usage scenarios and desired attributes for the interconnect as well as relevant interconnect topologies. In Sect. 14.3, we examine the interconnect architecture. We discuss the routing algorithms in Sects. 14.4–14.6. Section 14.7 discusses our prototyping and visualization environment, followed by the concluding remarks in Sect. 14.8.

14.2 Requirements for Tera-Scale Interconnects

Tera-scale many-core processors present opportunities that enable high levels of parallelism to address the demands of existing and emerging workloads. These workloads may create hot-spots, jitters, and congestion during communication among various blocks such as cores, on-die storage elements, memory controllers, and IO hubs. In this section, we first discuss a few example usage scenarios and their desired attributes, and then show how these would favor the use of a general-purpose interconnect.

14.2.1 Usage Scenarios

The on-chip interconnect architecture discussed here is targeted towards a general-purpose design that can meet the needs of different sub-systems in a highly integrated and highly parallel architecture. The challenge of the underlying architecture is to be competitive with application-specific designs. A few example scenarios that are potential targets for tera-scale architecture are presented below.

14.2.1.1 Resource-Efficient Data Center

The aggregate compute capacity of a tera-scale processor can be partitioned and virtualized to provide cloud computing services to multiple applications sharing its resources. This trend is also evident in the recent microserver products where large numbers of lower performance processors are densly packaged together to scale out workloads that do not need shared address spaces. We expect such approaches to be adapted to take advantage of energy and cost efficiencies provided by integration of large number of cores on a single die. An environment to support this should allow dynamic allocation and management of compute, memory, and IO resources with as much isolation between different partitions as possible. A large set of allocation and de-allocation of resources can create fragmentation that may not provide a clean and regular boundary between resources allocated for different purposes. The interconnection network bridging these resources should be flexible enough to allow such partitioning with high quality of service (QoS) guarantees and without causing undue interference between different partitions.

14.2.1.2 Scientific Computing

Scientific computing applications span the range from molecular dynamics to cosmic simulations, fluid dynamics, signal and image processing, data visualization, and other data and compute-intensive applications. These demanding applications are typically executed over large clusters of systems. The compute density per chip and energy per operation provided by tera-scale architecture can greatly enhance the capability of systems targeting this application domain. A typical design for this domain would be dominated by a large number of processing elements on chip with either hardware or software-managed memory structures and provision for very high off-chip memory and system bandwidth. An interconnect designed for such applications should provide high bandwidth across all traffic patterns, even patterns that may be adversarial to a given topology.

14.2.1.3 Visual Computing

Visual computing applications are similar to scientific computing applications in some aspects. However, these applications may additionally have real-time requirements including bounded-time and/or bandwidth guarantees. Also, different portions of an application such as AI, physical simulation and rendering have distinct requirements. This heterogeneity, reflected architecturally, would create topological irregularities and traffic hot-spots that must be handled to provide predictable performance for visual workloads.

14.2.1.4 Irregular Configurations

Cost and yield constraints for products with large numbers of cores may create a requirement for masking manufacturing defects or in-field failures of on-die components that in turn may result in configurations that deviate from the ideal topology of the on-chip interconnect. Another usage scenario that can create configurations that are less than ideal is an aggressive power-management strategy where certain segments of a chip are powered-down during periods of low utilization. Such scenarios can be enabled only when the interconnect is capable of handling irregular topologies with graceful performance degradation.

14.2.2 Desired Attributes

The on-chip interconnect for tera-scale architecture has to be designed keeping in mind the usage scenarios outlined above. Apart from the typical low-latency and high-bandwidth design goals, a flexible topology and predictable performance under a variety of loads are also essential.

Some of the main drivers for tera-scale architecture are the chip-level power and thermal constraints that have necessitated the shift from optimizing for scalar performance to optimizing for parallel performance. This shift has resulted in the integration of a higher number of relatively simpler cores on a chip instead of driving towards higher frequency and higher micro-architectural complexity. This trend also has implications for the on-chip interconnect. Due to the weak correlation of frequency with process technology generation, logic delay is becoming a diminishing component of the critical path, and wire delay is becoming the dominant component in the interconnect design. This implies that topologies that optimize for the reduction of wire delay will become preferred topologies for tera-scale architectures. Given the two-dimensional orientation of a chip, two-dimensional network topologies are obvious choices.

Another desirable attribute of an on-chip interconnect is the ability to scale-up or scale-down with the addition or reduction of processor cores and other blocks in a cost-performance optimal and simple manner, enabling the design to span several product segments. Based on the usage scenarios described at the beginning of this chapter, a latency optimized interconnect is critical for minimizing memory latency and ensuring good performance in a cache-coherent chip multi-processor (CMP). In addition, support is required for partitioning and isolation, as well as fault-tolerance, based on the envisaged usage scenarios. Due to these considerations, two-dimensional mesh and torus, and their variant topologies are good contenders for tera-scale designs. The design presented in this chapter assumes the 2D mesh and torus as primary topologies of interest and with support for some variations of the primary topologies to enable implementation flexibility.

Although on-chip wiring channels are abundant, shrinking core and memory structure dimensions and increasing numbers of agents on a die will put pressure on global wiring channels on the chip. This implies that wiring efficiency, i.e., the ability to effectively utilize a given number of global wires, will be an important characteristic of on-chip interconnect.

In order to support good overall throughput and latency characteristics with a modest buffering cost, our design assumes a buffered interconnect, based on wormhole switching [9] and virtual channel flow control [8]. It supports multiple

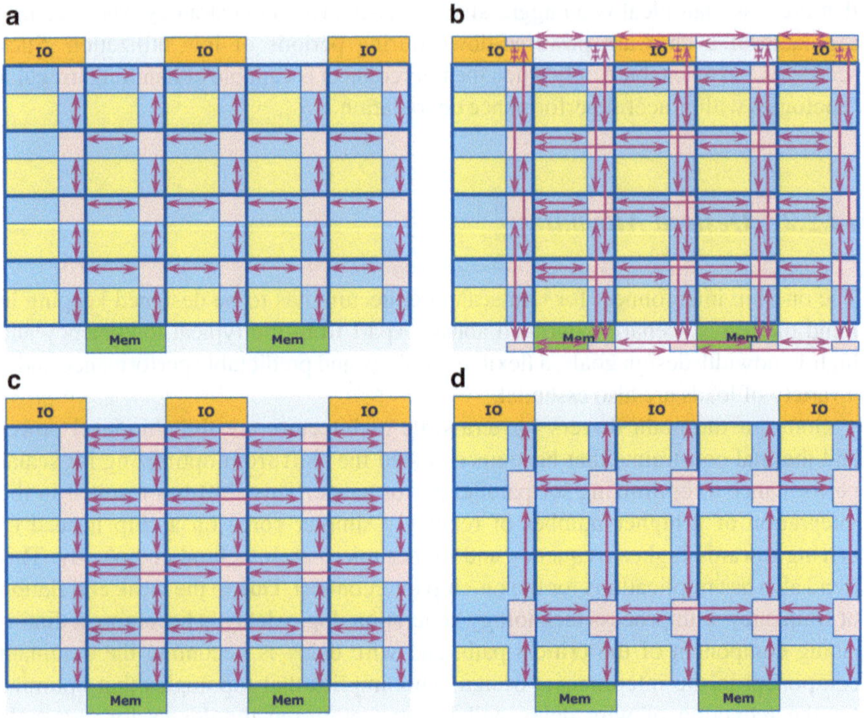

Fig. 14.1 Examples of 2D mesh and torus topology variants supported. (**a**) 2D mesh. (**b**) 2D torus. (**c**) Mesh-torus. (**d**) Concentrated mesh-torus

virtual channels to allow different types of traffic to share the physical wires. Some of the relevant high-level architectural features of the interconnect are discussed in Sect. 14.3. A more detailed discussion of these aspects is available in [2, 3].

14.2.3 Topological Aspects

Figure 14.1 depicts four on-chip interconnect topology options optimized for a CMP with the tiled modular design paradigm. All options are variants of 2-dimensional (2D) mesh or torus topologies. In option (a), each processor tile is connected to a 5-port router that is connected to the neighboring routers in both X and Y dimensions. IO agents and memory interfaces can be connected to the local port of the router or directly to an unused router port of a neighboring tile on the periphery of the chip. Option (b) depicts a 2D folded torus with links connecting routers in alternate rows and/or columns to balance wire delays. Compared to option (a), average hop count per packet is reduced at the expense of longer wires and the number of wires in the wiring channels is doubled for the same link width. In this particular example, more number of routers and links are needed to connect the peripheral devices. Option (c) shows a hybrid 2D mesh-torus topology with wraparound links in X dimension only by exploiting the fact that all peripherals are now located along the Y dimension. Compared to option (a), the same number of routers are used but the number of wires is doubled in the X dimension. Option (d) shows yet another variation of hybrid mesh-torus with two cores sharing one router, resulting in a concentrated topology that requires fewer routers for interconnecting the same number of cores. To enable this topology, either an extra port is required in every router to accommodate the additional core in a tile, or the two cores in a tile share a single local port through a mux and de-mux at the network interface.

The network topologies shown in Fig. 14.1 and many others can utilize the same basic router design. The tradeoff between different network topologies depends on the specific set of design goals and constraints of a given product. In other words substantial topology flexibility can be achieved through minor design modifications. In the next subsection, we discuss the micro-architecture and pipeline of one such router in detail.

14.3 Terascale Interconnect Architecture

14.3.1 Router Micro-architecture and Pipeline

The principal component of the 2D interconnect fabric is a pipelined, low-latency router with programmable routing algorithms that include support for performance isolation, fault tolerance, and dynamic adaptive routing based on network conditions.

Our router uses a non-blocking 5×5 full crossbar using wormhole switching and flit-level credit-based flow control. The functionality in our adaptive router includes most of the standard functionality that one can expect to see in a wormhole switched router [10].

The router has a three-stage pipeline with local arbitration (LA), global arbitration (GA), and switch traversal (ST) stages. The LA stage at each input port fairly selects one candidate packet request and presents it to the global arbitration (also known as switch arbitration) that occurs in the following clock. For each available output port, the GA stage grants that output port to one of potentially several input port requests selected in the previous cycle at the LA stage of input ports. Following the arbitration, the GA stage determines the crossbar configuration for switch traversal occurring in the following clock. The ST stage orchestrates the movement of the flits from the granted input ports to the corresponding output ports. In addition to the three pipe stages in the router, the link traversal (LT) stage drives a flit at the output port of an upstream router to the input port of a downstream router.

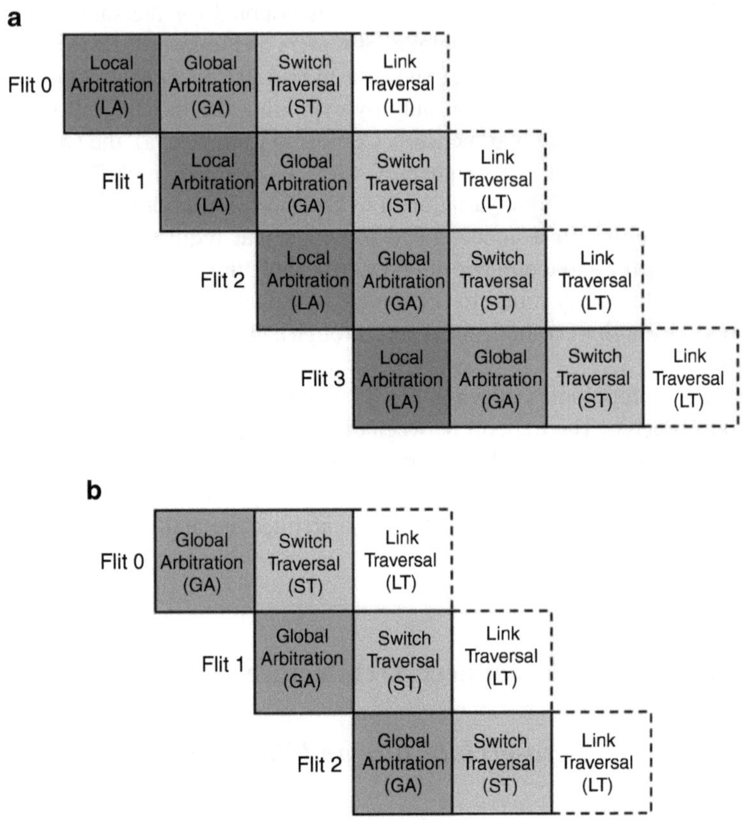

Fig. 14.2 Overview of the router pipeline. (**a**) Router pipeline under heavy traffic load. (**b**) Router pipeline under light traffic load

It should be noted that the LT stage is the boundary between a router and a link and is not considered as a stage in the router pipeline per se. The bulk of the LA stage operations (except some book keeping) is skipped entirely when a flit arrives at a previously idle input port with no queued up flits, thereby reducing the router latency under light traffic load to just two cycles. In such a case, the flit (or packet) proceeds directly to GA stage as it is the only candidate from that input port. The pipeline is reconfigured automatically according to the traffic conditions. Figure 14.2 captures the router pipeline stages under heavy and light traffic conditions.

Figure 14.3 shows the key functions performed in each stage. Of most relevance to us here is the route pre-compute functionality that is implemented in the GA or switch arbitration stage. We will discuss route computation in a later part of this subsection.

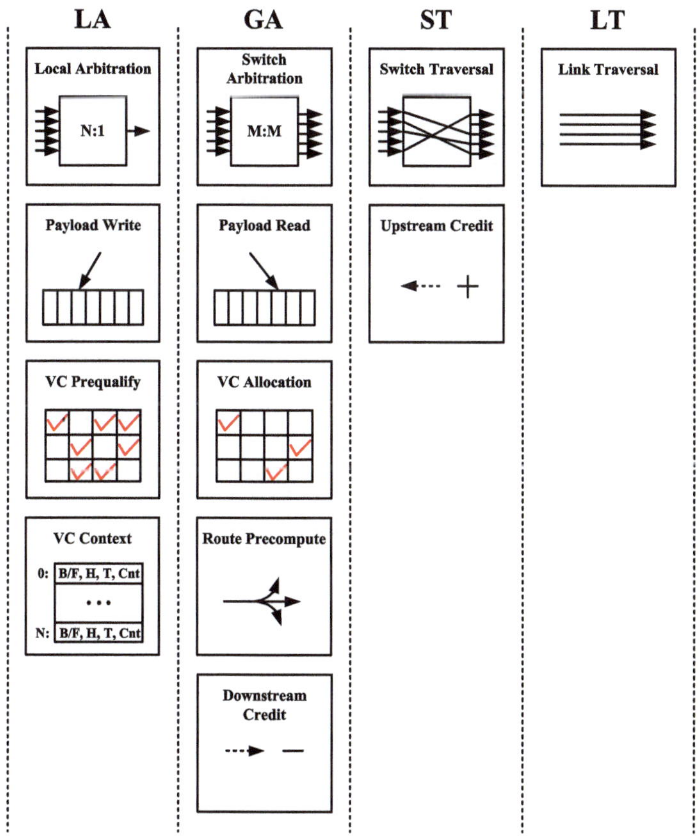

Fig. 14.3 Key functional blocks in the router pipeline

14.3.1.1 Virtual Channels and Buffer Management

Our router architecture relies on virtual channel flow control [8] both to improve performance and to enable support for deadlock-free routing with various flavors of deterministic, fault-tolerant and adaptive routing. The set of virtual channels (VCs) is flexibly partitioned into two logical sets: routing VCs and performance VCs. VCs are also logically grouped into virtual networks (VNs). VCs belonging to the same VN are used for message-class (MC) separation required by protocol-level MCs, but they use the same routing discipline. Routing VCs are also used for satisfying deadlock-freedom requirements of particular routing algorithms employed. Each routing VC is associated with one and only one MC with at least one reserved credit. Performance VCs belong to a common shared pool of VCs. They can be used by any MC at a given time both for adaptive or deterministic routing schemes. A VC is

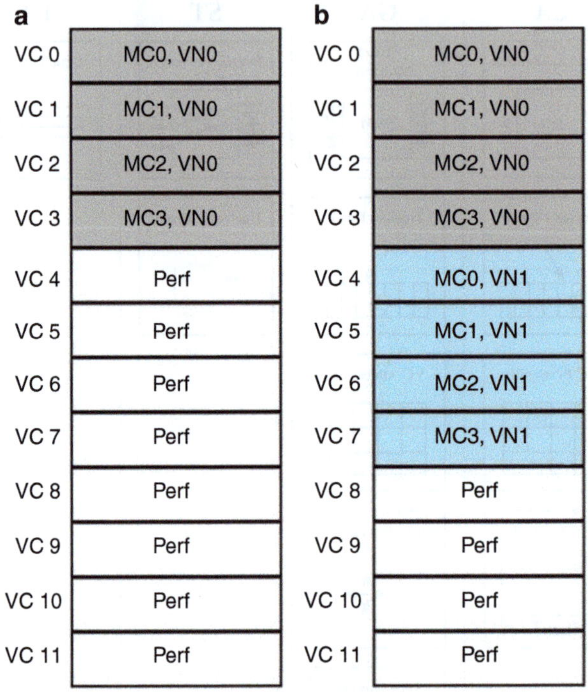

Fig. 14.4 Example mapping of virtual networks to virtual channels for 2D mesh and torus topologies. (**a**) 2D mesh. (**b**) 2D torus

used by only one message at any given time to manage design complexity and to ensure deadlock freedom for support of fully adaptive routing based on Duato's theory [13].

The number of VCs supported in a design is a function of design-time goals, area and power constraints. Figure 14.4 depicts examples of mapping of the supported VCs into routing VCs belonging to specific message-classes and VNs required for supporting minimal deadlock free XY routing algorithms for 2D mesh and torus topologies, as well as the pool of performance VCs. The example configurations assume a total of 12 VCs and 4 MCs; the mesh requires a single VN (VN0) for deadlock freedom, whereas the torus requires 2VNs (VN0, VN1).

A single shared buffer at each input port [19] is used to support flexibility and optimal usage of packet buffering resources with respect to performance, power, and area. The buffer is shared by all VCs, either routing VCs or performance VCs, at a port. Buffer slots are dynamically assigned to active VCs and linked lists are used to track flits belonging to a given packet associated with a VC. A free buffer list tracks buffer slots available for allocation to incoming flits.

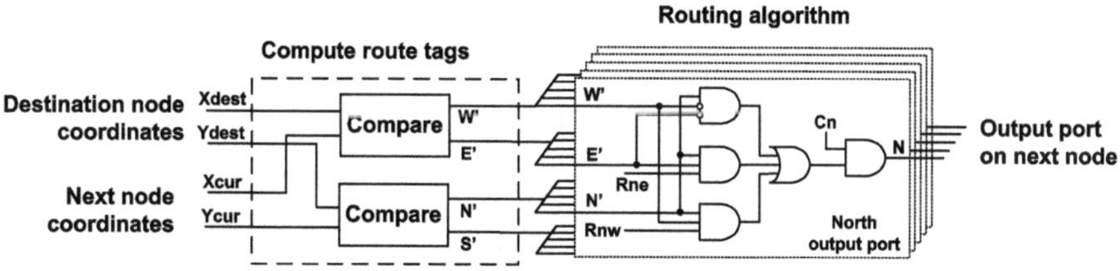

Fig. 14.5 Route pre-compute (Cn is north port connection bit; Rne and Rnw are turn-restriction bits for northeast and northwest turn, respectively)

14.3.1.2 Flow Control

The router uses credit-based flow control to manage the downstream buffering resources optimally. It tracks available input buffer space in the downstream router at the output side of the crossbar, through handshaking between two parts: upstream credit management and downstream credit management.

14.3.1.3 Route Computation

Routing determines which path a packet takes to reach its destination. We use a distributed routing scheme to select an output port at a given router that a packet must take to move towards its destination. For minimal adaptive routing, up to two distinct directions may be permitted based on the region a destination node falls into. The routing decision (i.e., output ports and VN choices permitted) at each router is based on the current input port and VN a packet belongs to, as well as on the destination address of the packet. Two different options are supported in our design: a compressed table-based distributed routing (TBDR) [14, 20] and logic based distributed routing (LBDR) [15] in order to enable a wide set of routing algorithms to be efficiently implemented. Compared to the LDBR scheme, the TBDR scheme uses a 9-entry table per router providing more routing flexibility with higher storage overhead.

The LBDR scheme uses a connection bit per output port and two turn-restriction bits. Different routing algorithms can be supported by setting appropriate turn restrictions. Our router architecture uses route pre-computation [10, 16] for the route decision of neighboring routers, thereby removing route computation from the critical path of the router pipeline. As shown in Fig. 14.5, it can be divided into two steps: (1) compute route tags based on packet destination, identifying the target quadrant and (2) determine the output port based on the selected routing algorithm. In an adaptive routing scheme, this can imply that the packet may have a choice of more than one output port towards its destination. We support up to two output port choices.

14.4 Flexible Deterministic and Adaptive Routing Support

14.4.1 Routing Algorithm Considerations

In this section we describe the support for various routing algorithms to enable a flexible, configurable, and adaptive interconnect, and we discuss the design implications.

14.4.1.1 Support for Multiple Routing Algorithms

The router architecture supports distributed routing wherein the subsets of the routing decisions are made at each router along the path taken by a given packet.

In two-dimensional networks like mesh and torus, given a source node, the set of shortest paths to the destination fall into one of four quadrants. With the LBDR framework [15], any turn model based routing algorithm [17] such as XY, west-first, odd-even [5], etc., can be implemented by setting the turn-restriction bits appropriately. For minimal adaptive routing, up to two distinct directions may be permitted based on the quadrant a destination node falls into. The routing decision (i.e., output ports and VN choices permitted) at each router is based on the current input port and VN a packet belongs to, as well as on the desired destination. For each VN, we support the flexible algorithms with very economical storage of only a few bits per port or with an alternative small 9-entry table [2].

Minimal path deterministic routing in mesh and torus and partially and fully adaptive minimal path routing algorithms, such as those based on the turn model [17], are supported. Our adaptive router architecture uses Duato's theory [13] to reduce the VC resource requirements while providing full adaptivity. Table 14.1 shows a comparison of the minimum number of VCs required to implement deadlock-free fully adaptive routing using the turn model versus one based on Duato's theory.

TBDR routing support also enables a deterministic fault-tolerant routing algorithm based on fault-region marking and fault-avoidance, such as in [4], as well as adaptive fault-tolerant routing algorithms [12]. Incomplete or irregular topologies caused by partial shutdown of the interconnect because of power-performance tradeoffs can be treated in a manner similar to a network with faults for routing re-configuration purposes.

Table 14.1 Minimum virtual channels required for fully-adaptive routing (Turn model v/s Duato's theory)

	VCs for 4 message classes	
Topology	Turn model	Duato's theory
2D mesh	8	5
2D torus	12	9

14.5 Pole Routing Algorithms for Load-Balancing

Pole routing is a novel two-stage routing algorithm for 2D meshes that supports regular, irregular and faulty mesh networks. The message routing is done in the first stage by sending it to predetermined intermediate destination called a pole node. In the second stage the message is forwarded from the pole node to the final destination node. Each routing stage uses a minimal deadlock-free routing algorithm in a mesh. However, the full route from the source to destination may or may not be minimal depending on the location of the pole with respect to the source-destination pair.

The selection of the pole node is done by the source node. Thus pole routing can be used as a hybrid source-controlled distributed routing algorithm and can be used in conjunction with table based or hardwired routing implemented in the fabric. The source can exercise control over load-balancing or have improved routability around faulty nodes by appropriate choice of pole location (pole placement) to route to a desired destination. We use minimal deterministic or partially adaptive routing algorithm based on the turn-model as the baseline routing algorithm for the "pre-pole" and "post-pole" stages. Each of these stages require just a single virtual channel (or virtual network) to guarantee deadlock freedom. However, in order to ensure that any cyclic channel dependence between pre-pole and post-pole phases does not arise, an additional virtual channel (virtual network) is required. This is because pre-pole to post-pole transition of channels by a message (at the pole router) may have to use a disallowed turn in the baseline routing algorithm.

Next we discuss a few different scenarios in which pole routing can be useful.

14.5.1 *Fault-Free Load Balanced Routing*

Load-balancing with pole routing can be done in several ways. Here we discuss static approaches for load balancing (i.e. we do not dynamically sense traffic conditions and then determine appropriate pole-placement). One simple and practical static approach is to look at a small number of pole placement options at a given source, say up to two options, for each destination and cycling between them. When a traffic pattern is static such an approach can lead to improved link utilization in the network leading to higher network throughput.

An example of this is shown for transpose traffic in a 2D mesh network in Fig. 14.6. In transpose traffic, a node S with coorinates (X,Y) sends messages only to destination node labeled D such that its coordinates are (Y,X). With deterministic XY-routing support in the network, the bottom left and top right corners of the network have the most congested links which limit the peak throughput. This can be see in Fig. 14.6a.

One option to get better load-balance for transpose traffic is for each source to alternately send messages to D choosing between pole-locations $P_0 = (X,X)$ and $P_1 = (Y,Y)$. While two virtual networks are required for pole-routing support

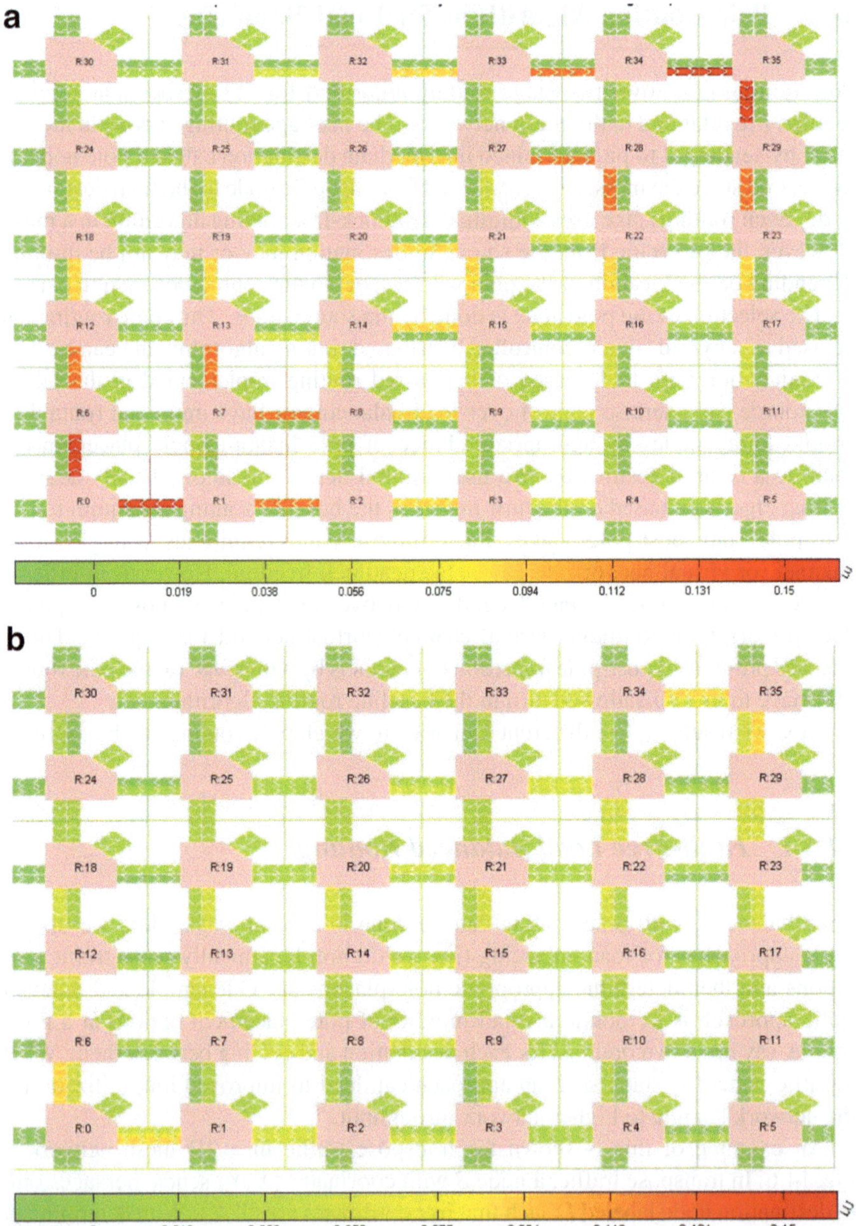

Fig. 14.6 Link utilization on a 6×6 mesh network with Transpose traffic. (**a**) XY deterministic routing algorithm causes congested (*red*) links at the corners of the mesh. (**b**) Load-balanced pole routing using appropriate pole placement reduces link utilizations of formerly congested links

compared to the baseline case, they both are still assumed to support XY-routing, however this specific choice of pole locations lead effectively to message being routed on an YX path (using pole P_0 and XY path using pole P_1). The value of this approach can be seen in Fig. 14.6b where the link utilizations are much lower at the same load compared to that with deterministic XY routing.

Clearly, there are other pole placement choices available for improving the network performance than the one discussed above. Later in this section, we also discuss optimal pole-placement heuristics. While we do not discuss dynamic load balancing, these approaches could also be coupled with a pole-routing based solution.

14.5.2 Fault-Tolerance Using Pole Routing

There are several well-known algorithms for implementing fault-tolerance in mesh networks given a known set of failed nodes in the network. Pole-routing provides yet another tool to implement fault-tolerance. Using pole routing provides a way to use simple baseline routing algorithm in the underlying network and then use appropriate pole-placement for fault avoidance.

For a given fault-free source-destination pair in the network, pole placement in the presence of network faults is done in such as manner as to avoid the faulty node both on the pre-pole and post-pole paths in the routing phases.

Examples, of fault-avoiding paths between source destination pairs are shown in Fig. 14.7. Note that, in order to present logically contiguous node IDs, the network interfaces may need to re-map logical IDs even though physical router and node are maintained in order to simplify routing in the network. This can be done in look-up tables maintained at network interfaces.

Fault-tolerant routing may need to take non-minimal paths in order to avoid faulty nodes. The presence of faults in the network and fault-avoidance also creates additional asymmetries in network link utilization. Appropriate pole placement can be used to minimize overall path latencies or even minimize network load imbalances caused by faults. We discuss heuristics for optimal pole-placement in the presence of faults in the following subsection.

14.5.3 Heuristics for Optimized Traffic Flows

We noted in the previous subsections that for routing a message between a given source-destination pair using pole routing there are multiple choices for pole placement. For a certain specified traffic patter it may be desirable to optimize routing performance by making suitable pole placement choices both in the no-fault cases as well as in cases where faults are present.

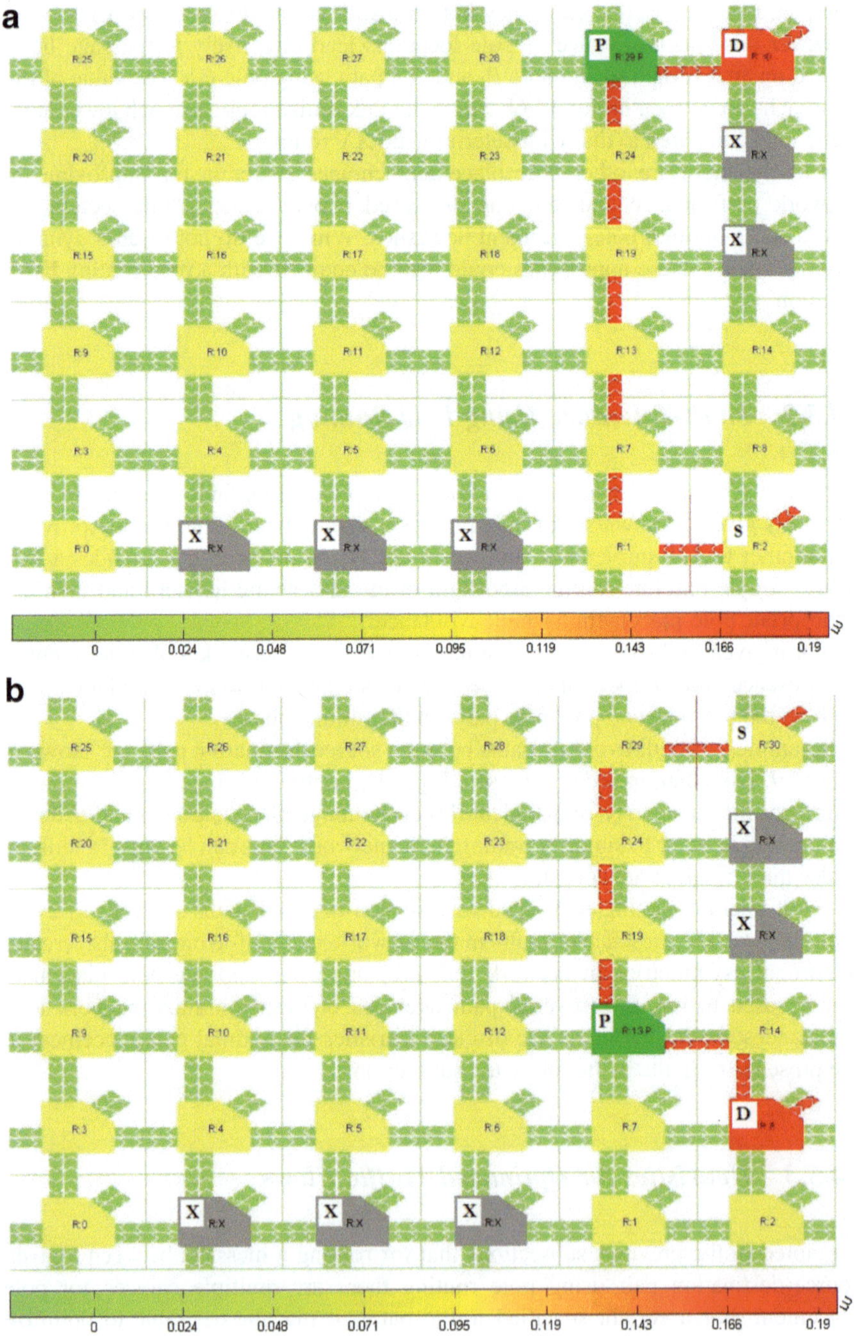

Fig. 14.7 Examples showing pole-placement (*P*) for fault-avoidance between two different source (*S*)-destination (*D*) pairs (**a**) ($S = 2,\ D = 30,\ P = 29$) (**b**) ($S = 30,\ D = 8,\ P = 13$). Nodes have been relabeled to skip faulty nodes. XY routing is the baseline routing algorithm assumed for pole routing

Here we present two heuristics to optimize pole placement for a given traffic pattern:

Min-hops the first heuristic attempts to minimize the overall network latency in terms of the average number of hops taken to route messages
Max-throughput the second heuristic attempts to maximize the overall throughput for the given traffic pattern

We also assume for the following discussion that a deterministic routing algorithm (such as XY routing) is used for both the pre-pole and post-pole routing virtual networks.

Min-hops heuristic: In this heuristic, for all fault-free source-destination pairs that are valid for a given traffic pattern, a pole location is picked such that it provides the least utilized shortest fault-free path. The key step of the heuristic are as follows:

1. Select a source-destination pair (S, D), from amongst all fault-free and valid source destination pairs for the given traffic pattern, at random.
2. Determine all valid pole positions that route a message from S to D using a minimal path (in a faulty network these should be minimal fault avoiding paths).
3. Pole a pole location P from amongst all candidates, such that the choice of P minimizes the utilization of the maximally used links amongst possible paths. If P is not unique, pick amongst pole location candidates at random.
4. Pick the next source-destination pair at random and repeat the process until all source destination pairs have been processed and all applicable pole locations have been chosen.

The above approach aims to minimize average number of hops for a given traffic pattern while simultaneously trying to improve link utilizations as well. In a network with faults present, utilizations of some links can be skewed quite adversely and this approach only helps somewhat. Figure 14.8a shows this behavior for uniform random traffic in a mesh with multiple faults.

Max-throughput heuristic: This heuristic attempts to balance network link utilization by making use of both minimal and non-minimal routes. As above, the source destination pairs are chosen at random. However, pole locations are picked from amongst all routes allowed by the routing algorithm (both minimal and non-minimal). Specifically, a pole location is chosen in such a manner that the link utilization of the maximally utilized link from amongst all candidate paths is minimized. If multiple choices of pole locations lead to the same value lowest value for the maximally utilized link, then the pole location with the shorter path length (number of hops) is chosen. Ties are broken by selecting a pole location at random from amongst equi-weight choices.

This heuristic is good for obtaining graceful degradation of performance (network throughput) in the presence of faults. Link utilizations for the same set of faults with the max-throughput heuristica can be significantly lower compared to the min-hops heuristics. This can be seen in Fig. 14.8b.

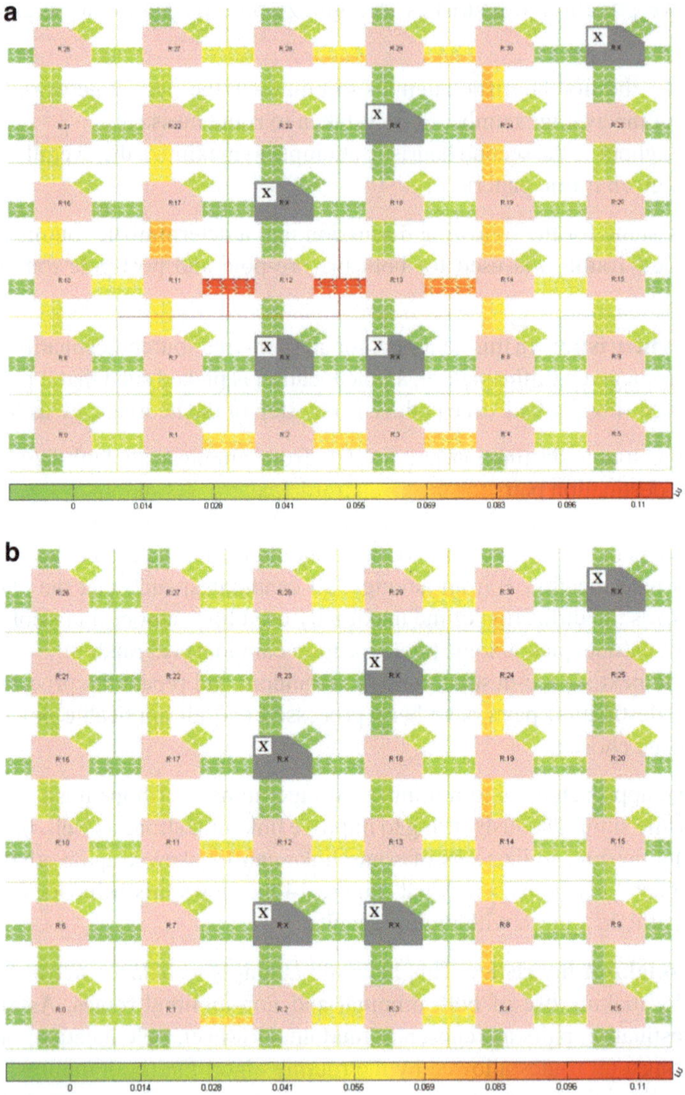

Fig. 14.8 Link utilization on a 6 × 6 mesh network with multiple faulty nodes and Uniform random traffic. (**a**) Pole placement heuristic for minimizing inter node hop count. (**b**) Pole placement heuristic to maximize the throughput for this traffic pattern. Non minimal routes may be chosen

14.6 Performance Isolation Through Routing

Large number of cores on a tera-scale processor may support multiple partitions such as with multiple virtual machines (VM) or with multiple tasks each being given a distinct set of processing and system resources. It is often desired that there be

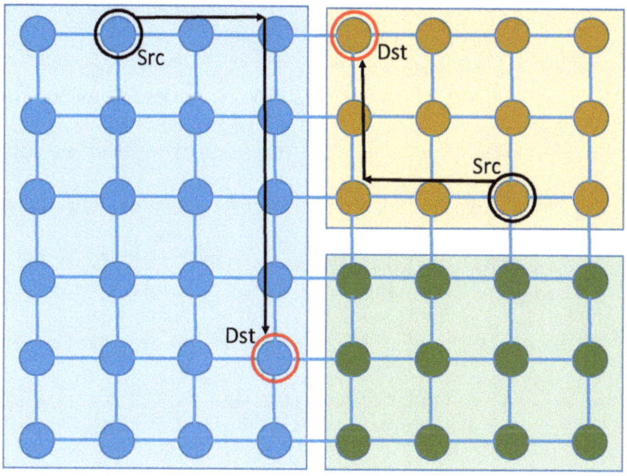

Fig. 14.9 Intrinsic performance isolation with rectangular/submesh shaped partitions

as much isolation maintained between different partitions as possible. As discussed earlier this isolation may be for performance reasons such as service level guarantees and/or security reasons.

We look at such isolation being provided through the interconnect and in particular through the routing algorithm(s) supported by the interconnect. In a mesh network, such isolation can be maintained simply by ensuring that each partition itself is rectangular, i.e. partitions are always submesh shaped. Cache memory banks, memory controllers and other system components should be contained within each partition as well. In such a scenario, the mesh interconnect with minimal routing algorithms intrinsically will ensure messages between cores and other system components belonging to a rectangular partition will be confined to the same partition. This is shown in Fig. 14.9.

However, the need to allow only rectangular partitions may underutilize resources, especially when such partitions are created and allocated and deallocated dynamically. In such an environment, one is likely to have rectangular and non-rectangular partitions operating side by side. The requirement to support (performance) isolation may still exist. A routing algorithm to support performance isolation in non-rectangular partitions is outlined below. The only restriction in this case is for each partition to be a connected partition (i.e. the underlying network graph of all nodes in the partition is a connected graph).

The routing algorithm to support performance isolation (i.e. that all traffic between nodes of the partition is contained within the partition) relies on the fact that every connected non-rectangular partition is made from a connected set of submeshes. Now, construct a new graph S constructed by placing a vertex in S for each component submesh of the partition and edge between corresponding vertices in S for adjacent submeshes in the partition. Given that the partition is a connected partition, S is a connected graph. A deadlock free hierarchical routing algorithm can now be constructed to route in the non-rectangular partition making use of deadlock free routing in S (using UP-DOWN routing for example) and deadlock free

```
// Source Node                  : S = Sm:Si
// Destination Node             : D = Dm:Di
// Intermediate Node            : X = Xm:Xi
// Node IDs have two parts : MeshID:IntrameshID
set X = S           // Xm = Sm, Xi=Si
while(X != D)       // Destination not reached
{
  if(Xm != Dm)      // Inter-submesh (Up-Down) route step
  {
    Use Virtual Network (VN) for Up-Down route
    Go from (Xm:Xi) 1-hop closer toward next submesh
  }
  else if (Xi != Di) // Xm == Dm, intra-submesh route
  {
    Use reserved VN(s) for intra-mesh route in Dm
    Route from (Dm:Xi) 1hop toward (Dm:Di); use mesh XY route
  }
  Update X with Node Id of node reached
}
Take egress channel at D // X == D, destination reached
```

Fig. 14.10 Algorithm sketch for performance isolation routing within a non-rectangular partition

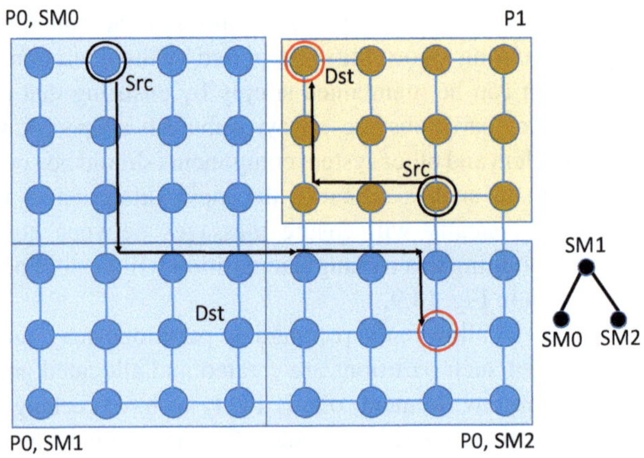

Fig. 14.11 Performance isolation with non-rectangular and rectangular shaped partitions

routing in meshes. Traversing in S represents traversing from a source submesh to the destination submesh which is done first and then intra-submesh routing delivers the message within the destination submesh to the final destination node. We use two distinct virtual networks one for inter-mesh traversal and a separate one for intra-submesh traversal at the destination submesh.

The algorithm sketch using a mix of pseudo-code and descriptive text is provided in Fig. 14.10.

The example mesh with an L-shaped partition and a rectangular partition showing performance isolation routing within each is shown in Fig. 14.11.

14.7 Environments for Prototyping, Debug and Performance Visualization

14.7.1 FPGA Emulation Prototyping

We have developed a full featured RTL implementation of the router using Verilog for the purpose of robust validation of the micro-architecture and design as well as to conduct a detailed performance characterization of the interconnect [7]. The larger goal of the interconnect prototyping effort is to have a robust interconnect which can then be interfaced to several production grade processor cores and their cache coherence protocol engines.

Our FPGA based emulation environment is highly configurable for various parameters including the number of MCs, performance and routing VCs and buffer sizes. Along with each router in the prototype, a network node also implements a network interface (NI) block for packet ingress and egress functionality as well as a synthetic traffic generator. Uniform random, transpose, bit-complement, and hotspot traffic patterns are currently supported by the traffic generator along with several additional configurable parameters for controlling injection rates, MCs and sizes. Various routing algorithms using programmable routing tables have been implemented including basic XY routing, turn model based routing, load balanced and adaptive routing, fault-tolerant routing and support for isolation of multiple partitions as well as support for mesh and torus topologies.

Table 14.2 summarizes the key features implemented in the RTL.

Emulator control and visualization software system enables one to initialize the network and run multiple experiments with various micro-architectural parameters, routing algorithm tables and traffic patterns. The software environment enables one to run each experiment for any given number of cycles after which a large array of performance counter values that can be recorded. Registered values include

Table 14.2 Key features of the emulation framework

Pipeline	2-cycle router pipeline with bypass support
Topology	2D mesh/torus
Routing algorithm	Hardwired routing/table-based routing
	Supported: East-Last/Odd-even/up-down/South-Last/West-First/ SR-Horizontal/SR-Vertical/X-Y
	Also support fault-tolerant/hierarchical routing (for performance isolation)
Traffic Patterns	Four traffic patterns are supported: uniform random/ transpose/bit-complement/hotspot
# nodes per board	2×3 MESH per board (2 nodes per FPGA and FPGAs are used for router design)
	In this work, 6 boards were used to form a 6×6 MESH
Other	Configurable network size, buffer sizes, virtual networks, number of performance VCs

number of injected and ejected packets, packet latencies split by MCs, buffer utilizations, bypass and arbitration success/failure rates per port, etc. A custom GUI for control and performance visualization is used to run experiments interactively and graphically render performance data in real time for each experiment.

14.7.2 Performance Simulation and Visualization Environment

Cycle accurate performance simulation played two distinct roles in our explorations: (a) validation of ideas through performance analysis; (b) validation of functionality before committing solutions to RTL and validation of RTL emulation at later stage. The first role is critical for understanding the complexity of packets flowing through the routers and experiencing the effect of many policies embedded in the micro-architecture of the router. Even though intuition of architects plays the key role in devising new router functionalities and micro-architecture, in most cases the validation of the idea through cycle by cycle simulator transforms the original intuition into new directions. In our effort we took two different approaches to modeling of the network and routers.

- Flexible pipeline with stage to stage abstraction API: Limited capability of adjusting pipeline stages such as combining stages into one.
- Event driven simulator: Every packet instigated hardware operation at each stage of the pipeline schedules the next operation in the future.

The first approach enabled the representation of multiple router pipelines with the same base code with simple modifications of the input pipeline configuration. However, validation of all possible pipeline configurations with changes in one was cumbersome. The second approach required bifurcations of the code for every pipeline of interest. It was easy to maintain for a specific pipeline but required reflection of new ideas in other live branches of the model. The appropriate choice is a function of the overall framework of the studies, i.e. simultaneous analysis of many alternative pipelines or detailed analysis of a few.

The second role is required since implementation of the idea in a higher level of abstraction in the cycle accurate performance simulation hashes out the details required for RTL implementation. In addition, debugging of the FPGA implementation of the RTL is a complex task and requires a reference point for comparison at every stage. Our performance simulator was made to be clock level accurate so that it could be used as the reference point for this debugging. The format and visualization of a subset of emulation/simulation output was made consistent across the two environments for ease of debugging, i.e. consistent set of APIs for configuration and visualizations. Multiple orders of simulation speed advantage of the simulator was useful in testing ideas before committing the changes to RTL.

14.7.3 Trace Driven Simulation Environment

The evaluation of interconnection topologies, architecture, and micro-architecture was carried in two different segments of servers and GPUs in our effort. Due to the differences in the nature of architecture and workloads in these two segments, the analysis tools became distinct even though the simulator and basic tools remained common. Homogeneity of the server processor architecture and limited number of agents, e.g. processor cores, cache controller, IO agents, and memory controllers, leads to regular structures and uniform connectivity in the network. On the other hand, GPU architectures are composed of larger number of agents with diverse set of functionalities and distinct set of flows both in terms of type and quantity and thus result in irregular agent connectivity to network.

In the server framework, coherence traffic was captured in a functionally correct parallel execution-driven environment of the target architecture under study. However, one expects that the execution of multi-threaded applications in such an environment to deviate from the desired execution in the target architecture due to the much higher level of abstraction and timing inaccuracies in the functional execution simulator which is mainly designed for speed. The main intention in this process is to capture the trace of an execution once and post process it for the performance simulation and analysis of the interconnect architecture many times. This approach enables performance analysis of large set of architecture and micro-architecture scenarios of the interconnection network by condensing the simulation time to a manageable level. The timing dependencies amongst coherence messages are captured in traces to enable accurate replay of those actions in the slightly different timing condition of the detailed underlying interconnection network simulation environment. Abstract models of coherent hardware agents behaviors using the underlying u-arch structures was used to capture a reasonably accurate timing behavior of coherent agents under queuing conditions. In other words, queueing behavior and timing was dictated by the underlying u-arch and flow back pressures while the relationship of coherency traffic was preserved adhering to the associated relationships embedded in the traces. IO traffic was modeled statistically, i.e. frequency of IO traffic, frequency of specific type of transactions, etc. Trace based simulation followed the initial studies based on pure statistical driven synthetic workloads, i.e. specific hit/miss ratios at different levels of private and shared caches.

In the GPU interconnect analysis traces of an older generation design was captured from the corresponding performance simulator, augmented appropriately to mimic the behavior of the future generation hardware alternatives, and then replayed in the interconnect model. The changes included potential increase in the ratios of number of agents of a particular type, change of hashing functions, filtering of certain traffic that could be avoided in the new architecture, introduction of new traffic, change of message formats, etc. The goal of this analysis was to identify potential bottlenecks in the interconnection network under consideration, optimized placement of various GPU agents in the network to better balance the traffic, analysis of alternative routing algorithms to remove bottlenecks, or enrich/reduce the configuration of the interconnect. Pattern of traffic in GPUs for various graphics workloads and for GPGPU compute is partly regular and can benefit from workload classification to identify worst case scenarios. The interdependence of messages in the GPU traces were analyzed and traces were pre-processed to help adherence to such dependencies in the replayed simulation on the interconnect simulator.

14.7.4 Trace Classification Through Clustering

Trace classification in servers was mainly done based on the coherence traffic to identify the primary nature of interconnect traffic such as second level cache miss traffic hitting the shared third level cache, dominant traffic of misses from third level cache diverted to memory controllers, IO traffic distribution, etc.

Due to large number of agents in the GPU architecture, the classification is more complex and has distinct phase characteristics. The structured computation and associated memory traffic in GPU workloads provided the opportunity to look across vast number of workloads, identify the traffic as mix of dominant traffic flows by small subsets of agents, and characterize the structure of such mix at different phases of the compute. Selection of the worst traffic pattern from the interconnect flow operation perspective can then be used to study a large selection of architecture and micro-architecture alternatives in the context of the interconnect topology, router design, and routing algorithms. Proper load balancing of the dominant traffic in the worst case scenarios and repeating the analysis enables effective interactive approach to topology and u-arch optimizations and demands fast simulation speeds consuming a large selection of workloads. The capability of automating the process of classification and selection of representative small phases of compute across all traces of every workload category provides a tremendous opportunity to focus on the interconnect design in an interactive mode.

14.7.5 GPU Interconnect Example

Figure 14.12 shows one example GPU interconnection network configuration (torus in horizontal dimension and mesh in the vertical dimension) with the associated GPU agents placements. Some agents such as caches have multiple interfaces connected to different routers to load balance the traffic through appropriate hashing schemes. Considering the regularity of GPU architectures, this can be a subset of a larger network encompassing it, i.e. extending the design in the vertical dimension. This particular solution is a function of the number of agents, their maximum injection and ejection bandwidth capabilities, real estate cost, power dissipation considerations, floor planning considerations, metal layer availability, modularity considerations across generations, priority of workloads from a business perspective, etc.

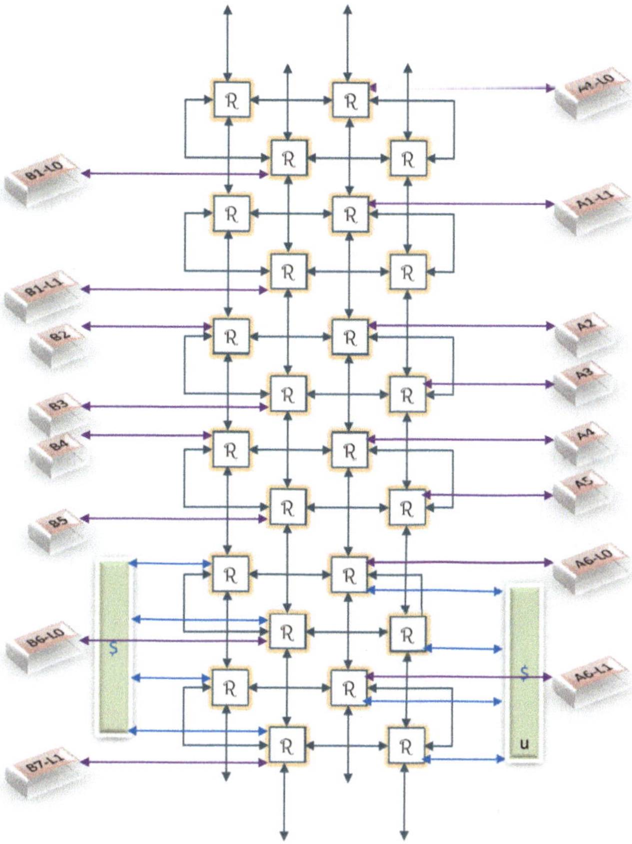

Fig. 14.12 Example 2D torus-mesh GPU interconnect

Figure 14.13 show a plot of pair-wise communication traffic mix for a selection of traces of different workloads, e.g. OpenCL, Direct X, etc. The x and y axis represent the various agent IDs and the Z dimension represents the traffic during the simulation time amongst a particular pair of agents. Dominant flows can be easily identified with this graph but the timing of such traffic would be missing, i.e. this is a cumulative plot across the whole simulation period.

Figure 14.14a, b represent the same set of data as in Fig. 14.13 as a function of time. The x dimension represents time and the stack of various colors at each point on the horizontal axis represent the traffic mix for that window of time represented at that single point. In this example, each point on the x axis represents an average over 50,000 simulation cycles. In general, one is interested in grouping regions that have very similar mix of pair wise communication traffic for a desired length of time and selecting the one(s) that have the largest volume within such mix. Such a subsetting approach will guarantee that the highest quantity of such distinct simultaneous

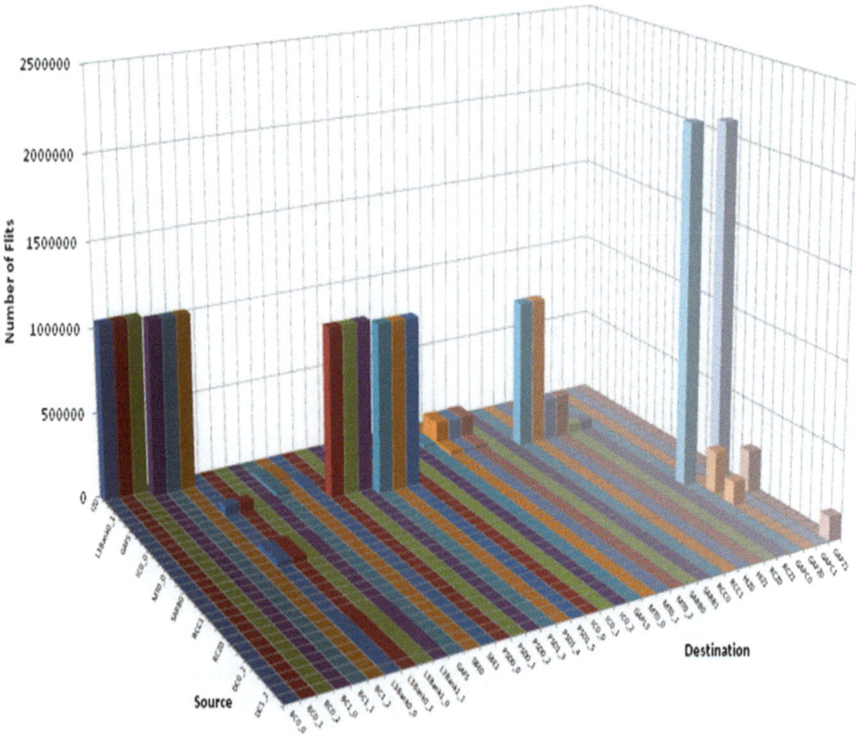

Fig. 14.13 Example plot from traffic visualization tool showing pair-wise traffic mix amongst various GPU nodes in the fabric

traffic flow is analyzed by the simulator. Proper classification of such distinct mixes and selection of highest volume of traffic amongst each mix will ensure that worse case interconnect flows are analyzed.

The color bands painted in the vertical direction in the background of the plots below represents the cluster classification of the pair-wise traffic. The mix of pair wise traffic in each cluster is the same but the overall magnitude of each member of that cluster can be different. By selecting a small number of traffic mixes with a pre-defined duration constraint, we can cover the simulation of all critical phases of all traces and reduce the simulation time by 2–3 orders of magnitude across thousands of traces.

The analysis of the representative phases requires additional support to expedite the interactive cycle of design decision, analysis, and required improvement. We developed a sophisticated interactive tool to examine the simulation replay of the traces, analyze the queueing effects on the links, visually identify the hot spots, and trace the link saturations directly to agents responsible for the traffic flow on those links. The color representation of the traffic levels in the interconnect links below quickly identifies potential hot spots for specific trace regions. Such trace

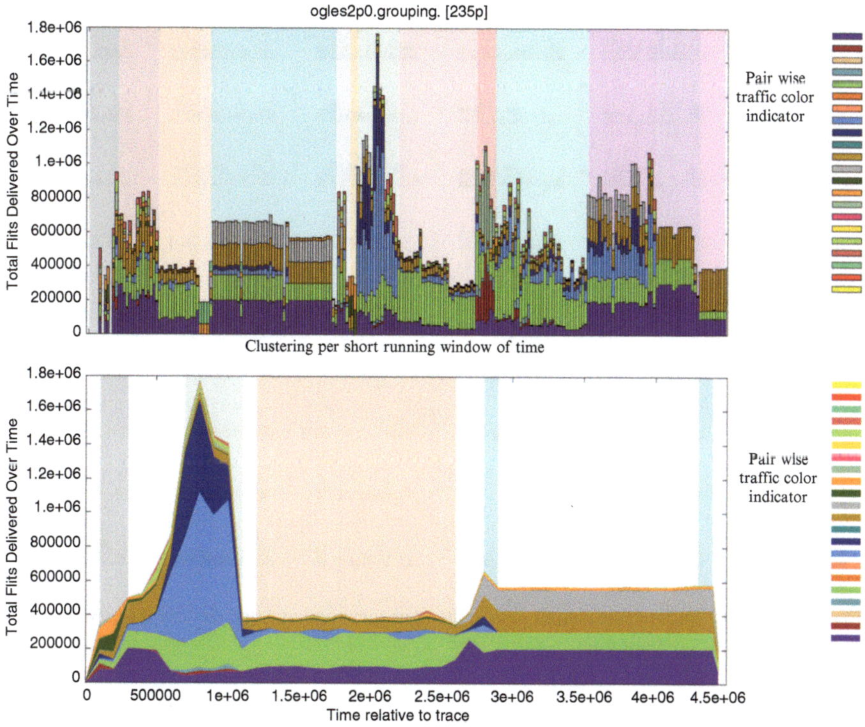

Fig. 14.14 GPU stacked pair-wise traffic over time with clustering to identify similar traffic phases

subsets can be automatically identified by our tool when link utilizations pass preset threshold levels. A tool view showing traffic analyses for multiple sets of traces simultaneously is shown in Fig. 14.15.

Figure 14.16 shows interactive aspects of the tool. For example left portions of the figure show the difference between injected traffic into the network (upper left plot) versus ejected traffic (lower left plot). By overlaying the ejected traffic line over the stacked of colored values of injection values one can identify potential queueing effects. A high level of queueing will show the quantum of deviation between the instantaneous injection and ejection. The user interface allows the architect to click on an interconnect link to see the originators of traffic flows through that link and the percentage contribution of each. Covering all scenarios which lead to queueing effects, one can identify the required changes required in agents' placement, routing algorithm tuning, etc. After reflection of such changes in the router performance model, the analysis can be repeated with minimal simulation time due to the huge computation savings enabled by clustering approach. The focus of the study will be limited to the small trace segments representing the highest level flows with distinct mixture of pair wise communications between agents. We found our tool to be extremely effectively in practice and shorten the interconnect design and optimization cycle enormously.

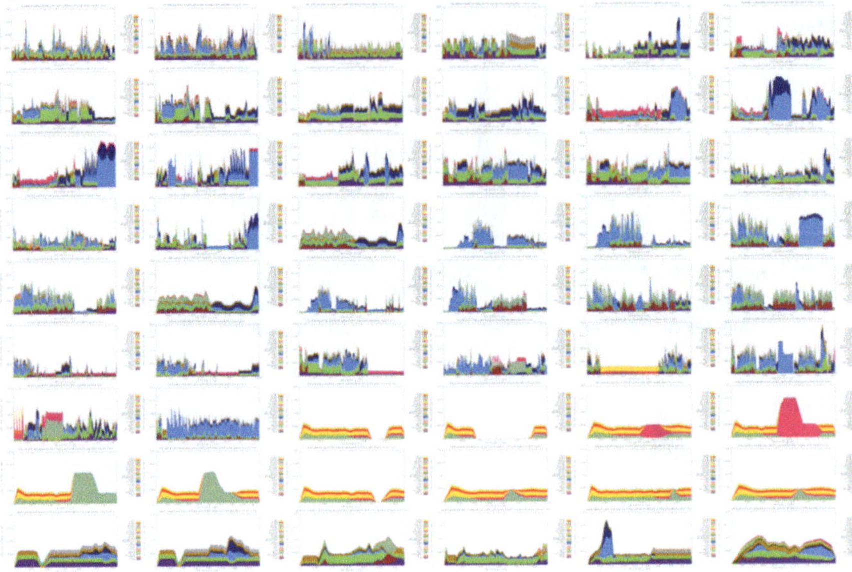

Fig. 14.15 GPU Traffic analysis tool with multi-trace view option helps identify traces which may stress interconnect configuration

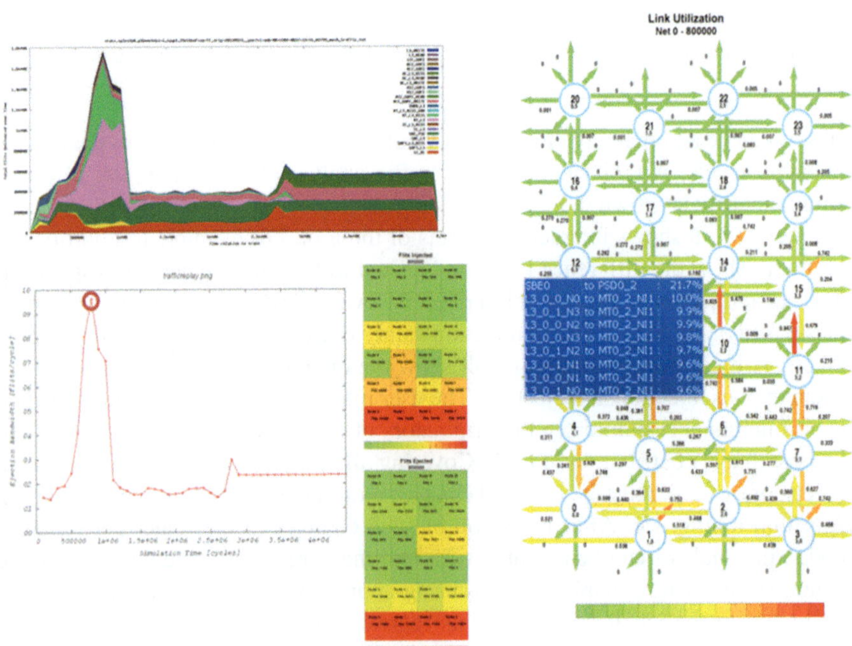

Fig. 14.16 Interactive traffic analysis and problem triage using our performance visualization tool

14.8 Summary and Concluding Remarks

Many interesting interconnection networks and variations have been explored over the last few decades in the academic circles and publications. The recent efforts in restricting the ideas to on-chip interconnects has allowed architects to make use of abundant wires, smaller clock synchronization overhead, lower error rate, and lower power dissipation in links compared to off-chip networks, where chip pin counts and power dissipation in the transceivers and links dominate design considerations. At the same time, the connectivity restriction to a planar substrate, aggressive power dissipation restrictions, and overall on-die design constraints dictated by the compute and storage elements on-die and by design teams introduces new additional requirements for practical solutions. A few such academic concepts have surfaced in some form or other in industrial implementations, but mostly in the extreme high-end and low volume part of the market. Otherwise, the adoption in the volume market has remained limited to simplistic instantiations of some of the elegant academic concepts. Better penetration of such interesting concepts into actual designs requires more effort on feasible micro-architecture and physical design implementations combined with creative validation techniques and tools to address the needs of industrial design teams that are extremely schedule driven and risk averse.

In this chapter, we have captured our experience with such efforts from the topology, architecture, routing algorithms, micro-architecture support, FPGA prototyping, performance simulation, debugging methods and tools. Unfortunately due to the scope and framework limitations, we have left out our physical implementation evaluations which is a critical part of the decision making loop in such implementations but is a tight function of the specific overall chip design, e.g. diversity of workload, overloading of network with independent traffic classes, sensitivity to latency, QoS requirements, validation resource constraints, number of agents, chip interconnect availability, power and real estate constraints, regularity of the design, reuse strategy across market segments, late binding of design, etc.

Micro-architecture and pipeline design is a critical element of the on-die design. The specific decisions made in the pipeline design are strong functions of the required bandwidth, latency, power and die space constraints, routing options, validation constraints, reusability in different segments, etc. The pipeline options discussed in this chapter are a subset of a much larger exploration. For example, latency tolerance of GPU environments provides flexibility in trading off frequency versus latency, its structured and more predictable traffic flow allows removing some of the routing flexibility requirements, its higher bandwidth requirements benefits from richer topologies imposing higher number of virtual channels. In short, the micro-architecture design and tuning within the vast set of design constraints is a complex task and requires a wide selection of architecture/design/power/debug tools at architects' disposal. Unfortunately, literature is full of apple to orange comparisons for specific ideas and designs due to distinct requirements and constraints in each project. Based on our experience, in many cases the real picture can turn out to be quite different from the suggested findings.

The routing algorithm section covered the benefits of the deterministic and adaptive approaches in realistic environments. Pole routing proved itself as a practical solution to address complex issues usually contained to specific hot spots and avoid a recourse to extreme complex solutions. Our real implementation and debugging of the Pole routing in a realistic and huge FPGA emulation of a complete system is a solid proof of its viability. At the same time, the adaptive routing solutions with Prof. Duato's theoretical basis has great potential but serious debugging and validation challenges. Our practical implementation and optimization of a practical adaptive routing mechanism offered an interesting solution but required major investment in definition, debugging hooks, tools, and simulator and FPGA debugging. Adaptive routing proved to address extreme hot spot conditions and avoid the required tuning of deterministic solutions and agent placements with a priori knowledge of diverse workload behaviors. A practical validation methodology is required prior to committing an adaptive routing for a general interconnection design.

Considering the wide spread usage of virtualization in the server domain and to some extent in the client domain, much higher level of integration on-die such as powerful GPUs on the processor die, introduction of massive number of cores and agents on-die such as network processors, etc., the sharing of interconnection network for distinct independent traffic is critical. The chapter of performance isolation provides practical trade-offs in providing network topology flexibility and adaptability versus the feature design complexity. We have been able to implement and emulate such capability in its full detail. However, one should pay close attention to the potential conflicting requirements of this capability against the many other desired routing optimization features. In some cases performance isolation may require disabling of other features, e.g. adaptive routing.

Emulation of complex features prior to design commitment is critical for complex networks. However, affordable FPGA emulation requires extensive tools and capabilities to make the effort manageable and worthwhile. The emulation section covers our experience with a 16 processor core emulation on a mesh/torus interconnect, porting of Linux, and actual execution of multi-threaded applications. Such effort required many creative solutions, including re-architecting the interrupt interconnect (APIC bus) on a 2D/2.5D network. In practice one can get away with a much simpler software stack for validation but capability of a full software stack allows one to evaluate the real interconnect effects under realistic workloads and systems. Emulation speeds are not fast enough to allow long execution of applications but fast enough for representative segments of workloads with prior sampling effort.

A fast, rich, and comprehensive cycle accurate simulator is the real required backbone of many routing, architectural, micro-architectural, and power analysis explorations. Creative visualization of the simulator results were absolutely critical for our FPGA emulation and design exploration across the board. Effective and creative visualization and interactivity of the tools allowed us to debug extremely complex adaptive features, address sharpening the latency distributions under complex u-arch techniques, and gain insight into potential new possible features.

Visualization was the key to quick optimization of our solutions in the GPU interconnect through topology modifications, routing optimizations, agent placement, traffic distribution, etc.

Acknowledgements Contributions and insights provided at various points in time by the following individuals are gratefully acknowledged: Donglai Dai, Dongkook Park, Andres Mejia, Gaspar Mora Porta, Roy Saharoy, Jay Jayasimha, Partha Kundu, Mani Ayyar and the late David James.

References

1. M. Azimi, N. Cherukuri, D.N. Jayasimha, A. Kumar, P. Kundu, S. Park, I. Schoinas, A.S. Vaidya, Integration challenges and tradeoffs for tera-scale architectures. Intel Technol. J. **11**(3), 173–184 (2007)
2. M. Azimi, D. Dai, A. Kumar, A. Mejia, D. Park, S. Saharoy, A.S. Vaidya, Flexible and adaptive on-chip interconnect for tera-scale architectures. Intel Technol. J. **13**(4), 62–79 (2009)
3. M. Azimi, D. Dai, A. Kumar, A.S. Vaidya, On-chip interconnect trade-offs for tera-scale many-core processors, in *Designing Network-on-Chip Architectures in the Nanoscale Era*, ed. by J. Flich, D. Bertozzi (Chapman & Hall/CRC, Boca Raton, 2011)
4. R.V. Bopanna, S. Chalasani, Fault-tolerant wormhole routing algorithms for mesh networks. IEEE Trans. Comput. **44**(7), 848–864 (1995)
5. G.M. Chiu, The odd-even turn model for adaptive routing. IEEE Trans. Parallel Distrib. Syst. **11**(7), 729–738 (2000)
6. G. Chrysos, Intel Xeon Phi coprocessor (codename Knights Corner). In: Hot Chips (2012)
7. D. Dai, A.S. Vaidya, S. Saharoy, S. Park, D. Park, H.L. Thantry, R. Plate, E. Maas, A. Kumar, M. Azimi, FPGA-based prototyping of A 2D mesh/torus on-chip interconnect, in *ACM/SIGDA International Symposium on Field-Programmable Gate Arrays (FPGA)*, 2010, p. 293. Abstract only
8. W.J. Dally, Virtual channel flow control. IEEE Trans. Parallel Distrib. Syst **3**(2), 194–205 (1992)
9. W.J. Dally, C.L. Seitz, Deadlock free message routing in multiprocessor interconnection networks. IEEE Trans. Comput. **36**(5), 547–553 (1987)
10. W.J. Dally, B. Towles, *Principles and Practices of Interconnection Networks* (Morgan Kaufmann, San Francisco, 2004)
11. S. Dighe, S. Vangal, P. Aseron, S. Kumar, T. Jacob, K. Bowman, J. Howard, J. Tschanz, V. Erraguntla, N. Borkar, V. De, S. Borkar, Within-die variation-aware dynamic-voltage-frequency scaling core mapping and thread hopping for an 80-core processor, in *IEEE International Solid-State Circuits Conference*, San Francisco, 2010, pp. 174–175
12. J. Duato, A theory of fault-tolerant routing in wormhole networks, in *International Conference on Parallel and Distributed Systems*, Hsinchu, 1994, pp. 600–607
13. J. Duato, A necessary and sufficient condition for deadlock-free adaptive routing in wormhole networks. IEEE Trans. Parallel Distrib. Syst. **6**(10), 1055–1067 (1995)
14. J. Flich, A. Mejia, P. López, J. Duato, Region-based routing: an efficient routing mechanism to tackle unreliable hardware in Network-on-Chips, in *International Symposium on Networks-on-Chip (NoCS-2007)*, Princeton, 2007
15. J. Flich, S. Rodrigo, J. Duato, An efficient implementation of distributed routing algorithms for NoCs, in *International Symposium on Network-on-Chips*, Newcastle, 2008
16. M. Galles, Spider: a high-speed network interconnect. IEEE Micro **17**(1), 34–39 (1997)
17. C.J. Glass, L.M. Ni, The turn model for adaptive routing. J. ACM (JACM) **41**(5), 874–902 (1994)

18. Intel Xeon Phi Coprocessor 5110P, Highly parallel processing to power your breakthrough innovations. Weblink, http://www.intel.com/content/www/us/en/processors/xeon/xeonphi-detail.html
19. Y. Tamir, G.L. Frazier, Dynamically-allocated multi-queue buffers for VLSI communication switches. IEEE Trans. Comput. **41**(6), 725–737 (1992)
20. A.S. Vaidya, A. Sivasubramaniam, C.R. Das, LAPSES: a recipe for high performance adaptive router design, in *Proceedings of the 5th International Symposium on High Performance Computer Architecture (HPCA'99)*, Orlando (IEEE Computer Society, Washington, DC, 1999), pp. 236–243

MIX
Papier aus verantwortungsvollen Quellen
Paper from responsible sources
FSC® C105338

If you have any concerns about our products,
you can contact us on
ProductSafety@springernature.com

In case Publisher is established outside the EU,
the EU authorized representative is:
Springer Nature Customer Service Center GmbH
Europaplatz 3, 69115 Heidelberg, Germany

Printed by Libri Plureos GmbH
in Hamburg, Germany